The Subnuclear Series • Volume 49

Proceedings of the International School of Subnuclear Physics

SEARCHING FOR THE UNEXPECTED AT LHC AND THE STATUS OF OUR KNOWLEDGE

THE SUBNUCLEAR SERIES*

Series Editor: ANTONINO ZICHICHI, *European Physical Society, Geneva, Switzerland*

16. 1978 THE NEW ASPECTS OF SUBNUCLEAR PHYSICS
17. 1979 POINTLIKE STRUCTURES INSIDE AND OUTSIDE HADRONS
18. 1980 THE HIGH-ENERGY LIMIT
19. 1981 THE UNITY OF THE FUNDAMENTAL INTERACTIONS
20. 1982 GAUGE INTERACTIONS: Theory and Experiment
21. 1983 HOW FAR ARE WE FROM THE GAUGE FORCES?
22. 1984 QUARKS, LEPTONS, AND THEIR CONSTITUENTS
23. 1985 OLD AND NEW FORCES OF NATURE
24. 1986 THE SUPERWORLD I
25. 1987 THE SUPERWORLD II
26. 1988 THE SUPERWORLD III
27. 1989 THE CHALLENGING QUESTIONS
28. 1990 PHYSICS UP TO 200 TeV
29. 1991 PHYSICS AT THE HIGHEST ENERGY AND LUMINOSITY: To Understand the Origin of Mass
30. 1992 FROM SUPERSTRINGS TO THE REAL SUPERWORLD
31. 1993 FROM SUPERSYMMETRY TO THE ORIGIN OF SPACE–TIME
32. 1994 FROM SUPERSTRING TO PRESENT-DAY PHYSICS
33. 1995 VACUUM AND VACUA: The Physics of Nothing
34. 1996 EFFECTIVE THEORIES AND FUNDAMENTAL INTERACTIONS
35. 1997 HIGHLIGHTS OF SUBNUCLEAR PHYSICS: 50 Years Later
36. 1998 FROM THE PLANCK LENGTH TO THE HUBBLE RADIUS
37. 1999 BASICS AND HIGHLIGHTS IN FUNDAMENTAL PHYSICS
38. 2000 THEORY AND EXPERIMENT HEADING FOR NEW PHYSICS
39. 2001 NEW FIELDS AND STRINGS IN SUBNUCLEAR PHYSICS
40. 2002 FROM QUARKS AND GLUONS TO QUANTUM GRAVITY
41. 2003 FROM QUARKS TO BLACK HOLES: PROGRESS IN UNDERSTANDING THE LOGIC OF NATURE
42. 2004 HOW AND WHERE TO GO BEYOND THE STANDARD MODEL
43. 2005 TOWARDS NEW MILESTONES IN OUR QUEST TO GO BEYOND THE STANDARD MODEL
44. 2006 THE LOGIC OF NATURE, COMPLEXITY AND NEW PHYSICS: FROM QUARK-GLUON PLASMA TO SUPERSTRINGS, QUANTUM GRAVITY AND BEYOND
45. 2007 SEARCH FOR THE "TOTALLY UNEXPECTED" IN THE LHC ERA
46. 2008 PREDICTED AND TOTALLY UNEXPECTED IN THE ENERGY FRONTIER OPENED BY LHC
47. 2009 THE MOST UNEXPECTED AT LHC AND THE STATUS OF HIGH ENERGY FRONTIER
48. 2010 WHAT IS KNOWN AND UNEXPECTED AT LHC
49. 2011 SEARCHING FOR THE UNEXPECTED AT LHC AND THE STATUS OF OUR KNOWLEDGE

Volume 1 was published by W. A. Benjamin, Inc., New York; 2–8 and 11–12 by Academic Press, New York and London; 9–10 by Editrice Compositori, Bologna; 13–29 by Plenum Press, New York and London; 30–49 by World Scientific, Singapore.

*For the complete list of titles in this series, please go to

http://www.worldscientific.com/series/ss

The Subnuclear Series • Volume 49

Proceedings of the International School of Subnuclear Physics

SEARCHING FOR THE UNEXPECTED AT LHC AND THE STATUS OF OUR KNOWLEDGE

Edited by

Antonino Zichichi

European Physical Society
Geneva, Switzerland

World Scientific

NEW JERSEY • LONDON • SINGAPORE • BEIJING • SHANGHAI • HONG KONG • TAIPEI • CHENNAI

Published by

World Scientific Publishing Co. Pte. Ltd.
5 Toh Tuck Link, Singapore 596224
USA office: 27 Warren Street, Suite 401-402, Hackensack, NJ 07601
UK office: 57 Shelton Street, Covent Garden, London WC2H 9HE

The Subnuclear Series — Vol. 49
SEARCHING FOR THE UNEXPECTED AT LHC AND THE STATUS OF OUR KNOWLEDGE
Proceedings of the International School of Subnuclear Physics 2011

ISBN 978-981-4522-50-2

In-house Editor: Rhaimie Wahap

Printed in Singapore by World Scientific Printers.

PREFACE

In June/July 2011, a group of 144 physicists of 21 countries met in Erice to participate in the 49th Course of the International School of Subnuclear Physics. The countries represented by the participants were: Australia, Brasil, Canada, France, Georgia, Germany, Hungary, India, Iran, Italy, Japan, Netherlands, Poland, Russia, Spain, Sweden, Switzerland, Turkey, Ukraine, United Kingdom and the United States of America.

The School was sponsored by the Academies of Sciences of Estonia, Georgia, Lithuania, Russia and Ukraine; the Chinese Academy of Sciences; the Commission of the European Communities; the European Physical Society; the Italian National Institute for Nuclear Physics; the Italian Ministry of Education, University and Scientific Research; the Sicilian Regional Government; the Weizmann Institute of Science; the World Federation of Scientists and the World Laboratory.

The programme of the School was focused on the latest results from LHC and on the highlights from all Laboratories. The theoretical lectures dealt with what could come out from fascinating topics such as extra-dimensions and Supersymmetry. This year's Course was the first of the three devoted to the celebrations of the 50th Anniversary of the School. Why three? Because Subnuclear Physics School needed three years for being established. In 1961 we started to discuss the bridge between University teaching and International Research Centers like CERN. In 1962 a formal paper for its existence was signed at CERN by Blackett, Rabi, Weisskopf and two young fellows, J.S. Bell and me. The first Course in Erice was held in 1963.

In each of the three years we will celebrate big discoveries. In 2011, the negative sign of the β–function and asymptotic freedom; in 2012, QCD and in 2013, SSB and Instantons.

An original feature of the School, introduced in 1996, is a series of special sessions devoted to "New Talents", with the aim of encouraging and promoting young physicists to achieve recognition at an international level. This is a serious problem in Experimental Physics, where collaborations count several hundreds of participants and it is almost impossible for young fellows to be known. Even if with much less emphasis, the problem exists also in Theoretical Physics. So it was decided to offer 47 young fellows the possibility to give an open presentation of the results of their studies, followed by a discussion. Among them 10 best "New Talents" have been selected for publication in this volume, and 4 were given an award. In particular, one award was assigned for an original presentation in theoretical physics, one for an original presentation in experimental physics, one for an original presentation in detector physics and one for an original presentation in mathematical physics.

During the organization and the running of the Course, I enjoyed the collaboration of my colleague and friend, Gerardus 't Hooft, who shared with me the Directorship of the School. I would like to thank him, together with the group of invited scientists and all the people who have contributed to the success of the Course.

I hope the reader will enjoy the book as much as the students attending the lectures and the discussion sessions, which, as every year, have been the focal point of this School activity. Thanks to the work of the Scientific Secretaries, the discussions have been reproduced as faithfully as possible. At various stages of my work I have enjoyed the collaboration of many friends whose contributions have been extremely important for the School and are highly appreciated. I thank them most warmly. A final acknowledgement to all those in Erice, Bologna and Geneva, who have helped me in so many occasions and to whom I feel very indebted.

Antonino Zichichi

Geneva, October 2011

CONTENTS

Hot Theoretical Topics

Seminars On Specialized Topics

HIGHLIGHTS FROM LABORATORIES

SPECIAL SESSIONS FOR NEW TALENTS

Testing strings at the LHC

I. Antoniadis*
Department of Physics,
CERN - Theory Division
CH-1211 Geneva 23, Switzerland

I discuss the main motivations and directions for physics beyond the Standard Model of elementary particles and I give a short introduction on string theory. This theory introduces a new physical constant, the string length, under which a new fundamental structure appears, changing drastically the laws of nature. In particular, lowering the string scale in the TeV region provides a theoretical framework for solving the so-called mass hierarchy problem: the apparent weakness of gravity can be accounted by the existence of large internal dimensions, in the submillimeter region, and transverse to a braneworld where our observed universe is confined. In these lectures, I describe the main properties of string phenomenology with high or low string scale. In the latter case, I discuss also its implications for new gravitational phenomena that can be observed at both particle colliders, and in non-accelerator experiments searching for new short range forces.

1. DEPARTURE FROM THE STANDARD MODEL

1.1. Why beyond the Standard Model?

Almost all research in theoretical high energy physics of the last thirty years has been concentrated on the quest for the theory that will replace the Standard Model (SM) as the proper description of nature at the energies beyond the TeV scale. Thus, it is important to know if this quest is justified or not, given that the SM is a very successful theory that has offered to us some of the most striking agreements between experimental data and theoretical predictions. We will start by briefly examining the reasons for leaving behind such a successful SM. Actually there is a variety of such arguments, both theoretical and experimental, that lead to the undoubtable conclusion that the Standard Model should be only an effective theory of a more fundamental one.

We briefly mention some of the most important arguments. Firstly, the Standard Model does not include gravity. It says absolutely nothing about one of the four fundamental forces of nature. Another is the popular "mass hierarchy" problem. Within the SM, the mass of the higgs particle is extremely sensitive to any new physics at higher energies and its natural value is of order of the Planck mass, if the SM is valid up to that scale. This is several orders of magnitude higher than the electroweak scale implied by experiment. The way to cure

*Electronic address: ignatios.antoniadis@cern.ch - *On leave from CPHT (UMR CNRS 7644) Ecole Polytechnique, F-91128 Palaiseau*

this discrepancy within the SM requires an incredible fine tuning of parameters. Also, the SM does not explain why the charges of elementary particles are quantized. Other equally important reasons include that the SM does not describe the dark matter or the dark energy of the universe. It does not explain the observed neutrino masses and oscillations and does not predict any gauge coupling unification, suggested by experiments.

Here, we describe some of these arguments in more detail, starting from the mass hierarchy problem. The mass of the higgs particle is very sensitive to high energy physics. For example, the main one loop radiative corrections from the virtual exchange of top quarks and the higgs particle itself are

$$\mu^2_{\text{eff}} = \mu^2_{\text{bare}} + \left(\frac{\lambda}{8\pi^2} - \frac{3\lambda_t^2}{8\pi^2}\right)\Lambda^2 + \cdots \tag{1}$$

where λ is the higgs quartic coupling, λ_t is the top Yukawa coupling and Λ is the ultraviolet (UV) cutoff, set by the scale where new physics appears. From the masses of all other particles and for reasons that we will explain later, we know that the higgs mass should be at the weak scale. Requiring the validity of the Standard Model at energies $\Lambda \gg \mathcal{O}(100)$GeV imposes an "unnatural" order by order fine tuning between the bare mass parameter μ^2_{bare} in the Lagrangian and the radiative corrections.

For example, for $\Lambda \sim \mathcal{O}(M_{\text{Planck}}) \sim 10^{19}$ GeV and a loop factor of the order of 10^{-2}, we get $\mu^2_{1-\text{loop}} \sim 10^{-2} \times 10^{38} = \pm 10^{36}$ (GeV)^2. Thus, we need $\mu^2_{\text{bare}} \sim \mp 10^{36}$ (GeV)$^2 + 10^4$ (GeV)^2. An incredible adjustment is then required, at the level of 1 part per 10^{32}: $\mu^2_{\text{bare}}/\mu^2_{1-\text{loop}} = -1 \pm 10^{-32}$. Even more, at the next and all higher orders, a new adjustment is required. The correction will be of the order of $\mathcal{O}(100)$ GeV only beyond 17 loops: $(10^{-2})^N \times 10^{38} \leq 10^4$ $\Rightarrow N \geq 17$ loops. If we want to avoid these unnatural adjustments we need $10^{-2}\Lambda^2 \leq 10^4$ (GeV)$^2 \Rightarrow \Lambda \leq 1$ TeV. The resolution of the mass hierarchy problem should reside within the energy reach of LHC!

Another open question is the charge quantization: all observed color singlet states have integer electric charges. The symmetry group of the SM is $SU(3) \times SU(2)_L \times U(1)_Y$ and the electric charge given by: $Q = T_3 + Y$. The representations of the symmetry group of all fields are:

$$\begin{aligned}
q &= (3,2)_{1/6} & \qquad q &= \begin{pmatrix} u_{2/3} \\ d_{-1/3} \end{pmatrix} \\
u^c &= (\bar{3},1)_{-2/3} & & \\
d^c &= (\bar{3},1)_{1/3} & & \\
\ell &= (1,2)_{-1/2} & \qquad \ell &= \begin{pmatrix} \nu_0 \\ e_{-1} \end{pmatrix} \\
e^c &= (1,1)_1 & &
\end{aligned}$$

where the subscripts denote the hypercharges in the left column and the electric charges in the right one. However, the SM does not tell us anything about this choice of the hypercharges, that guarantees the observed electric charge quantization.

Also, the experimental indication of gauge coupling unification is a feature of nature that asks for an explanation. The renormalization group evolution of the three SM gauge couplings $\alpha_i = \frac{g_i^2}{4\pi}$ is given by:

$$\frac{d\alpha_i}{d\ln Q} = -\frac{b_i}{2\pi}\alpha_i^2 \quad \Rightarrow \quad \alpha_i^{-1}(Q) = \alpha_i^{-1}(Q_0) - \frac{b_i}{2\pi}\ln\frac{Q}{Q_0} \tag{2}$$

where Q is the energy scale and b_i are the one-loop beta-function coefficients. The extrapolation of the low energy experimental data at high energies under the 'desert' assumption indicates an approximate unification of all couplings at energies of order $M_{GUT} \simeq 10^{15} - 10^{16}$ GeV. By doing a more precise analysis however, using the current experimental precision, on finds that the Standard Model fails to predict such a unification (see Fig. 1).

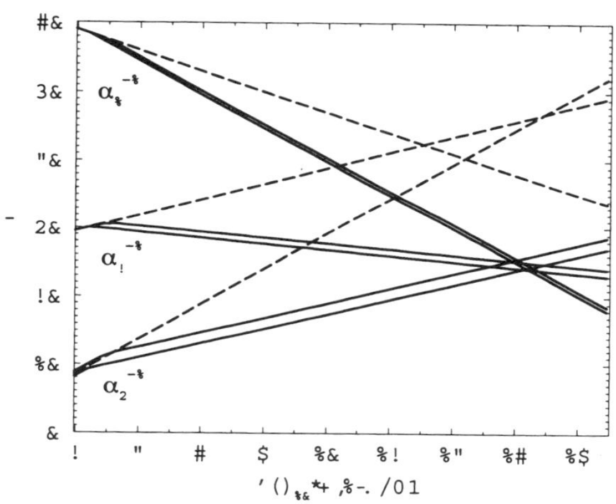

FIG. 1:

One more reason for going beyond the Standard Model is the existence of dark matter. Astronomical observations tell us that the ordinary baryonic matter is only a tiny fraction of the energy of the universe. There are observations that point towards the fact that a kind of non-luminous matter is out there and that is actually much more abundant than baryonic matter, consisting of around 25% of the total energy density of the universe. A natural explanation is that dark matter consists of a new kind of particles that are stable, massive at the electroweak scale and weakly interacting.

We could mention many other important reasons that drive us beyond the SM, like neutrino masses, dark energy and so on, but we will stop here. Even this brief and incomplete presentation shows that there are a lot of fundamental questions that the SM cannot address so the search for a new fundamental theory is definitely justified.

Modifying the SM though is not as easy as it might seem. If we try to identify the scale Λ of new physics by looking at what energies there are deviations from the SM predictions by observations, we will be disappointed. With the exception of dark matter and the higgs mass, all experimental indications suggest that Λ is very high. For example, the "see-saw" mechanism for the neutrino masses gives a very large scale: $m_\nu \sim \frac{v^2}{M} \sim 10^{-2}$ eV $\Rightarrow M \sim 10^{13}$

GeV, where v is the vacuum expectation value (VEV) of the SM higgs. Similarly, the scale of gauge coupling unification is even higher: $M_{\rm GUT} \sim 10^{16}$ GeV.

Even more, modifying the Standard Model in a way that solves the hierarchy problem is highly restricted by numerous experiments. Any new interactions could violate various Standard Model predictions that have been experimentally tested with good accuracy, such as lepton and baryon number conservation, absence of flavor changing neutral currents (FCNC), CP violation etc. For example, theories with baryon number violating operators like $\frac{1}{\Lambda_B^2}\left(\bar{q}\gamma^\mu u^c\right)\left(\bar{\ell}\gamma_\mu d^c\right)$ can lead to proton decay: $p = [uud] \to (e^+[\bar{d}d]) = e^+\pi^0, \quad uu \to e^+\bar{d}$. The bound from the Superkamiokande experiment on the proton lifetime of $\tau_p \geq 2.6 \times 10^{33}$ years highly restricts the allowed scale: $\Lambda_B \geq 10^{15}$ GeV. Similarly for the lepton number; in this case, the lowest dimensional operator that violates lepton number has dimension five: $\frac{1}{\Lambda_L}\left(\ell H^*\right)^2$, where H is the Higgs field. This operator leads in particular to Majorana neutrino masses. However, as we said before, neutrino masses are very low, $m_\nu = \frac{v^2}{\Lambda_L} \leq 0.1$ eV. With a VEV for the higgs at $\mathcal{O}(100)$ GeV, the scale Λ_L is also constrained to very high values: $\Lambda_L \geq 10^{14}$ GeV. Finally, FCNC operators like $\frac{1}{\Lambda_F^2}\left(\bar{d}_L\gamma_\mu s_L\right)^2$ are restricted by the measured value of $K_0 - \overline{K}_0$ mixing to be $\Lambda_F \geq 10^6$ GeV. It is very challenging to construct a model that provides us with a solution to the theoretical problems mentioned above and at the same time to comply with all experimental requirements.

1.2. Directions beyond the Standard Model

One way to encapsulate in the low energy regime the effects of a more fundamental theory is to use effective operators. These can be produced by integrating out heavy states in the fundamental theory. However, we can always write down generic effective operators even when that theory is unknown. We just require Lorentz symmetry and gauge invariance of the low energy effective action. These terms need not be renormalizable because by definition, the validity of an effective theory holds only up to the scale where the high energy physics appears.

Here we show an example of integrating out a heavy gauge boson of mass M

$$-\frac{1}{4}F_{\mu\nu}^2 - \frac{1}{2}M^2 A_\mu^2 + g\bar{\psi}\not{A}\psi \quad \Rightarrow \quad \frac{g^2}{M^2}\left(\bar{\psi}\gamma^\mu\psi\right)^2 \tag{3}$$

It gives rise to a six-dimensional non-renormalizable four-fermion effective operator. This is for instance the case in the popular Fermi theory, obtained by integrating out at low energies $E << m_W$ the heavy $SU(2)$ gauge bosons $W^\pm$:

$$\mathcal{L}_{\rm eff} = G_F\left(\bar{q}_L\gamma_\mu q_L + \bar{\ell}_L\gamma_\mu\ell_L\right)^2 + \cdots \tag{4}$$

with $G_F = g_2^2/m_W^2$.

Generally, effective field theory parametrizes our ignorance of the high energy physics by introducing local effective operators O_n^i of dim $(4+n)$

$$\mathcal{L}_{\rm eff} = \mathcal{L}_{\rm SM} + \sum_i \frac{c_n^i}{M^n} O_n^i \tag{5}$$

for $E \ll M$; M is the new physics scale where the heavy states become dynamical. If this is not too far away from the electroweak scale, the lowest dimensional (in powers of M) operators O_n^i can significantly affect the low energy physics. They can shift the mass of the higgs or violate symmetries of the Standard Model such as baryon number, lepton number, induce FCNC and important CP violation, etc. as we mentioned previously.

Effective field theory offers a systematic but blind way to explore the physics beyond the Standard Model. The alternative is to propose a totally new theory and study its consequences. There are several such proposals. They all cure some major weaknesses of the Standard Model but they also have their own drawbacks. Keeping the mass hierarchy as guidance, the following proposals exist so far:

1. Compositeness
2. Symmetry:
   - Supersymmetry
   - Higgs as pseudo-goldstone boson, little Higgs
   - conformal
   - Higgs as component of a higher-dimensional gauge field
3. Low UV cutoff:
   - low scale gravity ⇒ · large extra dimensions
     · warped extra dimensions
   - low string scale ⇒ · low scale gravity
     · ultra weak string coupling
   - large N degrees of freedom
   - higgsless models
4. Live with the hierarchy:
   - landscape of vacua, environmental selection
   - split supersymmetry

We will briefly describe only the first two and later we will focus on strings and extra dimensions.

The first proposal for physics beyond the SM is "compositeness". This idea postulates the existence of new strong dynamics at the energy range of TeV. The higgs field which breaks the electroweak symmetry is then a bound state of fermion bilinears just like the pions that break chiral symmetry in QCD. A concrete model realizing this idea is called "technicolor". However, most technicolor models have conflicts with electroweak precision measurements, leading in particular to the appearance of FCNC's, and are generally disfavored.

Another idea is to introduce a new symmetry. This provides a natural way to keep the mass of the higgs small and solve the mass hierarchy. After all, there are various examples where symmetries have been used to naturally obtain small masses for fermions and bosons. In the case of fermions, it is the chiral symmetry as it can be seen in the following simple Lagrangian

$$\mathcal{L}_F = i\bar{\psi}_L \not{D} \psi_L + i\bar{\psi}_R \not{D} \psi_R + m\left(\bar{\psi}_L \psi_R + \bar{\psi}_R \psi_L\right) \tag{6}$$

When $m = 0$ there is an enhanced symmetry $\psi_L \to e^{i\theta_1}\psi_L; \quad \psi_R \to e^{i\theta_2}\psi_R$ which breaks when the mass terms are included. Thus, imposing this symmetry on our model, one forbids any mass terms in the Lagrangian. Moreover, in their presence, since chiral symmetry is "softly" broken, radiative corrections behave like $\delta m \propto g^2 m$ where g is the gauge coupling (e.g. in QED with $\psi \equiv$ electron and $g \equiv e$). An alternative reason to understand why there is no $g^2\Lambda$ term, is because in a relativistic quantum field theory, any linear divergence cancels between the electron and positron contributions.

The corresponding symmetry for vector bosons is the gauge invariance. Consider for instance a massive abelian vector field:

$$\mathcal{L}_V = -\frac{1}{4}F_{\mu\nu}^2 - \frac{1}{2}m^2 A_\mu^2 \tag{7}$$

We see again that $m = 0$ respects $A_\mu \longrightarrow A_\mu + \partial_\mu \omega$. This is for instance how the gauge invariance of QED forbids any mass for the photon.

Is there a symmetry that can protect scalar masses, too? We can introduce for example a shift symmetry, implying that the Lagrangian should depend only through derivatives of the scalar field:

$$\phi \longrightarrow \phi + c \quad \Rightarrow \quad \mathcal{L}\left(\partial_\mu \phi\right) = \frac{1}{2}(\partial_\mu \phi)^2 + \frac{1}{\Lambda^4}\left[(\partial_\mu \phi)^2\right]^2 + \cdots \tag{8}$$

This is the case for a Goldstone boson without potential and in particular it does not hold when a quartic coupling is included. Scenarios like the "Little Higgs" take advantage of this property and treat the higgs as a pseudo-goldstone boson. In this case, the shift symmetry is broken by new gauge interactions, that generate quartic higgs interactions at higher orders. Another possibility is to use scale invariance:

$$x^\mu \longrightarrow a x^\mu \quad \varphi_d \longrightarrow a^{-d}\varphi(ax) \tag{9}$$

where d is the conformal dimension fo the field φ_d and $d = 1$ for a scalar field. This symmetry allows a quartic coupling but in a renormalizable theory is hardly broken by radiative corrections. In order to take advantage of this symmetry, the Standard Model should be embedded in a conformally invariant theory at the TeV scale.

Finally, we can use a different strategy by postulating a new symmetry that connects scalars to fermions or gauge fields:

$$\delta\phi = \xi\psi \quad \text{or} \quad \delta\phi = \epsilon_\mu A^\mu \tag{10}$$

Then the chiral or gauge invariance that protects the fermion or the gauge boson masses should also protect the masses of their scalar partners. This idea leads us to supersymmetry or to theories with extra dimensions, correspondingly. Indeed, the second case leads to the identification of the higgs field with an internal component of a higher-dimensional gauge field.

1.3. Advantages and problems of supersymmetry

We would like to conclude this introductory section by summarizing the main advantages and drawbacks of supersymmetry (SUSY). First of all, supersymmetry offers a solution to the hierarchy problem predicting a higgs mass that can be reached by colliders like LHC. Moreover, supersymmetry offers an excellent candidate for dark matter, the lightest supersymmetric particle. Another attractive feature is the gauge coupling unification a property that indicates that the theory might stay perturbative up to the GUT scale. On the conceptual level, we are given a framework where elementary scalars can be naturally included and treated on the same footing as fermions. Also, supersymmetry expanded our notion of spacetime including new dimensions of grassmannian nature. Last but not least, supersymmetry is a theory that makes specific predictions and can be verified or disproved experimentally in the real feature. If supersymmetry is true, there will be a rich spectrum of new particles within LHC reach.

On the other hand, one of the weak points is the fact that the MSSM without imposing phenomenological restrictions has far too many free parameters. Also, global symmetries like lepton and baryon number conservation are not automatic like in the Standard Model but need to be imposed, for instance by R-parity. Restrictions by hand also need to be imposed on the soft terms for suppressing dangerous FCNC processes. In addition there is a problem with the parameter μ. This is a SUSY mass parameter and a priori has nothing to do with SUSY breaking but its phenomenologically acceptable value is of the order of the soft breaking masses. Finally, there is some tension from experimental results in colliders, such as LEP2 and TEVATRON. The fact that they haven't seen any SUSY signatures yet has introduced already a fine tuning in the theory of the order of a few percent.

Because of these problems, several alternative theories have been proposed, that can also offer an explanation to the mass hierarchy, predicting new phenomena at the TeV scale. One radical idea, motivated by string theory, is based on extra dimensions (ordinary bosonic and not fermionic as in SUSY).

2. STRINGS AND EXTRA DIMENSIONS

In all physical theories, the number of dimensions is a free parameter fixed to three by observation, with one exception: string theory, which predicts the existence of six new spatial dimensions. This is the only known theory today that unifies the two great discoveries of 20th century: quantum mechanics, describing the behavior of elementary particles, and Einstein's General Relativity, describing gravitational phenomena in our Universe [1].

String theory replaces all elementary point-particles that form matter and its interactions with a single extended object of vanishing width: a tiny string. Thus, every known elementary particle, such as the electron, quark, photon or neutrino, corresponds to a particular vibration mode of the string (see Fig. 2). The diversity of these particles is due to the different properties of the corresponding string vibrations.

Until now, there is no experimental confirmation of string theory. No one has ever observed strings, not even indirectly, neither the space of extra dimensions where they live. The main arguments in its favor are theoretical, because it provides a coherent framework for unification of all fundamental interactions. For a long time, string physicists thought that strings were extremely thin, having the smallest possible size of physics, associated to the Planck length $\sim 10^{-35}$ meters.

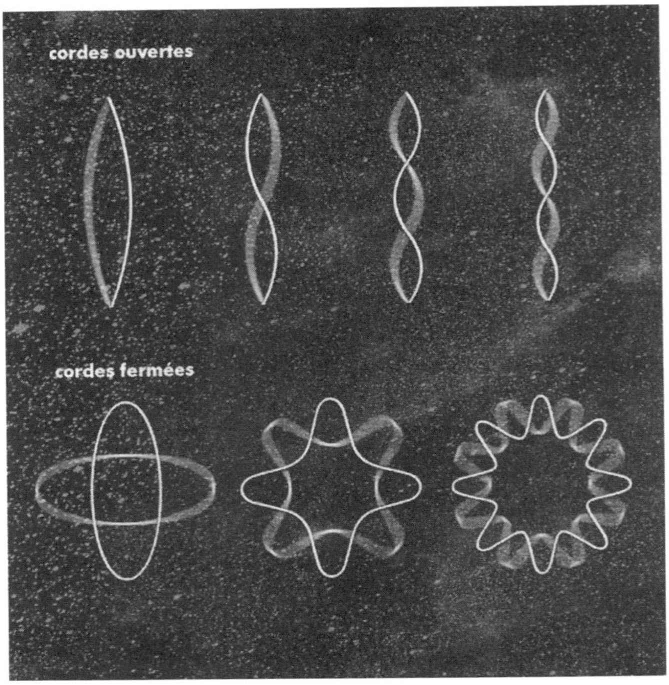

FIG. 2: *In string theory, the elementary constituent of matter is a miniscule string, having vanishing width but finite size. It can be open with free ends (upper part), or closed (lower part). Its vibration modes, like the ones shown above, correspond to various elementary particles.*

However, the situation changed drastically over the last decade. It has been realized that the "hidden" dimensions of string theory may be much larger than what we thought in the past and they become within experimental reach in the near future, together with the strings themselves [2, 3]. These ideas lead in particular to experimental tests of string theory that can be performed at TEVATRON and LHC.

The main motivation for these ideas came from considerations of the hierarchy problem that was described in the previous section. As mentioned already, a possible solution is provided by the introduction of supersymmetry. The appropriate and most convenient framework for low energy supersymmetry and grand unification is the perturbative heterotic string. Indeed, in this theory, gravity and gauge interactions have the same origin, as massless modes of the closed heterotic string, and they are unified at the string scale M_s. As a result, the Planck mass M_P is predicted to be proportional to M_s:

$$M_P = M_s/g\,, \tag{11}$$

where g is the gauge coupling. In the simplest constructions all gauge couplings are the same at the string scale, given by the four-dimensional (4d) string coupling, and thus no grand unified group is needed for unification. In our conventions $\alpha_{\rm GUT} = g^2 \simeq 0.04$, leading to a discrepancy between the string and grand unification scale $M_{\rm GUT}$ by almost two orders of magnitude. Explaining this gap introduces in general new parameters or a new scale, and the predictive power is essentially lost. This is the main defect of this framework, which remains though an open and interesting possibility [4].

On the other hand, a new idea has been proposed that solves the problem if the fundamental string length is fixed to $10^{-18} - 10^{-19}$ meters [3]. In this case, quantum corrections are controlled by the string scale, which is in the TeV region, and do not destabilize the masses of elementary particles. Moreover, it offers the remarkable possibility that string physics may be testable soon in particle colliders.

How is possible to lower the string scale from the traditional quantum gravity Planck scale to the TeV region without contradicting observations? in particular, what happens to the extra dimensions of string theory and why they have not been observed? A crucial role is played by the discovery of p-branes which are higher dimensional objects extended in p spatial dimensions, generalizing the notion of point-particle ($p = 0$) and string ($p = 1$). A main consequence of this discovery is that the string scale is in general a free parameter that can be dissociated from the Planck mass, if the universe is localized on a p-brane and does not feel all extra dimensions of string theory. The braneworld description of our universe separates the dimensions of space in two groups: those extended along our p-braneworld, called parallel, and those transverse to it.

If our universe has additional dimensions, we should observe new phenomena related to their existence. Why nobody has detected them until now? A possible answer was given in the beginning of 20th century by Kaluza and Klein [5]: because the size of the new dimensions is very small, in contrast to the size of the other three that we know, which is infinitely large. An infinite and narrow cylinder for example is a two-dimensional space, with one dimension forming a very small cycle: one can move infinitely far away along the axis, while one returns back at the same point when moving along the orthogonal direction (see fig. 3). If one of the three known dimensions of space was small, say of millimeter size, we would be flat and, while we could move freely towards left or right, forward or backward, it would be impossible to do more than a few millimeters up or down where space ends. It follows that extra dimensions along our universe escape observation if their size is less than 10^{-18} meters, requiring energies higher than those we dispose in present experiments to be explored [2, 6].

The next question is how we can detect their existence if we dispose sufficient energy to probe their size? The minimum energy required is given by their inverse size and is called compactification scale. The answer was given again by Kaluza and Klein: the motion of a particle in extra dimensions of finite size manifests to us as a tower of massive particles, called Kaluza-Klein excitations. If for instance the photon propagates along an extra compact dimension, one should observe a tower of massive particles with the same properties as the photon but with a mass that becomes higher as the size of the extra dimension is getting smaller. It follows, that for a size of order 10^{-18} meters, an energy of order of a few TeV would be sufficient to produce them.

The above analysis and bound does not apply though for transverse dimensions to our universe, since in this case we cannot send light or any form of observable matter to probe

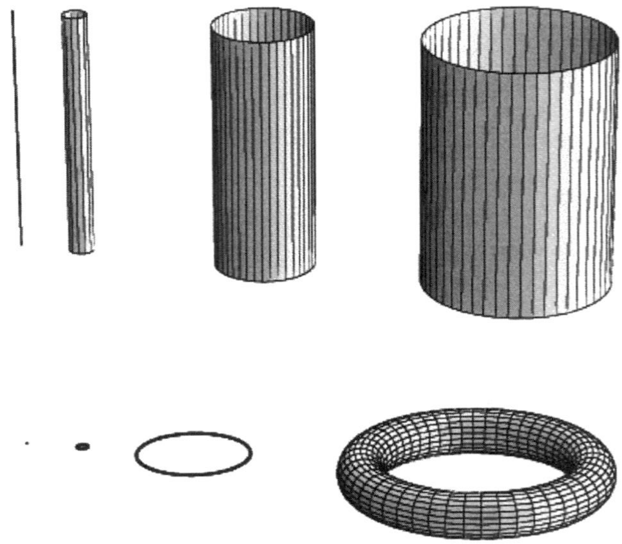

FIG. 3: *Possible forms of small extra dimensions of space. Far away they are unobservable, but at distances comparable to their size we start feeling their existence and exploring their shapes.*

their existence. The only way to communicate is through gravity which couples to any kind of energy density. However, our knowledge of gravity at short distances is much weaker than the other interactions, allowing for sizes of such "hidden" dimensions as large as almost a millimeter. This is roughly the shortest distance where Newton's law is tested in the laboratory.

The appropriate string perturbative framework that provides such a description and has been studied extensively in the more recent years is type I string theory with D-branes. Unlike in the heterotic string, gauge and gravitational interactions have now different origin. The latter are described again by closed strings, while the former emerge as excitations of open strings with endpoints confined on D-branes [7]. This leads to a braneworld description of our universe, which should be localized on a hypersurface, i.e. a p-brane extended in p spatial dimensions (see Fig. 4). Closed strings propagate in all nine dimensions of string theory: in those extended along the p-brane (parallel), as well as in the transverse ones. On the contrary, open strings are attached on the p-brane. Obviously, our p-brane world must have at least the three known dimensions of space. But it may contain more: the extra $d_{\|} = p - 3$ parallel dimensions must have a finite size, in order to be unobservable at present energies, and can be as large as $\text{TeV}^{-1} \sim 10^{-18}$ m. On the other hand, transverse dimensions interact with us only gravitationally and experimental bounds are much weaker: their size should be less than about 0.1 mm [8]. In the following, I review the main properties and experimental signatures of low string scale models [3, 9].

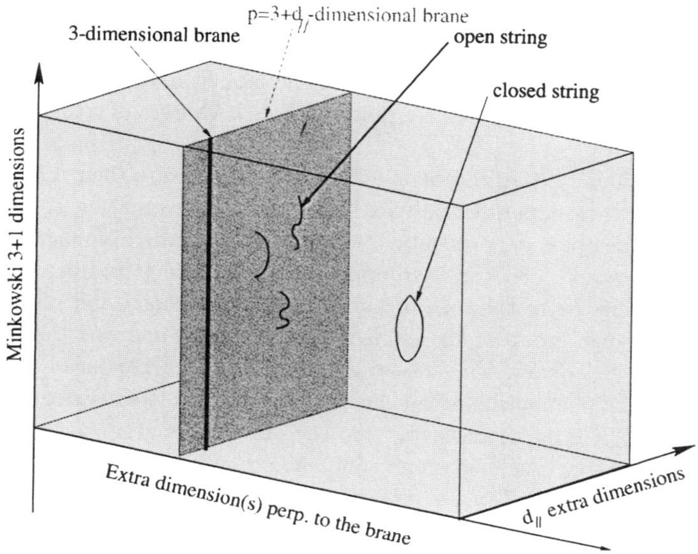

FIG. 4:

3. FRAMEWORK OF HETEROTIC STRING

Here, I discuss the general model independent properties and possible predictions of four-dimensional (4d) strings obtained upon compactification of the 10d perturbative Heterotic preserving $N = 1$ supersymmetry.[10] The gauge group can be as large as a group of rank 22.

3.1. Exotic representations and/or fractional electric charges

Every gauge group factor is characterized by a positive integer k, the level of the underlying 2d Kac-Moody algebra. This integer determines the corresponding tree-level gauge coupling constant, $g = g_4/\sqrt{k}$, while simultaneously it restricts the possible massless matter representations. For $k > 1$, large "exotic" representations are in general present in the spectrum, like color octets, $SU(2)$ triplets, etc. On the contrary, $k = 1$ constructions, which include generic free-fermionic models, guarantee that the only possible massless representations are fundamental and antisymmetric of unitary groups or vectors and spinors of orthogonal groups, which is consistent with the present experimental observations. However, in this case, the observed electric charge quantization cannot be imposed together with a value for the weak angle $\sin^2\theta_W = 3/8$ at the unification scale.[11, 12] Consequently, color singlet states with fractional electric charges are unavoidable, unless the weak angle is unacceptably small ($\sin^2\theta_W \leq 3/20$). In free-fermionic models a weaker charge quantization condition can always be imposed, namely that the electric charges of color singlet states are half-integers.

The lightest fractionally charged particle is stable. Generically their mass is in the TeV region, which seems to be experimentally excluded, since an estimation of their relic abundancies contradicts the upper experimental bounds by several orders of magnitude.[13] There are two possible ways out. The first is to make them superheavy and use inflation to suppress their abundancies as in the case of monopoles. The second and most natural is to confine them by a gauge group of the hidden sector, in the same way that QCD confines the fractionally charged quarks into integrally charged bound states.[14] This possibility may lead to interesting phenomenological consequences related to the existence of long-lived "crypto"-baryons.[15] It is also conceivable that the same gauge group is also responsible for dynamical supersymmetry breaking via gaugino and scalar condensation, in which case the confining scale must be in the TeV region.[16]

3.2. General properties of model building

It is by now known that string theory does not lead to a unique way to determine the vacuum. However, it provides a framework for model building with rules which are much more restrictive than in ordinary field theory. The simplest models correspond to $k = 1$ constructions and they are subject to several phenomenological constraints. On the one hand they have to provide solutions to basic outstanding problems, as the generation of the observed fermion mass hierarchies, and on the other hand they should not suffer from obvious diseases, as fast proton decay, or flavor changing neutral currents, etc. These may be imposed using discrete symmetries at the level of the Standard model, or proceeding through some grand unified group. In the second case ordinary GUT's should be modified, since $k = 1$ constructions have no higgses in the adjoint or in higher self-conjugate representations usually needed to break the GUT group.

In analogy with Calabi-Yau compactifications which lead to $E_6 \times E_8$ gauge symmetry, there is a class of free-fermionic 4d strings which lead to an intermediate step of $SO(10) \times U(1)^n \times SO(16)$. The gauge group is broken to a product of an "observable" with a "hidden" sector (modulo the $U(1)$ factors), $G_{\text{obs}} \times U(1)^n \times G_{\text{hid}}$, by 2d fermion number projections which correspond to the Wilson line breaking of E_6 in Calabi-Yau models. These projections also reduce the number of families to three. A general property of these constructions is that every fermion generation forms a spinor of $SO(10)$ instead of a **27** representation of E_6 and, thus, it contains just one more state, the right-handed neutrino, besides the known Standard model content. There are three distinct possibilities for G_{obs}, associated to the three different subgroups of $SO(10)$ which can be broken to the Standard model using higgses in low dimensional representations: the flipped $SU(5) \times U(1)$,[11, 14]the Pati-Salam left-right symmetric model $SU(4) \times SU(2)_L \times SU(2)_R$,[18]or directly the Standard-like model $SU(3) \times SU(2)_L \times U(1)_Y \times U(1)'$.[19]

The general strategy of this procedure, before introducing the breaking of space-time supersymmetry, is focusing two main issues: The gauge symmetry breaking and the fermion mass problem. The gauge symmetry breaking contains four steps:
(i) The breaking of the additional $U(1)$ factors. This is achieved because one particular linear combination $U(1)_A$ appears often to be anomalous at the tree-level. The anomaly is canceled by a generalization of the Green-Schwarz mechanism in four dimensions, implying that the axion is absorbed by the $U(1)_A$ gauge field which becomes massive at the one-loop level.[20] Furthermore, the corresponding $U(1)_A$ D-term is modified by an additional "constant" proportional to the anomaly $\mathrm{Tr}Q_A$:

$$D_A = \sum_i Q_A^i |\Phi_i|^2 + \frac{\mathrm{Tr}Q_A}{96\pi^2\alpha'} , \tag{12}$$

where Φ_i are the various scalars. As a result, some of the previously flat directions are destroyed, while the supersymmetric minimization conditions of F and D-flatness are satisfied by non vanishing VEV's of scalar fields which in general break all extra $U(1)$'s at a calculable scale M_A. For typical values of $\mathrm{Tr}Q_A \sim 1 - 100$, one finds a scale which is two to one orders of magnitude less than the string scale, $M_A \sim 10^{16} - 10^{17}$ GeV.
(ii) The breaking of the GUT-type group G_{obs} to the Standard model by a Higgs mechanism, at a scale $M_{\mathrm{GUT}} \sim 10^{16}$ GeV which could have the same or a different origin than M_A.
(iii) The condensation of the hidden group G_{hid} at some intermediate scale Λ_{cond}, which should also confine the fractionally charged states and guarantee the observed electric charge quantization. The hidden group may also be related with the dynamical breaking of space-time supersymmetry via gaugino condensation.[21]
(iv) The usual $SU(2) \times U(1)$ breaking which needs one pair of massless Higgs doublets. This is in general a problem in any construction, since vector-like representations tend to become massive.

Concerning the fermion masses, the strategy is the following:[14]
Only one generation has trilinear Yukawa couplings and aquires a mass at the lowest order in the α'-expansion. Because of the underlying $SO(10)$ symmetry one obtains $m_b = m_\tau$ at the unification scale, which leads to a successful prediction for the bottom to tau mass ratio at low energy, after taking into account the renormalization group evolution. Furthermore, the top Yukawa coupling λ_t is proportional to the gauge coupling at M_{st} implying that the top is in general heavy. In fact an upper bound can be obtained, corresponding to the limiting case $\lambda_t = \lambda_b = \lambda_\tau = \sqrt{2}g_4$ at M_{st} which is valid in the absence of any mixing:[14]

$$m_t \lesssim 180\mathrm{GeV} . \tag{13}$$

All other masses and mixings involving the other two generations appear in α'-corrections which correspond to non renormalizable terms in the effective superpotential. Consequently, the corresponding Yukawa couplings are naturally suppressed by powers of $M_A(\alpha')^{1/2}$, or $M_{\mathrm{GUT}}(\alpha')^{1/2}$, generated by the various scalar VEV's.

4. FRAMEWORK OF LOW SCALE STRINGS

In type I theory, the different origin of gauge and gravitational interactions implies that the relation between the Planck and string scales is not linear as (11) of the heterotic string. The requirement that string theory should be weakly coupled, constrain the size of all parallel dimensions to be of order of the string length, while transverse dimensions remain unrestricted. Assuming an isotropic transverse space of $n = 9 - p$ compact dimensions of common radius $R_\perp$, one finds:

$$M_P^2 = \frac{1}{g_s^2} M_s^{2+n} R_\perp^n \,, \qquad g_s \simeq g^2 \quad \text{for D-branes} \,, \tag{14}$$

where g_s is the string coupling. It follows that the type I string scale can be chosen hierarchically smaller than the Planck mass [3, 22] at the expense of introducing extra large transverse dimensions felt only by gravity, while keeping the string coupling small [3]. The weakness of 4d gravity compared to gauge interactions (ratio M_W/M_P) is then attributed to the largeness of the transverse space $R_\perp$ compared to the string length $l_s = M_s^{-1}$.

An important property of these models is that gravity becomes effectively $(4 + n)$-dimensional with a strength comparable to those of gauge interactions at the string scale. The first relation of Eq. (14) can be understood as a consequence of the $(4+n)$-dimensional Gauss law for gravity, with

$$M_*^{(4+n)} = M_s^{2+n}/g^4 \tag{15}$$

the effective scale of gravity in $4 + n$ dimensions. Taking $M_s \simeq 1$ TeV, one finds a size for the extra dimensions $R_\perp$ varying from 10^8 km, .1 mm, down to a Fermi for $n = 1, 2$, or 6 large dimensions, respectively. This shows that while $n = 1$ is excluded, $n \geq 2$ is allowed by present experimental bounds on gravitational forces [8, 23]. Thus, in these models, gravity appears to us very weak at macroscopic scales because its intensity is spread in the "hidden" extra dimensions. At distances shorter than $R_\perp$, it should deviate from Newton's law, which may be possible to explore in laboratory experiments (see Fig. 5).

The main experimental implications of TeV scale strings in particle accelerators are of three types, in correspondence with the three different sectors that are generally present: (i) new compactified parallel dimensions, (ii) new extra large transverse dimensions and low scale quantum gravity, and (iii) genuine string and quantum gravity effects. On the other hand, there exist interesting implications in non accelerator table-top experiments due to the exchange of gravitons or other possible states living in the bulk.

5. LARGE NUMBER OF SPECIES

Here, we open a parenthesis to describe that low scale gravity with large extra dimensions is actually a particular case of a more general framework, where the ultraviolate (UV) cutoff is lower than the Panck scale due to the existence of a large number of particle species coupled to gravity [24]. Indeed, it was shown that the effective UV cutoff M_* is given by

$$M_*^2 = M_P^2/N \,, \tag{16}$$

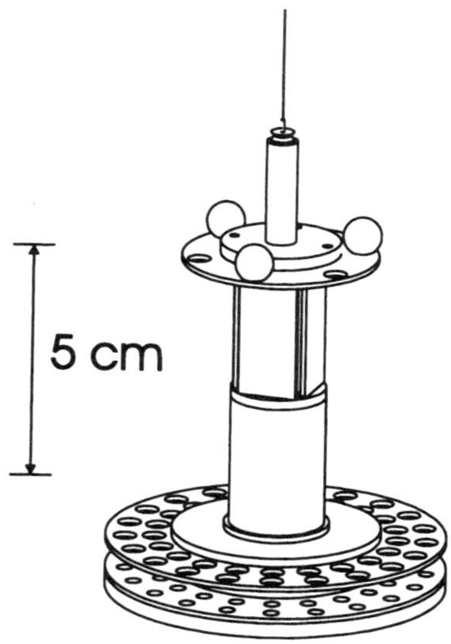

FIG. 5:

where the counting of independent species N takes into account all particles which are not broad resonances, having a width less than their mass. The derivation is based on black hole evaporation but here we present a shorter argument using quantum information storage [25]. Consider a pixel of size L containing N species storing information. The energy required to localize N wave functions is then given by N/L, associated to a Schwarzschild radius $R_s = N/LM_P^2$. The latter must be less than the pixel size in order to avoid the collapse of such a system to a black hole, $R_s \leq L$, implying a minimum size $L \geq L_{min}$ with $L_{min} = \sqrt{N}/M_P$ associated precisely to the effective UV cutoff $M_* = L_{min}$ given in eq. (16). Imposing $M_* \simeq 1$ TeV, one should then have $N \sim 10^{32}$ particle species below about the TeV scale!

In the string theory context, there are two ways of realizing such a large number a particle species by lowering the string scale at a TeV:

1. In large volume compactifications with the SM localized on D-brane stacks, as described in the previous section. The particle species are then the Kaluza-Klein (KK) excitations of the graviton (and other possible bulk modes) associated to the large extra dimensions, given by $N = R_\perp^n l_s^n$, up to energies of order $M_* \simeq M_s$.
2. By introducing an infinitesimal string coupling $g_s \simeq 10^{-16}$ with the SM localized on Neveu-Schwarz NS5-branes in the framework of little strings [26]. In this case, the particle species are the effective number of string modes that contribute to the black hole bound [27]: $N = 1/g_s^2$ and gravity does not become strong at $M_s \sim \mathcal{O}(\text{TeV})$.

Note the both TeV string realizations above are compatible with the general expression (14), but in the second case there is no relation between the string and gauge couplings.

6. WARPED SPACES

In these models, space-time is a slice of anti de Sitter space (AdS) in $d = 5$ dimensions while our universe forms a four-dimensional (4d) flat boundary [28]. The corresponding line element is:

$$ds^2 = e^{-2k|y|}\eta_{\mu\nu}dx^\mu dx^\nu + dy^2 \quad ; \quad \Lambda = -24M^3k^2\,, \tag{17}$$

where M, Λ are the 5d Planck mass and cosmological constant, respectively, and the parameter k is the curvature of AdS_5. The fifth coordinate y is restricted on the interval $[0, \pi r_c]$. Thus, this model requires two 'branes', a UV and an IR, located at the two end-points of the interval, $y = 0$ and $y = \pi r_c$, respectively. The vanishing of the 4d cosmological constant requires to fine tune the two tensions: $T = -T' = 24M^3k^2$. The 4d Planck mass is given by:

$$M_P^2 = \frac{1}{k}(1 - e^{-2\pi k r_c})M^3\,. \tag{18}$$

On the other hand, there is an exponential hierarchy of the mass scales in the IR relative to the UV brane due to the warped factor in the metric (17). One can then identify the electroweak scale $M_W \simeq M_P e^{-2\pi k r_c}$.

By studying the corresponding wave functions, one can show that 4d gravity is localized on the UV-brane, while KK gravitons on the IR. It follows that the main prediction of this model is the existence of spin-2 resonances at the TeV scale coupled with (TeV)-1 strength. Their masses are given by:

$$m_n = c_n k e^{-2\pi k r_c} \quad ; \quad c_n \simeq (n + \frac{1}{4})\text{for large } n \tag{19}$$

and they are weakly coupled for $m_n < M e^{-2\pi k r_c}$, *i.e.* for $k < M$. Viable models can be constructed when Standard Model gauge bosons propagate in the 5d bulk, fermions are localized near the UV brane, while the Higgs near the IR [29]. Using AdS/CFT correspondence, they are dual to strongly coupled 4d field theories with composite Higgs; their gauge coupling $g_{YM} \sim M/k$ is strong when spin-2 KK modes are weakly coupled in the gravity dual description.

7. EXPERIMENTAL IMPLICATIONS IN ACCELERATORS

7.1. World-brane extra dimensions

In this case $RM_s \gtrsim 1$, and the associated compactification scale $R_\parallel^{-1}$ would be the first scale of new physics that should be found increasing the beam energy [2, 6, 30]. There are several reasons for the existence of such dimensions. It is a logical possibility, since out of the six extra dimensions of string theory only two are needed for lowering the string scale, and thus the effective p-brane of our world has in general $d_\parallel \equiv p - 3 \leq 4$. Moreover, they can be used to address several physical problems in braneworld models, such as obtaining different SM gauge couplings, explaining fermion mass hierarchies due to different localization points of quarks and leptons in the extra dimensions, providing calculable mechanisms of supersymmetry breaking, etc.

The main consequence is the existence of Kaluza-Klein (KK) excitations for all SM particles that propagate along the extra parallel dimensions. Their masses are given by:

$$M_m^2 = M_0^2 + \frac{m^2}{R_\parallel^2} \quad ; \quad m = 0, \pm1, \pm2, \ldots \tag{20}$$

where we used $d_\parallel = 1$, and M_0 is the higher dimensional mass. The zero-mode $m = 0$ is identified with the 4d state, while the higher modes have the same quantum numbers with the lowest one, except for their mass given in (20). There are two types of experimental signatures of such dimensions [6, 31, 32]: (i) virtual exchange of KK excitations, leading to deviations in cross-sections compared to the SM prediction, that can be used to extract bounds on the compactification scale; (ii) direct production of KK modes.

On general grounds, there can be two different kinds of models with qualitatively different signatures depending on the localization properties of matter fermion fields. If the latter are localized in 3d brane intersections, they do not have excitations and KK momentum is not conserved because of the breaking of translation invariance in the extra dimension(s). KK modes of gauge bosons are then singly produced giving rise to generally strong bounds on the compactification scale and new resonances that can be observed in experiments. Otherwise, they can be produced only in pairs due to the KK momentum conservation, making the bounds weaker but the resonances difficult to observe.

When the internal momentum is conserved, the interaction vertex involving KK modes has the same 4d tree-level gauge coupling. On the other hand, their couplings to localized matter have an exponential form factor suppressing the interactions of heavy modes. This form factor can be viewed as the fact that the branes intersection has a finite thickness. For instance, the coupling of the KK excitations of gauge fields $A^\mu(x,y) = \sum_m A_m^\mu \exp i\frac{my}{R_\parallel}$ to the charge density $j_\mu(x)$ of massless localized fermions is described by the effective action [33]:

$$\int d^4x \sum_m e^{-\ln 16 \frac{m^2 l_s^2}{2R_\parallel^2}} j_\mu(x)\, A_m^\mu(x)\,. \tag{21}$$

After Fourier transform in position space, it becomes:

$$\int d^4x\, dy\, \frac{1}{(2\pi \ln 16)^2} e^{-\frac{y^2 M_s^2}{2\ln 16}} j_\mu(x)\, A^\mu(x,y)\,, \tag{22}$$

from which we see that localized fermions form a Gaussian distribution of charge with a width $\sigma = \sqrt{\ln 16}\, l_s \sim 1.66\, l_s$.

To simplify the analysis, let us consider first the case $d_\parallel = 1$ where some of the gauge fields arise from an effective 4-brane, while fermions are localized states on brane intersections. Since the corresponding gauge couplings are reduced by the size of the large dimension $R_\parallel M_s$ compared to the others, one can account for the ratio of the weak to strong interactions strengths if the $SU(2)$ brane extends along the extra dimension, while $SU(3)$ does not. As a result, there are 3 distinct cases to study [32], denoted by (t,l,l), (t,l,t) and (t,t,l), where the three positions in the brackets correspond to the three SM gauge group factors $SU(3) \times SU(2) \times U(1)$ and those with l (longitudinal) feel the extra dimension, while those with t (transverse) do not.

In the (t, l, l) case, there are KK excitations of $SU(2) \times U(1)$ gauge bosons: $W_{\pm}^{(m)}$, $\gamma^{(m)}$ and $Z^{(m)}$. Performing a χ^2 fit of the electroweak observables, one finds that if the Higgs is a bulk state (l), $R_{\parallel}^{-1} \gtrsim 3.5$ TeV [34]. This implies that LHC can produce at most the first KK mode. Different choices for localization of matter and Higgs fields lead to bounds, lying in the range $1 - 5$ TeV [34].

In addition to virtual effects, KK excitations can be produced on-shell at LHC as new resonances [31] (see Fig. 6). There are two different channels, neutral Drell–Yan processes

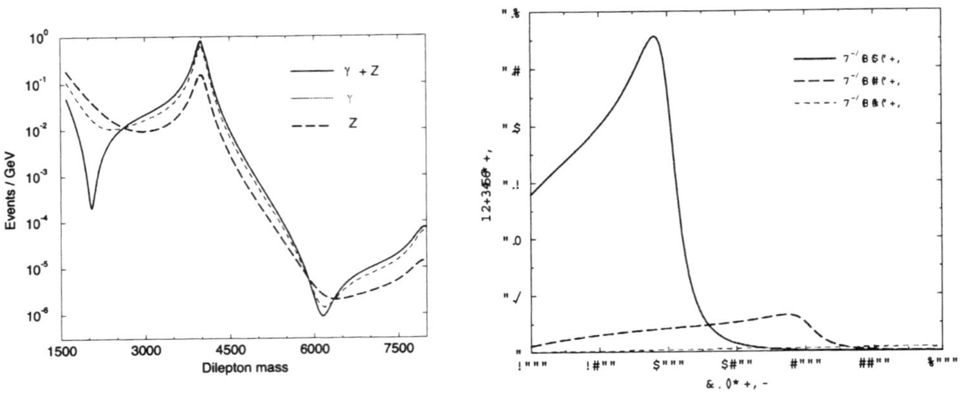

FIG. 6:

$pp \to l^+l^-X$ and the charged channel $l^{\pm}\nu$, corresponding to the production of the KK modes $\gamma^{(1)}, Z^{(1)}$ and $W_{\pm}^{(1)}$, respectively. The discovery limits are about 6 TeV, while the exclusion bounds 15 TeV. An interesting observation in the case of $\gamma^{(1)} + Z^{(1)}$ is that interferences can lead to a "dip" just before the resonance. There are some ways to distinguish the corresponding signals from other possible origin of new physics, such as models with new gauge bosons. In fact, in the (t, l, l) and (t, l, t) cases, one expects two resonances located practically at the same mass value. This property is not shared by most of other new gauge boson models. Moreover, the heights and widths of the resonances are directly related to those of SM gauge bosons in the corresponding channels.

In the (t, l, t) case, only the $SU(2)$ factor feels the extra dimension and the limits set by the KK states of $W^{\pm}$ remain the same. On the other hand, in the (t, t, l) case where only $U(1)_Y$ feels the extra dimension, the limits are weaker and the exclusion bound is around 8 TeV. In addition to these simple possibilities, brane constructions lead often to cases where part of $U(1)_Y$ is t and part is l. If $SU(2)$ is l the limits come again from $W^{\pm}$, while if it is t then it will be difficult to distinguish this case from a generic extra $U(1)'$. A good statistics would be needed to see the deviation in the tail of the resonance as being due to effects

TABLE I:

Experiment	$n=2$	$n=4$	$n=6$
Collider bounds			
LEP 2	5×10^{-1}	2×10^{-8}	7×10^{-11}
Tevatron	5×10^{-1}	10^{-8}	4×10^{-11}
LHC	4×10^{-3}	6×10^{-10}	3×10^{-12}
NLC	10^{-2}	10^{-9}	6×10^{-12}
Present non-collider bounds			
SN1987A	3×10^{-4}	10^{-8}	6×10^{-10}
COMPTEL	5×10^{-5}	-	-

additional to those of a generic $U(1)'$ resonance. Finally, in the case of two or more parallel dimensions, the sum in the exchange of the KK modes diverges in the limit $R_{\parallel}M_s >> 1$ and needs to be regularized using the form factor (21). Cross-sections become bigger yielding stronger bounds, while resonances are closer implying that more of them could be reached by LHC.

On the other hand, if all SM particles propagate in the extra dimension (called universal)[63], KK modes can only be produced in pairs and the lower bound on the compactification scale becomes weaker, of order of 300-500 GeV. Moreover, no resonances can be observed at LHC, so that this scenario appears very similar to low energy supersymmetry. In fact, KK parity can even play the role of R-parity, implying that the lightest KK mode is stable and can be a dark matter candidate in analogy to the LSP [35].

7.2. Extra large transverse dimensions

The main experimental signal is gravitational radiation in the bulk from any physical process on the world-brane. In fact, the very existence of branes breaks translation invariance in the transverse dimensions and gravitons can be emitted from the brane into the bulk. During a collision of center of mass energy $\sqrt{s}$, there are $\sim(\sqrt{s}R_\perp)^n$ KK excitations of gravitons with tiny masses, that can be emitted. Each of these states looks from the 4d point of view as a massive, quasi-stable, extremely weakly coupled (s/M_P^2 suppressed) particle that escapes from the detector. The total effect is a missing-energy cross-section roughly of order:

$$\frac{(\sqrt{s}R_\perp)^n}{M_P^2} \sim \frac{1}{s}\left(\frac{\sqrt{s}}{M_s}\right)^{n+2}. \tag{23}$$

Explicit computation of these effects leads to the bounds given in Table I. However, larger radii are allowed if one relaxes the assumption of isotropy, by taking for instance two large dimensions with different radii.

Fig. 7 shows the cross-section for graviton emission in the bulk, corresponding to the process $pp \to jet + graviton$ at LHC, together with the SM background [36]. For a given

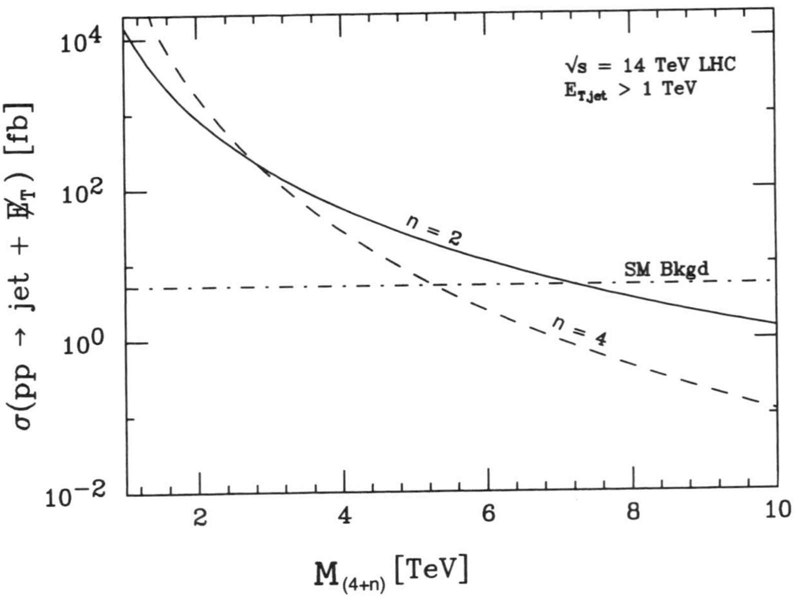

FIG. 7:

value of M_s, the cross-section for graviton emission decreases with the number of large transverse dimensions, in contrast to the case of parallel dimensions. The reason is that gravity becomes weaker if there are more dimensions because there is more space for the gravitational field to escape. There is a particular energy and angular distribution of the produced gravitons that arise from the distribution in mass of KK states of spin-2. This can be contrasted to other sources of missing energy and might be a smoking gun for the extra dimensional nature of such a signal.

In Table I, there are also included astrophysical and cosmological bounds. Astrophysical bounds [37, 38] arise from the requirement that the radiation of gravitons should not carry on too much of the gravitational binding energy released during core collapse of supernovae. In fact, the measurements of Kamiokande and IMB for SN1987A suggest that the main channel is neutrino fluxes. The best cosmological bound [39] is obtained from requiring that decay of bulk gravitons to photons do not generate a spike in the energy spectrum of the photon background measured by the COMPTEL instrument. Bulk gravitons are expected to be produced just before nucleosynthesis due to thermal radiation from the brane. The limits assume that the temperature was at most 1 MeV as nucleosynthesis begins, and become stronger if temperature is increased.

7.3. String effects

At low energies, the interaction of light (string) states is described by an effective field theory. Their exchange generates in particular four-fermion operators that can be used to extract independent bounds on the string scale. In analogy with the bounds on longitudinal extra dimensions, there are two cases depending on the localization properties of matter fermions. If they come from open strings with both ends on the same stack of branes, exchange of massive open string modes gives rise to dimension eight effective operators, involving four fermions and two space-time derivatives [33, 40]. The corresponding bounds on the string scale are then around 500 GeV. On the other hand, if matter fermions are localized on non-trivial brane intersections, one obtains dimension six four-fermion operators and the bounds become stronger: $M_s \gtrsim 2-3$ TeV [9, 33]. At energies higher than the string scale, new spectacular phenomena are expected to occur, related to string physics and quantum gravity effects, such as possible micro-black hole production [41, 42, 43]. Particle accelerators would then become the best tools for studying quantum gravity and string theory.

Direct production of string resonances in hadron colliders leads generically to a universal deviation from Standard Model in jet distribution [44]. In particular, the first Regge excitation of the gluon has spin 2 and a width an order of magnitude lower than the string scale, leading to a characteristic peak in dijet production; similarly, the first excitations of quarks have spin 3/2. The dijet cross-section is shown in Fig. 8 for LHC energies. The reason for the universal behavior is that tree N-point open superstring amplitudes involving at most two fermions and gluons are completely model independent from the details of the compactification, including the number of supersymmetries that are left unbroken in four dimensions (even if all are broken). Such tree-level amplitudes do not receive contributions from KK, string winding or closed string graviton modes, but are given as a universal sum over exchanges of open string Regge excitations lying in Regge trajectories with masses

$$M_n^2 = M_s^2 n \quad ; \quad n = 0, 1, \ldots \tag{24}$$

and maximal spin $n+1$.

The relevant partonic cross-sections for dijet production, involving at most two quarks are $|\mathcal{M}(gg \to gg)|^2$, $|\mathcal{M}(gg \to q\bar{q})|^2$, $|\mathcal{M}(q\bar{q} \to gg)|^2$ and $|\mathcal{M}(qg \to qg)|^2$, where g and q denote gluons and quarks, respectively. They can be obtained from the following two independent processes, up to crossing symmetries [45]:

$$\begin{aligned} |\mathcal{M}(gg \to gg)|^2 &= g^4 \left(\frac{1}{s^2} + \frac{1}{t^2} + \frac{1}{u^2} \right) \times \left[\frac{9}{4} \left(s^2 V_s^2 + t^2 V_t^2 + u^2 V_u^2 \right) - \frac{1}{3} \left(sV_s + tV_t + uV_u \right)^2 \right] \\ |\mathcal{M}(gg \to q\bar{q})|^2 &= g^4 \frac{t^2 + u^2}{s^2} \left[\frac{1}{6} \frac{1}{tu} (tV_t + uV_u)^2 - \frac{3}{8} V_t V_u \right] , \end{aligned} \tag{25}$$

where s, t, u are the usual Mandelstam variables in string mass units and

$$V_s = -\frac{tu}{s}\, B(t,u) = 1 - \frac{2}{3}\pi^2\, tu + \ldots \qquad V_t : s \leftrightarrow t \qquad V_u : s \leftrightarrow u\,. \tag{26}$$

The low energy expansion of these amplitudes reproduce the usual QCD expressions (upon inserting the color factors), while higher order terms describe the string corrections due to the

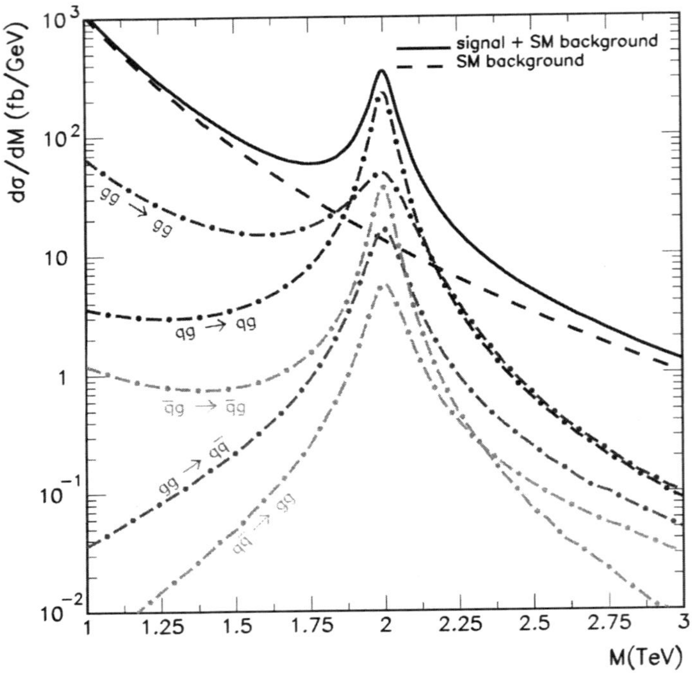

FIG. 8:

exchange of Regge excitations. Besides these amplitudes, there are also those involving four quarks, such as $|\mathcal{M}(q\bar{q} \to q\bar{q})|^2$ and $|\mathcal{M}(qq \to qq)|^2$. These are model dependent, because the details of the compactification do not decouple. The reason is that they involve four vertices containing twist fields that describe quark states arising from open strings stretched in brane intersections. However, taking into account QCD color factors, their contribution is suppressed because parton luminosities in proton-proton collisions at TeV energies favor at least one gluon in the initial state. As a result, the dominant contribution comes from the model independent cross-sections (25) leading to the effect of Fig. 8.

We finish this section with some comments related to the possible micro-black hole production. Independently on the unresolved issue of the convergence of string perturbation theory in the kinematic region relevant to micro-black hole formation, there is a simple argument showing that at least within the perturbative TeV string framework, the energy threshold for black hole production is far above the LHC reach. Indeed, a string size black hole has a horizon radius $r_H \sim 1$ in string units, while the d-dimensional Newton's constant behaves as $G_N \sim g_s^2$. It follows that the mass of a d-dimensional black hole is [46]:

$$M_{\rm BH} \sim r_H^{d/2-1}/G_N \simeq 1/g_s^2\,. \tag{27}$$

Thus, for a weakly coupled theory, this energy threshold is much higher than the string and the higher dimensional Planck scales M_s and M_* of eq. (15). Comparing this energy threshold with the mass of Regge excitations (24), one finds $n \sim 1/g_s^4$ which is actually compatible with the relation one obtains by identifying the black hole entropy $S_{\rm BH} \sim 1/G_N \sim 1/g_s^2$ with the perturbative string entropy $S_{\rm string} \sim \sqrt{n}$. Using now relation (14), and the value of the Standard Model gauge couplings $g_s \simeq g^2 \sim 0.1$, one finds that the energy threshold $M_{\rm BH}$ of micro-black hole production is about four orders of magnitude higher than the string scale, implying that one would produce 10^4 string states before reaching $M_{\rm BH}$.

8. SUPERSYMMETRY IN THE BULK AND SHORT RANGE FORCES

Besides the spectacular predictions in accelerators, there are also modifications of gravitation in the sub-millimeter range, which can be tested in "table-top" experiments that measure gravity at short distances. There are three categories of such predictions:
(i) Deviations from the Newton's law $1/r^2$ behavior to $1/r^{2+n}$, which can be observable for $n = 2$ large transverse dimensions of sub-millimeter size. This case is particularly attractive on theoretical grounds because of the logarithmic sensitivity of SM couplings on the size of transverse space [47], that allows to determine the hierarchy [48].
(ii) New scalar forces in the sub-millimeter range, related to the mechanism of supersymmetry breaking, and mediated by light scalar fields φ with masses [3, 49]:

$$m_\varphi \simeq \frac{m_{susy}^2}{M_P} \simeq 10^{-4} - 10^{-6}\ \text{eV}\,, \tag{28}$$

for a supersymmetry breaking scale $m_{susy} \simeq 1 - 10$ TeV. They correspond to Compton wavelengths of 1 mm to 10 μm. m_{susy} can be either $1/R_\parallel$ if supersymmetry is broken by compactification [49], or the string scale if it is broken "maximally" on our world-brane [3]. A universal attractive scalar force is mediated by the radion modulus $\varphi \equiv M_P \ln R$, with R the radius of the longitudinal or transverse dimension(s). In the former case, the result (28) follows from the behavior of the vacuum energy density $\Lambda \sim 1/R_\parallel^4$ for large $R_\parallel$ (up to logarithmic corrections). In the latter, supersymmetry is broken primarily on the brane, and thus its transmission to the bulk is gravitationally suppressed, leading to (28). For $n = 2$, there may be an enhancement factor of the radion mass by $\ln R_\perp M_s \simeq 30$ decreasing its wavelength by an order of magnitude [48].

The coupling of the radius modulus to matter relative to gravity can be easily computed and is given by:

$$\sqrt{\alpha_\varphi} = \frac{1}{M}\frac{\partial M}{\partial \varphi}\ ; \quad \alpha_\varphi = \begin{cases} \frac{\partial \ln \Lambda_{\rm QCD}}{\partial \ln R} \simeq \frac{1}{3} \ \text{for}\ R_\parallel \\ \\ \frac{2n}{n+2} = 1 - 1.5 \ \text{for}\ R_\perp \end{cases} \tag{29}$$

where M denotes a generic physical mass. In the longitudinal case, the coupling arises dominantly through the radius dependence of the QCD gauge coupling [49], while in the case of transverse dimension, it can be deduced from the rescaling of the metric which changes the string to the Einstein frame and depends slightly on the bulk dimensionality ($\alpha = 1 - 1.5$ for $n = 2 - 6$) [48]. Such a force can be tested in microgravity experiments and should be contrasted with the change of Newton's law due the presence of extra dimensions that is observable only for $n = 2$ [8, 23]. The resulting bounds from an analysis of the radion effects are [50]:

$$M_* \gtrsim 6\,\mathrm{TeV}\,. \tag{30}$$

In principle there can be other light moduli which couple with even larger strengths. For example the dilaton, whose VEV determines the string coupling, if it does not acquire large mass from some dynamical supersymmetric mechanism, can lead to a force of strength 2000 times bigger than gravity [51].

(iii) Non universal repulsive forces much stronger than gravity, mediated by possible abelian gauge fields in the bulk [37, 52]. Such fields acquire tiny masses of the order of M_s^2/M_P, as in (28), due to brane localized anomalies [52]. Although their gauge coupling is infinitesimally small, $g_A \sim M_s/M_P \simeq 10^{-16}$, it is still bigger that the gravitational coupling E/M_P for typical energies $E \sim 1$ GeV, and the strength of the new force would be $10^6 - 10^8$ stronger than gravity. This is an interesting region which will be soon explored in micro-gravity experiments (see Fig. 9). Note that in this case supernova constraints impose that there should be at least four large extra dimensions in the bulk [37].

In Fig. 9 we depict the actual information from previous, present and upcoming experiments [23, 48]. The solid lines indicate the present limits from the experiments indicated. The excluded regions lie above these solid lines. Measuring gravitational strength forces at short distances is challenging. The horizontal lines correspond to theoretical predictions, in particular for the graviton in the case $n = 2$ and for the radion in the transverse case. These limits are compared to those obtained from particle accelerator experiments in Table I.

9. ELECTROWEAK SYMMETRY BREAKING

Non-supersymmetric TeV strings offer also a framework to realize gauge symmetry breaking radiatively. Indeed, from the effective field theory point of view, one expects quadratically divergent one-loop contributions to the masses of scalar fields. The divergences are cut off by M_s and if the corrections are negative, they can induce electroweak symmetry breaking and explain the mild hierarchy between the weak and a string scale at a few TeV, in terms of a loop factor [55]. More precisely, in the minimal case of one Higgs doublet H, the scalar potential is:

$$V = \lambda(H^\dagger H)^2 + \mu^2(H^\dagger H)\,, \tag{31}$$

where λ arises at tree-level. Moreover, in any model where the Higgs field comes from an open string with both ends fixed on the same brane stack, it is given by an appropriate truncation of a supersymmetric theory. Within the minimal spectrum of the SM, $\lambda = (g_2^2 + g'^2)/8$, with g_2 and g' the $SU(2)$ and $U(1)_Y$ gauge couplings. On the other hand, μ^2 is generated at one loop:

$$\mu^2 = -\varepsilon^2\, g^2\, M_s^2\,, \tag{32}$$

where ε is a loop factor that can be estimated from a toy model computation and varies in the region $\epsilon \sim 10^{-1} - 10^{-3}$.

Indeed, consider for illustration a simple case where the whole one-loop effective potential of a scalar field can be computed. We assume for instance one extra dimension compactified

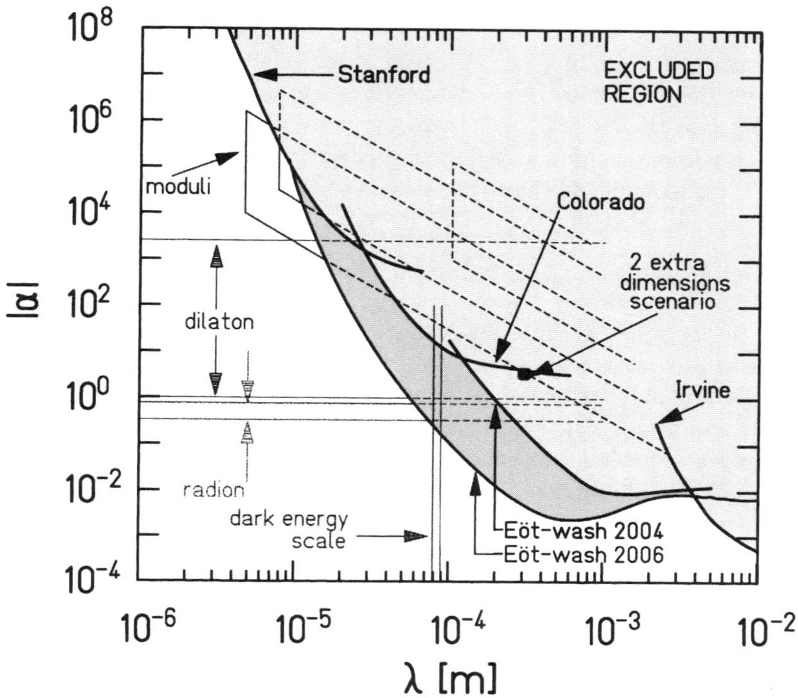

FIG. 9:

on a circle of radius $R > 1$ (in string units). An interesting situation is provided by a class of models where a non-vanishing VEV for a scalar (Higgs) field ϕ results in shifting the mass of each KK excitation by a constant $a(\phi)$:

$$M_m^2 = \left(\frac{m + a(\phi)}{R}\right)^2 , \tag{33}$$

with m the KK integer momentum number. Such mass shifts arise for instance in the presence of a Wilson line, $a = q \oint \frac{dy}{2\pi} gA$, where A is the internal component of a gauge field with gauge coupling g, and q is the charge of a given state under the corresponding generator. A straightforward computation shows that the ϕ-dependent part of the one-loop effective potential is given by [56]:

$$V_{eff} = -Tr(-)^F \frac{R}{32\,\pi^{3/2}} \sum_n e^{2\pi i n a} \int_0^\infty dl\ l^{3/2} f_s(l)\ e^{-\pi^2 n^2 R^2 l} \tag{34}$$

where $F = 0, 1$ for bosons and fermions, respectively. We have included a regulating function $f_s(l)$ which contains for example the effects of string oscillators. To understand its role we will consider the two limits $R >> 1$ and $R << 1$. In the first case only the $l \to 0$ region contributes to the integral. This means that the effective potential receives sizable contributions only from the infrared (field theory) degrees of freedom. In this limit we would have $f_s(l) \to 1$. For example, in the string model considered in [55]:

$$f_s(l) = \left[\frac{1}{4l}\frac{\theta_2}{\eta^3}(il + \frac{1}{2})\right]^4 \to 1 \qquad \text{for} \qquad l \to 0, \tag{35}$$

and the field theory result is finite and can be explicitly computed. As a result of the Taylor expansion around $a = 0$, we are able to extract the one-loop contribution to the coefficient of the term of the potential quadratic in the Higgs field. It is given by a loop factor times the compactification scale [56]. One thus obtains $\mu^2 \sim g^2/R^2$ up to a proportionality constant which is calculable in the effective field theory. On the other hand, if we consider $R \to 0$, which by T-duality corresponds to taking the extra dimension as transverse and very large, the one-loop effective potential receives contributions from the whole tower of string oscillators as appearing in $f_s(l)$, leading to squared masses given by a loop factor times M_s^2, according to eq. (32).

More precisely, from the expression (34), one finds:

$$\varepsilon^2(R) = \frac{1}{2\pi^2}\int_0^\infty \frac{dl}{(2\,l)^{5/2}}\frac{\theta_2^4}{4\eta^{12}}\left(il + \frac{1}{2}\right) R^3 \sum_n n^2 e^{-2\pi n^2 R^2 l}\,, \tag{36}$$

which is plotted in Fig. 10. For the asymptotic value $R \to 0$ (corresponding upon T-duality to a large transverse dimension of radius $1/R$), $\varepsilon(0) \simeq 0.14$, and the effective cut-off for the mass term is M_s, as can be seen from Eq. (32). At large R, $\mu^2(R)$ falls off as $1/R^2$, which is the effective cut-off in the limit $R \to \infty$, as we argued above, in agreement with field theory results in the presence of a compactified extra dimension [49, 57]. In fact, in the limit $R \to \infty$, an analytic approximation to $\varepsilon(R)$ gives:

$$\varepsilon(R) \simeq \frac{\varepsilon_\infty}{M_s\,R}\,, \qquad\qquad \varepsilon_\infty^2 = \frac{3\,\zeta(5)}{4\,\pi^4} \simeq 0.008\,. \tag{37}$$

The potential (31) has the usual minimum, given by the VEV of the neutral component of the Higgs doublet $v = \sqrt{-\mu^2/\lambda}$. Using the relation of v with the Z gauge boson mass, $M_Z^2 = (g_2^2 + g'^2)v^2/4$, and the expression of the quartic coupling λ, one obtains for the Higgs mass a prediction which is the Minimal Supersymmetric Standard Model (MSSM) value for $\tan\beta \to \infty$ and $m_A \to \infty$: $m_H = M_Z$. The tree level Higgs mass is known to receive important radiative corrections from the top-quark sector and rises to values around 120 GeV. Furthermore, from (32), one can compute M_s in terms of the Higgs mass $m_H^2 = -2\mu^2$:

$$M_s = \frac{m_H}{\sqrt{2}\,g\varepsilon}\,, \tag{38}$$

yielding naturally values in the TeV range.

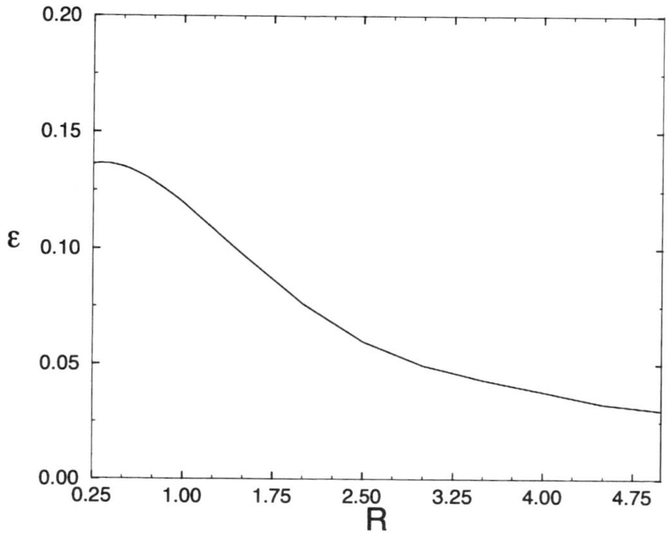

FIG. 10:

10. STANDARD MODEL ON D-BRANES

The gauge group closest to the Standard Model one can easily obtain with D-branes is $U(3) \times U(2) \times U(1)$. The first factor arises from three coincident "color" D-branes. An open string with one end on them is a triplet under $SU(3)$ and carries the same $U(1)$ charge for all three components. Thus, the $U(1)$ factor of $U(3)$ has to be identified with *gauged* baryon number. Similarly, $U(2)$ arises from two coincident "weak" D-branes and the corresponding abelian factor is identified with *gauged* weak-doublet number. Finally, an extra $U(1)$ D-brane is necessary in order to accommodate the Standard Model without breaking the baryon number [58]. In principle this $U(1)$ brane can be chosen to be independent of the other two collections with its own gauge coupling. To improve the predictability of the model, we choose to put it on top of either the color or the weak D-branes [59]. In either case, the model has two independent gauge couplings g_3 and g_2 corresponding, respectively, to the gauge groups $U(3)$ and $U(2)$. The $U(1)$ gauge coupling g_1 is equal to either g_3 or g_2.

Let us denote by Q_3, Q_2 and Q_1 the three $U(1)$ charges of $U(3) \times U(2) \times U(1)$, in a self explanatory notation. Under $SU(3) \times SU(2) \times U(1)_3 \times U(1)_2 \times U(1)_1$, the members of a family of quarks and leptons have the following quantum numbers:

$$\begin{array}{ll} Q & (\mathbf{3},\mathbf{2};1,w,0)_{1/6} \\ u^c & (\bar{\mathbf{3}},\mathbf{1};-1,0,x)_{-2/3} \\ d^c & (\bar{\mathbf{3}},\mathbf{1};-1,0,y)_{1/3} \\ L & (\mathbf{1},\mathbf{2};0,1,z)_{-1/2} \\ l^c & (\mathbf{1},\mathbf{1};0,0,1)_{1} \end{array} \tag{39}$$

The values of the $U(1)$ charges x, y, z, w will be fixed below so that they lead to the right hypercharges, shown for completeness as subscripts.

It turns out that there are two possible ways of embedding the Standard Model particle spectrum on these stacks of branes [58], which are shown pictorially in Fig. 11. The quark

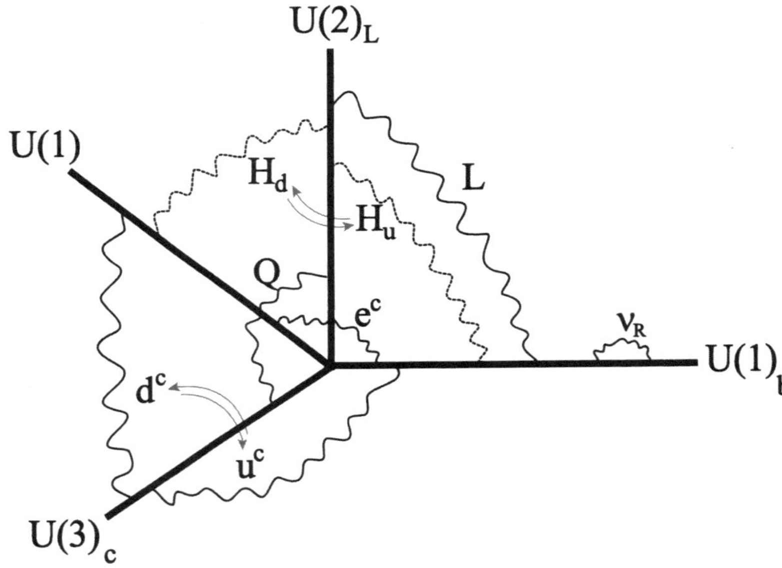

FIG. 11:

doublet Q corresponds necessarily to a massless excitation of an open string with its two ends on the two different collections of branes (color and weak). As seen from the figure, a fourth brane stack is needed for a complete embedding, which is chosen to be a $U(1)_b$ extended in the bulk. This is welcome since one can accommodate right handed neutrinos as open string states on the bulk with sufficiently small Yukawa couplings suppressed by the large volume of the bulk [60]. The two models are obtained by an exchange of the up and down antiquarks, u^c and d^c, which correspond to open strings with one end on the color branes and the other either on the $U(1)$ brane, or on the $U(1)_b$ in the bulk. The lepton doublet L arises from an open string stretched between the weak branes and $U(1)_b$, while the antilepton l^c corresponds to a string with one end on the $U(1)$ brane and the other in the bulk. For completeness, we also show the two possible Higgs states H_u and H_d that are both necessary in order to give tree-level masses to all quarks and leptons of the heaviest generation.

10.1. Hypercharge embedding and the weak angle

The weak hypercharge Y is a linear combination of the three $U(1)$'s:

$$Y = Q_1 + \frac{1}{2}Q_2 + c_3 Q_3 \quad ; \quad c_3 = -1/3 \text{ or } 2/3\,, \tag{40}$$

where Q_N denotes the $U(1)$ generator of $U(N)$ normalized so that the fundamental representation of $SU(N)$ has unit charge. The corresponding $U(1)$ charges appearing in eq. (39) are $x = -1$ or 0, $y = 0$ or 1, $z = -1$, and $w = 1$ or -1, for $c_3 = -1/3$ or $2/3$, respectively. The hypercharge coupling g_Y is given by [64]:

$$\frac{1}{g_Y^2} = \frac{2}{g_1^2} + \frac{4c_2^2}{g_2^2} + \frac{6c_3^2}{g_3^2}\,. \tag{41}$$

It follows that the weak angle $\sin^2\theta_W$, is given by:

$$\sin^2\theta_W \equiv \frac{g_Y^2}{g_2^2 + g_Y^2} = \frac{1}{2 + 2g_2^2/g_1^2 + 6c_3^2 g_2^2/g_3^2}\,, \tag{42}$$

where g_N is the gauge coupling of $SU(N)$ and $g_1 = g_2$ or $g_1 = g_3$ at the string scale. In order to compare the theoretical predictions with the experimental value of $\sin^2\theta_W$ at M_s, we plot in Fig. 12 the corresponding curves as functions of M_s. The solid line is the experimental

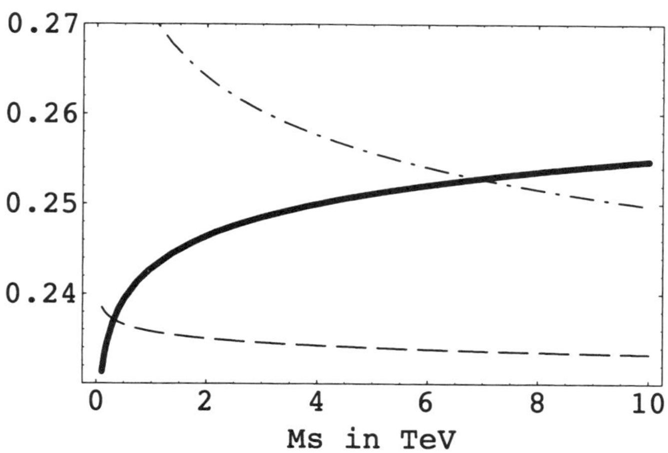

FIG. 12: The experimental value of $\sin^2\theta_W$ (thick curve), and the theoretical predictions (42).

curve. The dashed line is the plot of the function (42) for $g_1 = g_2$ with $c_3 = -1/3$ while the dotted-dashed line corresponds to $g_1 = g_3$ with $c_3 = 2/3$. The other two possibilities are not shown because they lead to a value of M_s which is too high to protect the hierarchy. Thus, the second case, where the $U(1)$ brane is on top of the color branes, is compatible with low energy data for $M_s \sim 6-8$ TeV and $g_s \simeq 0.9$.

From Eq. (42) and Fig. 12, we find the ratio of the $SU(2)$ and $SU(3)$ gauge couplings at the string scale to be $\alpha_2/\alpha_3 \sim 0.4$. This ratio can be arranged by an appropriate choice of the relevant moduli. For instance, one may choose the color and $U(1)$ branes to be D3 branes while the weak branes to be D7 branes. Then, the ratio of couplings above can be explained by choosing the volume of the four compact dimensions of the seven branes to be $V_4 = 2.5$ in string units. This being larger than one is consistent with the picture above. Moreover it predicts an interesting spectrum of KK states for the Standard model, different from the naive choices that have appeared hitherto: the only Standard Model particles that have KK descendants are the W bosons as well as the hypercharge gauge boson. However, since the hypercharge is a linear combination of the three $U(1)$'s, the massive $U(1)$ KK gauge bosons do not couple to the hypercharge but to the weak doublet number.

10.2. The fate of $U(1)$'s, proton stability and neutrino masses

It is easy to see that the remaining three $U(1)$ combinations orthogonal to Y are anomalous. In particular there are mixed anomalies with the $SU(2)$ and $SU(3)$ gauge groups of the Standard Model. These anomalies are cancelled by three axions coming from the closed string RR (Ramond) sector, via the standard Green-Schwarz mechanism [61]. The mixed anomalies with the non-anomalous hypercharge are also cancelled by dimension five Chern-Simmons type of interactions [58]. An important property of the above Green-Schwarz anomaly cancellation mechanism is that the anomalous $U(1)$ gauge bosons acquire masses leaving behind the corresponding global symmetries. This is in contrast to what would had happened in the case of an ordinary Higgs mechanism. These global symmetries remain exact to all orders in type I string perturbation theory around the orientifold vacuum. This follows from the topological nature of Chan-Paton charges in all string amplitudes. On the other hand, one expects non-perturbative violation of global symmetries and consequently exponentially small in the string coupling, as long as the vacuum stays at the orientifold point. Thus, all $U(1)$ charges are conserved and since Q_3 is the baryon number, proton stability is guaranteed.

Another linear combination of the $U(1)$'s is the lepton number. Lepton number conservation is important for the extra dimensional neutrino mass suppression mechanism described above, that can be destabilized by the presence of a large Majorana neutrino mass term. Such a term can be generated by the lepton-number violating dimension five effective operator $LLHH$ that leads, in the case of TeV string scale models, to a Majorana mass of the order of a few GeV. Even if we manage to eliminate this operator in some particular model, higher order operators would also give unacceptably large contributions, as we focus on models in which the ratio between the Higgs vacuum expectation value and the string scale is just of order $\mathcal{O}(1/10)$. The best way to protect tiny neutrino masses from such contributions is to impose lepton number conservation.

A bulk neutrino propagating in $4+n$ dimensions can be decomposed in a series of 4d KK excitations denoted collectively by $\{m\}$:

$$S_{kin} = R_\perp^n \int d^4x \sum_{\{m\}} \left\{ \bar{\nu}_{Rm} \partial\!\!\!/ \nu_{Rm} + \bar{\nu}^c_{Rm} \partial\!\!\!/ \nu^c_{Rm} + \frac{m}{R_\perp} \nu_{Rm} \nu^c_{Rm} + c.c. \right\} , \tag{43}$$

where ν_R and ν_R^c are the two Weyl components of the Dirac spinor and for simplicity we considered a common compactification radius $R_\perp$. On the other hand, there is a localized interaction of ν_R with the Higgs field and the lepton doublet, which leads to mass terms between the left-handed neutrino and the KK states ν_{Rm}, upon the Higgs VEV v:

$$S_{int} = g_s \int d^4x H(x) L(x) \nu_R(x, y=0) \quad \rightarrow \quad \frac{g_s v}{R_\perp^{n/2}} \sum_m \nu_L \nu_{Rm} , \tag{44}$$

in strings units. Since the mass mixing $g_s v / R_\perp^{n/2}$ is much smaller than the KK mass $1/R_\perp$, it can be neglected for all the excitations except for the zero-mode ν_{R0}, which gets a Dirac mass with the left-handed neutrino

$$m_\nu \simeq \frac{g_s v}{R_\perp^{n/2}} \simeq v \frac{M_s}{M_p} \simeq 10^{-3} - 10^{-2}\ \text{eV} , \tag{45}$$

for $M_s \simeq 1 - 10$ TeV, where the relation (14) was used. In principle, with one bulk neutrino, one could try to explain both solar and atmospheric neutrino oscillations using also its first KK excitation. However, the later behaves like a sterile neutrino which is now excluded experimentally. Therefore, one has to introduce three bulk species (at least two) ν_R^i in order to explain neutrino oscillations in a 'traditional way', using their zero-modes ν_{R0}^i [62]. The main difference with the usual seesaw mechanism is the Dirac nature of neutrino masses, which remains an open possibility to be tested experimentally.

Acknowledgments

This work was supported in part by the European Commission under the ERC Advanced Grant 226371 and the contract PITN-GA-2009-237920.

---

[1] M. Green, J. Schwarz and E. Witten, *Superstring Theory, Vols I and II*, Cambridge University Press, 1987.

[2] I. Antoniadis, *Phys. Lett.* B **246** (1990) 377.

[3] N. Arkani-Hamed, S. Dimopoulos and G. R. Dvali, *Phys. Lett.* B **429** (1998) 263 [arXiv:hep-ph/9803315]; I. Antoniadis, N. Arkani-Hamed, S. Dimopoulos and G. R. Dvali, *Phys. Lett.* B **436** (1998) 257 [arXiv:hep-ph/9804398].

[4] Fore a review, see e.g. K. R. Dienes, Phys. Rept. **287** (1997) 447 [arXiv:hep-th/9602045]; and references therein.

[5] T. Kaluza, Preuss. Akad. Wiss. (1921) 966; O. Klein, Z. Phys. **37** (1926) 895.

[6] I. Antoniadis and K. Benakli, *Phys. Lett.* B **326** (1994) 69.

[7] C. Angelantonj and A. Sagnotti, *Phys. Rept.* **371** (2002) 1 [Erratum-ibid. **376** (2003) 339] [arXiv:hep-th/0204089].

[8] D. J. Kapner, T. S. Cook, E. G. Adelberger, J. H. Gundlach, B. R. Heckel, C. D. Hoyle and H. E. Swanson, Phys. Rev. Lett. **98** (2007) 021101.

[9] For a review see e.g. I. Antoniadis, *Prepared for NATO Advanced Study Institute and EC Summer School on Progress in String, Field and Particle Theory, Cargese, Corsica, France (2002)*; and references therein.

[10] For a recent review see e.g. H. P. Nilles, S. Ramos-Sanchez and P. K. S. Vaudrevange, AIP Conf. Proc. **1200** (2010) 226 [arXiv:0909.3948 [hep-th]].

[11] X. Wen and E. Witten, *Nucl. Phys.* B 261 (1985) 651.

[12] A. Schellekens, *Phys. Lett.* B 237 (1990) 363.

[13] G. Athanasiu, J. Atick, M. Dine and W. Fischler, *Phys. Lett.* B 214 (1988) 55.

[14] I. Antoniadis, J. Ellis, J. Hagelin and D.V. Nanopoulos, *Phys. Lett.* B 231 (1989) 65.

[15] J. Ellis, J.L. Lopez and D.V. Nanopoulos, *Phys. Lett.* B 245 (1990) 375 and 247 (1990) 257.

[16] I. Antoniadis and K. Benakli, *Phys. Lett.* B 295 (1992) 219.

[17] I. Antoniadis, J. Ellis, J. Hagelin and D.V. Nanopoulos, *Phys. Lett.* B 194 (1987) 231; 205 (1988) 459 and 208 (1988) 209.

[18] I. Antoniadis and G.K. Leontaris, *Phys. Lett.* B 246 (1989) 333; I. Antoniadis, G.K. Leontaris and J. Rizos, *Phys. Lett.* B 216 (1990) 165.

[19] A.E. Faraggi, D.V. Nanopoulos and K. Yuan, *Nucl. Phys.* B 335 (1990) 347; A.E. Faraggi, *Phys. Lett.* B 278 (1992) 131; *Nucl. Phys.* B 387 (1992) 239.

[20] M. Dine, N. Seiberg and E. Witten, *Nucl. Phys.* B 289 (1987) 585.

[21] J.-P. Derendinger, L.E. Ibáñez and H.P. Nilles, *Phys. Lett.* B 155 (1985) 65; M. Dine, R. Rohm, N. Seiberg and E. Witten, *Phys. Lett.* B 156 (1985) 55.

[22] J. D. Lykken, *Phys. Rev.* D **54** (1996) 3693 [arXiv:hep-th/9603133].

[23] J. C. Long and J. C. Price, *Comptes Rendus Physique* **4** (2003) 337; R. S. Decca, D. Lopez, H. B. Chan, E. Fischbach, D. E. Krause and C. R. Jamell, *Phys. Rev. Lett.* **94** (2005) 240401; R. S. Decca et al., arXiv:0706.3283 [hep-ph]; S. J. Smullin, A. A. Geraci, D. M. Weld, J. Chiaverini, S. Holmes and A. Kapitulnik, arXiv:hep-ph/0508204; H. Abele, S. Haeßler and A. Westphal, in 271th WE-Heraeus-Seminar, Bad Honnef (2002).

[24] G. Dvali, arXiv:0706.2050 [hep-th]; Int. J. Mod. Phys. A **25** (2010) 602 [arXiv:0806.3801 [hep-th]]; G. Dvali and M. Redi, Phys. Rev. D **77** (2008) 045027 [arXiv:0710.4344 [hep-th]]; R. Brustein, G. Dvali and G. Veneziano, JHEP **0910** (2009) 085 [arXiv:0907.5516 [hep-th]].

[25] G. Dvali and C. Gomez, Phys. Lett. B **674** (2009) 303.

[26] I. Antoniadis and B. Pioline, *Nucl. Phys.* B **550** (1999) 41; I. Antoniadis, S. Dimopoulos and A. Giveon, JHEP **0105** (2001) 055 [arXiv:hep-th/0103033]; I. Antoniadis, A. Arvanitaki, S. Dimopoulos and A. Giveon, arXiv:1102.4043.

[27] G. Dvali and D. Lust, arXiv:0912.3167 [hep-th]; G. Dvali and C. Gomez, arXiv:1004.3744 [hep-th].

[28] L. Randall and R. Sundrum, *Phys. Rev. Lett.* **83** (1999) 4690 and *Phys. Rev. Lett.* **83** (1999)

3370.
[29] For a review see e.g. R. Contino and A. Pomarol, Comptes Rendus Physique **8** (2007) 1058; and references therein.
[30] K. R. Dienes, E. Dudas and T. Gherghetta, Phys. Lett. B **436** (1998) 55 [arXiv:hep-ph/9803466]; Nucl. Phys. B **537** (1999) 47 [arXiv:hep-ph/9806292].
[31] I. Antoniadis, K. Benakli and M. Quirós, *Phys. Lett.* B **331** (1994) 313 and *Phys. Lett.* B **460** (1999) 176; P. Nath, Y. Yamada and M. Yamaguchi, *Phys. Lett.* B **466** (1999) 100 T. G. Rizzo and J. D. Wells, *Phys. Rev.* D **61** (2000) 016007; T. G. Rizzo, *Phys. Rev.* D **61** (2000) 055005; A. De Rujula, A. Donini, M. B. Gavela and S. Rigolin, *Phys. Lett.* B **482** (2000) 195;
[32] E. Accomando, I. Antoniadis and K. Benakli, *Nucl. Phys.* B **579** (2000) 3.
[33] I. Antoniadis, K. Benakli and A. Laugier, *JHEP* **0105** (2001) 044.
[34] P. Nath and M. Yamaguchi, *Phys. Rev.* D **60** (1999) 116004; *Phys. Rev.* D **60** (1999) 116006; M. Masip and A. Pomarol, *Phys. Rev.* D **60** (1999) 096005; W. J. Marciano, *Phys. Rev.* D **60** (1999) 093006; A. Strumia, *Phys. Lett.* B **466** (1999) 107; R. Casalbuoni, S. De Curtis, D. Dominici and R. Gatto, *Phys. Lett.* B **462** (1999) 48; C. D. Carone, *Phys. Rev.* D **61** (2000) 015008; A. Delgado, A. Pomarol and M. Quirós, *JHEP* **1** (2000) 30.
[35] G. Servant and T. M. P. Tait, *Nucl. Phys.* B **650** (2003) 391.
[36] G. F. Giudice, R. Rattazzi and J. D. Wells, *Nucl. Phys.* B **544** (1999) 3; E. A. Mirabelli, M. Perelstein and M. E. Peskin, *Phys. Rev. Lett.* **82** (1999) 2236; T. Han, J. D. Lykken and R. Zhang, *Phys. Rev.* D **59** (1999) 105006; K. Cheung and W.-Y. Keung, *Phys. Rev.* D **60** (1999) 112003; C. Balázs *et al.*, *Phys. Rev. Lett.* **83** (1999) 2112; L3 Collaboration (M. Acciarri *et al.*), *Phys. Lett.* B **464** (1999) 135 and **470** (1999) 281: J. L. Hewett, *Phys. Rev. Lett.* **82** (1999) 4765.
[37] N. Arkani-Hamed, S. Dimopoulos and G. Dvali, *Phys. Rev.* D **59** (1999) 086004.
[38] S. Cullen and M. Perelstein, *Phys. Rev. Lett.* **83** (1999) 268; V. Barger, T. Han, C. Kao and R. J. Zhang, *Phys. Lett.* B **461** (1999) 34.
[39] K. Benakli and S. Davidson, *Phys. Rev.* D **60** (1999) 025004; L. J. Hall and D. Smith, *Phys. Rev.* D **60** (1999) 085008.
[40] E. Dudas and J. Mourad, Nucl. Phys. B **575** (2000) 3 [arXiv:hep-th/9911019]; S. Cullen, M. Perelstein and M. E. Peskin, *Phys. Rev.* D **62** (2000) 055012; D. Bourilkov, *Phys. Rev.* D **62** (2000) 076005; L3 Collaboration (M. Acciarri *et al.*), *Phys. Lett.* B **489** (2000) 81.
[41] P. C. Argyres, S. Dimopoulos and J. March-Russell, Phys. Lett. B **441** (1998) 96 [arXiv:hep-th/9808138]; T. Banks and W. Fischler, arXiv:hep-th/9906038.
[42] S. B. Giddings and S. Thomas, *Phys. Rev.* D **65** (2002) 056010 [arXiv:hep-ph/0106219]; S. Dimopoulos and G. Landsberg, *Phys. Rev. Lett.* **87** (2001) 161602 [arXiv:hep-ph/0106295].
[43] P. Meade and L. Randall, arXiv:0708.3017 [hep-ph].
[44] L. A. Anchordoqui, H. Goldberg, D. Lust, S. Nawata, S. Stieberger and T. R. Taylor, Phys. Rev. Lett. **101** (2008) 241803 [arXiv:0808.0497 [hep-ph]].
[45] D. Lust, S. Stieberger and T. R. Taylor, Nucl. Phys. B **808** (2009) 1 [arXiv:0807.3333 [hep-th]].
[46] G. T. Horowitz and J. Polchinski, Phys. Rev. D **55** (1997) 6189 [arXiv:hep-th/9612146].
[47] I. Antoniadis, C. Bachas, *Phys. Lett.* B **450** (1999) 83.
[48] I. Antoniadis, K. Benakli, A. Laugier and T. Maillard, *Nucl. Phys.* B **662** (2003) 40 [arXiv:hep-

ph/0211409].

[49] I. Antoniadis, S. Dimopoulos and G. Dvali, *Nucl. Phys.* B **516** (1998) 70; S. Ferrara, C. Kounnas and F. Zwirner, *Nucl. Phys.* B **429** (1994) 589.

[50] E. G. Adelberger, B. R. Heckel, S. Hoedl, C. D. Hoyle, D. J. Kapner and A. Upadhye, Phys. Rev. Lett. **98** (2007) 131104.

[51] T. R. Taylor and G. Veneziano, *Phys. Lett.* B **213** (1988) 450.

[52] I. Antoniadis, E. Kiritsis and J. Rizos, *Nucl. Phys.* B **637** (2002) 92.
I. Antoniadis, K. Benakli and A. Laugier, *Nucl. Phys.* B **631** (2002) 3.

[53] I. Antoniadis and M. Tuckmantel, *Nucl. Phys.* B **697** (2004) 3.

[54] I. Antoniadis, M. Tuckmantel and F. Zwirner, *Nucl. Phys.* B **707** (2005) 215 [arXiv:hep-ph/0410165].

[55] I. Antoniadis, K. Benakli and M. Quirós, *Nucl. Phys.* B **583** (2000) 35.

[56] I. Antoniadis, K. Benakli and M. Quiros, *New Jour. Phys.* **3** (2001) 20.

[57] I. Antoniadis, C. Muñoz and M. Quirós, *Nucl. Phys.* B **397** (1993) 515; I. Antoniadis and M. Quirós, *Phys. Lett.* B **392** (1997) 61; A. Pomarol and M. Quirós, *Phys. Lett.* B **438** (1998) 225; I. Antoniadis, S. Dimopoulos, A. Pomarol and M. Quirós, *Nucl. Phys.* B **544** (1999) 503; A. Delgado, A. Pomarol and M. Quirós, *Phys. Rev.* D **60** (1999) 095008; R. Barbieri, L. J. Hall and Y. Nomura, *Phys. Rev.* D **63** (2001) 105007.

[58] I. Antoniadis, E. Kiritsis and T. N. Tomaras, *Phys. Lett.* B **486** (2000) 186; I. Antoniadis, E. Kiritsis, J. Rizos and T. N. Tomaras, *Nucl. Phys.* B **660** (2003) 81.

[59] G. Shiu and S.-H. H. Tye, *Phys. Rev.* D **58** (1998) 106007; Z. Kakushadze and S.-H. H. Tye, *Nucl. Phys.* B **548** (1999) 180; L. E. Ibáñez, C. Muñoz and S. Rigolin, *Nucl. Phys.* B **553** (1999) 43.

[60] K. R. Dienes, E. Dudas and T. Gherghetta, *Nucl. Phys.* B **557** (1999) 25 [arXiv:hep-ph/9811428]; N. Arkani-Hamed, S. Dimopoulos, G. R. Dvali and J. March-Russell, *Phys. Rev.* D **65** (2002) 024032 [arXiv:hep-ph/9811448]; G. R. Dvali and A. Y. Smirnov, *Nucl. Phys.* B **563** (1999) 63.

[61] A. Sagnotti, *Phys. Lett.* B **294** (1992) 196; L. E. Ibáñez, R. Rabadán and A. M. Uranga, *Nucl. Phys.* B **542** (1999) 112; E. Poppitz, *Nucl. Phys.* B **542** (1999) 31.

[62] H. Davoudiasl, P. Langacker and M. Perelstein, Phys. Rev. D **65** (2002) 105015.

[63] Although interesting, this scenario seems difficult to be realized, since 4d chirality requires non-trivial action of orbifold twists with localized chiral states at the fixed points.

[64] The gauge couplings $g_{2,3}$ are determined at the tree-level by the string coupling and other moduli, like radii of longitudinal dimensions. In higher orders, they also receive string threshold corrections.

CHAIRMAN: I. ANTONIADIS

Scientific Secretaries: M. Schröder, A. Zein Assi

DISCUSSION I

- *Cerchiai:*

1. How do you solve the problem of fractional charges in GUTs?
2. Can you explain in more detail orbifold GUTs?

- *Antoniadis:*

1. By making them superheavy and using inflation to suppress their flux, or by assuming an extra gauge group that confines them in integral charged states in the same way that QCD confines the fractional charged quarks into integral charged nucleons.

2. Orbifold are obtained from smooth manifolds upon quotienting them by a discrete symmetry group. For example, consider a circle S1 of radius R and the symmetry Z_2 : y → -y, where y is an extra dimension. Then, a simple orbifold is $S1/Z_2$, in which you identify y with -y. It forms an interval from 0 to πR. The endpoints 0 and πR are fixed points because they remain invariant under Z_2. At the fixed points one obtains usually chiral fermions. Orbifold GUTs can be studied as effective theories without knowing the full machinery of string theory.

- *Biassoni:*

Is the problem of flavor (LFV, baryon number violation, FCNC) addressed in a natural way, i.e. without stringent requirements on the mass matrix, in extra dimensions or string models?

- *Antoniadis:*

The lepton and baryon number conservations can be assured by imposing R-parity. However, FCNC cannot be addressed by discrete symmetries but you must impose some structure of the soft terms. One needs a theory of soft terms. Actually, in extra dimensions or string theories, you have the same problem as in supersymmetry (SUSY).

- *Alba:*

1. How can we check experimentally the type of geometry of possible extra dimensions?
2. How can we distinguish between geometric and algebraic (Gepner) approaches to string compactification?
3. Is it possible to find out whether there is another brane near our brane?

- *Antoniadis:*

1. Experimentally we can only obtain limits of their sizes but it is very difficult to deduce their geometries.
2. If the size of extra dimensions is large compared to the string length, then we can distinguish between different geometries. In contrast, if the size of the extra dimensions is comparable to the string scale, then classical geometry breaks down, different geometries can give rise to equivalent string physics. Gepner models describe this regime.
3. Depending on the model, we could feel that brane via gravity or gauge interactions.

- *Burda:*

Could you say a few words about TeV-scale gravity?

- *Antoniadis:*

This will be discussed in the next lecture.

- *Aidala:*

Is there any existing theory in which the fractional charges of quarks and the integer charges of all directly observable states (not just in QCD) come out naturally?

- *Antoniadis:*

There exist theories with a gauge group factor additional to QCD that confines fractional charged states into integral charged bound states. Such models have been used to obtain dark matter candidates or to break SUSY.

- *Vlasov:*

Could you comment on the naturalness problem in SUSY? SUSY is introduced in order to explain 5 – 10 parameters (hierarchy problem) but it has itself many more parameters related to the breaking.

- *Antoniadis:*

Yes, I agree, we have this problem. The newly introduced parameters have to be of the same order if the hierarchy problem is to be solved. But they have to satisfy additional constraints imposed by low energy physics, which are not automatic.

- *Miyazaki:*

String Theory needs SUSY. Is it possible that extra-dimensions coincide with SUSY at TeV scale? Are there any possibilities of KK-gravitino?

- *Antoniadis:*

String theory needs SUSY on the world-sheet, not on space-time. It might be that SUSY is broken at the string scale and in that case there will be no low-energy manifestation of SUSY. If the string scale is high (like GUT scale), then string theory could provide a framework for SUSY GUTs and SUSY-breaking scale should be generated at low-energy. But it could also be that SUSY is broken at a high scale. However, if the string scale is around the TeV, then there is no need for SUSY to solve the hierarchy problem anymore (SUSY-breaking scale doesn't need to be different than the string scale).

- *Minkowski:*

Let us invoke a reflex of a baby when shown a mirror: it seeks the picture on the other side of the mirror. What happens with the brane-world for scales much beyond the string scale?

- *Antoniadis:*

At some energies, of the order of string scale, the field theory description of particle physics breaks and one has to use a string description.

- *Kepuladze:*

What is the explanation of the smallness of extra small dimensions? Unless it is explained, it moves the hierarchy problem from one point to another.

- *Antoniadis:*

For SUSY to solve the hierarchy problem, it must be broken near the electroweak scale. This solves the technical aspect of stability of quantum corrections, but we need still to understand how this scale is generated (dynamically). The same logic applies in extra-dimensions. If the string scale is around the TeV and there are large extra-dimensions (with respect to the string length), the technical aspect of the hierarchy is again solved but one should still explain its dynamical aspect, and this is related to the moduli stabilization problem: the size of an extra-dimension is provided by a vacuum expectation value of some scalar field (with flat potential) called modulus. Some solutions (generate potentials with minima hierarchically different than the string scale) are provided by introducing fluxes.

- *Krylov:*

Do you know any expected signals from LHC or other experiments which can be described only by models with extra dimensions?

- *Antoniadis:*

Yes. It will be discussed in tomorrow's lecture.

- *Mager:*

Talking about extra-dimensions, the 3D space is very special in terms of the confined structures. Take for example knot theory from which we know that all knots may be solved in 4D (and higher). Do you preserve these structures when adding dimensions? If yes, how?

- *Antoniadis:*

These are generally related to topological properties of the space. For instance, in compactifications of extra-dimensions on non-trivial manifolds, the topological properties of the latter are reflected to properties of the massless spectrum and its effective field theory, like the number of chiral fermions. Also, there is a special class of topological theories that translate the topological properties of the compactification manifold into real-world properties (spectrum, etc.).

- *Badziak:*

Recently, there has been a lot of interest in F-theory GUTs. What are the advantages of these theories as compared to orbifold GUTs?

- *Antoniadis:*

F-theory GUTs share good features of orbifold GUT models and of intersecting-brane models. The main disadvantage is that F-theory is very difficult to study in a systematic way at the quantum level because of its non-perturbative nature. Only some limits of F-theory are known using effective field theory.

- *Csanàd:*

If you expect Kaluza-Klein modes to appear at LHC energy scales, wouldn't that mean too large extra-dimensions?

- *Antoniadis:*

There are two types of scales for K-K modes. One is the TeV scale (inverse TeV size for the extra dimensions seen by Standard Model gauge interactions); these K-K modes are produced in LHC as heavy resonances that may be detected in di-lepton or di-jet distributions for example. On the other hand, there are other K-K excitations associated to millimeter size extra-dimensions of gravitational nature; these have very small masses (meV for the graviton for example), they interact gravitationally and one cannot detect them directly but give rise to missing energy signal.

- *Rinaldi:*

What are the mathematical tools used to compute spectra in String Theory? And can we compare the results of these tools to the numbers obtained from experimental results?

- *Antoniadis:*

The main mathematical tool is perturbation theory on the world-sheet. String theory is studied from the first quantization point of view (no Field Theory analog for Strings), and the surfaces replace the trajectories of particles. Spectra, cross-sections, etc. are extracted by Conformal Field Theory techniques (2D world-sheet embedded in space-time). The numbers we get out of the calculation cannot be easily compared with experiments because there are arbitrary unknown parameters besides the string tension, associated to the details of the compactification manifold. These parameters have to be fixed and the comparison with the experiments becomes model-dependant.

CHAIRMAN: I. ANTONIADIS

Scientific Secretaries: A. Agostinelli, M. Franchini

DISCUSSION II

- *Inguglia:*

a. Are there any cosmological signatures of your model?
b. How can you understand if you have a neutralino signature rather than a γ^* signal?
c. Does the existence of the γ^* affect some SM processes?

- *Antoniadis :*

a. The cosmological signatures depend on the evolution of the size of the extra dimensions and at what time of the evolution their size was fixed. An example of signatures is the graviton flux which implies in particular that the re-heating temperature has to be low due to nucleosynthesis constraints.
b. The difference between the two signatures is the spin of the particle associated to the missing energy. This gives rise to a different angular distribution for the observed particles or jets. Thus a study of the angular distribution can differenciate the two models, but this requires a lot of statistics that probably cannot be reached in the first LHC run.
c. No, it is stable because it is odd under Kaluza-Klein parity and is a candidate for dark matter.

- *Burda:*

If I remember properly a few years ago people discussed a problem of localization of gauge theory on a brane. Does this problem affect this picture or not?

- *Antoniadis:*

No. Localization in string theory is a consequence of Dirichelet boundary conditions for the end-points of the open strings. These are forced to be localized at a fixed position, defining a brane. Actually, one can show that D-branes have a width given by the string scale.

- *Manoni:*

Some extra dimension models can benefit from indirect searches for new physics such as rare B decays mediated by loops in which Kaluza-Klein excitations may enter. Is there any indirect search that can be related to your model?

- *Antoniadis:*

Yes but such searches are model dependent. Usually, to have rare decays, one needs a model in which some global symmetry of the Standard Model is violated and a corresponding quantum number is not conserved.

- *Alba:*

a. Why do you have 6 separate dimensions in slide 27?
b. You told us about the difference in observation between supersymmetric and extra dimension scenario. Could you comment on possibility having combined scenario?

- *Antoniadis:*

a. Because here I am in the framework of string theory with total 10 space-time dimensions, and our observed world is 4-dimensional, so we have 6=10-4 extra.

b. Some people consider such scenarios for various reasons, e.g. the so-called little hierarchy. We then have at the same time supersymmetric particles and Kaluza-Klein excitations.

- *Miyazaki:*

In case of UED/SUSY do we need ILC to determine the spin? Is it possible in LHC to determine this spin?

- *Antoniadis:*

Yes, LHC is possible to determine the spin but probably in the second higher luminosity run. Of course a linear collider such as ILC can also determine the spin much better, but we are not yet there.

- *de With:*

Is there a mass scalar for the Kaluza-Klein particle at which they no longer solve the hierarchy problem as it happens for SUSY particles?

- *Antoniadis:*

In principle there are constraints, but it is more difficult to extract a precise bound in relation to the degree of fine-tuning. This is partly because there is no simple minimal model of extra dimensions like the MSSM for supersymmetry, and partly because the effective ultraviolet cutoff is at the TeV. I remind you that in MSSM, there is already some tension between the experimental bounds on the superparticle masses and the degree of fine-tuning. The point is that we believe that supersymmetry is not there only for hierarchy but also for the unification of gauge couplings and therefore we expect that the supersymmetric SM should be valid up to the GUT scale. Thus, the effective cutoff is high and the logarirthmic corrections cancel the loop factor suppression I used in the fine-tuning discussion and make the problem more severe.

- *Sisman:*

In slide 29 (concerning gravitational radiation in the bulk in form of missing energy), you emphasize on the spin dependence of the missing energy, but also there will be missing energy for every process and missing energy will depend on the energy of LHC, is it right?

- *Antoniadis:*

Yes, in principle, there is missing E_t for any process because any process can radiate gravitons, but if the Center of Mass energy is below the Quantum Gravity scale, there is a power suppression from the coupling.

- *Vlasov:*

Do the extra dimensions participate in cosmological expansion?

- *Antoniadis:*

It depends on the time at which the size of the extra dimensions was fixed. If the corresponding scalar potential had occurred at the string scale, there is no room for cosmological evolution. If on the contrary, it occurs at lower energies, the extra dimensions can grow for instance together with our three dimensions for some time.

- *Rinaldi:*

Kaluza-Klein parity is the remnant of Poincaré symmetry in the extra dimensions. However it is usually imposed by hand. Do you think that this is a problem for UED models?

- *Antoniadis:*

UED models are based on the assumption that KK-parity is an exact symmetry of the model. This is indeed an extra condition imposed by hand, like the R-parity in supersymmetry. In practice, it is not there generically since in a string embedding of the SM usually some particles are localized in brane intersections and KK parity is lost.

- *Blankenburg:*

How strong are the LEP indirect bounds on these models?

- *Antoniadis:*

For one extra-dimension, the stronger bound is 3-4 TeV in the localized case, while the weaker bound is a few hundreds GeV in UED models. In general, it is somewhere in the middle, depending on the localization properties of quarks and leptons in a given model.

- *Mager:*

Can you comment on micro gravity measurements? What are their upper bounds?

- *Antoniadis:*

The micro gravity experiments test Newton laws at short distances and look for new short-range forces. In both cases, the deviations are parameterized by adding a Yukawa term to the Newtonian potential, depending on two parameters: the wave length associated to the mass of the exchanged particle and its strength relative to gravity. In the case of extra dimensions, the wavelength corresponds to their size (mass of the first KK excitation) while the strength is of order gravity up to a calculable numerical constant depending on the internal geometry. As I told you, for forces of strength comparable to gravity the bound is about 50 μm, but for forces stronger than gravity the bound is at much shorter distances.

- *'t Hooft:*

You mention that SM particles can propagate in extra-dimensions in that case you obtain a higher dimensional SM which is not renormalizable which means that the cutoff is not very far away?

- *Antoniadis:*

Yes that's right. For longitudinal extra dimensions, their size is of the order of the string length so the cutoff is of the same order (but it may differ by a factor of two or so, which is important for experimental searches of KK modes).

R. BOUSSO

"Entropy Bounds and the Holographic Principle"

The Slides of the Lecture can be found at:

http://www.ccsem.infn.it/issp2011/

CHAIRMAN: R. BOUSSO

Scientific Secretaries: N. Krupina, A. Sadofyev

DISCUSSION I

- Burda:

For ordinary matter entropy behaves as S~A^(3/4), for black holes S~A, the question is about this gap, do I understand properly, that gap filling by consideration of matter coupled to gravity in curved background?

- Bousso:

So there is a gap of a factor of area to the one quarter between the entropy, which we can actually obtain in spherical symmetric systems like this, and a bound S is less than A, and that is a little bit puzzling. I mean at least from the understanding point of view, if we have an entropy bound that can be weird, if you can saturate it well, you can say that you can saturate it with a BH, but why is there a jump between ordinary matter and Bhs? There are two examples in which this gap can be filled, but in both cases you really need to use covariant entropy bound, and not just such weak statement about how the entropy inside some sphere of radius a is less than A, you have to use this light-sheets. One example is when you make a BH very slowly. So you have a BH, and now you feed it with quanta very slowly. The way to do this is to add the most entropy to the BH while increasing its mass and its area as little as possible. You always add quanta that have wavelengths of the order of radius of the BH and which are massless. If they have any longer wavelengths and they, of course, add even less mass to the BH, but they probably want to go in. And if they have shorter wavelength, then you have one beta transformation BH, but more energy than you needed. And so if you keep building up to BH like that one quantum at the time, then the actual matter entropy that crosses the horizon is of order the area of the BH at the end. And the horizon of BH is a light-sheet of its final BH horizon area surface. That was one example. The other example is open FRW geometries, and, of course, you can cut out pieces of matter type geometries. So it turns out that in open FRW universe, which is curvature dominated early, the matter entropy can be of order of the matter entropy on unobserved past light cone, which consists of two light-sheets, can be of order the maximum area under past light cone. And it is actually the lower one of this maximum area I thought that is starting point for two acceptable light-sheets, because of area decreases in its future and it was a big bang in its past and it's on the past directed light-sheet the entropy is the order of area. So the gap can be filled with ordinary matter.

- Burda:

What can you say about two strange examples - white holes and "monsters"?

- Bousso:

Concerning white holes, the covariant entropy bound is time reversal invariant, I mean there is no time direction, which is preferred, and I guess, if you consider time reversal collapse solution of a BH, you won't change the fact that bound is satisfied. Regarding monsters, monsters are as you can make something that is almost the BH, so the spheres get smaller and smaller, and you think you get into horizon, but then there is a large piece where you have a long throat and you don't have an actual horizon, and you can store a lot of matter in that long throat. So the surface area of that throat can be smaller than the actual entropy of all the stuff that is in it in a special sense. If I take an instant of time and look at all of entropy that can be larger than area, but if I follow light-rays, they will not propagate to the top of the throat before they hit a singularity, in the end this geometry does form a BH. And the light-rays are cut off by the future or the past singularity, depending which direction you pick before they propagated all the way down to the end of the throat, so they don't see all of the entropy, so the bound is upheld.

- Aidala:

What generates the entropy of a BH?

- Bousso:

That's a deep question to which we still don't have a general answer. We only understand the case of a supersymmetric BH of zero temperature.

- Palhares:

Where does the Boltzmann constant come from in this D brane scenario?

- Bousso:

The Boltzmann constant comes from the relation between energy and temperature.

- de With

Does the fact that the total area of all BH never decreases (proven by Hawking) contradict the idea of Hawking radiation and BH evaporation?

- Bousso:

It does, however to prove that the total area of all BH never decreases, Hawking used the null energy condition, which is violated by quantum fluctuations and BH evaporation is a quantum process.

- Alba:

Is entropy of BH completely quantum phenomenon? Should entropy classically be equal zero?

- Bousso:

The answer to the first part of the question is yes, it is a quantum phenomenon, but the answer to the second part of the question is no. If you lead hbar tend to zero, the entropy of the BH does not go to zero and goes to infinity, if you hold the area fixed in some other units. The entropy is given by S~A/(G*hbar), so if you turn to quantum mechanics, if you take hbar is zero, the entropy actual goes to infinity. Of course you don't care very much about that, because BH becomes completely non radiating, by the fact that the temperature goes to zero, that is true. It doesn't conflict with having nonzero entropy, and an example is sypersymmetric BHs, which have zero temperature, but finite entropy, even in finite values of hbar.

- Campoleoni:

Is the holographic principle a peculiarity of gravity or something that can emerge in many classical non-linear systems?

- Bousso:

It seems to be a peculiarity of gravity, for instance in the limit of vanishing Newton constant, the entropy bound becomes trivial.

CHAIRMAN: R. BOUSSO

Scientific Secretaries: A. Festanti, T. C. Sisman

DISCUSSION II

- *Cerchiai:*

Could you tell more about constructing holographic screens in arbitrary spacetimes?

- *Bousso:*

First of all, let me try to describe the question a bit better. After the discovery of the AdS/CFT, it seemed like that you just need to find the holographic dual for the spacetime you are interested in, then you will have the nonperturbative quantum theory of gravity. However, it was quickly noticed that in realistic cosmologies, you did not know what you mean by the boundary of the spacetime on which the theory supposed to be defined. The simplest example is that a close universe manifestly has no boundary, so it is not the case that the holographic theory lives at a boundary separated by an infinite distance. I think what is fruitful in this problem of trying to find a definition of the quantum gravity for arbitrary spacetimes is to make a distinction between the question of exactly what this theory is and how it should be formulated, and a separate question which is how many degrees of freedom it should have or more generally whether the holographic principle can be a guide in finding that theory. What the statement here is that you can always do something that is analogues to finding a holographic screen in AdS. In AdS, for every time slice represented as a horizontal disk, see Fig. 1, there is a boundary on the rim of the disk with enough area to encode all the information on the disk.

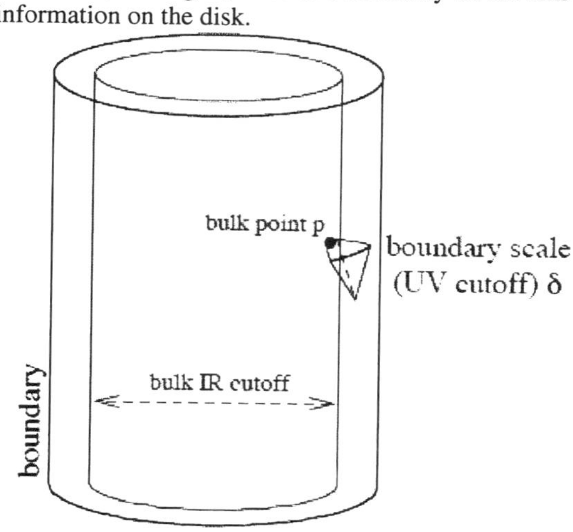

Figure 1: AdS spacetime where the time is in the vertical direction.

If one tries to apply this idea for a closed universe, first thing to notice is that the screen of interest cannot be at infinity, since there is no infinity. But, we can construct a screen that has enough area to encode everything that happens in the spacetime, and such a construction is not unique. Consider the closed FRW universe whose Penrose diagram given in Fig. 2.a, and has the slicing of the spacetime in the light cones as shown in Fig.2.b. Then, a sphere on the apparent horizon will be the largest sphere on the corresponding (null) slice whose area bounds the entropy in both parts of the (null) slice by use of covariant entropy bound.

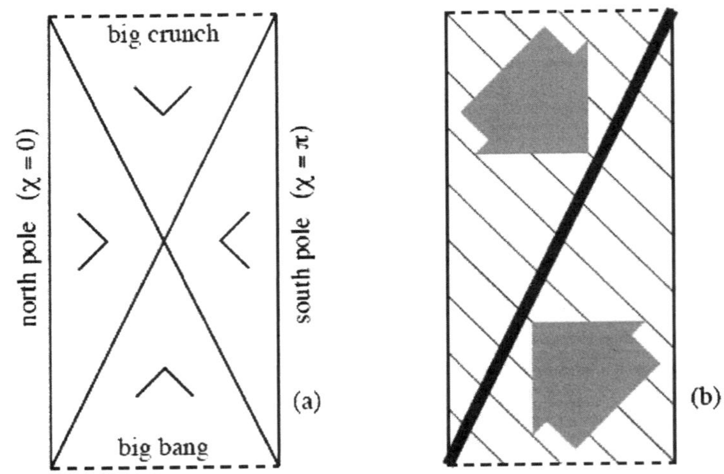

Figure 2: a) Penrose diagram of the closed FRW universe, b) Slicing of the spacetime into light cones.

Therefore, we can describe all of the information about what happens on this slice by as many bits as this area in Planck units. In that sense, one can holographically encode the information on a time slice of the spacetime on a surface that sits within the spacetime. You can do it for every one of the slices. If this construction is applied to the AdS space, it would simply reproduce the statement that the screen is the boundary of the AdS space. But, it is general enough to be applied to arbitrary spacetimes, and it tells you that there is always a holographic screen that you can find. What it does not tell you is what the theory is that holographically describes the spacetime. But, it tells you that it is possible to think of the spacetime as being encoded in a hologram with only degrees of freedom involved in the area.

- Alba:

Bekenstein bound is $S \leq 2\pi MR/\hbar$. How is it consistent with $S \leq A/4G\hbar$?

- Bousso:

Heuristic derivation of Bekenstein bound and the bound in terms of the area assumed that we are dealing with weakly gravitating systems. In other words, we are dealing with the systems where the mass, M, is much less than the radius, R, divided by Newton's constant, G. By just substituting this assumption into Bekenstein bound, one gets that S is much less than R^2 which is the area, A. One must not take that M is scaling like R^3, precisely because of the gravitational back reaction. It is true that as long as gravity is incredibly weak M scales like R^3, but you have to also assume that the density is very small. The mass of a black hole only scales like R.

- Alba:

Can you comment on the effect of quantum fluctuations of spacetime on the construction of the covariant entropy bound?

- Bousso:

Covariant entropy bound should be constructed with semiclassical gravity that is Einstein's classical gravity coupled to the expectation value of the stress tensor in the quantum field theory describing the matter. It involves the quantities such as the area of the surfaces which you do not expect to be infinitely sharply defined in a full quantum theory of gravity. So, it is a statement about how classical gravity/geometry has to emerge from whatever the underlying theory is. It is not directly speaking the language of that underlying theory. That is not unusual, similarly in the equivalence principle case you are noticing something that is curious and seems very coincidental. The hope is that this statement tells you something about how to write the new theory where this is no more coincidental. The new theory is going to be much richer, and the original language that makes you to ask this question may not even be present in the new theory. That is exactly the case for the equivalence principle leading to general relativity. So, again this is a statement about something that we already see happening in the limit of classical gravity where you do not worry about the quantum fluctuations of the geometry, and it needs an explanation. It is just very strange behavior of the effective theory that we worked with.

- *de With:*

In which situations are there boundaries which have more than two light sheets in non-expanding directions?

- *Bousso*:

For a two-dimensional surface, there are four null directions away from it, see Fig. 3, and there are only two independent values of the expansion which are calculated locally. For example, if the expansion in the direction of F_2 is negative, it will be a light sheet; however, F_3 will not be a light sheet, since it has an expansion value opposite of the expansion of F_2. If F_2 has zero expansion, then F_3 will also be an allowed light sheet. Therefore, if one of the two independent values of expansion is zero, then there are three light sheets; if both of them are zero, then one will have four light sheets.

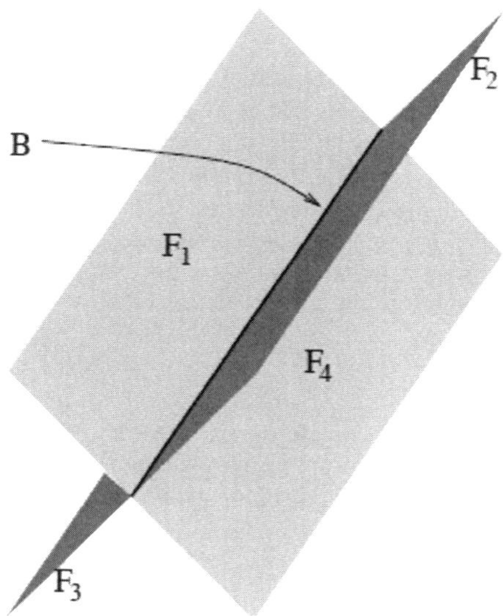

Figure 3: Spatial surface B bounding (2+1 D) null hypersurfaces generated by a congruence of null geodesics perpendicular to B.

For example, any surface located at the apparent horizon of a flat FRW universe will have three allowed light sheets, see Fig 4. In general, if you are at the boundary of two or more areas with different wedges, you will have three or four light sheets in non-expanding directions.

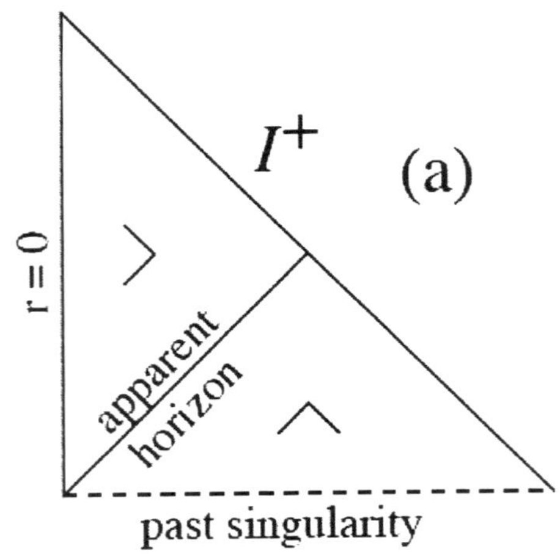

Figure 4: Penrose diagram of flat FRW universe.

- Rothkopf:

Please elaborate on the statement: AdS/CFT is a complete, non-perturbatve quantum theory of gravity. Usually I have heard about a $N \to \infty$ CFT limit without any fluctuations being mapped to a completely classical gravity in higher dimensions.

- Bousso:

If N is strictly infinity, then this corresponds to a classical theory in which there are no fluctuations of the metric in the bulk. But, that is a special limit of the correspondence which is claimed to be much more general statement and supposed to be valid at finite N. So, you need N to be large and you need the 't Hooft coupling to be large to have a bulk regime where the gravity is classical is approximately valid. But, as long as the N is finite, quantum fluctuations will be correctly encoded in the CFT. The reason that I am saying it is a completely nonperturbative definition is the fact that CFT is a theory which we understand completely. If you have a big enough computer you can calculate anything you want and that is the definition of the entire theory in the bulk.

- Aidala

Going from a big bang to a big crunch, the past and the future singularities have to be quiet different, with no or very little entropy at the past singularity and large entropy at the future singularity. What does this mean for cyclic cosmological models?

- Bousso:

It means that cyclic cosmological models don't make sense. Time shouldn't be privileged-spacetime coordinate, and there should be no way of transmitting information from the "earlier" universe to the "later" one. But we still really don't understand spacelike singularities.

- Burda:

Before the question, I would like to make a comment about AdS/CFT correspondence. As I remember, it is commonly believed that for finite N there is a correspondence between CFT which is N=4 super-Yang-Mills and string theory on preliminary given curved spacetime namely on AdS_5xS^5. In principle, string theory could be a theory of quantum gravity, but who knows?

- *Bousso:*

I am not sure what is the distinction you drawing out. The AdS/CFT correspondence is a conjecture that is supported by enormous amount of evidence. It is evidence for the conjecture that the CFT manages to reproduce stringy physics that is supposed to be there at the perturbative level at the relatively low energies where the classical background make sense. That was a requirement for the consistency of this conjecture. We could have studied perturbative string theory in an AdS5xS^5 background before, and said we have a quantum theory of gravity, but it would only be a perturbative quantum gravity theory. We did not have a nonperturbative completion. In particular, to say that it is a completion of perturbative string theory in AdS_5xS^5, it had better to reproduce the string states in that perturbative theory as a special case. But, it also knows about the nonperturbative things like black holes that we are not able to describe in perturbative string theory.

- *Burda:*

The holographic principle in the form of the covariant entropy bound looks so beautiful and general, almost as the principle of least action, so the question is: how do you think it is possible to obtain some dynamical principle which will give the dynamics of the matter from the covariant entropy bound? For example, Verlinde proposed the principal of maximal entropy and he told that we can obtain classical Einstein gravity from this principle, but his arguments are a bit hand-wavy. I want to learn whether it is possible to find a more elegant way to obtain the dynamics of gravity by use of covariant entropy bound or not?

- *Bousso:*

Well, this is why I gave these lectures. I am sort of hoping that you come up with ideas of trying to understand how this bound could come out from the deeper level. I think there is actually slightly less handwaving version of what Verlinde has said that appeared earlier by Ted Jacobson some time ago in the 90's. It is an interesting attempt to derive Einstein's equations from black hole thermodynamics applied to tiny little Rindler horizons at every point in space. It is very intriguing, and I wish it had gone somewhere, but I think nobody knows how to go further. I think that to go further we have to understand better how the result is tangent that is expressed in the entropy bound: the tangent between on one hand the very small number of degrees of freedom that we naively expected and the other hand the appearance of perfectly good locality which has made quantum field theory such a successful pursue. In a way, the AdS/CFT correspondence has not really helped us that much with that. It just told us how huge the tangent actually is. It is incredibly difficult to understand locality; at least, at scales much smaller than the AdS curvature scale in the dual boundary theory. Locality is still completely obscure that enough special cases, where there are a lot of work, that people convince themselves reconstruct things on smaller scales than the AdS scale, but the general problem it has not really thrown so much light on it. I think that is one of the problems we really need to focus on, we need to understand how the result is in conflict. In the case of the AdS, what is so nice about the CFT is that it has exactly the right number of degrees of freedom as demanded by the bound. May be it would be even better if you had more degrees of freedom but could explicitly identify some of them as pure gauge, so that you could have enough degrees of freedom that you can mimic locality but at the same time you could also see that the true number of degrees of freedom obeys the holographic bound. There might be different gauges you could choose. This is very handwaving, but the problem is really important, and it deserves attention.

- *Inguglia:*

In a paper published last year whose title is "Cosmology with torsion: an alternative to cosmic inflation", it is claimed that the universe that we are living in may be considered as a black hole in another universe. What do you think about this idea?

- *Bousso:*

I am not familiar with that paper. I probably have to look at the paper to comment on.

Quantum Gravity Needs Supersymmetry

Sergio Ferrara[1,2] and Alessio Marrani[1]

1 *Physics Department, Theory Unit, CERN, CH 1211, Geneva 23, Switzerland*
sergio.ferrara@cern.ch
alessio.marrani@cern.ch

2 *INFN - Laboratori Nazionali di Frascati, Via Enrico Fermi 40,00044 Frascati, Italy*

Abstract

We report on some old and new results on the quantum aspects of four-dimensional maximal supergravity, and its hypothetical ultraviolet finiteness.

1 From Supersymmetry...

If supersymmetry is used in *string theory* (see *e.g.* [1, 2]), then it is possible to construct five seemingly consistent (M-theory descendant [3]; see *e.g.* [4] for a list of Refs.) theories in $D = 10$ space-time dimensions (see *e.g.* [5]). However, due to the large possible choice of compactification manifolds

$$\mathcal{M}_{10} \longrightarrow \mathcal{M}_4 \times X_6, \tag{1.1}$$

a multitude of $D = 4$ theories arise (it can be classified *e.g.* by a statistical approach to the vacua [6]), and it is yet a mystery how string theory could select the physical vacuum. Further models involving physical degrees of freedom living in the bulk and on branes give very interesting possibilities (see *e.g.* [7, 8, 9, 10]).

On the other hand, if supersymmetry is used in *field theory*, then one encounters the opposite situation. For *rigid* theories (*i.e.*, neglecting gravity), one can build nice but unrealistic models, explore AdS/CFT duality, and study strongly coupled gauge theories. However, when gravity is included, *at least* in the cases with $\mathcal{N} \leqslant 16$ supercharges, it seems that one fails to unify supersymmetry, gravity and quantum theory.

In spite of these drawbacks, as we will shortly report in the next Sections, there are some indications that for some $D = 4$ special theories of supergravity, namely $\mathcal{N} = 8$ and possibly $\mathcal{N} = 5, 6$, "miracles" occur due to unexpected cancellations in multi-loop perturbative calculations [11, 12, 13, 14, 15, 16, 17, 20, 21].

Arguments based on M-theory compactifications [22, 23, 24] and E_7 invariance [20] prevent counterterms only up to a finite loop order. On the other hand, analysis based on possible divergences in the supefield light-cone formulation may indicate the absence of counterterms at all loops [14, 17].

In October 1976, S.F. was invited by John Schwarz to visit Caltech and to give a seminar on the construction of the first supersymmetric theory of gravity in $D = 4$ (named $\mathcal{N} = 1$ *supergravity* [18, 19]); at that time, the first extended ($\mathcal{N} = 2$) supergravity had just been completed [25]. Murray Gell-Mann, during an enlightening conversation in his office, remarked that if $D = 4$ higher-$\mathcal{N}$ supergravities existed, then $\mathcal{N} = 8$ would be the supergravity theory with maximal supersymmetry [26].

Today, after 35 years, we are still struggling with $\mathcal{N} = 8$ *maximal* supergravity, its connections to superstring and M- theory, its hypothetical perturbative finiteness and its non-perturbative completion(s).

2 ...to Maximal Supergravity...

As anticipated by Gell-Mann, $\mathcal{N} = 8$ supergravity [27, 28] is the theory with the largest possible amount of supersymmetry for particles with spin $s \leqslant 2$ in $D = 4$ (namely, no higher spin fields in the massless spectrum). Indeed, in supersymmetric gravity theories with $\mathcal{N}$-extended supersymmetry, the massless particle content is given by

$$\binom{\mathcal{N}}{k} \equiv \frac{\mathcal{N}!}{k!\,(\mathcal{N}-k)!} \text{ particles of } \textit{helicity } \lambda = 2 - \frac{k}{2}, \tag{2.1}$$

where $k_{\max} = \mathcal{N}$, and $\mathcal{N} \leqslant 8$ if $|\lambda| \leqslant 2$ is requested.

One possible approach to maximal supergravity is to consider it as it comes from M-theory restricted to the massless sector. The problem is that this theory, even if preserving maximal $\mathcal{N} = 8$ supersymmetry (corresponding to $32 = 8 \times 4$ supersymmetries), is *not* uniquely defined, because of the multiple choice of internal compactification manifolds and corresponding duality relations:

$$\begin{array}{lll} \mathbf{I}.\ M_{11} \longrightarrow M_4 \times T_7 & (GL^+(7,\mathbb{R}) \text{ and } SO(7) \text{ manifest}); & \\ \mathbf{II}.\ M_{11} \longrightarrow AdS_4 \times S^7 & (SO(8) \text{ manifest, } \textit{gauged}); & (2.2) \\ \mathbf{III}.\ M_{11} \longrightarrow M_4 \times T_{7,\mathcal{R}} & (SL(8,\mathbb{R}) \text{ and } SO(8) \text{ manifest}), & \end{array}$$

where T_7 is the 7-torus and S^7 is the 7-sphere. $T_{7,\mathcal{R}}$ denotes the case in which, according to Cremmer and Julia [27], the dualization of 21 vectors and 7 two-forms makes $SL(8,\mathbb{R})$ (in which $GL^+(7,\mathbb{R})$ is maximally embedded) manifest as maximal non-compact symmetry of the Lagrangian. Note that in case **III** one can further make $E_{7(7)}$ (and its maximal compact subgroup $SU(8)$) manifest *on-shell*, by exploiting a Cayley transformation supplemented by a rotation through $SO(8)$ gamma matrices on the vector 2-form field strengths [27, 29].

The fundamental massless fields (and the related number $\sharp$ of degrees of freedom) of M-theory in $d = 11$ flat space-time dimensions are [30]

$$\begin{array}{llll} g_{\mu\nu}\ (\textit{graviton}): & \sharp = \frac{(d-1)(d-2)}{2} - 1, & \text{in } d = 11 : \sharp = 44; & \\ \Psi_{\mu\alpha}\ (\textit{gravitino}): & \sharp = (d-3)2^{(d-3)/2}, & \text{in } d = 11 : \sharp = 128; & (2.3) \\ A_{\mu\nu\rho}\ (\textit{three-form}): & \sharp = \frac{(d-2)(d-3)(d-4)}{3!}, & \text{in } d = 11 : \sharp = 84. & \end{array}$$

Because a $(p+1)$-form ("Maxwell-like" gauge field) A_{p+1} couples to p-dimensional extended objects, and its "magnetic" dual B_{d-p-3} couples to $(d-p-4)$-dimensional extended objects, it follows that the fundamental (massive) objects acting as sources of the theory are $M2$- and $M5$-branes.

In the formulation **III** of (2.2) [27], the gravitinos ψ_I and the gauginos χ_{IJK} respectively have the following group theoretical assignment[1] (I in $\mathbf{8}$ of $SU(8)$):

$$\text{theory } \mathbf{III}\ [27]: \left\{ \begin{array}{l} \psi_I : \underset{\mathbf{8}}{SO(7)} \subset \underset{\mathbf{8}_s}{SO(8)} \subset \underset{\mathbf{8}}{SU(8)}; \\ \\ \chi_{IJK} : \underset{\mathbf{8}+\mathbf{48}}{SO(7)} \subset \underset{\mathbf{56}_s}{SO(8)} \subset \underset{\mathbf{56}}{SU(8)}. \end{array} \right. \tag{2.4}$$

On the other hand, the 70 scalar fields arrange as

$$\text{theory } \mathbf{III}\ [27]: \underset{(\sharp=70)}{s = 0\ \textit{dofs}} : \underset{\mathbf{1}+\mathbf{7}+\mathbf{21}+\mathbf{35}}{SO(7)} \subset \underset{\mathbf{35}_v+\mathbf{35}_c}{SO(8)} \subset \underset{\mathbf{70}}{SU(8)}, \tag{2.5}$$

[1] As evident from (2.4), we use a different convention with respect to [31] (see *e.g.* Table 36 therein). Indeed, we denote as $\mathbf{8}_v$ of $SO(8)$ the irrep. which decomposes into $\mathbf{7}+\mathbf{1}$ of $SO(7)$, whereas the two spinorial irreps. $\mathbf{8}_s$ and $\mathbf{8}_c$ both decompose into $\mathbf{8}$ of $SO(7)$. The same change of notation holds for $\mathbf{35}$ and $\mathbf{56}$ irreps..

where **70** is the rank-4 completely antisymmetric irrep. of $SU(8)$, the maximal compact subgroup of the U-duality group $E_{7(7)}$ (also called $\mathcal{R}$-symmetry). It follows that scalars parameterize a non-compact coset manifold $\frac{G}{SU(8)}$. Indeed, the $SU(8)$ under which both the scalar fields and the fermion fields transform is the *"local"* $SU(8)$, namely the stabilizer of the scalar manifold. On the other hand, also a *"global"* $SU(8)$ ($\mathcal{R}$-symmetry group) exists, under which the vector 2-form self-dual/anti-self-dual field strengths transform. Roughly speaking, the physically relevant group $SU(8)$ is the diagonal one in the product $SU_{\text{local}}(8) \times SU_{\text{global}}(8)$ (see also discussion below).

Remarkably, there exists an *unique* simple, non-compact Lie group with real dimension $70 + 63 = 133$ and which embeds $SU(8)$ as its *maximal* compact subgroup: this is the real, non-compact split form $E_{7(7)}$ of the exceptional Lie group E_7, thus giving rise to the symmetric, rank-7 coset space

$$\frac{E_{7(7)}}{SU(8)/\mathbb{Z}_2}, \tag{2.6}$$

which is the scalar manifold of $\mathcal{N} = 8$, $D = 4$ supergravity ($\mathbb{Z}_2$ is the kernel of the $SU(8)$-representations of even rank; in general, spinors transform according to the double cover of the stabilizer of the scalar manifold; see *e.g.* [32, 33]).

$E_{7(7)}$ acts as electric-magnetic duality symmetry group [34], and its maximal compact subgroup $SU(8)$ has a chiral action on fermionic as well as on (the vector part of the) bosonic fields. While the chiral action of $SU(8)$ on fermions directly follows from the chirality (complex nature) of the relevant irreps. of $SU(8)$ (as given by Eq. (2.4)), the chiral action on vectors is a crucial consequence of the electric-magnetic duality in $D = 4$ space-time dimensions. Indeed, this latter allows for *"self-dual / anti-self-dual"* complex combinations of the field strengths, which can then fit into complex irreps. of the stabilizer H of the coset scalar manifold G/H itself. For the case of maximal $\mathcal{N} = 8$ supergravity, the relevant chiral complex irrep. of $H = SU(8)$ is the rank-2 antisymmetric **28**.

Note that if one restricts to the $SL(8,\mathbb{R})$-covariant sector, the chirality of the action of electric-magnetic duality is spoiled, because the maximal compact subgroup of $SL(8,\mathbb{R})$, namely $SO(8)$, has not chiral irreps.

Composite (sigma model G/H) anomalies can arise in theories in which G has a maximal compact subgroup with a *chiral* action on bosons and/or fermions (see *e.g.* [35, 36, 13]). Surprising cancellations among the various contributions to the composite anomaly can occur as well. An example is provided by $\mathcal{N} = 8$, $d = 4$ supergravity itself, in which standard anomaly formulæ yield the remarkable result [36, 13]

$$3Tr_{\mathbf{8}}X^3 - 2Tr_{\mathbf{28}}X^3 + Tr_{\mathbf{56}}X^3 = (3 - 8 + 5)\,Tr_{\mathbf{8}}X^3 = 0, \tag{2.7}$$

where X is any generator of the Lie algebra $\mathfrak{su}(8)$ of the *rigid* (*i.e. global*) $SU(8)$ group ($\mathcal{R}$-symmetry). In light of the previous considerations, the first and third contributions to (2.7) are due to fermions: the 8 gravitinos ψ_A and the 56 spin-$\frac{1}{2}$ fermions χ_{ABC}, respectively, whereas the second contribution is due to the 28 chiral vectors. Note that, for the very same reason, the *local* $SU(8)$ (stabilizer of the non linear sigma-model of scalar fields), under which only fermions do transform[2], would be *anomalous* [35]. In an analogous way, in [36] it was discovered that $\mathcal{N} = 6$ and $\mathcal{N} = 5$ "pure" supergravities are *composite anomaly-free*, whereas $\mathcal{N} \leqslant 4$ theories are not.

[2]Also scalar fields transform under *local* $SU(8)$, but they do not contribute to the composite anomaly, because they sit in the *self-real* (and thus *non-chiral*) rank-4 antisymmetric irrep. **70** of $SU(8)$.

3 ...and its UV Perturbative Finiteness?

At homotopical level, the following holds:

$$E_{7(7)} \cong (SU(8)/\mathbb{Z}_2) \times \mathbb{R}^{70}, \tag{3.1}$$

implying that the two group manifolds have the same De Rham cohomology. This is a key result, recently used in [13] to show that the aforementioned absence of $SU(8)$ current anomalies yield to the absence of anomalies for the non-linearly realized $E_{7(7)}$ symmetry, thus implying that the $E_{7(7)}$ continuous symmetry of classical $\mathcal{N} = 8$, $d = 4$ supergravity is preserved at all orders in perturbation theory (see *e.g.* [11, 14, 37, 15, 16, 17, 20]). This implies the perturbative finiteness of supergravity at least up to seven loops; Bern, Dixon *et al.* explicitly checked the finiteness up to four loops included [11] (computations at five loops, which might be conclusive, are currently in progress; for a recent review, see *e.g.* [21]).

A puzzling aspect of these arguments is that string theory certainly violates continuous $E_{7(7)}$ symmetry at the perturbative level, as it can be easily realized by considering the dilaton dependence of loop amplitudes (see *e.g.* [20]). However, this is not the case for $\mathcal{N} = 8$ supergravity. From this perspective, two (perturbatively finite) theories of quantum gravity would exist, with 32 local supersymmetries; expectedly, they would differ at least in their non-perturbative sectors, probed *e.g.* by black hole solutions. String theorists [38, 39, 40] claim that $\mathcal{N} = 8$, $d = 4$ supergravity theory is probably not consistent at the non-perturbative level. From a purely $d = 4$ point of view, their arguments could be overcome by excluding from the spectrum, as suggested in [37], black hole states which turn out to be singular or ill defined if interpreted as purely four-dimensional gravitational objects. Inclusion of such singular states (such as $\frac{1}{4}$-BPS and $\frac{1}{2}$-BPS black holes) would then open up extra dimensions, with the meaning that a non-perturbative completion of $\mathcal{N} = 8$ supergravity would lead to string theory [38]. Extremal black holes with a consistent $d = 4$ interpretation may be defined as having a Bertotti-Robinson [41] $AdS_2 \times S^2$ near-horizon geometry, with a non-vanishing area of the event horizon. In $\mathcal{N} = 8$ supergravity, these black holes are[3] $\frac{1}{8}$-BPS or non-BPS (for a recent review and a list of Refs., see *e.g.* [44]). The existence of such states would in any case break the $E_{7(7)}(\mathbb{R})$ continuous symmetry, because of Dirac-Schwinger-Zwanziger dyonic charge quantization conditions.

The breaking of $E_{7(7)}(\mathbb{R})$ into an arithmetic subgroup $E_{7(7)}(\mathbb{Z})$ [45] would then manifest only in exponentially suppressed contributions to perturbative amplitudes (see *e.g.* the discussion in [13], and Refs. therein), in a similar way to instanton effects in non-Abelian gauge theories. Indeed, as for the $SL(2,\mathbb{R})$ symmetry of Type IIB superstrings in $D = 10$ [22], the continuous symmetry is believed to be broken down to discrete $E_{7(7)}(\mathbb{Z})$ by non-perturbative effects like instantons [46]. The breaking of $E_{7(7)}(\mathbb{R})$ to $E_{7(7)}(\mathbb{Z})$ is instrumental to setting a uniform mass-gap, given by Planck Mass $M_{Pl} = \sqrt{\hbar c/G_N} = 1.22 \times 10^{19} GeV/c^2$, everywhere inside the moduli space $E_{7(7)}/SU(8)$, for all regular BPS states with $\mathcal{I}_4 \neq 0$ where [47, 48]

$$\mathcal{I}_4 = \mathbb{K}_{MNPQ} Q^M Q^N Q^P Q^Q \tag{3.2}$$

[3] We also remark that these are the only black holes for which the *Freudenthal duality* [42, 43] is well defined.

is the *quartic* Cartan invariant of $E_{7(7)}$ [27] and Q^M is a 56-dimensional vector of 'bare' quantized electric and magnetic charges with respect to the 28 $U(1)$ gauge groups.

In [37], after taking into account charge quantization and assuming that the perturbative theory be UV finite to all orders, the plausibility of a *non-perturbative* completion of genuinely $D = 4$, $\mathcal{N} = 8$ supergravity, only including regular black hole states with $\mathcal{I}_4 \neq 0$ and excluding all singular states with $\mathcal{I}_4 = 0$, was proposed. This proposal has some analogy with $\mathcal{N} = 4$ super Yang-Mills (SYM) theory in $D = 4$, decoupled from gravity and other stringy interactions. Even after including non-perturbative effects, $\mathcal{N} = 4 = 4$ SYM in $D = 4$ should not be thought of as a compactification of Type I or Heterotic strings, that contain the same massless states but differ by the massive completion, but rather in terms of the AdS/CFT correspondence [49]. Analogously, pure $D = 4$ $\mathcal{N} = 8$ supergravity, including *regular* non-perturbative states, may be disconnected from toroidal compactifications of Type II superstrings, that unavoidably give rise to 1/2 BPS states with $\mathcal{I}_4 = 0$. The fact that all known $\mathcal{N} = 8$ supergravity perturbative amplitudes could be expressed in terms of $\mathcal{N} = 4$ SYM amplitudes in the superconformal phase, where the latter enjoys 32 supersymmetries (16 of Poincaré type plus 16 superconformal), might be more than an analogy in this respect.

The conjectured UV finiteness of $\mathcal{N} = 8$ supergravity, associated with continuous $E_{7(7)}$ symmetry, has been questioned by Green, Ooguri, Schwarz in [50], where non-decoupling of BPS states from four-dimensional $\mathcal{N} = 8$ supergravity was discussed. The main conclusion of [50] was that the $\mathcal{N} = 8$ supergravity limit of string theory does not exist in four dimensions, irrespective of whether or not the perturbative approximation is free of UV divergences. String theory adds to the 256 massless states of four-dimensional $\mathcal{N} = 8$ supergravity an infinite tower of states, such as Kaluza-Klein momenta and monopoles, wound strings and wrapped branes.

As mentioned, classical solutions of the $\mathcal{N} = 8$ version of non-linear Einstein equations including stable, zero temperature, extremal, BPS and non-BPS charged black holes [51], should play a crucial role in defining the quantum features of the theory. For appropriate choices of the charges, these black holes can be viewed as smooth solitons interpolating between flat Minkowski space-time at infinity and Bertotti-Robinson $AdS_2 \times S^2$ geometry [41] near the horizon. The asymptotic values of the scalar fields are largely arbitrary and determine the ADM mass M [52] for given charges Q^M. Thanks to the *attractor mechanism* [53], their near-horizon values are determined in terms of the charges. The entropy of a black hole in $\mathcal{N} = 8$ supergravity is related to the horizon area by the Bekenstein-Hawking formula

$$S_{BH} = \frac{1}{4} A_H = \pi \sqrt{|\mathcal{I}_4|}, \tag{3.3}$$

and $\mathcal{N} = 8$ attractors were studied in [54].

Within this framework, the ADM mass of an $\mathcal{N} = 8$ extremal black hole depends on its charges and on the asymptotic values of the scalar fields, both transforming under $E_{7(7)}$ symmetry:

$$M_{BH} = M_{ADM}(Q, \phi). \tag{3.4}$$

A manifestly $E_{7(7)}$ covariant expression for the mass is related to the maximal eigenvalue of the central charge matrix [45],[55]

$$M^2_{ADM}(Q, \phi) \geq \mathrm{Max}_i\{|Z_i(Q, \phi)|^2\}, \tag{3.5}$$

where $Z_i(Q,\phi), i = 1,2,3,4$, are the four (skew) eigenvalues [56] of the 'dressed' central charge matrix $Z_{AB}(Q,\phi)$. Indeed, the positive hermitian matrix $\mathcal{H}_A{}^B = Z_{AC}\bar{Z}^{BC}$ has four *real positive eigenvalues* $|Z_i(Q,\phi)|^2 = \lambda_i$ which, without any loss of generality, can be put in decreasing order $\lambda_1 \geq \lambda_2 \geq \lambda_3 \geq \lambda_4$. The BPS condition requires that the ADM mass be exactly equal to the largest eigenvalue of the central charge matrix, i.e. it saturates the bound (3.5). On the other hand, non-BPS extremal black holes have a mass which is strictly larger than the largest eigenvalue of the central charge matrix; in such a case $\mathcal{I}_4 < 0$, and the non-BPS extremal black hole geometry is regular and its mass is never zero : $S_{BH,non-BPS} = \pi\sqrt{-\mathcal{I}_4}$. Thus, it follows that for regular black holes ($\mathcal{I}_4 \neq 0$) the ADM mass is bounded from below as a function in the moduli space; no "massless black holes" can exist in this case, contrarily to the $\mathcal{N} = 2$ [57, 40] (and $\mathcal{N} = 4$ [37]) cases.

In the SU(8) covariant 'dressed' central charge basis, the quartic invariant (3.2) in the area/entropy formula (3.3) reads

$$\mathcal{I}_4(Q,\phi) = Tr[(Z\bar{Z})^2] - \frac{1}{4}[Tr(Z\bar{Z})]^2 + 8\mathrm{Re}Pf(Z), \tag{3.6}$$

where $Pf(Z)$ denotes the *Pfaffian* of Z_{AB}; it should be remarked that each $SU(8)$ invariant term in this expression depends on the moduli, but nevertheless the total expression is moduli independent due to $E_{7(7)}$ symmetry.

As a final comment, it should be stressed that if the perturbative theory has some - yet unknown - UV divergences [23, 24], the analysis of the massless black hole solutions and the proposal put forward in [37] may require modifications. However, if there are no UV divergences in perturbation theory, corrections to the analysis of the space-time properties of these states are not expected. In particular, the states in [50] have been shown to be singular in [37]. There are two reasonable options: *i)* these states may be consistently excluded from the four-dimensional theory and therefore do not affect UV properties of $\mathcal{N} = 8$, $D = 4$ supergravity; *ii)* they can be proven to be required in $D = 4$ and affect the perturbative theory.

While some evidence for the plausibility of the first option was given in [37], a conclusive answer has not been given yet.

4 Conclusive Remarks

If $\mathcal{N} = 8$, $D = 4$ supergravity turns out to be UV perturbatively finite, according to Bern *et al.* [11, 21], it is not only due to maximal supersymmetry and to perturbatively unbroken $E_{7(7)}$ symmetry, but also to other reasons.

One of these is the *"double copy"* structure [58], which implies a relation, not only kinematical but also dynamical, between the square of the $\mathcal{N} = 4$ super Yang-Mills amplitudes and $\mathcal{N} = 8$ amplitudes. At loop level, the "double copy" properties of amplitudes have been extended to supergravity theories with $\mathcal{N} \geqslant 4$; in this case, one copy is given by $\mathcal{N} = 4$ Yang-Mills and the other copy is an $\mathcal{N} = 0, 1, 2$ Yang-Mills gauge theory, thus giving rise to $\mathcal{N} = 4, 5, 6$ supergravity theories. From the analysis of divergences, one is led to conclude that $\mathcal{N} = 6$ and $\mathcal{N} = 5$ supergravity may be UV finite (if $\mathcal{N} = 8$ is), while $\mathcal{N} = 4$ probably is not. It is worth remarking that these results are in agreement with the "composite anomaly" arguments [36, 13] for which $\mathcal{N} = 5, 6$ do not exhibit duality anomalies, while $\mathcal{N} = 4$ does.

Another interesting aspect [21] which should be implied by UV finiteness of $N = 8, 6, 5$ supergravity in $D = 4$ dimensions is that their *gauged* versions should be possibly UV finite, as well. Roughly speaking, this is related to the fact that *gauging* may be regarded as a spontaneous soft breaking of an unbroken gauge symmetry, and UV properties should not be affected by such a spontaneous breaking, as it happens in the Standard Model of electro-weak interactions.

We have already commented on the difficulties and subtleties related to the question of whether a point-like non-perturbative completion of $\mathcal{N} = 8$ supergravity exists. Single-centered BPS black hole states preserving a large fraction of supersymmetry (1/2 or 1/4) are singular in Einstein theory of gravity, and thus physically unacceptable. However, such states may be *"confined"*, *i.e.* they may only exist as building blocks of multi-centered black hole configurations; the viability and physical meaning of such a phenomenon are currently unexplored issues. On the other hand, from a superstring theory perspective such singular states are just an indication that the fundamental theory is ill-defined, and that extra dimensions *and/or* a *"non-local"* structure with minimal length open up in the non-perturbative quantum regime.

Acknowledgments

Enlightening discussions with Zvi Bern and Renata Kallosh are gratefully acknowledged.

The work of S.F. is supported by the ERC Advanced Grant no. 226455, *"Supersymmetry, Quantum Gravity and Gauge Fields"* (*SUPERFIELDS*).

References

[1] J. H. Schwarz (Ed.) : *"The First 15 Years of Superstring Theory"* (2 Vols.), World Scientific, 1985.

[2] J. Polchinski, *"String Theory"*, Cambridge University Press, 1998.

[3] E. Witten, *String theory dynamics in various dimensions*, Nucl. Phys. **B443**, 85 (1995), hep-th/9503124.

[4] M. J. Duff, *M theory (The Theory formerly known as strings)*, Int. J. Mod. Phys. **A11**, 5623 (1996), hep-th/9608117.

[5] M.B. Green, J.H. Schwarz and E. Witten, *"Superstring Theory"*, (2 Vols.), Cambridge University Press, 1987.

[6] M.R. Douglas, *The Statistics of String / M-Theory Vacua*, JHEP **0305**, 046 (2003). M.R. Douglas and F. Denef, *Distributions of Flux Vacua*, JHEP **0405**, 072 (2004). *Distributions of Nonsupersymmetric Flux Vacua*, JHEP **0503**, 061 (2005).

[7] L. Randall and R. Sundrum, *Out of This World Supersymmetry Breaking*, Nucl. Phys. **B557**, 79 (1999). *A Large Mass Herarchy from a Small Extra Dimension,*

Phys. Rev. Lett. **83**, 3370 (1999). *An Alternative to Compactification*, Phys. Rev. Lett. **83**, 4690 (1999).

[8] I. Antoniadis, *From Extra-Dimensions : Multiple Brane Scenarios and their Contenders*, Int. J. Mod. Phys. **A25**, 5817 (2010). I. Antoniadis, M. Atkins and X, Calmet, *Brane World Models need Low String Scale*, JHEP **1111**, 039 (2011).

[9] G.R. Dvali, G. Gabadadze and M. Porrati, *4-D Gravity on a Brane in 5-D Minkowski Space*, Phys. Lett. **B485**, 208 (2000). G. R. Dvali, G. Gabadadze, M. Kolanovic and F. Nitti, *The Power of Brane Induced Gravity*, Phys. Rev. **D64**, 084004 (2001).

[10] N. Arkani-Hamed, S. Dimopoulos, G.R. Dvali and N. Kaloper, *Infinitely Large New Dimensions*, Phys. Rev. Lett. **84**, 586 (2000). N. Arkani-Hamed, S. Dimopoulos, N. Kaloper and R. Sundrum, *A Small Cosmological Constant from a Large Extra Dimension*, Phys. Lett. **B480**, 193 (2000).

[11] Z. Bern, J. J. Carrasco, L. J. Dixon, H. Johansson and R. Roiban, *The Ultraviolet Behavior of $\mathcal{N}=8$ Supergravity at Four Loops*, Phys. Rev. Lett. **103**, 081301 (2009).

[12] G. Bossard, P.S. Howe and K.S.Stelle, *The Ultra-Violet Question in Maximally Supersymmetric Field Theories*, Gen. Rel. Grav. **41**, 919 (2009).

[13] G. Bossard, C. Hillmann and H. Nicolai, *$E_{7(7)}$ Symmetry in Perturbatively Quantised $\mathcal{N}=8$ Supergravity*, JHEP **1012**, 052 (2010).

[14] R. Kallosh, *On UV Finiteness of the Four Loop $\mathcal{N}=8$ Supergravity*, JHEP **0909**, 116 (2009).

[15] P. Vanhove, *The Critical Ultraviolet Behaviour of $\mathcal{N}=8$ Supergravity Amplitudes*, `arXiv:1004.1392 [hep-th]`.

[16] L. J. Dixon, *Ultraviolet Behavior of $\mathcal{N}=8$ Supergravity*, Lectures presented at ISSP 2009, Aug 29 - Sep 7 '09, Erice (Italy), `arXiv:1005.2703 [hep-th]`.

[17] R. Kallosh, *The Ultraviolet Finiteness of $\mathcal{N}=8$ Supergravity*, JHEP **1012**, 009 (2010).

[18] D. Z. Freedman, P. van Nieuwenhuizen and S. Ferrara, *Progress Toward a Theory of Supergravity*, Phys. Rev. **D13**, 3214 (1976).

[19] S. Deser and B. Zumino, *Consistent Supergravity*, Phys. Lett. **B62**, 335 (1976).

[20] N. Beisert, H. Elvang, D. Z. Freedman, M. Kiermaier, A. Morales and S. Stieberger, *$E_{7(7)}$ Constraints on Counterterms in $\mathcal{N}=8$ Supergravity*, Phys. Lett. **B694**, 265 (2010).

[21] Z. Bern, J. J. Carrasco, L. Dixon, H. Johansson and R. Roiban, *Amplitudes and Ultraviolet Behavior of $\mathcal{N}=8$ Supergravity*, `arXiv:1103.1848 [hep-th]`.

[22] M. B. Green and P. Vanhove, *D-instantons, strings and M-theory*, Phys. Lett. B **408**, 122 (1997).

[23] M. B. Green, J. G. Russo and P. Vanhove, *String theory dualities and supergravity divergences*, JHEP 1006 (2010), 075.

[24] J. Bjornsson and M. B. Green, *5 loops in 24/5 dimensions*, JHEP **1008** (2010), 132.

[25] S. Ferrara and P. Van Nieuwenhuizen, *Consistent Supergravity with Complex Spin 3/2 Gauge Fields*, Phys. Rev. Lett. **37**, 1669 (1976).

[26] M. Gell-Mann, P. Ramond and R. Slansky, *Color Embeddings, Charge Assignments, and Proton Stability in Unified Gauge Theories*, Rev. Mod. Phys. **50**, 721 (1978). *Complex Spinors and Unified Theories*, Conf.Proc. C790927, 315 (1979).

[27] E. Cremmer and B. Julia, *The $SO(8)$ Supergravity*, Nucl. Phys. B **159**, 141 (1979).

[28] B. de Wit and H. Nicolai, *$\mathcal{N}= 8$ Supergravity*, Nucl. Phys. B **208** (1982) 323.

[29] C. M. Hull and N. P. Warner, *The Structure of the Gauged $\mathcal{N}= 8$ Supergravity Theories*, Nucl. Phys. **B253**, 650 (1985). C. M. Hull and N. P. Warner, *The Potentials of the Gauged $\mathcal{N}= 8$ Supergravity Theories*, Nucl. Phys. **B253**, 675 (1985).

[30] E. Cremmer, B. Julia and J. Scherk, *Supergravity Theory in Eleven-Dimensions*, Phys. Lett. **B76**, 409 (1978).

[31] R. Slansky, *Group Theory for Unified Model Building*, Phys. Rep. **79**, 1 (1981).

[32] I. Yokota, *Subgroup $SU(8)/\mathbb{Z}_2$ of compact simple Lie group E_7 and non-compact simple Lie group $E_{7(7)}$ of type E_7*, Math. J. Okoyama Univ. **24**, 53 (1982).

[33] P. Aschieri, S. Ferrara and B. Zumino, *Duality Rotations in Nonlinear Electrodynamics and in Extended Supergravity*, Riv. Nuovo Cim. **31**, 625 (2008).

[34] M. K. Gaillard and B. Zumino, *Duality Rotations for Interacting Fields*, Nucl. Phys. **B193**, 221 (1981).

[35] P. Di Vecchia, S. Ferrara and L. Girardello, *Anomalies of Hidden Local Chiral Symmetries in Sigma Models and Extended Supergravities*, Phys. Lett. **B151**, 199 (1985).

[36] N. Marcus, *Composite Anomalies in Supergravity*, Phys. Lett. **B157**, 383 (1985).

[37] M. Bianchi, S. Ferrara and R. Kallosh, *Perturbative and Non-Perturbative $\mathcal{N}= 8$ Supergravity*, Phys. Lett. **B690**, 328 (2010). M. Bianchi, S. Ferrara and R. Kallosh, *Observations on Arithmetic Invariants and U-Duality Orbits in $\mathcal{N}= 8$ Supergravity*, JHEP **1003**, 081 (2010).

[38] M. Green, H. Ooguri and J. H. Schwarz, *Nondecoupling of Maximal Supergravity from the Superstring*, Phys. Rev. Lett. **99**, 041601 (2007).

[39] N. Arkani-Hamed, F. Cachazo and J. Kaplan, *What is the Simplest Quantum Field Theory?*, JHEP **09**, 016 (2010).

[40] T. Banks, *Why I don't Believe $\mathcal{N}$= 8 SUGRA is Finite*, talk at the Workshop *"Supergravity versus Superstring Theory in the Ultraviolet"*, PennState Univ, PA USA, August 27-30 2009.

[41] B. Bertotti, *Uniform Electromagnetic Field in the Theory of General Relativity*, Phys. Rev. **116**, 1331 (1959). I. Robinson, *A solution of the Maxwell-Einstein equations*, Bull. Acad. Pol. Sci. Ser. Sci. Math. Astron. Phys. **7**, 351 (1959).

[42] L. Borsten, D. Dahanayake, M. J. Duff and W. Rubens, *Black Holes admitting a Freudenthal Dual*, Phys. Rev. **D80**, 026003 (2009).

[43] S. Ferrara, A. Marrani and A. Yeranyan, *Freudenthal Duality and Generalized Special Geometry*, Phys. Lett. **B701**, 640 (2011).

[44] L. Borsten, D. Dahanayake, M. J. Duff, S. Ferrara, A. Marrani and W. Rubens, *Observations on Integral and Continuous U-Duality Orbits in $\mathcal{N}$= 8 Supergravity*, Class. Quant. Grav. **27**, 185003 (2010).

[45] C. M. Hull and P. K. Townsend, *Unity of superstring dualities*, Nucl. Phys. B **438**, 109 (1995). *Enhanced gauge symmetries in superstring theories*, Nucl. Phys. B **451**, 525 (1995).

[46] M. Bianchi, M. B. Green, S. Kovacs and G. Rossi, *Instantons in supersymmetric Yang-Mills and D-instantons in IIB superstring theory*, JHEP **9808**, 013 (1998).

[47] R. Kallosh and B. Kol, *E(7) Symmetric Area of the Black Hole Horizon*, Phys. Rev. D **53**, 5344 (1996).

[48] M. Cvetic and D. Youm, *Dyonic BPS saturated black holes of heterotic string on a six torus*, Phys. Rev. **D53**, R584 (1996), `hep-th/9507090`. M. Cvetic and A. A. Tseytlin, *General class of BPS saturated dyonic black holes as exact superstring solutions*, Phys. Lett. **B366**, 95 (1996), `hep-th/9510097`.

[49] J. M. Maldacena, *The large N limit of superconformal field theories and supergravity*, Adv. Theor. Math. Phys. **2**, 231 (1998) [Int. J. Theor. Phys. **38**, 1113 (1999)].

[50] M. B. Green, H. Ooguri and J. H. Schwarz, *Nondecoupling Supergravity from the Superstring*, Phys. Rev. Lett. **99** (2007) 041601.

[51] S. Ferrara, G. W. Gibbons and R. Kallosh, *Black holes and critical points in moduli space*, Nucl. Phys. B **500** (1997) 75.

[52] R. Arnowitt, S. Deser and C. W. Misner: *Canonical Variables for General Relativity*, Phys. Rev. **117**, 1595 (1960).

[53] S. Ferrara, R. Kallosh and A. Strominger, *$\mathcal{N}$= 2 extremal black holes*, Phys. Rev. D **52**, 5412 (1995).

[54] S. Ferrara and R. Kallosh, *On $\mathcal{N}$= 8 attractors*, Phys. Rev. D **73**, 125005 (2006).

[55] A. Ceresole, R. D'Auria, S. Ferrara and A. Van Proeyen, *Duality transformations in supersymmetric Yang-Mills theories coupled to supergravity*, Nucl. Phys. B **444**, 92 (1995).

[56] L. K. Hua, *On the Theory of Automorphic Functions of a Matrix Variable. I: Geometrical Basis*, Amer. J. Math. **66**, 470 (1944). C. Bloch and A. Messiah, *The Canonical Form of an Antisymmetric Tensor and its Application to the Theory of Superconductivity,* Nucl. Phys. **B39**, 95 (1962). B. Zumino, *Normal Forms of Complex Matrices*, J. Math. Phys. **3**, 1055 (1962).

[57] A. Strominger, *Massless Black Holes and Conifolds in String Theory*, Nucl. Phys. **B451**, 96 (1995).

[58] Z. Bern, C. Boucher-Veronneau and H. Johansson, $\mathcal{N} \geqslant 4$ *Supergravity Amplitudes from Gauge Theory at One Loop*, Phys. Rev. **D84**, 105035 (2011), `arXiv:1107.1935 [hep-th]`.

CHAIRMAN: S. FERRARA

Scientific Secretaries: B.L. Cerchiai, A. Yeranyan

DISCUSSION

- Bousso:

Why should there be a mechanism for selecting a unique vacuum, when we have a plausible scenario (eternal inflation) for creating a very large number of vacua?

- Ferrara:

Originally in string theory, as a theory of everything, it was expected that there would be a unique vacuum related to a Calaby-Yau manifold and where generation was related to the Hodge numbers. Nowadays there is a change of perspective, a quantum field theory vacuum is not necessarily so simple and could be analyzed statistically.

- Borsten:

What are the other four-dimensional supergravities with vanishing chiral anomaly?

- Ferrara:

Along with the N=8 theory, according to an argument by Z. Bern and collaborators, also N=5 and N=6 theories would have vanishing chiral anomaly. The anomaly argument should not work for theories with N≤4, i.e. when there isn't a unique multiplet, but also matter multiplets. Of course, one could have particular matter multiplets for which the chiral anomaly vanishes, as in the standard model. Usually the chiral anomaly comes from fermion loops. However, due to the particular nature of electric-magnet duality, in this case spin 1 bosons are chiral and contribute to the anomaly. Hillman-Bossard-Nicolai have shown by explicit computation of Feynman graphs that this cancellation of anomaly occurs for certain combinations, as in the N=8 theory.

- Borsten:

Has anyone checked whether there are U-duality invariant counterterms in N=5 and N=6 theories?

- Ferrara:

No, because N=5 and N=6 theories are complicated theories which can be obtained from string theory only by reducing the number of supersymmetries, while the N=8 theory is easier, because it has the same number of supersymmetries as in 10 or 11 dimensions.

- Sibiryakov:

If N=8 in d=4 supergravity is finite, does it have any implications for N=1 D=11 supergravity?

- Ferrara:

Supergravity in 10 and 11 dimensions is certainly not finite. Finiteness of the N=8 supergravity theory in 4 dimensions is due to special properties of the theory in 4 dimensions. In particular, to certain special properties of the electric-magnetic duality in 4 dimensions, allowing e.g. cancellation of the chiral anomaly. In higher dimensions certain counterterms exist, which vanish in 4 dimensions. According to Z. Bern et al. divergencies of the N=8 supergravity are not those of Einstein theory, but those of N=4 Super-Yang-Mills theory, which is a finite theory.

- *Sibiryakov:*

Is it possible to introduce a cosmological constant in N=8 in d=4 supergravity?

- *Ferrara:*

So far all the perturbative studies of N=8 supergravity have been performed in flat space. It is possible to consider deformations of the theory corresponding to the introduction of a cosmological constant, but these are not expected to change the ultraviolet properties of the theory.

- *Sibiryakov:*

Even if N=8 in d=4 supergravity turns out to be finite, the interactions still become strong at Planckian energies. So, the theory is inevitably non-perturbative in the ultraviolet. How is this reconciled with perturbative finiteness?

- *Ferrara:*

The theory definitely needs a non-perturbative completion. Some think that this will require string theory. Alternatively, the non-perturbative effects may be of the same kind as instanton effects in gauge theories. If this non-perturbative completion exists, it should break the continuous E7 symmetry to a discrete subgroup because of black holes charge quantization. For sure, it should explain several things, e.g. the inclusion of black holes in the theory.

- *Minkowski:*

What happens to the $SO(8)_L$ x $SO(8)_R$ gauge structure of the theory? Can the uncanceled version of the anomaly of supergravity be revealed by coupling it to an external structure?

- *Ferrara:*

The SO(8) in supergravity is ungauged, because the theory lives in flat space. The SO(8) group is embedded in the $E_{7(7)}$ electric-magnetic duality. An "external current" can display the anomaly in an unusual way, because not only fermions are chiral in this theory, but bosons as well.

- *Cerchiai:*

What is Renata Kallosh's argument why N=8 d=4 supergravity should be finite even beyond 7 loops?

- *Ferrara:*

It is a technical computation and it is not sure whether it is correct. But the argument is similar to the one for the Born-Infeld action. If you impose electric magnetic duality on a lagrangian with higher order terms, then it poses such severe constraints on the theory that it fixes almost everything. If R. Kallosh's argument is correct, then it would mean that electric magnetic duality is enough to guarantee perturbative finiteness of the theory, which in this context is the E7 symmetry.

- *Mouslopoulos:*

You said that one can reproduce the area law for black holes in the case of extremal black holes. What happens for non-extremal ones?

- *Ferrara:*

Of course, you can recover the same result also for non-extremal black holes. However, in the case of extremal black holes there is an elegant way of obtaining the result almost without doing any explicit computation. In some sense these states are like the elementary stable constituents of the theory. Using the electric-magnetic E_7 duality group one can compute the entropy as its unique quartic invariant.

- *Bousso:*

If N=8 supergravity is finite and string theory is not needed for completion, then D-branes and strings cannot explain the microstates of black holes. But then do you have any idea of what could be responsible for them?

- *Ferrara:*

Denef, Moore, Sen et al. have been able to count microstates essentially by counting BPS states. There is a counting in supergravity based on classifying representations of the duality groups of BPS states. This calculation includes large black holes as composite of small constituents.

Composite Weak Bosons, Leptons and Quarks

Harald Fritzsch
University of Munich
Faculty of Physics,
Arnold Sommerfeld Center for Theoretical Physics,

Abstract

The weak bosons are interpreted as composite particles. They consist of two fundamental fermions, bound by a confining gauge interaction. The mass scale of this new interaction is determined. A new neutral isoscalar weak boson X must exist. Its mass is expected to be less than 1 TeV. It will decay mainly into quark and lepton pairs and into two or three weak bosons. Above 1 TeV excitations of the weak bosons must exist, which mainly decay into pairs of weak bosons. Leptons and quarks consist of a fermion and a scalar. Two scalars form bound states, which are either neutral in charge or have charge 2/3 e.

The Standard Model of particle physics is known since 1973. It combines the gauge theory of the electroweak interactions (ref.(1)) and quantum chromodynamics (ref.(2)). In the theory of the electroweak interactions the weak bosons are considered as elementary gauge bosons. Their masses and the masses of the leptons and quarks are due to the spontaneous breaking of the gauge symmetry (ref.(2)). In the field theory of the strong interactions the hadronic masses are due to the field energy of the confined gluons and quarks inside the hadrons (confinement mechanism). In the absence of the quark masses the masses of the baryons and mesons are proportional to the QCD mass scale Λ_c.

Thus in the Standard Model there are two different ways to introduce masses. They are either due to a spontaneous symmetry breaking (the masses of the weak bosons, of the leptons and of the quarks) or due to the confinement mechanism in QCD (e.g. the masses of the nucleons).

The masses of the hadrons are certainly due to the confinement mechanism. It is not yet known, whether the mechanism of spontaneous symmetry breaking is realized in nature. The masses of the weak bosons, the leptons and the quarks might also be due to a confinement mechanism - this possibility will be discussed here.

Many years ago the ρ -mesons were considered as elementary gauge bosons. Their masses can be generated by a spontaneous symmetry breaking. One needs a complex doublet of scalar fields. After the symmetry breaking the ρ -mesons are massive, and one massive scalar "Higgs" particle is present. Today we know that this theory is not realized in nature. In the theory of quantum chromodynamics the ρ -mesons are quark-antiquark bound states, and the masses of the ρ -mesons are due to the field energies of the quarks and gluons inside the mesons, as the masses of the nucleons.

In this paper the weak bosons, the leptons and the quarks are interpreted as composite particles. The constituents of the weak bosons and of the leptons and quarks are bound by a new confining gauge interaction. The masses of the weak bosons and of the leptons and quarks are due to the confinement mechanism - they can be calculated, once the details of the theory are known. Note that the present limit on the size of the electron, the muon and of the light quarks is about 10^{-17} cm.

We denote the constituents as "haplons" - the Greek translation of "simple" is "haplos". The new confining gauge theory is denoted as QHD. The QHD mass scale is given by a mass parameter Λ_h. This mass scale determines the masses of the weak bosons. A theory of this type was proposed in 1981 (ref.(4), see also ref.(5,6,7,8,9)).

Two types of lefthanded spin 1/2 haplons are needed as constituents of the weak bosons, denoted by α and β. The doublet h of the weak isospin group $SU(2)$ is given by the two lefthanded haplon fields:

$$h = \begin{pmatrix} \alpha \\ \beta \end{pmatrix} \tag{1}$$

The electric charges of the haplons are determined by the requirement that the charge difference must be one and the sum of the charges should vanish. The electric charges in units of e are:

$$h = \begin{pmatrix} +\frac{1}{2} \\ -\frac{1}{2} \end{pmatrix} \tag{2}$$

The three weak bosons have the following internal structure:

$$\begin{aligned} W^+ &= \overline{\beta}\alpha\,, \\ W^- &= \overline{\alpha}\beta\,, \\ W^3 &= \frac{1}{\sqrt{2}}\left(\overline{\alpha}\alpha - \overline{\beta}\beta\right)\,. \end{aligned} \tag{3}$$

We expect that in a composite model the structure of the spectral functions of the weak currents is similar to the structure of the spectral functions in hadronic physics. At energies of the order of the QCD mass scale the spectral functions of the hadronic currents are dominated by the lowest states, the ρ-mesons. At higher energies excited mesons are present. At energies larger than the QCD mass scale the spectral functions are given by the continuum of quark-antiquark states.

Analogously we expect that at the energy of the order of the QHD mass scale the spectral functions of the weak currents are dominated by the lowest states, the weak bosons. At higher energies excited weak bosons are present. At energies larger than the QHD mass scale there would be a continuum of states as in QCD.

In strong interaction physics the universality of the couplings of the ρ-mesons to the hadrons follows from the current algebra and the dominance of the matrix elements of the vector currents by the ρ-mesons. In the Standard Model the universality of the weak coupling constants is due to the gauge invariance. In our composite model of the weak bosons it follows from the algebra of the weak currents and the dominance of the weak currents by the weak bosons (ref.(6)).

In the absence of electromagnetism and the quark masses the three ρ-mesons are degenerate in mass. If the electromagnetic interaction is introduced, the neutral ρ-meson changes its mass due to a mixing with the photon. The mass shift, caused by this mixing, can be calculated. It depends on a mixing parameter μ, which is determined by the electric charge, the decay constant F_ρ and the mass of the ρ-meson:

$$\mu = e\frac{F_\rho}{M_\rho} \tag{4}$$

The decay constant is measured to about 220 MeV, which gives $\mu \approx 0.09$. The mass shift is about 3 MeV. There is also a contribution due to the isospin breaking, caused by the quark masses. These mass shifts are too small to be measured in the experiments.

Analogously in QHD the three weak bosons are degenerate in mass in the absence of electromagnetism. If the electromagnetic interaction is introduced, the mass of the neutral boson increases due to the mixing with the photon. This mass shift has been measured in the LEP-experiments: 10.79 ± 0.03 GeV.

In the Standard Model the mixing between the neutral weak boson and the photon is generated by scalar fields. In our model it is caused by the dynamics, like the mixing between the photon and the neutral ρ-meson in QCD. The mixing of the weak bosons is described by a mixing parameter m, which is determined by the decay constant of the weak boson F_W, defined in analogy to the decay constant of the ρ-meson in QCD (ref.(6)):

$$\langle 0 \left| \frac{1}{2} \left(\overline{\alpha}\gamma_\mu \alpha - \overline{\beta}\gamma_\mu \beta \right) \right| Z \rangle = \varepsilon_\mu M_W F_W \, . \tag{5}$$

In the Standard Model the mixing is described by the weak mixing angle. In QHD the mixing is a dynamical phenomenon, and the mixing parameter m is given by the decay constant of the W-boson F_W:

$$m = e \frac{F_W}{M_W} \tag{6}$$

In the Standard Model the mixing parameter m is determined by the weak mixing angle (ref.(6)):

$$\sin \theta_w = m \tag{7}$$

The mass difference between the Z-boson and the W-boson is given by the mixing parameter m and the W-mass:

$$M_Z^2 - M_W^2 = M_W^2 \left(\frac{m^2}{1 - m^2} \right) . \tag{8}$$

We shall use the following experimental values:

$$\begin{aligned} M_W &= 80.4 \text{ GeV} , \\ M_Z &= 91.9 \text{ GeV} , \\ F_W &= 124 \text{ GeV} , \\ \sin^2\theta_W &= 0.2315 , \\ \alpha &= \frac{e^2}{4\pi} \approx \frac{1}{128.9} , \\ e &= 0.3122 , \\ m &= 0.481 \end{aligned} \tag{9}$$

In strong interaction physics the decay constant of the ρ-meson meson and the QCD mass parameter Λ_c are proportional. The decay constant of the ρ-meson is measured to about 220 MeV, with a large error. The QCD mass parameter Λ_c has also been measured: 217 ± 25 MeV. Thus both parameters are about equal.

We expect a similar connection between the decay constant of the weak boson and the QHD mass parameter Λ_h. If the QHD gauge group would be SU(3), the ratio of Λ_h and Λ_c would be given by the measured ratio of the decay constants:

$$\frac{\Lambda_h}{\Lambda_c} \approx \frac{F_W}{F_\rho} \simeq 560. \tag{10}$$

In that case Λ_h would be about 0.12 TeV. For larger groups Λ_h would be larger, but it should be less than 0.5 TeV. Thus the mass scales of QCD and QHD differ by about three orders of magnitude.

If the weak bosons consist of the two haplons α and β, there must exist a second neutral weak boson X, which is an isoscalar and has the internal structure:

$$X = \frac{1}{\sqrt{2}}\left(\overline{\alpha}\alpha + \overline{\beta}\beta\right) . \qquad (11)$$

This boson is not present in the Standard Model, and its mass must be very large. The present lower limit on the mass of the X-boson is about 0.8 TeV (see also ref.(9)).

In strong interaction physics the mass of the ρ-meson and of the ω-meson are nearly the same, and one would expect that the mass of the X–boson is about the same as the mass of the Z-boson, but this is excluded by the experiments. The fact that the X–boson, if it exists, is much heavier than the Z-boson, might be related to the QHD analogy of the gluonic anomaly of QCD. The latter implies that the mass of the η'-meson is different from zero in the chiral limit, while the masses of the π-mesons and of the η-meson vanish.

In QHD the isospin singlet axial vector current also has an anomaly, and this might be the reason why the X-boson is very heavy. The theory of QHD is a confining chiral gauge theory, and low mass pseudoscalar bosons do not exist. The anomalous divergence of the singlet axial vector current might increase the mass of the X-boson. But details about the dynamics of chiral gauge theories are not yet known. For our further discussion we shall assume a mass of 0.8 TeV for the X-boson.

The X-boson would couple to the leptons and quarks with about the same strength as the Z-boson. Thus it can easily be produced at the LHC by quark-antiquark-fusion. The cross section for the production of Z-bosons at the LHC is estimated to about 60 nb. If the X-boson has a mass of 0.8 TeV, the cross section for its production at the LHC should be about 0.8 nb.

An important decay mode of the X-boson is the decay into lepton pairs, e.g. into muon pairs. The partial width for this decay can be estimated by comparing it with the decay of a charged weak boson into a muon and the muon antineutrino, which has a partial width of about 220 MeV. Using this result, we can calculate the partial width for the leptonic decay of the

X-boson with a mass of 0.8 TeV:

$$\Gamma(X \to \mu^+\mu^-) \approx 1.2\, GeV. \tag{12}$$

The X-boson will decay primarily into lepton pairs and quark pairs. We expect the following relations to hold between the branching fractions of the various decays:

$$\begin{aligned}
&Br(X \to e^+e^-) \cong Br(X \to \nu_e\nu_e) \\
&Br(X \to \nu_e\nu_e) \cong Br(X \to \nu_\mu\nu_\mu) \cong Br(X \to \nu_\tau\nu_\tau) \\
&Br(X \to e^+e^-) \cong Br(X \to \mu^+\mu^-) \cong Br(X \to \tau^+\tau^-) \\
&3Br(X \to e^+e^-) \cong Br(X \to \overline{u}u) \cong Br(X \to \overline{d}d) \\
&Br(X \to \overline{u}u) \cong Br(X \to \overline{c}c) \cong Br(X \to \overline{t}t) \\
&Br(X \to \overline{d}d) \cong Br(X \to \overline{s}s) \cong Br(X \to \overline{b}b)
\end{aligned} \tag{13}$$

The X-boson can also decay into weak bosons. We expect:

$$\Gamma(X \to W^+W^-) \cong \Gamma(X \to ZZ) \simeq \Gamma(X \to \mu^+\mu^-) \tag{14}$$

For the decays of the X - boson into three and four weak bosons the following relations for the partial widths are expected:

$$\begin{aligned}
&\Gamma(X \to W^+W^-Z) \cong \Gamma(X \to ZZZ) \\
&\Gamma(X \to W^+W^-W^+W^-) \cong \Gamma(X \to W^+W^-ZZ) \\
&\Gamma(X \to W^+W^-W^+W^-) \cong \Gamma(X \to ZZZZ)
\end{aligned} \tag{15}$$

We introduce the parameters a, b and c:

$$\begin{aligned}
&\Gamma(X \to W^+W^-) = a \cdot \Gamma(X \to \mu^+\mu^-) \\
&\Gamma(X \to W^+W^-Z) = b \cdot \Gamma(X \to \mu^+\mu^-) \\
&\Gamma(X \to W^+W^-W^+W^-) = c \cdot \Gamma(X \to \mu^+\mu^-)
\end{aligned} \tag{16}$$

The parameter "a" should be of the order of one. The parameters "b" and "c" are expected to be smaller than one. As an example we set a=1, b=0.5, c=0.3 and estimate the total width of the X-boson. There are three decay channels for the charged leptons, three channels for the neutrinos and 18 channels for the quark-antiquark pairs (including the color degree of freedom). The decays into weak bosons are added, using the parameters above. Other decays of the X - boson are expected to be small and are neglected. Then we find for the total width:

$$\Gamma(X \to all) \approx 35\ GeV. \tag{17}$$

The best way to observe the X - boson in the collisions at the LHC is to find the decays into muon pairs and into electron-positron pairs. Once it has been found, one can search for the decays into quark-antiquark pairs. Two narrow quark jets should be observed with an invariant mass given be the mass of the X-boson.

The QHD mass scale is three orders of magnitude larger than the QCD mass scale. In strong interaction physics complexities arise at the energy of 1 GeV and above. Analogously there should be complexities due to the QHD dynamics at the energy of 1 TeV and above.

In strong interaction physics there exist excited states of the vector mesons. Analogously we expect that there exist excited states of the weak bosons, with masses of the order of 1 TeV and higher. These states can have the total angular momentum 1,2,3 etc. The excited state of a charged weak boson can decay into a charged weak boson and a Z-boson or a photon. The excited state of a Z-boson will decay mainly into two weak bosons. Decays of the excited weak bosons into quark pairs or leptons pairs are expected to be suppressed.

The weak bosons couple universally to the leptons and quarks, as the ρ-mesons to the nucleons. Inside the ρ-meson are the same quarks as inside the nucleons, and this leads to the universality of the coupling parameters. Analogously we expect form the universality of the weak coupling parameters that inside a lepton and quark the haplons α and β are also present. A bound state model of the weak bosons requires that the leptons and quarks are also composite systems.

The simplest model of composite leptons and quarks is the one discussed in ref.(4). Besides the fermions α and β four scalar haplons are needed, one scalar for the leptons, denoted by l, and three scalars for the three colors of the quarks, denoted by r, g and b:

$$h(fermion) = \begin{pmatrix} \alpha \\ \beta \end{pmatrix} \tag{18}$$

$$h(scalar) = (l,\ r,\ g,\ b) \tag{19}$$

Both the fermions and the scalars transform according to the fundamental representation of the QHD gauge group. Thus bound states of the fermions and the scalars exist - the lowest states would be the observed leptons and quarks. The lowest leptons have the structure:

$$\begin{aligned} \nu &= (\alpha l) \\ e^- &= (\beta l) \end{aligned} \tag{20}$$

The structure of the up and down quarks (with red color) is given by:

$$u = (\alpha r)$$
$$d = (\beta r)$$

(21)

In such a model the first generation of leptons and quarks would describe the ground states of the fermion-scalar bound states, the second and third generation must be dynamical excitations - the electron is the ground state of the charged leptons, the muon and the tau-lepton are excitations. Likewise the u-quark is the ground state of the up-quarks, the c-quark and the t-quarks are excitations. Compared to the QHD mass scale the masses of the observed leptons and quarks are essentially zero. The number of nearly massless bound states could be related to the rank of the QHD gauge group. Three generations might be obtained, if the gauge group is SU(3).

In principle the masses of the leptons and quarks can be calculated, but details of the bound state structures of a confining chiral gauge theory are not yet known.

The quarks in a meson can have a non-zero orbital angular momentum. In QHD we also expect excited states of the leptons and quarks with non-zero orbital angular momenta at energies above 1 TeV. For example, there should exist excitations of the muon with spin 3/2, spin 5/2 etc. The excited muon with spin 3/2 would decay mainly into a muon and a photon or a Z-boson. At energies above 1 TeV there should exist also quarks with spin 3/2, spin 5/2 etc. These quarks would mainly decay into a normal quark, emitting a gluon, a photon or a Z-boson.

In the model, discussed here, there should also exist bound states of the scalars above 1 TeV. A leptonic scalar and its antiparticle will form a bound state S:

$$S = (\bar{l}l) \tag{22}$$

This state would decay mainly into a neutrino and an antineutrino or into a charged lepton and its antiparticle. Decays into pairs of weak bosons are strongly suppressed. The branching ratio for the decay into a muon pair would be about 0.17.

A bound state T, consisting of a leptonic scalar and a red scalar, has the electric charge 2/3:

$$T = (\bar{l}r) \tag{23}$$

This colored scalar with the electric charge 2/3 would decay into a u-quark or a c-quark or a t-quark and an antineutrino, or into a d-quark or a s-quark or a b-quark and a positively charged lepton.

We expect that the first signal of the new substructure of the weak bosons is the discovery of the X-boson at the Large Hadron Collider at CERN.

This paper was written during my visit at the ECT* in Trento in July 2011. I am grateful to Prof. A. Richter for the hospitality.

References

[1] S. L. Glashow and J. C. Ward, Phys. Lett. 13 (1964) 168;
S. Weinberg, Phys. Rev. Lett. 19 (1967) 1264;
A. Salam, in: Elementary Particle Theory, p. 367, Stockholm 1969.

[2] H. Fritzsch and M. Gell-Mann, Proceedings of the XVI Int. Conf. on High Energy Physics, Chicago 1972, Vol. 2
H. Fritzsch, H. Leutwyler and M. Gell-Mann, Physics Lett. 47B (1973) 365.

[3] P. Higgs, Phys. Rev. Lett. 13 (1964) 508;
R. Brout and F. Englert, Phys. Rev. Lett. 13 (1964) 321;
G. S. Guralnik, C. R. Hagen, and T. W. B. Kibble, Phys. Rev. Lett. 13 (1964) 585.

[4] H. Fritzsch and G. Mandelbaum, Phys. Lett. B102 (1981) 319;
Phys. Lett. B 109 (1982) 224;
X. Calmet and H. Fritzsch, Phys. Lett. B496 (2000), 161.

[5] R. Barbieri, R. Mohapatra and A. Masiero, Phys. Lett. B 105 (1981) 369.

[6] H. Fritzsch. D. Schildknecht and R. Kogerler, Phys. Lett. B 114 (1982) 157.

[7] L. F. Abbott and E. Farhi, Phys. Lett. B 101, 69 (1981).

[8] T. Kugo, S. Uehara and T. Yanagida, Phys. Lett. B 147, 321 (1984).

[9] S. Uehara and T. Yanagida, Phys. Lett. B 165, 94 (1985).

[10] U. Baur and K. H. Schwarzer, Phys. Lett. B 180, 163 (1986).

CHAIRMAN: H. FRITZSCH

Scientific Secretaries: E.Rinaldi, S. Biondini

DISCUSSION

- Alba:

Since the W-bosons are composite objects, should we expect the Standard Model to have current-current interactions between the constituent fermions?

- Fritzsch:

The W boson consists of a fermion and anti-fermion just as the rho meson consists of a quark and anti-quark pair. As a first approximation, as far as I know. I just get the Standard Model. However, deviations from the SM can be seen at higher energies.

- Alba:

So, is the QHD a fundamental theory or just an effective theory?

- Fritzsch:

It is a model that reproduces the SM at low energies. QHD is a more fundamental theory than the SM in the sense that the W boson are made from more fundamental particles called haplons. The effects due to the composite nature of the W bosons can be seen only at very high energies and should be discovered soon by the LHC.

- Alba:

Why is Lambda_h of order of 1 TeV?

- Fritzsch:

Lambda_h cannot be determined precisely. However I can calculate the order of magnitude using the decay constant of the W bosons. The order of magnitude for Lambda_h still depends on the choice of the gauge group of QHD. If it is SU(3), than Lambda_h should be approximately 0.2 TeV.

- Alba:

Have you considered models where the W bosons are made up by more than 2 fermions?

- Fritzsch:

No, I have never considered those models. In my model I just have fermion and anti-fermion. Adding more fermions just make things more complicated. The model should be as simple as possible.

- Alba:

QHD is in the strongly coupled regime where there is confinement. How do you calculate things in this regime?

- Fritzsch:

I can make calculations of decay and mixing constant, as I did for the Z boson, but that's about it. It's a confining theory like QCD and the calculations are very difficult at low energy.

- *Bertolucci:*

Why is the cross section for a 500 GeV composite X boson of the order of 2 nb?

- *Fritzsch:*

I used the Z production cross section that has already been calculated and I scaled that up to the energy of 500 GeV. The X production cross section strongly depends on the mass of this new X boson.

- *Burda:*

What kind of fermions are these haplons? Are they Majorana or Dirac fermions?

- *Fritzsch:*

I assume them to be Dirac fermions.

- *Burda:*

Does QHD have four-fermion interaction? What about their non renormalizability?

- *Fritzsch:*

I don't care about non-renormalizability so much. This is a bound state model, but QHD, is renormalizable. The SM comes out just as a first approximation even if without a Higgs boson and this can cause some problems with renormalizability. However, the QHD theory is renormalizable whereas, the SM is just an effective theory.

- *Burda:*

What are the masses of these new fermions (haplons)? Are there any new fundamental parameter introduced by QHD?

- *Fritzsch:*

I assume the new fermions to be massless. All the masses in the SM can in principle be calculated. The only new parameter would be the coupling constant of the gauge group of QHD, on the scale Lambda_h.

- *Burda:*

About oscillations of W and gamma, did you consider only uncharged modes?

- *Fritzsch:*

I just considered oscillations, due to QED, between the neutral Z and the photon. This mixing lifts up the mass of the Z boson with respect to the masses of the W charged bosons.

- *Inguglia:*

Is there any reason why the mass difference in the X-Z system is so big when compared to the \rho-\omega system?

- *Fritzsch:*

There could be the possibility that this mass difference is due to an anomaly in QHD. This anomaly could be similar to the QCD anomaly that enhances the mass of the eta-prime with respect to the pi meson. This is just a hypothesis because I could not do any calculation.

- *Galakhov:*

Should we add something special to the QCD Lagrangian to the break left right symmetry?

- *Fritzsch:*

No, the parity violating nature of the W bosons just follows from the left handed nature of the haplons.

- *Galakhov:*

Can we consider the scalar constituents of leptons in QHD as supersymmetric partners of the fermions?

- *Fritzsch:*

I thought about it. The problem is that the quantum numbers are different. The fermionic haplons have charge plus and minus one half, whereas the scalar haplons in leptons and quarks have charge minus one half or one sixth. Therefore they cannot be superpartners of each other.

- *Galakhov:*

Will radial excitations produce a lot of W-like families?

- *Fritzsch:*

I will discuss this in my next lecture. There are excited W bosons, with spin one half or spin three half, but their masses are very high, of the order of one TeV. There are also excited leptons and excited quarks. For example LHC could observe a spin three half muon at the TeV scale.

- *Galakhov:*

Does your model have any relation to Technicolor theories?

- *Fritzsch:*

Well, there are some similarities, but I don't know the details very well.

- *Morozov:*

What is QHD analog of pi meson?

- *Fritzsch:*

In QHD there is no pi meson, because QHD is a chiral gauge theory. Scalar and pseudo-scalar mesons have very high masses because they come from p-wave excitations of haplons. They don't exist as ground states.

- *Biassoni:*

Are there any other observables, other than the presence of the X boson, that can show the W compositeness?

- *Fritzsch:*

I don't know any. The only way is to see the other particles predicted by my theory. You cannot measure the form factor of the W boson because it is an unstable particle. It is extremely difficult.

- *Preghenella:*

Could your theory explain neutrino oscillations?

- *Fritzsch:*

Neutrino oscillations come because neutrinos have a mass. But my theory does not concern the mass of neutrinos. It might be calculated in the future though.

- *Vlasov:*

What is the ratio of mass, contributed by sea haplons and valence haplons?

- *Fritzsch:*

The haplons are considered massless in my theory.

- *Vlasov:*

What is the interaction that keeps the haplons together? And what is the carrier of this force?

- *Fritzsch:*

Haplons interact via the QHD gauge bosons that are the analog of the gluons.

- *Vlasov:*

Will they contribute to the mass of the W bosons?

- *Fritzsch:*

For sure. The mass of W bosons is coming out from QHD, from the dynamics of the QHD gauge bosons.

- *Inguglia :*

Since quarks and leptons are composite in your model, do you expect radial excitations in this sector as well? And how can these excitations be observed if you think that we haven't even observed a single quark yet?

- *Fritzsch:*

We have observed the quarks. You can see them clearly at SLAC, and in future experiments you might see haplons in the same way inside the W. Excitations of leptons and quarks can decay in their ground states plus gamma rays. This can be observed at the LHC.

- *Inguglia :*

So in your model there are a lot of new particles. I am really scared.

- *Fritzsch:*

There are actually infinitely many new particles. In strong interactions, many years ago, people had seen the Delta, the nucleons, the pi mesons. Now, at high energies, we know that the SM predicts infinitely many hadrons. In my model, as well, there are infinitely many of these composite new particles.

- *Manoni:*

From the experimental point of view, how can you distinguish the decay of an X boson from the decay of other heavy BSM particles?

- *Fritzsch :*

It is a very simple boson and it should decay in the simplest way. For example, it could decay in lepton anti-lepton pairs or quark-quark pairs. It can also decay into 2 Z bosons that will then decay into dimuon pairs, but this signal will be very difficult to observe. On the other hand, the direct dimuon decay can be easily seen at LHC.

- *Barkowski:*

What is the ultimate number of fundamental particles in your model? And how are they divided into generations?

- *Fritzsch:*

Without counting the gauge bosons of the QHD gauge group, the fundamental particles of my model are just the 2 fermionic haplons, alpha and beta, and 4 scalars, one leptonic scalar and 3 coloured scalars. In total I have 6 particles.

- *Barkowski:*

All the SM particles are made of these 6 fundamental particles?

- *Fritzsch:*

Yes. Leptons are made of 1 fermionic haplon and 1 leptonic scalar. Quarks are made of 1 haplon and 1 coloured scalar.

- *Barkowski:*

What is the difference between the lepton and quark generations?

- *Fritzsch :*

They are just dynamical excitations of each other. But it is very complicated, and I could not do any specific calculation so far.

- *Alba :*

Does QHD have a hierarchy problem?

- *Fritzsch:*

I don't know. I don't understand the hierarchy problem in the SM either.

- *Alba :*

What is the main reason to introduce haplons? Does it solve any other problem ot the SM?

- *Fritzsch:*

It might solve many problems in the future, if we will be able to make specific calculations. The only thing I mentioned already, is that I don't like the fact that the SM has two different ways of generating masses. My model essentially solves this problem by using just the QCD-like mechanism to generate the masses.

- *Galakhov:*

Would the loop corrections make the constituent scalars in leptons very heavy and the correspondent leptons very heavy too?

- *Fritzsch:*

I don't know. This is a specific theoretical question. At the moment all my fundamental constituents are massless.

ITP-UU-11/40
SPIN-11/31

The Negative β Function: from the Standard Model to Quantum Gravity

Gerard 't Hooft

Lectures presented at the International School of Subnuclear Physics, 49th course: Searching for the Unexpected at LHC and Status of our Knowledge, Erice, Sicily, June 2011.

Institute for Theoretical Physics
Utrecht University
and
Spinoza Institute
Postbox 80.195
3508 TD Utrecht, the Netherlands
e-mail: g.thooft@uu.nl
internet: http://www.phys.uu.nl/~thooft/

Abstract

After a short Introduction on beta functions and asymptotic freedom (Lecture 1), the concept of Black Hole Complementarity is introduced, and in connection with that, the importance of local conformal invariance and its spontaneous breakdown (Lecture 2). The emergence of anomalies is discussed, and it is explained that requiring these to cancel out has important implications for model building. In Lecture 3 it is outlined what role the beta functions may play in black hole physics, leading to constraints that make coupling constants, masses, and even the cosmological constant all computable once the algebraical symmetries of a particle model are fixed, although explicit, physically interesting solutions of the resulting equations have not yet been found.

June 22, 2011

1 Lecture 1: The Negative β function

1.1 The scale transformation

Coupling constants, g or λ, in the lagrangian of a 4-dimensional quantum field theory often appear to be dimensionless. This means that, under a scaling transformation,

$$x^\alpha \to x^\alpha/\mu\ ; \qquad \varphi(x) \to \mu\,\varphi(x)\ ; \qquad g \to g\ , \quad \lambda \to \lambda\ , \tag{1.1}$$

the action integral, $\int \mathrm{d}x\mathcal{L}(x)$ remains unchanged. However, the renormalization procedure in the theory requires some cut-off Λ, and this is not scale-invariant. Consequently, when renormalized quantum corrections to a vertex amplitude are calculated, one finds that the total amplitude does change, and this can be interpreted by saying that the coupling "constant" has an anomalous dimension [1][2][3][4]. Upon scaling, see Fig. 1, the coupling strength g changes according to, for instance,

$$\frac{\mu\mathrm{d}}{\mathrm{d}\mu}g(\mu) = \beta(g) = b_3 g^3 + b_5 g^5 + \cdots\ , \tag{1.2}$$

a series with coefficients b_i that can be computed order by order using Feynman diagrams.[1]

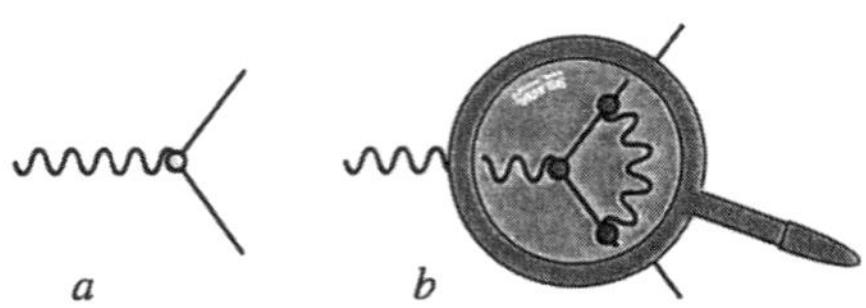

Figure 1: A vertex at large distance scales (a) is resolved in terms of bare verices at smaller distance scales (b). This gives a renormalization of the vertex.

The simplest diagram, contributing to b_3, is sketched in Fig. 1. The fact that the canonical dimension of g vanishes is reflected in the absence of a term $b_1 g$ in Eq. (1.2). Concentrating on the dominant term, we write

$$\begin{aligned}
\frac{\mu\mathrm{d}}{\mathrm{d}\mu}\left(\frac{1}{g^2}\right) &= -2b_3 + \cdots\ ,\\
\frac{1}{g^2(\mu)} &= \frac{1}{g_0^2} - 2b_3\log\mu\ + \cdots\ ;\\
g^2(\mu) &= \frac{g_0^2}{1-2b_3 g_0^2\log\mu} + \cdots\ .
\end{aligned} \tag{1.3}$$

[1]In more general cases, there are several coupling paramters, so that these equations act in higher dimensional spaces.

We see that, if the dominant coefficient b_3 happens to be positive, the coupling strength $g(\mu)$ explodes[2] at large values for μ (that is, at very tiny distance scales), whereas, if b_3 happens to be negative, it slowly runs to zero there, while now at very large scales (small μ) the strength might explode.

All renormalizable quantum field theories known before 1970 had the feature that the coefficients corresponding to b_3 (there could be several of such coefficients, referring to different coupling parameters) were positive. In fact, one could argue that the contributions of virtual particles in the propagators of observable fields, *had* to be positive. The argument would roughly go as follows. If the bare propagator of a field reads

$$P_0(k) = \frac{1}{k^2 + m^2 - i\varepsilon} , \tag{1.4}$$

then the virtual particles in the higher loop corrections should add up to give as the renormalized propagator

$$P(k) = \int_0^\infty \mathrm{d}\mu \frac{\varrho(\mu)}{k^2 + \mu^2 - i\varepsilon} ; \qquad \varrho(\mu) \geq 0 . \tag{1.5}$$

This is the so-called Källen-Lehmann representation of the propagator. Since $\rho(\mu)$ represents the contribution of intermediate virtual particle states, one can prove that it has to be positive. Thus, at high energies the propagator should contribute more strongly than at low energies, relative to the bare propagator. From here to show that the effective coupling should increase may seem easy to conclude. In any case, it is true for all theories with only scalar fields, spinnors and *Abelian* vector fields such as the Maxwell field. These happened to be the only theories known and understood before 1970.

So why does the argument fail in the case that non-Abelian fields are present [5][6][7]? One obvious reason could be that, while the scalar and spinor fields are gauge-invariant, the non-Abelian vector field does not correspond to a physically directly observable particle. The argument that $\varrho(\mu) \geq 0$ only holds for gauge-invariant amplitudes. The discovery that $\beta(g)$ can easily become negative nevertheless came as a surprise.

In Fig. 2 the running of coupling strengths in various theories is displayed. For a conventional field theory to make sense, the coupling parameters must be small enough to permit perturbative expansions. In view of this, we observe that n on-renormalizable theories (a) have a much more limited domain of validity than renormalizable ones (b). An asymptotically free theory (c) may be well-defined everywhere, because its large distance behavior ($\mu \to -\infty$) should follow from integrating out its effective field equations, while the other theories still depend on perturbative expansions.

Various explanations have been put forward concerning the physical origin of the sign switch for β in non-Abelian gauge theories. Since the color electric charges are non linear, and therefore not additively conserved, the dual of these charges, the color magnetic charges, cannot be constrained to vanish under all circumstances; this is related

[2]More precisely, it becomes so large that the perturbative expression is no longer valid; instead of exploding, it could reach a fixed point. Often, it is impossible to decide which of these two cases one is dealing with.

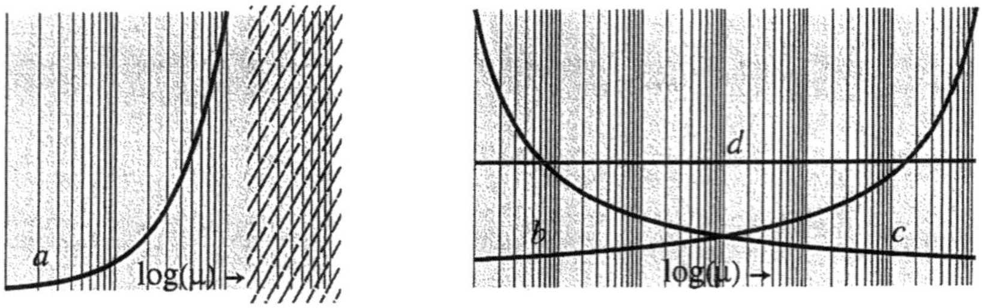

Figure 2: *a*) Running coupling strength for a non-renormalizable theory; in the shaded region the theory is ill-defined. *b*) Running coupling for a renormalizable theory that is not asymptotically free. *c*) Asymptotically free theory. *d*) Conformally invariant theory. The horizontal scale is logarithmic in μ.

to the inevitable presence of magnetic monopoles in purely non-Abelian gauge theories with Higgs mechanism. Once we have magnetic charges, we can imagine that virtual magnetic charges can be involved with an effective screening mechanism for color magnetic charge. As the product of the electric charge unit g_e and the magnetic charge unit g_m is fixed by Dirac's condition for magnetic monopoles, $g_e \cdot g_m = 2\pi$, screening of magnetic charge at large distance implies a negative screening effect for colored electric charges. In practical calculations, one finds that the dominant contribution to the negative sign of the β function comes from the fairly large natural magnetic moments of the virtual vector particles.

1.2 The computation of the β function: the background field method

Suppressing details such as indices, the lagrangian for a pure non Abelian gauge theory looks as

$$\mathcal{L} = -\tfrac{1}{2}\partial A^2 - g[AA\partial A] - g^2[AAAA] \,. \tag{1:6}$$

The cut-off dependent renormalization procedure renormalizes the vector fields A and the coupling strength g multiplicatively:

$$A \to Z_A\, A\,; \qquad g \to Z_g\, g\,. \tag{1.7}$$

Thus, the entire lagrangian renormalizes as follows:

$$\mathcal{L} \to -\tfrac{1}{2}Z_A^2\partial A^2 - Z_g Z_A^3\, g[AA\partial A] - Z_g^2 Z_A^4\, g^2[AAAA] \,. \tag{1.8}$$

Should gauge invariance then not dictate that $Z_g Z_A \stackrel{?}{=} 1$? This, in general, turns out not to be the case. The reason is that the field A is not a gauge invariant observable.

To compute how the coupling strength g scales, one first determines Z_A^2 by computing how the 2-point function scales, then one determines $z_g Z_A^3$ from the 3-point function. This leads to expressions for Z_A and Z_g separately, which one can check by determining $Z_g^2 Z_A^4$ from the 4-point function.

This is how the β functions were first computed, but it can be done in a smarter way, by introducing a background field[8]. Let us write

$$A = A^{\text{back}} + A^{\text{quant}} , \tag{1.9}$$

allowing us to couple artificial sources J only to A^{back}, so that only the background components of the fields describe the external lines. A^{back} is taken to be a classical field, while the functional integrals only have A^{quant} as functional integration variables. Then, we consider two distinct types of gauge transformations, both being symmetries of the lagrangian:

$$\text{I}: \quad A^{\text{back}} \to A^{\text{back}} ; \qquad A^{\text{quant}} \to \Omega(A^{\text{back}} + A^{\text{quant}}) - A^{\text{back}} + \Omega\partial\Omega^{-1} ; \tag{1.10}$$

$$\text{II}: \quad A^{\text{back}} \to \Omega A^{\text{back}} + \Omega\partial\Omega^{-1} ; \qquad A^{\text{quant}} \to \Omega A^{\text{quant}} . \tag{1.11}$$

For the functional integral, it is imperative that the type I gauge invariance is fixed by a gauge condition, *but we can keep the type* II *invariance!* This is achieved by choosing the gauge fixing term

$$\mathcal{L}_C = -\tfrac{1}{2}(D_\mu^{\text{back}} A_\mu^{\text{quant}})^2 , \tag{1.12}$$

where D_μ^{back} is the covariant derivative containing only the background fields. This leads to a ghost lagrangian

$$\mathcal{L}_{\text{ghost}} = -\varphi^* D_\mu^{\text{back}}(D_\mu^{\text{back}} + A_\mu^{\text{quant}})\varphi , \tag{1.13}$$

and this procedure guarantees that all one-loop diagrams with only background field external lines are gauge-invariant, so they will correspond to a formalism with $Z_g Z_A = 1$. Therefore, computing the 2-point diagrams suffices to determine Z_g. One finds that all one-loop 2-point diagrams containing only scalar and spinor fields inside the loop give positive contributions to β, but the vector fields in the loop, and the ghost field, give a minus sign. The vector field contributions dominate by a big margin (typically a factor 12 compared to the others).

1.3 Scaling behavior of the general renormalizable theory

We write the lagrangian of the most general renormalizable quantum field theory as

$$\mathcal{L} = -\tfrac{1}{4}G_{\mu\nu}^a G_{\mu\nu}^a - \tfrac{1}{2}(D_\mu\varphi_i)^2 - V(\varphi) - \bar\psi(\gamma D + (S_i + i\gamma^5 P_i)\varphi_i)\psi , \tag{1.14}$$

where φ_i and $\bar\psi$, ψ are in general in reducible representations of the gauge group, D_μ is the gauge covariant derivative, $V(\varphi)$ is a gauge-invariant quartic scalar potential, and

S_i and P_i are matrices in terms of the fermion flavor indices. Everything must be gauge invariant and the theory must be anomaly free[9][10].

The covariant derivatives contain the hermitean representation matrices T^a_{ij}, $U^{La}_{\alpha\beta}$ and $U^{Ra}_{\alpha\beta}$:

$$D_\mu\varphi_i \equiv \partial_\mu\varphi_i + iT^a_{ij}A^a_\mu\varphi_j \; ; \tag{1.15}$$

$$D_\mu\psi_\alpha \equiv \partial_\mu\psi_\alpha + i(U^{La}_{\alpha\beta}P^L + U^{Ra}_{\alpha\beta}P^R)A^a_\mu\psi_\beta \; ; \quad P^{L,R} \equiv \tfrac{1}{2}(1\pm\gamma^5) \; . \tag{1.16}$$

The gauge coupling constant(s) g are assumed to be included in these matrices T and U. The operators P^L and P^R are projection operators for the left- and right handed chiral fermions.

The group structure constants f^{abc} are also assumed to include a factor g, and they are defined by

$$[T^a, T^b] = if^{abc}T^c \; ; \quad [U^{La}, U^{Lb}] = if^{abc}U^{Lc} \; ; \quad [U^{Ra}, U^{Rb}] = if^{abc}U^{Rc} \; . \tag{1.17}$$

Casimir operators C_g, C_s and C_f will be defined as

$$f^{apq}f^{bpq} = C^{ab}_g \, , \quad \mathrm{Tr}\,(T^aT^b) = C^{ab}_s \, , \quad \mathrm{Tr}\,(U^{La}U^{Lb} + U^{Ra}U^{Rb}) = C^{ab}_f \; . \tag{1.18}$$

All these algebraical numbers were defined such that they are either real or hermitean.

All beta functions are given by writing down how the entire lagrangian (1.14) runs as a function of the scale μ. Since the diagrams always have the same mathematical form, the resulting expression can be written in terms of the algebraical coefficients of the theory. One finds:[11]

$$\mu\frac{\partial}{\partial\mu}\mathcal{L} = \beta(\mathcal{L}) = \frac{1}{8\pi^2}\Delta\mathcal{L} \, , \tag{1.19}$$

$$\Delta\mathcal{L} = -\tfrac{1}{4}G^a_{\mu\nu}G^b_{\mu\nu}(\tfrac{11}{3}C^{ab}_g - \tfrac{1}{6}C^{ab}_s - \tfrac{2}{3}C^{ab}_f) - \Delta V - \bar\psi(\Delta S_i + i\gamma^5\Delta P_i)\varphi_i\psi \; . \tag{1.20}$$

Here,

$$\begin{aligned}\Delta V \;=\; & \tfrac{1}{4}V^2_{ij} - \tfrac{3}{2}V_i(T^2\varphi)_i + \tfrac{3}{4}(\varphi T^aT^b\varphi)^2 \; + \\ & \varphi_iV_j\mathrm{Tr}\,(S_iS_j + P_iP_j) - \mathrm{Tr}\,(S^2+P^2)^2 + \mathrm{Tr}\,[S,P]^2 \, ,\end{aligned} \tag{1.21}$$

where $S = S_i\varphi_i$, $P = P_i\varphi_i$, $V_i = \partial V(\varphi)/\partial\varphi^i$, $V_{ij} = \partial^2V(\varphi)/\partial\varphi_i\partial\varphi_j$, etc.

It is convenient to define the complex matrices W as

$$W = S + iP \, , \qquad \tilde W = S - iP \, , \tag{1.22}$$

so that

$$\begin{aligned}\Delta W_i \;=\; \Delta S_i + i\Delta P_i \;=\; & \tfrac{1}{4}W_k\tilde W_kW_i + \tfrac{1}{4}W_i\tilde W_kW_k + W_k\tilde W_iW_k \\ & - \tfrac{3}{2}(U^R)^2W_i - \tfrac{3}{2}W_i(U^L)^2 + W_k\mathrm{Tr}\,(S_kS_i + P_kP_i) \; .\end{aligned} \tag{1.23}$$

2 Lecture 2: Complementarity in black holes

At first sight, the relation between the general beta function, as discussed in the previous lecture, and the notion of black hole complementarity seems to be a distant one at best, but we shall see how conformal symmetry may play a crucial role in black holes, and with that, the beta function. This second lecture is about black holes.

2.1 Introduction

As soon as we approach the Planck scale, the spectrum of elementary particle states is accompanied by black holes. In classical gravity, black holes can be seen as legitimate solutions of Einstein's equations, describing the final states of implosion processes. At first sight, they are just like soliton solutions of other field theoretical systems, so one expects that, just as is the case with solitons, their properties are completely determined by the local field equations. Black holes should peacefully coexist with elementary particles and obey the laws of quantum mechanics.

This, however, does not seem to be the case. There is a serious clash with unitarity and/or locality[12]. How can that be?

Deeply hidden in the inside, there is a space-time singularity. This singularity is *not*, or at least not directly, the cause of our problem. It is shielded from any observer by an event horizon, so, in a sense, it can be compared to singularities in the complex plane, being harmless for physical observers. The cause of the problem is the horizon, which requires space-time transformations of a type not earlier considered in particle physics. But there should be no such clashes, and we shall argue that, insisting on this requirement, leads to further constraints on how particles, including gravitons and black holes, interact. The remaining parts of these lectures are intended to show that constants such as the finestructure constant α and m_e/m_P, will not be freely adjustable but completely computable – in principle.

A horizon is a stationary region of space-time, but what is to be seen as a *time translation* for a distant observer, amounts to a *Lorentz transformation* to a local observer. Consequently an astronaut diving into a black hole will be seen by a distant observer as being more and more Lorentz contracted as time proceeds. The distant observer will never see the astronaut actually pass the horizon. The astronaut him (or her-) self will have a quite different experience. (S)he will pass right through the horizon, staying unaffected for a short period of time, before having a deadly encounter with the singularity.

This non-trivial and non-stationary transformation of the time variable has severe consequences when one considers quantized relativistic particles and fields there. Creation and annihilation operators are defined as the fourier coefficients in the time direction, and therefore distant observers and local observers require different definitions of these operators. Consequently, their definition of the *vacuum state*, as being the one state on which all annihilation operators (but none of the creation operators) give zero, is different. One generally assumes that the local observer sees a vacuum, or at least a state only containing limited numbers of particles. That's enough to derive that distant observers see an incessant stream of quantized particles, leaving the horizon as black body radiation, associated to a given temperature: *Hawking radiation* at the Hawking temperature [13][14].

The fact that this radiation is thermal is the real problem. It cannot be described as a single, pure quantum state. This means that we still do not know how to describe a black hole, regardless whether it is radiating or not, as a single quantum state. The question is then: how do we construct a Fock space for black holes? How do we record *Nature's book keeping system* near a horizon? This is our problem. No quantum theory of gravity can be called satisfactory if it leaves this question unanswered.

2.2 Black hole complementarity: CPT invariance in black holes

If an observer entering the horizon would be able to detect the Hawking particles, (s)he would pass through the entire sequence of Hawking particles emitted by the hole in its entire life, which is essentially infinitely many. Instead, the black hole solution assumes that only the ingoing matter will play a role there, so, by construction, the Hawking particles are unobservable for an ingoing observer. His or her Hilbert space, however, does include other particles that enter into the black hole.

Conversely, an outside observer does detect the Hawking particles, but not the particles inside. *Black hole complementarity* [15] is the principle that these two observers should not disagree about the outcomes of experiments; this is a kind of relativity principle: there should exist a dictionary translating the observations of one observer in terms of the terminology of the other.

A strongly related question is about CPT invariance. Parity (P) and charge conjugation (C) symmetry are fairly obvious symmetries for a black hole, but what about time reversal invariance?

When Hawking had derived the fact that black holes radiate away matter, he noted that a black hole may decay completely, and subsequently asked himself how this would affect its Penrose diagram. A Penrose diagram is a picture in a conveniently chosen coordinate frame, such that the local light cones stand upright as in Minkowski space. Since a local observer experiences a singularity, while distant observers see it decay completely, he concluded that the Penrose diagram must be as pictured in Fig. 3a. However, particle physicists might insist that, since the Schwarzschild solution is symmetric under time reversal, as is all of quantum field theory, also the Penrose diagram should be CPT symmetric, whereas the CPT transform of 3a is Figure 3b. How is this to be explained?

Figure 4a shows the simplest case of a collapsing black hole, in coordinates that are convenient for an observer going in: a single spherical shell of matter collapses, a case where we can easily write the solution of Einstein's equation in closed form. The dotted line is the horizon. All points behind it are unobservable for the outside observer. Now we first observe that we have the freedom to distort the coordinates such that all points on the horizon are in the infinite future (Fig. 4b). Since the decaying black hole, in some

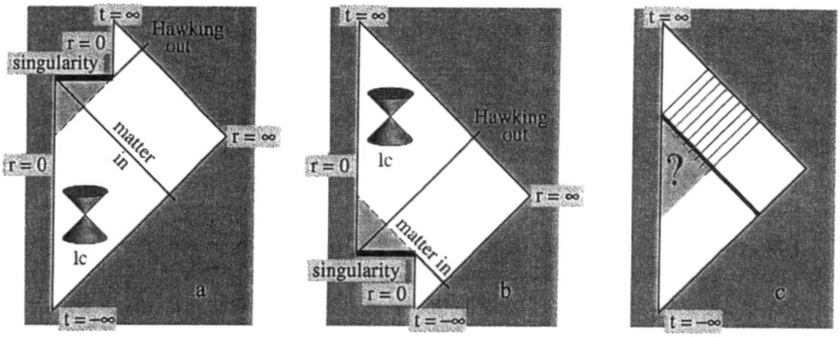

Figure 3: Penrose diagrams are renditions of spacetime in terms of coordinates chosen such that locally the light cones are oriented as in Minkowski spacetime. "lc" indicates the orientation of the local light cones. *a*) Penrose diagram of decaying black hole as proposed by Hawking. *b*) Its CPT reversal. Black holes appear not to be CPT covariant. *c*) See later in text.

sense, might be seen as the CPT reverse of the collapsing black hole, one might suspect that its future is the time reverse of Fig. 4b. So, one might suggest that a decaying black hole could be represented as in Fig. 4c. Transforming back, this would imply that Hawking radiation originates in a more expanded region than in 4a, as given in Fig. 4d. The Penrose diagram of that is the one shown in Fig. 3c.

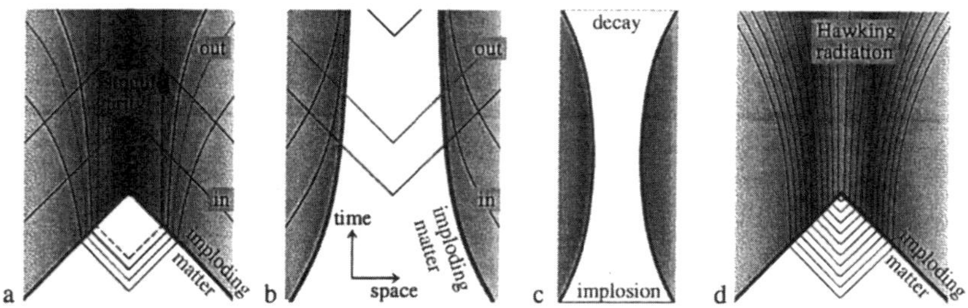

Figure 4: The decaying black hole in various coordinate frames, with horizontally the space coordinates, vertically time. Blue coloring indicates Riemann curvature; see text.

It must be noted, however, that this picture would imply a violation of Einstein's equation in the region of the question mark. This, of course, is why the Penrose diagram is usually not pictured this way. It requires a modification of Nature's laws. It will be seen as an essential modification, even if it all happens behind the horizon.

Let us first introduce an *extreme version* of black hole complementarity. Note, that Hawking radiation is thermal; therefore, *all* configurations of out going radiation are possible, though suppressed by the usual Boltzmann factor $e^{-E/kT}$. Since every final state ends up with the same total energy, this Boltzman factor is the same for all configurations, but the total probabilities are multiplied by the volume of phase space. This means that there is a certain probability that the out going particles are in the state of a single expanding shell of matter, the CPT reverse of the in-state. The probability of this is small only because its phase space is very small compared to that of the usually expected thermal radiation. But it is not zero, and so we can consider this special case. According to complementarity, observers going into the hole only observe the ingoing shell. Similarly, the external observers only see the out going shell. The ingoing shell has left no trace. Viewed in this manner, black hole complementarity merely states that there exists a scattering matrix relating in-states to out-states, just like in any other scattering process.

The special case of a single shell of Hawking radiation is CPT symmetric. Now the two pictures, as experienced by the two observers in the conventional formalism, do appear to have two regions in common: the exterior, both to be described by the Schwarzschild metric, and a *diamond shaped region* inside the horizon of both cases (the diamond region in Fig. 4a and the question mark region in Fig. 3c). So, we must postulate that the region as depicted in Fig. 4a can be mapped onto a similar region in the CPT reversed frame.

We now take a bold step[16][17][18]. We suspect that the mapping in question is a space-time mapping, and it should respect causality. In other words, if two events are causally connected in one picture, they should be connected similarly in the other picture, or, they should have the same local light cones. The light cone at a space-time point x is given by

$$\mathrm{d}s^2 = g_{\mu\nu}(x)\mathrm{d}x^\mu \mathrm{d}x^\nu = 0 \ . \tag{2.1}$$

This means that the two regions must have the same space-time metric $g_{\mu\nu}$, *apart from a possible scalar overall factor.* What the regions must have in common is the normalized metric:

$$\hat{g}_{\mu\nu} = g_{\mu\nu}/(-\det(g))^{1/4} \ ; \qquad \det(\hat{g}) = -1 \ . \tag{2.2}$$

2.3 Black hole complementarity and local conformal invariance

This subsection of the written version of the Erice lecture notes deviates from the original lecture given, because a new finding clarifies the situation much better. Originally, the conformal transformation

$$x^\mu \to y^\mu = \frac{-\varrho^2 x^\mu}{|x^2|} \ , \qquad (\mathrm{d}y^\mu)^2 = \alpha^2(x)(\mathrm{d}x^\mu)^2 \ ; \qquad \alpha(x) = \varrho^2/|x^2| \tag{2.3}$$

was applied in the small flat region that imploding and exploding shells of matter have in common. But it is not a priori clear where the origin of the transformation should be placed. Instead, the more general black hole has no vacuum region deep inside. The more elegant argument now goes as follows.

Under a general, infinitesimal local conformal transformation,

$$g_{\mu\nu}(x) \to (1 + 2\lambda(x))g_{\mu\nu}(x) \ , \tag{2.4}$$

the Ricci tensor is found to transform as

$$R_{\mu\nu}(x) \;\rightarrow\; R_{\mu\nu}(x) - 2D_\mu\partial_\nu\lambda(x) - g_{\mu\nu}D^2\lambda(x) \;, \tag{2.5}$$

so that the Einstein tensor $G_{\mu\nu}(x)$ transforms as:

$$G_{\mu\nu}(x) = 8\pi G_N T_{\mu\nu}(x) \;\rightarrow\; G_{\mu\nu}(x) - 2\left(D_\mu\partial_\nu\lambda(x) - g_{\mu\nu}D^2\lambda(x)\right) + \mathcal{O}(\partial\lambda)^2 \;, \tag{2.6}$$

where G_N is Newton's constant. We see that energy momentum conservation is maintained:

$$D^\mu T_{\mu\nu} = 0 + \mathcal{O}(\partial\lambda^2) \;. \tag{2.7}$$

Furthermore, flat space-time only transforms into flat space-time iff $\lambda(x)$ is linear in x, that is, a global conformal transformation. Indeed, as one can easily check, an infinitesimal global conformal transformation, which is an inversion with its origin far away, has

$$\lambda(x) \rightarrow \; a^\mu x_\mu + b \;. \tag{2.8}$$

However, more general local conformal transformations lead to non vanishing contributions to the energy-stress-momentum tensor $T_{\mu\nu}$. Or, they do not transform the vacuum into the vacuum, the vacuum is not invariant under these transformations. This is a situation very similar to the Higgs mechanism. We now say that, in canonical gravity, local conformal symmetry is broken by a Higgs mechanism.

It is now proposed that, in Eq. (2.2), only $\hat{g}_{\mu\nu}(x)$ is locally observable. We have a local symmetry (2.4), and we have to *fix the gauge.* A more or less standard way to do this is to consider a dilaton field $\omega(x)$, as we shall do in Eq. (3.3),and postulate it to take a trivial value. Thus, the *standard gauge* will be

$$\omega(x) \;=\; 1 \;. \tag{2.9}$$

But we can fix the gauge in different ways. In view of its non-trivial transformation law, we can use *one* of the ten components of the energy-stress-momentum tensor $T_{\mu\nu}(x)$, demanding it to vanish. Taking radial lightcone coordinates,

$$r + 2M\log(r - 2M) \pm t \;=\; r^\pm \;, \tag{2.10}$$

the *black hole gauge* is defined to be

$$T_{--}(x) \;=\; 0 \;, \tag{2.11}$$

and finally the *white hole gauge* is

$$T_{++}(x) \;=\; 0 \;. \tag{2.12}$$

Thus, the imploding black hole is characterized by the fact that local observers see no outgoing Hawking particles, while its time-reversed image, a white hole, is defined by having no locally observable particles going in.

We claim that now everything comes together. The particles produced by an exploding black hole are related to the ones going in by a local gauge transformation, the one transforming our states from the gauge (2.11) to the same states in the gauge (2.12). We do note that there are also transverse components in the energy-stress-momentum tensor, such as T_{-i} and T_{+i}, where i is a transverse index, and these we cannot also tune to zero. Although these components are related to T_{--} and T_{++} by continuity equations, questions concerning their behavior for generic black holes do have to be addressed.

Let us emphasize once again why we limit ourselves to *only* conformal transformations of the metric $g_{\mu\nu}(x)$, rather than more general transformations: this is because we think it is mandatory to keep the light cones where they are, so that the causal ordering of events is not affected.

3 Lecture 3: the local conformal group in canonical quantum gravity

The Einstein-Hilbert action of the gravitational part of a generally covariant theory reads

$$S^{\mathrm{EH}} = \int \mathrm{d}^4x \, \frac{\sqrt{-g}}{16\pi G_N}(R - 2\Lambda) \, , \tag{3.1}$$

where for later convenience we already added a possible cosmological term -2Λ, and G_N is Newton's constant. The total reparametrization invariant action is written as

$$S^{\mathrm{total}} = S^{\mathrm{EH}} + \int \mathrm{d}^4x \sqrt{-g}\mathcal{L}^{\mathrm{matt}} \, , \tag{3.2}$$

where the matter Lagrangian $\mathcal{L}^{\mathrm{matt}}$ is written in a generally covariant manner using the space-time metric $g_{\mu\nu}(x)$. To study conformal invariance, it is useful to split the metric tensor $g_{\mu\nu}(x)$ as follows:

$$g_{\mu\nu}(x) \stackrel{\mathrm{def}}{=} \omega^2(x)\, \hat{g}_{\mu\nu}(x) \, , \tag{3.3}$$

where $\hat{g}_{\mu\nu}(x)$ may be subject to some additional constraint such as

$$\det(\hat{g}) = -1 \, , \tag{3.4}$$

besides imposing a gauge condition for each of the $n = 4$ coordinates. Condition (3.4) is not at all necessary; it is just one of the possible choices for fixing the gauge concerning the following local gauge transformation,

$$\hat{g}_{\mu\nu}(x) \rightarrow \alpha^2(x)\, \hat{g}_{\mu\nu}(x) \, , \qquad \omega(x) \rightarrow \alpha^{-1}(x)\, \omega(x) \, . \tag{3.5}$$

This transformation, which is *not* a coordinate transformation, will be referred to as a *local conformal transformation.* We note that a special way to fix the gauge here would simply be the standard gauge,

$$\omega(x) \rightarrow 1 \, , \tag{3.6}$$

in which case the conventional theory is reobtained. On the other hand, however, one could note that, if $\alpha(x)$ is taken to be constant, this transformation is a scale transformation for the modified distance unit $\mathrm{d}\hat{s}^2 = \hat{g}_{\mu\nu}\mathrm{d}x^\mu \mathrm{d}x^\nu$. Furthermore, if $\alpha(x) = \varrho^2/x^2$, where ϱ is a fixed constant, we can transform a flat $\hat{g}_{\mu\nu}$ back into a flat space-time metric if it is combined with the space-time transformation

$$x^\mu \to \varrho^2\, x^\mu / x^2 \; ; \tag{3.7}$$

this will always be referred to as a *special* (or global) *conformal transformation.*

The standard gauge (3.6) could also be interpreted as the 'unitarity gauge' while a conformal Higgs mechanism takes place: the "vacuum expectation value" of ω is one.

Leaving the problem of fixing the gauge aside, we rewrite the functional integral as

$$\begin{aligned} \int \mathcal{D}g_{\mu\nu} e^{i(S^{\mathrm{EH}}+S^{\mathrm{matt}})} &= \int \mathcal{D}\hat{g}_{\mu\nu} e^{iS^{\mathrm{eff}}(\hat{g})} \; ; \\ e^{iS^{\mathrm{eff}}(\hat{g})} &= \int \mathcal{D}\omega e^{i(S^{\mathrm{EH}}(\hat{g},\omega)+S^{\mathrm{matt}}(\hat{g},\omega))} \; . \end{aligned} \tag{3.8}$$

Since S^{eff} has been obtained by integrating over ω, one might expect it to be entirely conformally invariant. If that were so, the integration over the remaining field $\hat{g}_{\mu\nu}$ would be renormalizable, or, in a sense, even finite. For two reasons, this is not so. The first is that there are conformal *anomalies*. The second reason is that the $\hat{g}_{\mu\nu}$ propagator might develop unphysical poles at zero mass: there would be ghost particles around, possibly violating unitarity.

3.1 Working out the action

In terms of $\hat{g}_{\mu\nu}$ and $\omega(x)$, the Einstein Hilbert action (3.1) turns out to be

$$S^{\mathrm{EH}} = \int \mathrm{d}^4x \tfrac{1}{16\pi G_N} \sqrt{-\hat{g}}(\omega^2 \hat{R} + 6\hat{g}^{\mu\nu}\partial_\mu\omega\partial_\nu\omega - 2\omega^4\Lambda) \; . \tag{3.9}$$

Let's first consider the example that the matter fields consist of a single real scalar field, and, for convenience we write its kinetic part in a conformally invariant way:

$$\mathcal{L}^{\mathrm{matt}}(\omega, \hat{g}, \phi) = -\tfrac{1}{2}\hat{g}^{\mu\nu}\partial_\mu\phi - \tfrac{1}{12}\hat{R}\,\phi^2 - \tfrac{1}{2}m^2\omega^2\phi^2 \; . \tag{3.10}$$

It is now convenient to rescale ω:

$$\begin{aligned} \omega = \sqrt{\tfrac{16\pi G_N}{12}}\,\tilde{\omega} \; ; \qquad \mathcal{L} \;=\; \mathcal{L}^{\mathrm{EH}} + \mathcal{L}^{\mathrm{matt}} \;=\; \\ +\tfrac{1}{2}\hat{g}^{\mu\nu}\partial_\mu\tilde{\omega}\partial_\nu\tilde{\omega} + \tfrac{1}{12}\hat{R}\tilde{\omega}^2 - \tfrac{2}{9}\pi G_N \Lambda\tilde{\omega}^4 - \tfrac{1}{2}\hat{g}^{\mu\nu}\partial_\mu\phi\partial_\nu\phi - \tfrac{1}{12}\hat{R}\,\phi^2 - \tfrac{2}{3}\pi\, G_N m^2\,\tilde{\omega}^2\phi^2 \; . \end{aligned} \tag{3.11}$$

Note that 1) the scalar field and the field $\tilde{\omega}$ emerge in exactly the same way in this lagrangian, apart from a sign switch. This sign switch reflects the well-known fact that gravity is unstable if Wick rotated to Euclidean spacetime. As for the Feynman rules, this sign switch has little effect; one is free to replace

$$\tilde{\omega} = i\eta \ , \tag{3.12}$$

at which the η field contributes just as an ordinary scalar field (apart from having imaginary odd couplings).

And 2) there is no kinetic part for the field $\hat{g}_{\mu\nu}$. It only occurs as metric field for the other fields. This means that the Euler Lagrange equations can only give

$$T^{\rm total}_{\mu\nu} = T^{(\omega)}_{\mu\nu} + T^{\rm matt}_{\mu\nu} = 0 \ , \tag{3.13}$$

at least classically. Note that $T^{(\omega)}_{\mu\nu}$ is just Einstein's tensor $-G_{\mu\nu}$; Eq. (3.13) is just Einstein's equation. For Eq. (3.13) to hold, the "wrong" sign of the ω field is now crucial.

3.2 Conformal anomalies

For $S^{\rm eff}$ in Eq. (3.8) to be indeed conformally invariant, as it was supposed to be by construction, the conformal anomalies should cancel out. Now those terms in the effective action that depend explicitly on the curvature of the background $\hat{g}_{\mu\nu}$ space, are somewhat more difficult to handle. They come from the divergent one-loop diagrams with external $\hat{g}_{\mu\nu}$ lines, see Fig. 5. After integrating over the ω fields, all these anomalies depend on the background curvature in the same way, as we now shall explain.

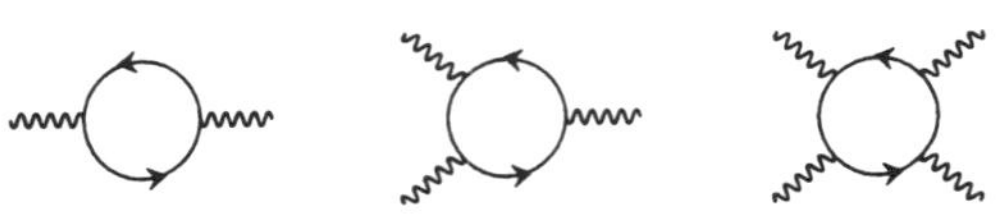

Figure 5: Divergent diagrams with external lines describing the curved background metric $\hat{g}_{\mu\nu}$. The wiggly lines are gravitons; the closed loops are matter or ω fields.

3.2.1 Lessons in conformal invariance

In 4 dimansional space-time, the Riemann curvature $R^{\mu}_{\ \nu\alpha\beta}$ has 20 mutually independent components – the other components are then fixed by symmetry relations:

$$R_{\mu\nu\alpha\beta} = -R_{\mu\nu\beta\alpha} = R_{\alpha\beta\mu\nu} \ ; \qquad R^{\mu}_{\ \nu\alpha\beta} + R^{\mu}_{\ \alpha\beta\nu} + R^{\mu}_{\ \beta\nu\alpha} = 0 \ . \tag{3.14}$$

Its contraction, the Ricci tensor

$$R_{\mu\nu} = R_{\nu\mu} = R^{\alpha}_{\ \mu\alpha\nu} \ , \tag{3.15}$$

has 10 independent components. The remaining 10 components of the Riemann tensor form the Weyl tensor, which is the traceless part of the Riemann tensor,

$$W_{\mu\nu\alpha\beta} = R_{\mu\nu\alpha\beta} + \tfrac{1}{2}(g_{\mu\beta}R_{\nu\alpha} - g_{\mu\alpha}R_{\nu\beta} - g_{\nu\beta}R_{\mu\alpha} + g_{\nu\alpha}R_{\mu\beta}) + \tfrac{1}{6}(g_{\mu\alpha}g_{\nu\beta} - g_{\nu\alpha}g_{\mu\beta})R \,. \tag{3.16}$$

Under a general conformal transformation, the Ricci tensor transforms in a non-homogeneous way, see Eq. (2.5), but it is easy to argue that the Weyl tensor must transform trivially:

$$W_{\mu\nu\alpha\beta} \to \alpha^2(x) W_{\mu\nu\alpha\beta} \,, \tag{3.17}$$

so that the following action integral is completely invariant:

$$\int \mathrm{d}^4x\sqrt{-g}\, W_{\mu\nu\alpha\beta}W^{\mu\nu\alpha\beta} \to \int \mathrm{d}^4x\sqrt{-g}\, W_{\mu\nu\alpha\beta}W^{\mu\nu\alpha\beta}\alpha^{4-8+4} \,. \tag{3.18}$$

According to the definition (3.16),

$$W_{\mu\nu\alpha\beta}W^{\mu\nu\alpha\beta} = R_{\mu\nu\alpha\beta}R^{\mu\nu\alpha\beta} - 2R_{\mu\nu}R^{\mu\nu} + \tfrac{1}{3}R^2 \,. \tag{3.19}$$

Now the following is a topological invariant, closely related to the instanton action $\frac{1}{4}\int \mathrm{d}^4x F_{\mu\nu}\tilde{F}^{\mu\nu}$ in gauge theories; it is called the Pontryagin action:

$$\tfrac{1}{4}\int \mathrm{d}^4x\,(-g)^{-1/2}\varepsilon^{\mu\nu\alpha\beta}\varepsilon^{\kappa\lambda\gamma\delta}\,R_{\mu\nu\kappa\lambda}R_{\alpha\beta\gamma\delta} = \int \mathrm{d}^4x\sqrt{-g}\,(R^2_{\mu\nu\alpha\beta} - 4R^2_{\mu\nu} + R^2) \,. \tag{3.20}$$

Adding the Pontryagin action does not modify the classical or quantum equations of motion, provided we confine ourselves to topologically flat space-time, *i.e.* spacetime without wormholes. Even our black hole is topologically flat[3], so here it is legitimate to subtract (3.20) from the Weyl action (3.18), (3.19):

$$\int \mathrm{d}^4x\sqrt{-g}\, W_{\mu\nu\alpha\beta}W^{\mu\nu\alpha\beta} = 2\int \mathrm{d}^4x\sqrt{-g}\,(R_{\mu\nu}R^{\mu\nu} - \tfrac{1}{3}R^2) \,, \tag{3.21}$$

which makes it easier to handle, in practice. Since all functional integrations performed earlier were in a conformally invariant setting, we expect all divergences due to diagrams such as Fig. 5 to be proportional to $R^2_{\mu\nu} - \frac{1}{3}R^2$.

Detailed calculations give the coefficients of this anomaly [19]–[30]. Assuming, besides the ω field, N_0 real scalar fields, $\frac{1}{2}N_{1/2}$ Dirac spinors (or $N_{1/2}$ Majorana fields), N_1 real vector fields, $N_{3/2}$ gravitinos and N_2 massless spin 2 fields, one finds a divergent contribution in the effective, divergent lagrangian, in n space-time dimensions,

$$S^{\text{eff}} = 2C\int \mathrm{d}^nx\sqrt{-\hat{g}}\Big(\hat{R}^{\mu\nu}\hat{R}_{\mu\nu} - \tfrac{1}{3}\hat{R}^2\Big) \,, \qquad \text{where} \tag{3.22}$$

$$C = \frac{1}{16\pi^2(4-n)}\left(\tfrac{1}{120}(1+N_0) + \tfrac{1}{40}N_{1/2} + \tfrac{1}{10}N_1 - \tfrac{233}{720}N_{3/2} + \tfrac{53}{45}N_2\right) \,. \tag{3.23}$$

[3] In more refined versions of our theory we may well feel the need to consider topologically non-trivial spacetimes, in which case this argument will have to be refined.

This diverges as $n \to 4$, and in practice $1/(4-n) \to \frac{1}{2}\log(\Lambda^2/k^2)$, where Λ is a momentum space cut-off. The cut-off violates conformal invariance, and the theory fails to be conformally invariant in $n \neq 4$ dimensions. Thus, we have here a conformal anomaly. The problem is that all renormalizable fields add to this anomaly with the same sign. Only gravitinos could cancel the anomaly out, but for that we need a supergravity theory; this goes beyond our present ambitions.

This means that the curvature anomaly is as yet not removed, and we do not know how to handle it properly. Therefore, we simplify the discussion to *flat background spaces*, or more precisely: $\hat{g}_{\mu\nu}$ is kept flat; we do allow for non-trivial ω fields.

3.3 Conformal anomalies in the scale transformations

There is another anomaly, which can be cured, and we consider it now: the scale anomaly. Let us consider some matter fields in a flat background $\hat{g}_{\mu\nu}$ and a somewhat simplified notation:

$$\begin{aligned}\mathcal{L} &= -\tfrac{1}{4}G_{\mu\nu}G_{\mu\nu} + \tfrac{1}{2}D_\mu\tilde{\omega}^2 - \tfrac{1}{6}\tilde{\kappa}\Lambda\tilde{\omega}^4 - \tfrac{1}{2}D_\mu\varphi^2 - \tfrac{1}{8}\lambda\varphi^4 \\ &\quad -\tfrac{1}{2}(\tilde{\kappa}^2 m_s^2)\tilde{\omega}^2\varphi^2 - \tfrac{1}{6}(\tilde{\kappa}g_3)\tilde{\omega}\varphi^3 - \overline{\psi}(\gamma D + \tilde{\kappa}m_d\tilde{\omega} + Y\varphi)\psi\ , \end{aligned} \tag{3.24}$$

where $Y = S + i\gamma^5 P$, g_3 is a possible cubic coupling term for the scalar fields φ and $\tilde{\kappa}^2 = \frac{1}{6}\kappa^2 = \frac{4}{3}\pi G_N$. Note that all parameters here, including $\tilde{\kappa}^2\Lambda$, $\tilde{\kappa}g_3$ and the masses $\tilde{\kappa}m$ are dimensionless.

We now need complete conformal invariance (when $\tilde{\omega}$ transforms together with the conformal transformation just as the φ fields)[4], but we have the familiar scale anomaly. Thus, we could require to have it cancel out:

$$\frac{\mu \mathrm{d}}{\mathrm{d}\mu}(\tilde{\kappa}^2\Lambda, g, \lambda, Y, \tilde{\kappa}g_3, \tilde{\kappa}m_s, \tilde{\kappa}m_d, \cdots) = \vec{\beta}(\tilde{\kappa}^2\Lambda, g, \lambda, Y, \tilde{\kappa}g_3, \tilde{\kappa}m_s, \tilde{\kappa}m_d, \cdots) = 0\ . \tag{3.25}$$

Particularly in the presence of (only) non-Abelian gauge fields, it is quite conceivable that there are solutions to these equations, but, since there are as many equations as there are unknowns, these solutions are isolated points in parameter space, or, all coupling parameters, all masses and the cosmological constant Λ are determined by imposing this requirement.

Where did this result come from? Did we not simply consider conventional, canonical quantum gravity? The answer to this is that we imposed a special condition on the ω field: *it should transform just as any other scalar field under conformal transformations.* In "ordinary" canonical gravity, ω is always kept at, or close to, one. So fractional powers of ω when $n \to 4$ are accepted there. Here, we do not accept that. If ω is an ordinary scalar field, the bare lagrangian should be polynomial in ω even at $n \neq 4$. In particular, as $\omega \to 0$, there should be no (logarithmic) branch cut. Small ω means small distances, se we actually imposed a new condition at the very small distance scale, where

[4]Note, that this really is merely a coordinate transformation in terms of canonical gravity.

ordinary canonical quantum gravity simply runs out of control. This condition is where our constraints come from. We think it is worth-while to consider this condition, and how it constrains Standard Model parameters.

3.4 Finding solutions of the equations $\vec{\beta} = 0$

The equations are quartic in some of the coupling parameters citeGtHcompl, and they are complicated. Indeed, no non-trivial solution has yet been found [31] (the one existing solution, $\mathcal{N} = 4$ super-Yang-Mills [32] is a bit too trivial to be physically interesting; it would have vanishing masses and a vanishing cosmological constant). Instead, we would search for a solution where all parameters are small, so that a perturbative approach can be tried. This is possible. In principle, the procedure would have to be an iterative one with three steps in the iteration:

First, consider the gauge coupling strength. We need its β function to 2 loop order [33][34]:

$$\beta(g^2) = \tfrac{1}{24\pi^2}(\tfrac{1}{2}C_s + 2C_f - 11C_g)g^4 + K(Y/g)g^6 \ . \tag{3.26}$$

Here, K depends on the Yukawa couplings Y – it so happens that, at this order, the quartic scalar self coupling does not yet enter. Now suppose that we have a fairly large gauge group of the form $SU(N_g)$ or $SO(N_g)$. Then, C_g is of order N_g, but C_s and C_f can be adjusted so that the first coefficient is small. The g^6 term, however, will in general be of order N_g^2. Thus, by carefully picking our gauge group and the scalar and spinor representations, we can find a solution where

$$g^2 = \mathcal{O}(1/N_g^2) \ . \tag{3.27}$$

At large N_g, even $g^2 N_g$ stays small, so perturbation theory makes sense.

The next step is then to consider the Yukawa couplings, including the Dirac masses, if allowed by our symmetries. They are determined by imposing that Eq. (1.23) vanishes. In general, this is an equation of the form

$$c_1(S,P)^3 - c_g\, g^2(S,P) = 0 \ , \tag{3.28}$$

where the first term is cubic in the Yukawa couplings and the second one linear. This often gives solutions where $(S,\,P) = \mathcal{O}(g)$, which justifies that we considered the term Kg^6 in Eq. (3.26) to be of order g^6. We do have to check at this point whether the relative signs of the two-loop terms and the 1 loop terms in Eq. (3.26) emerge as desired.

Finally, we have to find the zeros of the beta function for the scalar couplings. This is the most complicated part. In general, according to Eq. (1.21), they take the form

$$\beta(\lambda) \propto \lambda^2,\ g^2\lambda,\ g^4,\ (S,P)^4 \ . \tag{3.29}$$

so that a solution of the form $\lambda \propto g^2$ is quite conceivable.

4 Conclusions

We find that, investigating the scaling behavior of the matter system in a gravitational field, constraints are found that may make the Standard Model couplings computable. A drawback is that this procedure in general will produce only masses and couplings that are of order one at the Planck scale. This includes the cosmological constant. Finding cases where these numbers come in a hierarchy of different scales will be very difficult. Thus, what we have not solved is the *hierarchy problem*: why do the ratios of coupling strengths include such large dimensionless numbers? In theories with $SU(N_g)$ gauge couplings, we can make them of order $1/N_g$ or $1/N_g^2$ but this is clearly not enough. Perhaps instanton effects can be invoked that yield numbers of the order of $e^{\pm 1/N_g^2}$, which would bring us closer to interesting solutions. We have not (yet?) been able to use instantons in this way.

Even so, the approach advocated here may open up new allies for further research. Perhaps our research of a topic as esoteric as black hole quantum mechanics is finally paying off: we are re-approaching the domain of the Standard Model.

References

[1] E.C.G. Stueckelberg and A. Peterman, Helv. Phys. Acta **26** (1953) 499.

[2] C.G. Callan, *Phys. Rev.* **D2** (1970) 1541.

[3] K. Symanzik, Commun. Math. Phys. **16** (1970) 48; *ibid.* **18** (1970) 227.

[4] K.G. Wilson, *Phys. Rev.* **D3** (1971) 1818; K.G. Wilson and M.E. Fisher, *Phys. Rev. Lett.* **28** (1972) 240.

[5] G. 't Hooft, remarks made after K. Symanzik's presentation in the Marseilles Conf. on Yang-Mills fields, 19-23 June 1972, C.P. Korthals Altes et al, eds.

[6] D.J. Gross and F. Wilczek, "Ultraviolet behavior of Non-Abelian Gauge Theories", *Phys. Rev. Lett.* **30** (1973) 1343.

[7] H.D. Politzer, "Reliable Perturbative Results for Strong Interactions?", *Phys. Rev. Lett.* **30** (1973) 1346.

[8] B.S. DeWitt, *Phys. Rev. Lett.* **12** (1964) 742; *Phys. Rev.* **160** (1967) 1113; ibid. **162** (1967) 1195,1239.

[9] S.L. Adler, *Phys. Rev.* **177** (1969) 2426; J.S. Bell and R. Jackiw, *Nuovo Cim.* **60A** (1969) 47;

[10] S.L. Adler and W.A. Bardeen, *Phys. Rev.* **182**; W.A. Bardeen, *Phys. Rev.* **184** (1969) 1848.

[11] G. 't Hooft, *The birth of asymptotic freedom*, Nucl. Phys. **B254** (1985) 11; id., *The Conceptual Basis of Quantum Field Theory*, in Handbook of the Philosophy of Science. Philosophy of Physics, Vol. Eds. J. Butterfield and J. Earman, 2007 Elsevier B.V., p. 661.

[12] G. 't Hooft, "Quantum Structure of a Black Hole", *Nucl. Phys.* **B256** (1985) 727; "The black Hole Horizon as a Quantum Surface", Physica Scripta **T36** (1991) 247.

[13] S.W. Hawking, "Particle Creation by Black Holes", Commun. Math. Phys. **43** (1975) 199, Erratum ibid. **46** (1976) 206; J.B. Hartle and S.W. Hawking, "Path Integral derivation of black hole radiance", Phys.Rev.**D13** (1976) 2188.

[14] G. 't Hooft, "The scattering matrix approach for the quantum black hole: an overview", J. Mod. Phys. **A11** (1996) pp. 4623-4688. gr-qc/9607022.

[15] L. Susskind, L. Thorlacius and J. Uglum, 1993, "The Stretched Horizon and Black Hole Complementarity", arXiv:hep-th/9306069.

[16] G. 't Hooft, *Quantum Gravity without Space-time Singularities or Horizons*, Erice School of Subnuclear Physics 2009, to be publ.; arXiv:0909.3426.

[17] G. 't Hooft, *Probing the small distance structure of canonical quantum gravity using the conformal group*, arXiv:1009.0669[gr-qc].

[18] G. 't Hooft, *The Conformal Constraint in Canonical Quantum Gravity*, arXiv: 1011.0061[gr-qc].

[19] G. 't Hooft and M. Veltman, *One Loop Divergences in the Theory of Gravitation*, Ann. Inst. Henri Poincaré, **20** (1974) 69.

[20] D.M. Capper and M.J. Duff, *Conformal Anomalies and the Renormalizability Problem in Quantum Gravity*, Phys. Lett. **53A** 361 (1975).

[21] M.J. Duff, *Twenty Years of the Weyl Anomaly*, Talk given at the Salamfest, ICTP, Trieste, March 1993, arXiv:hep-th/9308075.

[22] S. Solodukhin, private communication.

[23] S. Deser and A. Schwimmer, *Geometric Classification of Conformal Anomalies in Arbitrary Dimensions*, Phys.Lett. **B309** (1993) 279, arXiv:hep-th/9302047 (12 Feb 1993).

[24] S. Deser, *Closed Form Effective Conformal Anomaly Actions in* $D \geq 4$, Phys.Lett. **B479**:315, 2000, arXiv:hep-th/9911129 (23 Feb 2000).

[25] E.S. Fradkin and A.A. Tseytlin, *Conformal Anomaly in Weyl Theory and Anomaly Free Superconformal Theories*, Phys. Lett. **134B** 187 (1984).

[26] B. Hasslacher and E. Mottola, *Asymptotically free quantum gravity and black holes*, Phys. Letters **B 99**, 221 (1981).

[27] S. Deser and P. van Nieuwenhuizen, *One-loop divergences of quantized Einstein-Maxwell fields*, Phys. Rev. **D10** 401 (1974);

[28] S. Deser and P. van Nieuwenhuizen, *Nonrenormalizability of the Quantized Einstein-Maxwell System*, Phys. Rev. Lett. **32**, no 5, 245 (1973); id., *Nonrenormalizability of the quantized Dirac-Einstein system*, Phys. Rev. **D10** 411 (1974); P. van Nieuwenhuiozen and J.A.M.Vermaseren, *One loop divergences in the quantum theory of supergravity*, Phys. Lett. **65 B** 263 (1976).

[29] F. Englert, C. Truffin and R. Gastmans, *Conformal Invariance in Quantum Gravity*, Nucl. Phys. **B117** (1976) 407.

[30] N.D. Birrell, and P.C.W Davies, *Quantum Fields in Curved Space*, (Cambridge University Press, New York, 1982).

[31] C. Wever, work in progress.

[32] L. Brink, J.H. Schwarz and J. Scherk, *Supersymmetric Yang-Mills Theories*, ERDA Research and development report, CALT-68-574 (1976); M. Sohnius *Introducing supersymmetry*, Physics Reports **128** (2-3) (1985) 39-204.

[33] D.R.T. Jones, Nucl. Phys. **B75** (1974) 531; W.E. Caswell, Phys. Rev. Lett. **33** (1974) 244; R. van Damme, *Phys. Lett.* **110B** (1982) 239; *2-loop renormalization of the gauge coupling and the scalar potential for an arbitrary renormalizable field theory*, Nucl. Phys. **B227** (1983) 317.

[34] A.G.M. Pickering, J.A. Gracey and D.R.T. Jones, *Three loop gauge beta-function for the most general single gauge-coupling theory*, arXiv:hep-ph/0104247.

CHAIRMAN: G. 't HOOFT

Scientific Secretaries: A. Morozov, A. Vlasov

DISCUSSION I

- *Galakhov:*

Do you believe in the existence of an elementary scalar field which has not been observed yet?

- *'t Hooft:*

First of all, this is not quite true, because three of the four components of the Higgs field have already been observed – the three longitudinal parts of the $W^{\pm}$ and Z. So only the fourth one is still missing, the one that should confirm that the $W^{\pm}$ and Z mass really are due to the Higgs mechanism. So in this sense we do have scalar fields already in the Standard model. Renormalizability requires the presence of these scalar fields, otherwise we cannot have any symmetry breaking mass terms. The general belief is, I would think, that models such as a Standard model are not absolute truths. I mean, as always they are an approximation of something else, and what this something else exactly is we do not know. We could think of a superstring theory or loop quantum gravity, for example. The "true theory" would be something very very fancy, a model that may only become more definite in some distant future. In the meantime we are using models, such as a Standard model as a temporary description of the best of our present knowledge, Thus, the Standard model is not an ultimate truth, it represents today's knowledge as well as possible. Today we see that we want to use renormalizable theories, because that gives you a finite number of adjustable parameters, a model that is most economic in describing everything we know. Non renormalizable theories are not economical in this sense. You have millions of coupling constants, which you cannot compute, cannot test and so on. And now the question is: Should such models contain scalar fields? To the best of our knowledge today the answer is yes, why not? A scalar field just happens to have no fancy properties, such as Dirac field's anticommutation or the association of gauge groups such as in the case of the vector fields. But there is no good reason to say that scalar fields don't exist, no good principle would lead to such a conclusion. Of course we know very well that experiments so far have not revealed explicitly the presence of scalar fundamental particles, but there is no reason why you should not expect them. Sometimes, naturalness is used as an argument. Well, any Standard model as a whole lacks some amount of naturalness anyway. So the present state of my understanding is: yes, the theory that works best is not a completely natural theory. That means that probably scalar fields do play a role at some point. Therefore, the answer to your question is yes I believe that scalar fields are necessary ingredients to describe our present knowledge of particles. If in due time any more superior theory of nature will be found, where for instance superstrings would get a natural place, then of course you will no longer talk about scalar fields; you are then talking about strings or something like that.

- *Galakhov:*

Can we describe gravity as some effective theory?

- *'t Hooft:*

What is fundamental about general relativity is most of all the invariance under the diffeomorphisms, *i.e.*, reparametrizations of the coordinates. That is fundamental. How to implement that in terms of an Einstein-Hilbert action is something else. Einstein's field theory for gravity works classically very well, and the first attempts to quantize the theory directly give approximate results. What I mean is that we do expect the existence of gravitons, and we do think we know how those gravitons will interact in a first approximation with other kinds of matter. But it is not a closed theory: if you want to do infinitely accurate calculations you discover that you need to renormalize the amplitudes that you find, and this is not possible. So the theory has to be replaced by something fancier and the basic difficulty today is that we have no good way of writing down any theory that allows you to do absolutely accurate calculations at all levels. Of course you have the so called candidate theories, such as the superstring scenario. I would call that a scenario; it is the beginning of a theory, but it is still unfinished. It is quite possible that it will be a more fundamental theory, but string theory has its own weaknesses as well. Whether it could be on its way towards a final answer, we really don't know. Should we expect a fundamental theory of gravity? I think yes. But how will it be embedded in the theory of the other interactions? I do not believe that gravity can be emergent. Past experience tell us that if a problem is as complex as the one we are facing today, a more likely the answer will be some sort of further unification. Probably, matter as such is not fundamentally different from gravity; matter and gravity should form one combined system of objects that we are trying to describe and so on, and so forth. So the best answer to our present problems of gravity will be a unified system. Whether it will contain branes or strings remains to be seen.

- *Burda:*

How is the background field method related to the so-called Wilsonian approach to renormalization?

- *'t Hooft:*

The Wilsonian approach is in a sense more ambitious than our approach to scaling. Wilson aims to compute the entire beta-function in a situation where perturbation theory does not work because the interactions are strong. Wilson's approach is to look at the coarse-grained theory that you get by integrating out all degrees of freedom at small distances. This introduces a cutoff like as in a lattice theory. In my approach I stay well within the perturbation theory, both $\boldsymbol{A}^{\text{background}}$ and $\boldsymbol{A}^{\text{quantum}}$ occur at all scales. So, at this stage, there is no large distance – short distance separation. The only reason for the splitting I discussed is the need to have the gauge constraint depend differently on what is "background" and what is the "quantum field". It is just a trick. The trick was explained to me in the early seventies by Honerkamp, but actually it was used earlier by Bryce DeWitt in his early theories of gravity. The trick is to say that you have two types of gauge transformations, obtained by splitting the gauge transformation of any field between background and quantum. The fields are split into two parts, and this splitting can be defined arbitrarily. The background field basically describes the external lines of the diagrams. Then you have the gauge transformation of type II which transforms the background field. It also acts on a quantum field homogeneously. On the other hand, gauge transformations of type I leave the background unchanged, but only act on the quantum fields. A smart way to choose the gauge is such that you fix type I but you keep invariance with respect to type II. This is not exactly what Wilson had in mind when he was splitting the fields in small-distance and large-distance scales, that is a different, efficient way of doing calculations.

- *Burda:*

What is your opinion about the recent model proposed by Verlinde, entropic gravity? If it is true, the gravity is not a fundamental force and we should not try to quantize it.

- *'t Hooft:*

It is more appropriate question for the next lecture, but nevertheless. I think that Verlinde may have captured some ideas which I also was trying to propose. The world at the Planck scale is chaotic and extremely complex. And so yes, there are lots of fluctuations and statistical phenomena taking place there. And those could eventually give rise to what is called the gravitational force. In general I agree, but it will be great problem, if you have an idea like that, how to make it work, how to turn it into a more satisfactory model of gravity. Because you can have very good idea, but if you do not know how to implement it, pursue it, then the idea turns out not to be so good. And I still have to wait and see, how Verlinde will implement this idea and make it work.

- *Shöfbeck:*

When will we make progress on the non-perturbative aspects of QCD?

-'t *Hooft:*

Well, there has been a considerable amount of progress in the past. Originally QCD was formulated as a perturbative theory just like all the others. But it was realized that particularly in large distance behavior, if we bring quarks large apart to make some vortex between them, we have fundamentally non-perturbative behavior. At first we were completely mystified with this situation but gradually the answer has been obtained and we have a general rough idea of what is going on in terms of topology, in terms of phase transitions and so on. So the rough idea about how QCD behaves non-perturbatively is very well understood now. Quantitavely the questions are a lot harder. I think the lattice theory of QCD is a very beautiful answer, because it allows you to go to the large distance limit. In principle, without making any major truncation or modification of the theory, you can use the lattice, because you can then make that lattice very fine-grained, put the whole thing on a computer and just wait. Formally we know that if you wait long enough you will get the correct theory on the computer. So you can in principle do all the calculations. That's a nice fundamental answer, but in practice not good enough, because it converges too slowly, so if you want to have accurate results you have to wait for an infinite amount of time to get reasonable results. So we want to do something smarter than that. And now the question is what else can we do? 1/N-expansions simplify things a little bit, but also don't give the answer. We would try to make effective theories of all sorts and my hope is that someday someone will come with a smart effective theory of hadrons, explaining how hadrons can be reexpressed in terms of the quarks and gluons, but now in systematic way. What has been done today is that you describe the pion, the proton and so on, and also the vortex between quarks, in a sort of effective way by using a lot of hand waving. Then we do get a rough picture. When the rough picture doesn't quite give the answers we want, we try to repair it here, repair it there. That is certainly a very pragmatic way, it sounds ugly, but in practice it works very well. A similar result is jet phenomenology. Experimentalists work with jets every day, a jet is basically a way to describe essentially non-perturbatively the final states of QCD. QCD says a single quark comes out and the quark will do something, it hadronizes, whatever that means. It was discovered, a very clever discovery, that these quarks actually exhibit their presence in the form of a jet. And that is a non-perturbative statement, which works very well in practice, and phenomenologically it is a very nice way. However, I hope that one day someone will come with a more systematic, but very fast convergent way of dealing with the features of QCD without hand waving in between arguments. But we have not come that far. I understand your question very well and I wish that in the future a better procedure will be discovered. I do not know whether that will happen at all. In principle we have the lattice approach. Maybe the computers will become so much faster that lattice theory will be a satisfactory way to handle QCD. Or at least one could keep the errors of the calculations under control, and that is a very nice aspect of lattice QCD. I don't know for sure, I can't see in the future that well, so I don't know what will happen, whether something better than the lattice will ever be discovered.

- *Sibiryakov:*

You mentioned in your lecture, that the apparent contradiction between the negative sign of beta function and unitarity is removed by the fact that the gauge boson propagator is not gauge-invariant. My question is: is it possible to see the asymptotic freedom by looking at gauge-invariant or covariant observables? How is it reconciled with the Kallen-Leman representation?

- 't Hooft:

This is an interesting point. However, gauge-invariant observables are necessarily composite, and you can leave out the mass, because the mass is an infrared feature of the theory; in the ultraviolet there are only massless particles. Massless vector particles themselves are not gauge-invariant. You have to make composite objects. Also for composite objects the propagator satisfies some Beta-like equation, but there the situation is totally different. Unitarity does not work the same way for Bethe-Salpeter equations (the equations obeyed by composite objects), which people try to solve. Composite things are gauge-invariant, but the argument concerning asymptotic freedom using unitarity would break down in gauge theories. This is because the only propagator which we have is the gauge-dependent one. There is a subtlety that you can, of course, make asymptotically free theories also with scalar fields present. I don't know whether the scalars are allowed to be singlet representations of the gauge group, presumably not, I have to check that. Generally speaking, all elementary excitations of the theory are either gauge-dependent or composite, so the counterargument just does not hold.

- Alberte:

It seems that the sign of the beta function is gauge dependent or at least depends on the choice of the gauge fixing. This type II gauge fixing seems unnatural, at least if one considers perturbations of the metric.

-'t Hooft:

This argument I gave does not apply to gravity yet. My entire lecture this morning was about standard renormalizable field theories in a flat background. I wasn't talking about gravity. In case of gravity reality is much more complicated. In fact the situation is much worse for gravity, because not only it is not asymptotically free, it is not even perturbatively scale invariant. QCD in perturbation expansion in the lowest order terms is completely scale invariant. As you know this is not so for the quantum corrections because the higher order loops require a cutoff and that generates scale dependence in QCD. In gravity although, even the lowest order theory has a scale in it, Newton's constant is not dimensionless. If you put Planck's constant and the speed of light equal to one, Newton's constant has a dimension of a mass raised to the power minus two. So that theory does not have scale invariance at all already at the level of the tree diagrams. Consequently, that theory in the far ultraviolet explodes in a way comparable to the old weak interaction theory due to Fermi which has a direct interaction of the four fermions. That theory also exploded in the not so far ultraviolet. And we know that that theory of course cannot work very well beyond a fraction of a TeV.

- Alberte:

Does it really mean that the sign of the beta function depends on the gauge?

- 't Hooft:

No, the beta function itself is certainly gauge invariant for gauge invariant quantities. The coupling constant g itself is gauge invariant, so the beta function for that must be a gauge invariant expression. It does not depend on how you fix the gauge. Even though in the procedure to calculate it you do fix the gauge first and then calculate it. At first sight you expect gauge dependent answers, but when all the dust settles you discover that no matter how you fix the gauge you get the same expression for the beta function. But note that for every effective coupling constant you have to define how to subtract it. If that subtraction becomes gauge dependent then there will be some effect at higher orders. The fact that it scales down to become asymptotically free cannot depend on a gauge choice. The higher order terms might, just because the theory for the coupling constant subtraction might depend on a gauge choice.

- *Alberte:*

But what about your argument that the probability in the Lehmann representation can become negative because it is gauge non-invariant?

-*'t Hooft:*

You can say that the gauge non-invariant particles contain ghosts at all stages, so they are not physically observable particles. Many of the ghosts have the wrong sign in front of the metric, so they contribute negatively to the probabilities. And that is why in the propagators for these theories sometimes do generate negative metric. In the diagrams I illustrated, the dotted line is a ghost field. Its direct contribution to the beta function is very small. It is only one twelfth out of the eleven twelfth that is due to the ghost in the background gauge. The main contribution comes from the diagram at the left that is the vector particle itself. So it was very surprising that even though the contribution of ghosts at first sight seems to be very small, nevertheless they dominate in a big way in the beta function. Curiously, the vector diagram in the lower left gives the biggest contribution to the beta with the wrong sign. the fact that this effect of the ghost could be so big is something of a surprise. But I think the answer to that is that the Lehmann representation doesn't work for gauge dependent quantities.

- *Krylov:*

You said that the quantum field theories with a big coupling constant are well-defined. What did you mean?

- *'t Hooft:*

Quantum field theories with large coupling constants in principle are not so well-defined, I cannot remember whether I have said something about that. In theories with small coupling constants the perturbation theory expansions are well-defined, an example is QED. In QED, the constant is alpha = 1/137, and the first four or five terms give us excellent precision, despite the fact that the entire series are divergent. They begin to diverge at the hundredth term or so, but nobody is able to calculate diagrams with a hundred loops in them. If the constant alpha were of order unity, one would not be able to use perturbation theory. You can use the Feynman diagram technique in theories with large coupling constants, only when they are asymptotically free, such as QCD. We think that we understand very well the behavior of the theory at small distances, and by extrapolation we can use the perturbation technique at larger distances where it actually breaks down. The lattice theory is a good example of this. What you can't use, is lattice theory in momentum space, that is, you cannot look what happens in the theory at very small distances starting from the large distance behavior. So, if at small distances the coupling constant grows, there is nothing much you can do.

- *Aidala:*

You said that the theories of spins 0, 1/2 and 1 are privileged because they are renormalizable, and spin 2 and higher are not renormalizable. What about the hypothetical particle of spin 3/2?

-*'t Hooft:*

Well, that's the same problem; this particle is not normally renormalizable. The spin 3/2 occurs in supergravity, but it is not renormalizable. Supergravity is not renormalizable, just as usual gravity. The presence of a spin 3/2 particle in supersymmetric theories automatically implies the presence of a spin 2 particle; there are no known supersymmetric theories which stop at spin 3/2.

- *Sibiryakov:*

I want to say a short comment to your statement on a supergravity. Recently there has been a lot of progress in theories of supergravity, which can be not only renormalizible, but even finite.

- *'t Hooft:*

Those were the subjects of beautiful talks by Zvi Bern during previous years of this school. It is very intriguing, but I frankly don't quite understand what it means. Because it doesn't take away the fact that Newton's constant has a scale. A suspicion could be that we can reformulate quantum gravity in such way that there is a new constant that has no scale. Just like saying that if we have a theory that refuses to produce ghosts, maybe it is because it is actually a Higgs theory of a renormalizable type in some hidden jacket, which you do not see. Those results seem to indicate that perhaps supergravity or other theories of gravity are actually renormalizable theories in disguise. That would be absolutely beautiful, but I have never seen, how such an algorithm works in detail. Only if scientists can implement such a theory and describe how it behaves in the ultraviolet I would accept it.

- *Yeletskikh:*

Many physicists and most textbooks consider gauge symmetry as a fundamental property of Nature. But isn't it just a mathematical trick to enable the vector fields in the theory?

-*'t Hooft:*

This question is closely related with one of the previous questions about scalar fields. The question was whether scalar fields really exist or not. Vector particles require gauge invariance, scalars do not require that, and now you are asking whether gauge-invariance itself is the fundamental property of nature. I believe that the Standard Model is only today's understanding of nature. It is the theory, that describes the phenomena observed today and it agrees with experiments. Yes, it requires gauge-invariance. Conceivably, someday the Standard Model will be replaced by some more fundamental theory such as string theory or something even fancier and completely unknown. Similarly, gauge-invariance may be also a relic of some other property. I like to think that quantum mechanics itself is not a fundamental theory, but just the machine which is being used today to describe events observed today. The underlying theory could have dissipation, or information loss, in it. And I found that, in that case, we will have equivalence classes of states, being classes of states with equal information. In such a theory, gauge invariance gets a natural place, giving us equivalence classes of states. I would identify an information equivalence class with a gauge equivalence class. Whether this makes sense or not, only the distant future can tell us.

- *Rinaldi:*

What do you think about walking Technicolor theories, where the Beta-function behaves differently in different energy regions?

-*'t Hooft:*

Of course, these can be possible, you mention more complicated situations, where the coupling constant is flat in one region and in some other domain the theory becomes asymptotically free or whatever it can be. That is, of course, possible. Then you will have a theory that behaves differently at different scales, these possibilities are also investigated by scientists.

- *Rinaldi:*

But the question was what do you think about these theories?

-*'t Hooft:*

In such theories the naturalness question always arises: where do these several scales come from? This is a very deep fundamental question. It is similar to the question "Why is the Universe so big and its structure is so complex?" Even if we would have the theory of everything, which gives us all the equations, it might still not be understood how this theory could produce a Universe as complex as ours. Our Universe is complex because we have many different scales in it. Any simple theory is bound to give you only few characteristic scales, and it is difficult to invent a theory that will give us all the varieties of those. This question remains completely unanswered today, except the one suggestion called "the anthropic principle", a very ugly principle. This principle just says that the Universe is so big and complex not because our theories are complicated, but because we are complicated. We sit in the Universe and we look at it and try to understand it, and this has to be described by a very complicated process.

CHAIRMAN: G. 't HOOFT

Scientific Secretaries: P. Burda, D. Galakhov

DISCUSSION II

- Yeletskikh:

Due to presence of interaction in the boost transformation operator, the boost transformation can change number of particles in the system. So, maybe, one doesn't need a special transformation for objects nearby the black hole.

-'t Hooft:

If you make a Lorenz transformation the vacuum is, of course, invariant. But for other transformations of general form the quantum vacuum state is not invariant, because if the time in one frame is not linearly related to the time variable in another frame then the creation and annihilation operators get mixed up. The vacuum is defined as the only state, which, when acted upon by annihilation operators, gives zero; it is therefore different for observers who use different annihilation operators. For the conformal transformation a more drastic thing happens. In these transformations the matter is created classically. We can present a conformal transformation of the interior of the black hole into the exterior of an exploding black hole; at the edges it is no longer a global conformal transformation. It is a more general transformation which as a result transforms the Hawking particles in one frame into physically observable particles in another frame.

- Alba:

Theories with nom-zero beta-function have problems either at small scales or at the large ones, might it mean that a real fundamental theory should be conformal?

-'t Hooft:

No, I don't think so. One could argue that the fundamental theory should be asymptotically free. The fact that the interaction explodes at large distances just means that perturbation theory is not applicable at large distance scales. You should use something else like lattice-QCD. So QCD is one of the best quantum field theories to my mind. It is asymptotically free and is better defined than QED or even conformal theories, because their small distant structure is less transparent.

- Alba:

You have told us that due to a special conformal transformation it is possible to make non-trivial map. How it is possible to make a non-trivial change with the aid of variable changes?

-'t Hooft:

You should distinguish between the coordinate transformations and what I called a general conformal transformation. The general conformal transformation leaves the coordinates unchanged. This is not a normal invariance of the quantum field theory. Only conformal field theories are invariant under this transformation. What I propose is to consider such a transformation where you multiply the metric by an arbitrary factor. Sometimes you can undo this transformation by a coordinate transformation, but only in special cases. Thus you have to distinguish these two sets of transformations. I would like also to say that a conformal theory, the theory where a rescaled metric is the only relevant quantity, is a five dimensional gauge theory, it contains four coordinate transformations and the fifth one, which is a multiplication of the original metric by a conformal factor. This will be considered in detail in the next lecture.

- *Alba:*

What is a meaning of a letter "*W*" in your talk?

-*'t Hooft:*

You have left-handed and right-handed fermions. The symbol *Y* represents a sum of scalar and pseudo-scalar interactions. So the left fermions will be coupled with $W = S + i\,\gamma^5 P$ and you should have a hermitian conjugated term in the Lagrangian I denoted *W*-tilde. In the next lecture I will distinguish between complex conjugation and a change of sign in front of " *i* " (because *S* will not always be hermitian).

- *Alba:*

Is a singularity on the Penrose diagram for the Hawking radiation real or just a coordinate one?

-*'t Hooft:*

This is a real singularity. The only apparent coordinate singularity is a horizon. The presented singularity is physically real, at this point the Riemann curvature becomes singular.

- *Galakhov:*

To which statistics should the black holes satisfy being described as elementary particles?

-*'t Hooft:*

The black holes you normally encounter in the gravity theory are macroscopic black holes. It is difficult to tell definitely if one of those is bosonic or fermoinic because it is a large non-local system and it is difficult to find two identical black holes to discuss their statistics. But for a very tiny black hole as an eigen state of the Hamiltonian this question is well-defined. We know that a black hole will emit both bosons and fermions. So as it emits a fermion it should experience a transition from boson to fermion or back, thus both statistics should be represented in a spectrum of black holes.

- *Palhares:*

To what extent, do you think, it is proven that QCD is a confining theory and what is your criterion for this?

-*'t Hooft:*

The question of the mathematical mechanism of confinement is a very interesting question. Conversely to asymptotic freedom in the ultraviolet, in the infrared limit the interaction force becomes extremely large, but this does not necessarily mean confinement. It could be that some fixed point occurs or that something else happens. Generally ultraviolet asymptotic freedom does not guarantee the presence of confinement. One should find a topological description of what is really happening. You should consider Wilson loop integrals and the analogous dual loops I have introduced some time ago. When these loops envelop vortices connecting two quarks then their amplitudes test the presence of these vortices. There are various options of how these loops behave in the infrared limit. If all the light states in QCD are massive then correlation functions should behave in the infrared limit in the exponential way with these masses as a factor.in the exponent. So you should consider the infrared limit of the correlators and confinement will occur in the case when the correlators give a particular exponential behavior. In terms of loops this can be described in that way that the Wilson loop and the 't Hooft loop when intertwined obey a non-trivial commutation rule. There are different options consistent with this loop behavior: Higgs mechanism, confinement, deconfinement. The latter ones are in some sense electro-magnetic dual of the Higgs mechanism. There could be a very complicated Coulomb mode as well. To summarize, confinement is a well-defined phase transition in QCD, in the confinement phase two quarks are connected by a stable vortex.

- *Palhares:*

Confinement is well-defined for pure gauge theories without quark fields, what about QCD?

- *'t Hooft:*

For real QCD, the definition of confinement is more subtle, because quarks are in the fundamental representation. Quark-antiquark pairs can confine, but due to quark-antiquark back reaction can lose each other and then connect to new quarks created by this back reaction. There is still confinement, because every quark is bound to another quark. If you have a scalar in the fundamental representation then the situation is more delicate, the quark can go out, bound to this scalar. This will be a quite exotic object, a quark-scalar bound state; they might be observed in the future. Now if these extra scalar states would be very light the definition of confinement will be more confusing, since in the dual analogous Higgs model the phase transition will be very soft without sharp transition points.

- *Campbell:*

You said that the CPT-invariant black hole has no horizon, I'm wondering if you can consider such black holes in classical large mass limit and maybe find things which are significantly different from the Einstein equations obeyed by black holes near the classical horizon, or they are indistinguishable?

- *'t Hooft:*

The point is that outside the horizon I always talk about ordinary black holes. Which are discribed by just the Schwarzschild solution. Inside the horizon I'll have an option, I will explain more about it tomorrow. The option is to multiply the metric inside the horizon by a conformal factor. And in ordinary physics that conformal factor is dictated by convention, by the fact that we are in, or very close to, the vacuum state, and we want to keep things that way, so in all the reference frames we have a particular choice of this conformal factor, and with that choice we have complete Einstein solutions, which means that we have a Schwarzschild solution also in the interior, or the Kerr-Newman solution in the more general case, so you come back to the original situation. But if you choose this conformal factor in more exotic ways, which is formally allowed, like it is formally allowed by gauge invariance to choose the Higgs field to be not in (1,0) direction, but in another direction, though you usually don't do that. Similarly here you have a normally defined choice for the ω, but you might decide to choose it differently, and then the solution becomes more general.

- *Campbell:*

But, if I remember properly, you said that there is still no causal separation anymore. Was that right, or I misunderstood something?

- *'t Hooft:*

The question, actually, is about how an observation of one observer is related to observations by another observer. In the standard picture there is no such a relation. Because in the classical black hole description there is no Hawking radiation at all, so there is no evaporation, there is no other exploding black hole on which you can map, at all. Now with Hawking radiation in the classical limit, which is singular, even though Hawking radiation is very weak (a black hole has a lifetime of many times the age of the universe before it eventually decays), then this transformation becomes extremely singular and not useful in practice. But if you consider a tiny black hole, then these transformations make more sense, and they can transform spacetime to another spacetime;

you can get observers at large exploding black holes, and then you get these alternative possibilities for the description of the black hole interior.

- Aidala:

Can the Hawking radiation be completely thermal? As many processes with the black hole horizon, I assume that you will have some finite equilibration time. Do people understand that equilibration process completely or they are working on it?

-'t Hooft:

Well, while a black hole is formed, matter forming this black hole is usually at a very high temperature, much higher than the Hawking temperature. Thus, after formation, the state of the black hole near its horizon is not the vacuum one. When you form a black hole you are in the interior of very dense matter. But in general there is no need to consider that question, because things cool down rapidly and you end up at equilibrium. But, of course, a static black hole, which has lived for very long time, by definition has a fixed temperature. Actually, I think, that we have a problem only when we have matter interacting with the background metric, where we have a complicated system of equations. This imploding stuff also emits radiation, but this radiation dies off and radiates away from the black hole, and what you are left with is just that thermal thing. In practice, such calculations can be done, it's one of the questions Kip Thorne wrote a whole book about.

- Csanàd:

Would it be possible to observe any remainder of an evaporated black hole or the radiation itself?

-'t Hooft:

Well, black holes that are so small that the Hawking radiation is of any significance, not only have not been observed, it is very unlikely they will be observed. Astrophysicists can imagine processes where black holes with a mass of few suns are produced, It is very difficult to imagine an astrophysical processes leading to a black hole with the mass of the sun itself, and lighter black holes appear to be impossible to make. We don't know any astrophysical mechanism to make tiny black holes. And only black holes of the size of millimeter or smaller start to radiate any substational amount of the Hawking radiation, and this has not been seen in any astrophysical observation. The point is that you would have to make them so small, that the Hawking radiation becomes more important than the radiation emitted by things captured by the black hole, and for that you need to make them so small, that it will be very difficult to observe them. In principle, one thing that could be observed is a black hole that ends its life by emitting the last amount of Hawking radiation, because those will be very violent explosions, that could be visible many light years away. Actually, after such events were first thought of, people began to search for flashes of light far away that could be associated to decaying black holes. The light flash of an exploding black hole should be very well visible, but unfortunately there is no good cosmological scenario for the creation of a large number of such tiny black holes, and we do not know where to look for them, because before the decay they are totally invisible in the sky. So it will be a tough problem to do experiments with tiny black holes.

- Sibiryakov:

Could you comment more on this logical step from the complementarity principle to this conformal invariance, because why wouldn't it be satisfactory to have just an S-matrix to describe black hole creation and evaporation from the point of view of the outside observer and maybe some limited amount of the local Green functions, or something, for an observer who falls inside a black hole?

- 't Hooft:

For a long time I have indeed been thinking about this question, that somehow you have to have a scattering matrix expression for the black hole. The big problem is how to compute this scattering matrix. It is not the question how to compute Hawking radiation, because that we can do, and it is always thermal. But that is not described by a unitary scattering matrix, so the question was how to modify some non-unitary object into a unitarity one. And turning around this question millions of times, I ended up with the picture discussed now, that you can somehow find a reformulation of quantum field theory, such that within that reformulated framework I will have black holes, which nevertheless can be described as an ordinary scattering process, and then the scattering matrix should be a well-defined object. It was a way to rescue the local physics, in terms of which the scattering matrix could be constructed. This was the fundamental idea; it seems to me what I have now it is a better approach than the earlier ones, and I will pursue that further.

- Dorosti:

You introduce a transformation such that in the case of the Hawking radiation we will have ordinary particles, if I understood well, and I was wondering in that case will you be able to observe black holes, since we can have some other processes, like creation of jets, that also can accelerate particles from the black holes, and, probably, they would also cover a very large energy spectrum. Could these ordinary particles be emitted from the black hole. According to what you tell us, this transformation have a kind of characteristic in energy spectrum or something special, so maybe we can distinguish them?

- 't Hooft:

Hawking particles behave just as in a thermal system, so they can come literally in all the states of Hilbert space, and the probability for any particular one of the emerging states is controlled by the Boltzman factor. So if you look at very energetic objects, they are very strongly suppressed by the exponential of minus E / kT, while if you look at light objects, such as neutrinos, photons, and gravitons, they all are allowed if the energy is sufficiently small (in fact, the energy must be so small that the wavelength is of the order of the Schwarzschild radius of the black hole), then anything goes, so the spectrum is extremely defused, in contrast with what you may see in the accretion disk or some other more classical phenomena, where the energy spectrum is not thermal at all, but oriented in certain directions.

- Agocs:

I'm not an expert in the general relativity, so I don't understand why is the conformal transformation usable if it is depends on the choice of coordinates?

- 't Hooft:

You have to distinguish the conformal transformations from the coordinate transformations, although normally you mix them up. You may say that the *global* conformal transformation is a special case of a coordinate transformation. I mean, the transformation of x to $1/x$ is certainly a very special coordinate transformation, and those are presented as usual coordinate transformations in general relativity. I consider the transformations when the metric tensor gets a multiplicative factor in front of it without transformation of the coordinates. And that is a new kind of transformation, that transforms one physical situation into an apparently different physical situation. Now a special class of conformal transformations is formed by the coordinate transformations such as x to $1/x$, so there is an overlap between the general conformal and the general coordinate transformations. I would say that in the most generic case I have a five dimensional gauge group. The coordinate transformations for the four dimensional coordinates and, in addition to that, the set of multiplications of the metric tensor with spacetime depended functions ω is the fifth component in the gauge theory. That is where this picture differs from the ordinary general relativity, when you just have the four dimensional coordinate transformations, which include automatically a very special set of global conformal transformations.

- *Kaminska:*

As a person also not working in the gravity I would naively wonder why really people worry about what is going on below the event horizon and about the singularity. It would be like worrying about Landau pole, that you mentioned in the previous lecture. For the black hole which is far away the classical picture looks okay, but then if we get closer we get this fluctuations, we get quantum gravity effects, so can we really treat this classical solution as reasonable background for our considerations?

-*'t Hooft:*

Well, the observation that we make is that these solutions to the Einstein equations exists. And according to these solutions nothing happens when you go through the horizon. But then you may say that there is some kind of censorship, that if you cross the horizon, then from that moment onwards, I should shut you out from the physical world and you are no longer there. In principle, that is a valid description, but then you are losing some important piece of information, and furthermore it could even be possible that we are falling into a black hole right now, the horizon could pass through us, without us noticing. Now these are the usual words that are used against what you have said just now. For example, in a gauge theory I always can say that I choose just one gauge and never consider any gauge transformations, well, you can do this, but then you loose much of the mathematical control over the theory, similarly here, if you want to have a better mathematical control to understand what your theory says it is useful to consider what happens beyond the horizon, even if distant observers will not be able to see that. This is just useful to make further predictions. But, of course, if you do not want to make such predictions, you always can say, that I should not look behind the horizon. Well, history told us, that this is not the right way, if you want to make progress.

- *Kepuladze:*

How the non gauge invariant interaction of vector field will affect the running of coupling constant?

-*'t Hooft:*

Well, you need gauge invariance for renormalizability, but you can always ask: what about nonrenormalizable theories? Well, these theories have a very limited domain of applicability, since when you go to high energies the effective interaction quickly runs to become so strong that you cannot compute anything. As you know, in quantum field theory you are basically convicted to perturbative expansions, even if you don't like to do them. So if a coupling constant becomes strong, then you are running outside the domain where you can do any calculation at all. This is a very pragmatic, but very important argument why I prefer not to look for nonrenormalizable theories of elementary particles. In a renormalizable theory you are not allowed to put anything in front of the Yang-Mills term in the action and not allowed to break gauge invariance, and you are forced to limit yourself to the models I spoke about. Of course you may consider a nonrenormalizable theory, for instance, to describe effective interactions, for example, the four-fermion theory of weak interactions. And if you are not interested in very highly accurate higher order corrections, then this is a very useful way for describing weak interactions, and many people do this for very good reasons. But, if you want to be more accurate and construct a theory with bigger domain of applicability than you have to consider renormalizable theories.

CHAIRMAN: G. 't HOOFT

Scientific Secretaries: P. Dunin-Barkowski, O. Ptitsyna

DISCUSSION III

- *Tapia Takaki:*

Yesterday when you discussed the implosion and decay picture, there was a region there where Einstein equations cannot be calculated. You said this was one of the weakest points of this picture. Can you calculate something there?

-*'t Hooft:*

Consider the Penrose diagram according to the conventional theory. As far as Einstein's equations go, there's nothing wrong with it. Physically, however, if we want to do quantum black holes, the singularity there is a big nuisance because that singularity there appears to absorb states without giving anything in return. So, to square that with unitarity is a big problem. And that's why I say while this is nice to describe macroscopic black holes; the outside world won't ever notice what happens within the lower triangle that reaches the singularity, so that part is basically unphysical, but there's nothing wrong in adding it, because it allows you to do your calculations better. But I say for quantum physics this is not good enough. I want a mapping that maps the interior of the black hole to an outside region which is accessible to the outside observers, and that's where I need my conformal transformation. And then you get the other picture I described. This is not really a Penrose diagram. This is the picture where I said that Einstein's equations were not really properly obeyed. And in my theory this would have been very nice and general relativity would have agreed with the picture except on those two spots. There, where delta peaks arise in the curvature that are not due to matter, those delta peaks come about from my conformal transformations that turn inside matter into outgoing matter and that's just because the vacuum state is not unique. So these transformations transform the vacuum into something else. I think that's in principle also all right, although it is unconventional; it's not the way general relativity is normally formulated.

- *Tapia Takaki:*

But then you cannot calculate equations of motion in these places?

- *'t Hooft:*

I don't say you cannot calculate equations of motion, I say if you calculate it, you'll see some quite anomalous looking peaks in the energy-momentum tensor. In principle you can calculate it essentially using the conventional Hawking procedure, but, more to the point, if I would have a theory which is completely invariant under local conformal transformations, this would be just a totally regular representation of what is going on, and, in principle, as soon as you have the dynamical details of such a theory, you should be able to calculate everything, which includes not only Hawking particles that are going out of the black hole but even writing all this in terms of unitary evolution operator. And that was the main shortcoming of the conventional theory.

- *Zein Assi:*

Today you've mentioned the result of the contribution of a scalar and a spinor and a vector to the trace anomaly. I was a bit confused because I know that the contribution of a vector comes with the minus sign compared to the contribution of a scalar and of a fermion. And this is like, for example, if I remember correctly, if we plug in the right degrees of freedom of an any conformal hypermultiplet, then we obtain zero. But I was wondering why the sign was the same also for the vector in your case?

- *'t Hooft:*

This is not the scale anomaly, this is the conformal anomaly when the background curvature is unequal to zero. So this is another conformal anomaly quite distinct from the scale anomaly. And in this particular case scalars as well as the vectors all give the same sign. This calculation was there in the literature, but I redid it, and after all dust settled I found agreement with what's in the literature. So these coefficients are actually well known. The first three of these are, indeed, positive. The negative one comes from the gravitino. That's not my calculation, so I cannot really guarantee its correctness, but that's what I've found in the literature. And that has a minus sign, so maybe it is possible to remedy the problem if you add gravitinos, but that would force you to adopt the supergravity scenario where somehow the supersymmetry is broken in a way that's not yet understood to match with low energy theories. There could be some sort of a program there, although I would not know how to implement it. So this anomaly is quite distinct from the scale anomaly, also the coefficients are quite different. The scale anomaly occurs in a flat background, while this anomaly doesn't arise in a flat background. In a flat background you have the scale anomaly from indeed spin one, the additional scalar that I add which is part of the metric tensor (the conformal factor) and the conventional scalar fields and then, of course, you have at the very end the spinor fields. And that is the conventional scaling anomaly. And I said: alright, I don;t know what to do with this curvature dependent anomaly, that's still under discussion, maybe something new is needed which has not yet been discovered. But I do know what to do with the scale anomaly and that's just to demand that the theory with that ω field included has vanishing beta function.

- *de With:*

At the end of your talk you told us that you have been looking at several test models to search for a set of real values for the coupling constant for which all beta functions are zero but that you have not succeeded yet. Which test models you have considered so far and why do you think such a set of coupling constants will be eventually found for these models?

- *'t Hooft:*

There is one example which I don't like but which is an existence proof, and that is $\mathcal{N} = 4$ super Yang-Mills theory. That theory is known to have a completely vanishing beta function. It contains scalars, spinors and vectors. They form an $\mathcal{N} = 4$ supermultiplet, the biggest supermultiplet you can have in flat spacetime with spin one, spin one half and spin zero and it has a vanishing beta. That is an existence proof. So one possible project would be to start with this model and try to extend it into more general models which are more interesting from a physical point of view where you would have mass terms. The ω field should not be in a nontrivial representation of the gauge group. $\mathcal{N} = 4$ supergravity doesn't have this, so it doesn't really have this ω field. The ω field basically decouples, the theory will not only be conformally invariant, it will also have vanishing ω couplings which means also vanishing cosmological constant. That may sound interesting, the cosmological constant in the real world is extremely small, so maybe we are very very close to a supergravity theory. But that's at least an existence proof, so we shouldn't give up hope.

- *Alba:*

As you told us today, there is some kind of divergence in this theory and in order to regularize it we should introduce some scale parameter and so our theory will not be conformal. But the constraint of the theory to be conformal is crucial for our theory. What should we do with this? And also when we consider supergravity instead of gravity to cancel divergences, what will be the special conformal transformation?

- *'t Hooft:*

Clearly, what you're thinking of is to extend the whole picture to the supergravity scenario and that's not really my expertise; I have some doubts that it's really the way to go, but I do see some very important features now in supersymmetry. In particular I've just mentioned $\mathcal{N} = 4$ super Yang-Mills theory which has vanishing beta-function. It is known that some conformal versions of supergravity also have this cancellation of the anomaly I've discussed. But conformal supergravity has its own problems because it's not evidently unitary. In fact, the problems have not been solved, because this effective conformal gravity would possibly generate Landau-like ghosts. Conformal supergravity also appears to have unphysical particles in it. Ghostlike features in it seem to be dangerous. So such questions still have to be addressed. Also, of course, the question is how come that we don't see any trace of supersymmetry. Even now, LHC is still working on it, but so far they haven't found any trace of supersymmetry. So I have some reservations against supersymmetry. Not that I say that it's forbidden, I'm just trying not to refer prematurely to supergravity and supersymmetry scenarios. I might well be forced in that direction, in which case I would like to see how that happens and when that happens.

- *Alba:*

As far as I understand, the idea is to fix all parameters by demanding the beta function to be equal to zero. And my question is: let's consider two dimensional Liouville theory which is conformal and has one parameter, the central charge. The beta function of the theory is equal to zero from the very beginning. So what will we obtain after introducing some analog to this special conformal transformation for Liouville theory?

- *'t Hooft:*

You're quite right in mentioning the 2D case, but I do think that two dimensions in case of gravity is very very different from four. Even three dimensions is again a quite different theory. So, dimensions 2, 3 and 4 give three entirely different gravity theories where all these arguments might not carry through so easily, so if you look at the effective Lagrangian I had, there is a mass term and a quartic interaction term which is typically good in four dimensions. If you go to 2 or 3 or 5 dimensions, it all looks completely different. So in two dimensions the worldsheet of the string can be parametrized by any kind of parameters that you like and so the theory naturally becomes invariant under reparametrizations, also under conformal transformations. So there's a lot of experience with conformal transformations from that area. Maybe indeed one should look at the experience one has from there to see what else is possible. We are very accustomed to anomalies in conformal transformations from precisely that theory as well because also in string theory there is a big issue about cancelling of the anomalies, which gives you the number of dimensions of string theory. So this is not the first time in physics when anomalies are playing a decisive role about how to proceed with theory.

- *Alba:*

Your theory from the very beginning was reparametrization invariant and will it be still reparametrization invariant after introducing the ω field?

- *'t Hooft:*

The ω field was there already, it has not been introduced, it is just a parametrization of things that already existed. The metric tensor exists in the theory of quantum gravity as a 10 dimensional variable, you have to functionally integrate over all its components. That includes its conformal term. To reach this expression I have added no new physics, so it's the same mathematical theory, all of the fields were already there. I just split the metric tensor into a hat tensor and the ω part. I haven't changed the canonical theory, in that respect nothing new had been done. The new thing that has been done is to demand that ω behaves just like a scalar field. It looks like a scalar field from a distance, but actually the canonical theory would generate logarithmic singularities at $\omega \to 0$. And then I say that's unphysical, because when the other scalar field go to zero nothing happens. So why should there be such a singularity? Demanding this singularity to be absent is the really important difference with the canonical theory. That's where physics comes in.

- *Galakhov:*

You have decomposed the metric tensor into conformal factor and g hat, and how technically you are going to perform computations with g hat? I mean it looks like some kind of gauge fixing for g, shouldn't you introduce some Faddeev-Popov ghosts?

- *'t Hooft:*

Precisely. The answer is that with the ω field in there, in a sense I turn the theory into a five dimensional gauge theory. So the four usual gauge degrees of freedom are the four coordinates for the coordinate reparametrization of general relativity, those are the four standard transformations which you do in general relativity. I add a fifth one which affects the conformal term of the metric tensor. Then you can gauge fix that one. The obvious thing to do is to gauge fix omega to be equal to one, then you're back at where you were before. But life becomes more interesting if you don't involve ω directly into your gauge fixing but you gauge fix in some other way. This is much like the Higgs theory, where you could use the Higgs as a spurion saying it is a field in the (1,0) direction always and ever, and nothing in the theory differs from a theory with no Higgs. But now demanding the Higgs to be a dynamical object tells you that there must be a Higgs particle. In the same sense this theory differs from the conventional theory. I promote omega to a full-fledged quantum field with dynamical properties just like the scalar field and not in a passive way as in the standard theory. And then you can fix the gauge differently, so my theory then turns to a five dimensional gauge theory.

- *Galakhov:*

And can we then apply some technique similar to what can be done, for example, with four-fermion action?

- *'t Hooft:*

Yes, I think that conventional gravity shows some resemblance with the four-fermion interaction theory of the old weak interactions which is not renormalizable. I see some similarity here, that without constructing this ω field quantum gravity looks hopelessly non-renormalizable. Now, introducing this omega field turning gravity into a five dimensional gauge theory rather than a four dimensional one appears to cure some of its problems. Unfortunately, of course, not all of its problems, there are still problems there, I'm not claiming to have solved the quantum gravity problem. But it seems to be in a somewhat better shape if you rearrange the fields this way.

- *Burda:*

My question is about this final equation with vanishing beta functions. As I understand, we now all together should try to solve this equation. My question is why can't we use already calculated examples of beta functions for various parts of standard model just to end up with algebraic equations for the parameters? Why should we recalculate all the loops from the very beginning?

- *'t Hooft:*

Of course we should not recalculate things that have already been calculated. But the Standard Model is probably only the tip of an iceberg. One generally believes that there is much much more under water and LHC will bring the water level down a bit to expose a little bit more of this iceberg, but there will be still much more hidden. And all those hidden degrees of freedom will also contribute to the beta functions. Of course, we could try the Standard Model itself, but it will be presumably premature. It will be a much better exercise to try to get some experience on the general behavior of the beta functions for all theories. And that's why I advocate using this algebra, indeed, the algebra you were thinking about; it implies that we don't have to compute Feynman diagrams anymore because we've done that and now we know that the outcome of that is just a number of coefficients. We can use those coefficients. We can substitute any gauge group, any set of representations for the scalars and the spinors and find immediately what the beta function is. The problem here is one of bookkeeping because whatever you do, you end up with very lengthy expressions. And probably the way to go is to find some smart computer program that can handle all these expressions. One thing to do would be to take a large class of gauge groups, a large class of representations and even a larger class of all these mutual interactions these things can have with each other and see if we can write down a general equation and can we then smoothen or simplify this procedure so as to get a whole list of beta functions and of possible models where this procedure would work. So far no single entry of the list had been found. So, what I will be interested in, first of all, is to find the simplest possible example which will very probably generate a very large cosmological constant and no real hierarchy yet of interaction parameters, which means that such a universe will be as small as the Planck scale. Not very interesting at all, it will show nothing lighter than about a Planck mass, but this is a start to do some exercises and see if we can find the next to simplest model which will have a little more structure, perhaps a little of hierarchy and it might indicate even the way of how to proceed from there and eventually build up a more general model. And hopefully that will be a project for ten generations of physicists after me (and after you?) who will generate exactly the right model which boils down to the Standard Model at low energies. That would be my vision of a possible program.

- *Burda:*

So, you propose to move step by step? And can you specify what can the next step look like? What theory should I take precisely and try to calculate this equation?

- *'t Hooft:*

My next step would be to familiarize with this algebra. This algebra is for the general renormalizable quantum field theory, we have to make the constraint that anomalies cancel. Then we have this effective set of beta functions, the Feynman diagrams have already been computed, there is only the algebra that survives. These all contain the group Casimirs, all one has to do is to write down lists and tables of Casimirs and of interactions between representations of the theory and get some familiarity with this algebra. For instance, my answer to one of the previous questions is that supersymmetry simplifies here a lot because in case of super-models much fewer terms survive and $\mathcal{N} = 4$ just has one single vanishing beta function. That's the easiest case of all. $\mathcal{N} = 1$ and $\mathcal{N} = 2$ theories also have expressions of this sort but a lot simpler. Although I think that we should get familiarity with thsituation entire quantum field theory.

- *Burda:*

What do you think about the last Maldacena paper where he proposed to consider conformal gravity which on the one hand contains all of the solutions to Einstein equations but there are also additional solutions and on the other hand this theory of conformal gravity is finite. So he proposed to compute everything in conformal gravity theory and after that, by imposing boundary conditions in the future he can get rid of these additional solutions of the conformal gravity.

- *'t Hooft:*

The problem of that, of course, is unitarity or other physical demands of the theory. Conformal gravity is a theory which has four derivatives in its Lagrangian, therefore it doesn't even look like anything unitary. It doesn't even look like having a proper vacuum. If my theory is right, saying that the conformal transformations transform the vacuum states to other states, if you look at conformally invariant formulations, you might have a problem of how to identify the vacuum state. With that comes the problem of how to identify positive energy states, because if you don;t know what the vacuum is, you don't know what the positive and negative energy particles are. In fact, energy is an ill-defined concept in a conformal theory because energy is the canonical conjugate to time but we have these conformal transformations in time. So we have to reconsider energy, we have to reconsider unitarity, there is hell of a lot to do before such a program may work. But maybe Maldacena is in a position to give it a try. I think that it is very hard, but I think, I can agree with him to a large extent that this is a possible scenario for reaching an improved understanding.

- *Sisman:*

You start with locality and unitarity and to satisfy unitarity you want to relate in and out states for a black hole and you relate them by having causality as the basic principle. To have this causality you use conformal transformations. Is this picture correct?

- *'t Hooft:*

Yes, you basically described what I was trying to tell you in the lectures.

- *Sisman:*

Can you clarify the point where you said that states connected by conformal transformations have different vacua?

- *'t Hooft:*

I'm describing conformal transformations as transformations which do not leave the vacuum invariant. Basically you can see that in two ways. One is the way I introduce with black holes saying that one conformal transformation which maps x into 1/x can map an imploding black hole onto an exploding black hole, so that one black hole which first implodes and later explodes can be described in one representation of the theory, by one set of coordinates, and causality is still maintained. The person who observes the imploding black hole has a different conformal frame than the observer looking at the exploding black hole, but there is a natural conformal transformation which then takes the whole thing to one single coordinate frame. You have to realize that these conformal transformations affect the matter content of spacetime. At first sight all you have in mind is the Hawking particles which look different in the different frames. But then you say, wait a minute, if I perform a nontrivial conformal transformation, that is a conformal factor which depends in a complicated, generic way on spacetime, then the Ricci tensor is not invariant, which means if you start with a vacuum or anything close to the vacuum, and you make such an exotic conformal transformation, you end up in a state which has lots and lots of particles. So, this means that states containing lots of particles can be obtained from states of the vacuum by a conformal transformation. This is a very nontrivial idea which you have to work at a lot further to see to what extent it makes sense; it seem to make a lot of sense to me to say that the vacuum is related to states containing particles using these transformations.

- *Sisman:*

What is the situation with the sign of the ω field term in Lagrangian?

- *'t Hooft:*

The omega field has the sign opposite to that of all other scalar fields, but I can easily cure this by rotating the omega field integration contour towards the imaginary direction; then it behaves just like an ordinary scalar field

- *Alberte:*

Why is it possible to split the integration measure as a product of the conformal factor and the transformed metric? Should there not be some Jacobian?

- *'t Hooft:*

Fortunately we have a lot of experience with that in ordinary quantum field theory, particularly in the gauge theories where we often go from one gauge to another gauge and that involves a nontrivial Jacobian. The measure transforms in a nontrivial way and you have to take that into account. The answer that was basically found by Faddeev and Popov and then smoothened a lot by various researchers is now called the Faddeev-Popov contribution that turns the whole thing into an invariant measure. So, now we know what to do with gauge theories, we have the Faddeev-Popov determinant and the Lagrangian ensures that the measure over the field degrees of freedom is the same before and after the transformation. Same thing here: if you split up the metric into its conformal part and the remainder, then you're bound again to encounter a problem with the measure, the Jacobian of the transformation. Again the answer will be: add the appropriate Faddeev-Popov terms in the Lagrangian, which I have not shown here explicitly. But as soon as someone will do explicit calculations, one has to write down the five Faddeev-Popov fields which are there in this now five dimensional gauge theory.

- *Inguglia:*

What is the practical meaning of the Landau ghost particle?

- *'t Hooft:*

It's certainly not a physical particle. Physical particles are such that if you compute the probability of producing such a particle, you get some positive number. Now that positive number is usually the square of an amplitude. And if you get indeed the absolute square of an amplitude, it's guaranteed to be positive. That's how it works in this conventional Schroedinger theory of scattering. Ghosts have the property that they basically give amplitudes which contain additional minus signs such that if you try to compute the probability of making a ghost, more often than not you will get some minus sign, which means a negative probability, and that is physically meaningless. These particles are not physical objects. They are auxiliary things, useful for the calculations, but you should leave them out. Then the problem is to prove that being restricted only to physical particles, the theory is a unitary one and obeys all the positivity constraints that you have. Sometimes ghosts also come with negative energies, but either if you have a particle with negative energy, or a particle with negative probability, they give rise to similar problems.

Effective actions for high energy processes in QCD and in quantum gravity

L.N. Lipatov *
Petersburg Nuclear Physics Institute,
Gatchina, 188300, St.Petersburg, Russia

Abstract

The Gribov approach to the hadron-hadron high energy scattering is based on an effective field theory for Pomerons. We construct similar effective actions for gluons and gravitons in QCD and in gravity. The BFKL approach in QCD is reviewed and presented in the form of the gauge-invariant effective theory for the interactions local in the particle rapidities. An analogous generally covariant action for gravity is formulated in terms of effective currents satisfying the Hamilton-Jacobi equation. It gives a possibility to construct various effective vertices and graviton j-plane trajectory. In the $N = 4$ extended supersymmetric gauge model such effective action in the 10-dimensional anti-de-Sitter space can be used for the description of the Pomeron interactions.

1 Gribov calculus

In the high energy kinematics

$$s = 4E^2 >> -t = \vec{q}^2 \tag{1}$$

the t-channel partial wave expansion for the elastic scattering amplitude with the definite signature $p = \pm 1$ has the form

$$A^p(s,t) = s \int_{a-i\infty}^{a+i\infty} \frac{d\omega}{2\pi i} \left((-s)^\omega - p s^\omega\right) f^p_\omega(t) \,. \tag{2}$$

In the pole approximation the t-channel partial wave is given below

$$f^p_\omega(t) = \frac{\gamma^2(t)}{\omega - \omega_p(t)} \,, \tag{3}$$

where ω_p is the Regge trajectory assumed to be linear

$$\omega_p(t) = \Delta - \alpha' \vec{q}^2 \,. \tag{4}$$

In this model the amplitude has the power-like form [1]

$$A^p_{Regge}(s,t) = \xi_p(t)\, s^{1+\omega_p(t)}\, \gamma^2(t) \,, \; \xi_p(t) = e^{-i\pi\omega_p(t)} - p \,. \tag{5}$$

S. The poles generate the cut singularities in the ω-plane [2]. For example, in the simple case of the two reggeon exchange we have for the Mandelstam contribution

$$A^p_{Mand}(s,t) = \xi_p\, s \int \frac{d^2k}{(2\pi)^2} \Phi^2(k, q-k) s^{\omega_{p_1}(-k^2)} s^{\omega_{p_2}(-(q-k)^2)} \,, \tag{6}$$

where according to the Gribov signature conservation rule one obtains

$$p = p_1\, p_2 \,. \tag{7}$$

V. Gribov, I. Pomeranchuck and K. Ter-Martirosyan argued that the Mandelstam cuts appear as a result of an analytic continuation of the multi-particle t-channel unitarity relation [3]

$$\Im_t f_\omega(t) \sim \sum_n \int d\Omega_n |f^{(n)}_\omega|^2 \,. \tag{8}$$

To describe an aproximately constant behaviour of the total cross-sections σ related to the elastic amplitudes by the optical theorem a special j-plane singularity - Pomeron with vacuum quantum numbers is introduced

$$\sigma = \frac{1}{s} \Im A(s.0) \,, \; A(s,t) \approx is\, s^{\Delta - \alpha'_P q^2} \,, \tag{9}$$

where the intercept Δ is small.

The produced and virtual particles at high energies are separated in clusters of particles with approximately the same rapidities y_r

$$0 < y_1 < y_2 < ... < y_k < \ln s \,, \; y_k - y_{k-1} \gg 1 \,. \; y_k = \frac{1}{2} \ln \frac{E_k + |p_k|}{E_k - |p_k|} \,. \tag{10}$$

A cluster in the Gribov picture interacts with other clusters by the exchange of the Pomerons. The Pomeron Green function in a linear approximation for its trajectory has a non-relativistic form

$$G_0 = \frac{1}{E + \Delta - \frac{q^2}{2m}}\,, \quad E = -\omega\,, \quad \alpha' = \frac{1}{2m}\,. \tag{11}$$

Pomeron interactions are described by an effective 2+1 dimensional Gribov action [4]

$$S = \int dy\, d^2\rho \left(\phi^*(\partial_y - \Delta)\phi + \frac{1}{2m}\,|\partial_\mu \phi|^2) + i\lambda(\phi^*\phi^2 + \phi\phi^{*2}) + \ldots \right)\,, \tag{12}$$

where ϕ is a complex scalar field.

2 Elastic and production amplitudes in QCD

The Born scattering amplitude in the high energy limit of QCD has the form

$$M_{AB}^{A'B'}(s,t)|_{Born} = g\,T^c_{A'A}\,\delta_{\lambda_{A'}\lambda_A}\,\frac{2s}{t}\,g\,T^c_{B'B}\,\delta_{\lambda_{B'}\lambda_B}\,, \tag{13}$$

corresponding to the conservation of helicities λ of the colliding particles A and B. In the leading logarithmic approximation (LLA)

$$\alpha_s \ln s \sim 1\,, \quad \alpha_s = \frac{g^2}{4\pi} \ll 1 \tag{14}$$

it is multiplied by the factor

$$M_{AB}^{A'B'}(s,t) = M_{AB}^{A'B'}(s,t)|_{Born}\, s^{\omega(t)}\,, \tag{15}$$

where the gluon Regge trajectory in LLA is given below

$$\omega(-|q|^2) = -\frac{\alpha_s N_c}{4\pi^2}\int d^2k\,\frac{|q|^2}{|k|^2|q-k|^2} \approx -\frac{\alpha_s N_c}{2\pi}\,\ln\frac{|q^2|}{\lambda^2}\,. \tag{16}$$

In the multi-Regge kinematics, where the pair energies $\sqrt{s_k}$ of the produced gluons are large in comparison with momentum transfers q_i and transverse components $k_{r\perp}$ of the final particle momenta the amplitude has the factorized form [5]

$$M_{2\to 2+n} \sim \frac{s_1^{\omega_1}}{|q_1|^2}\,gT^{d_1}_{c_2c_1}\,C(q_2,q_1)\,\frac{s_2^{\omega_2}}{|q_2|^2}\ldots gT^{d_n}_{c_{n+1}c_n}\,C(q_{n+1},q_n)\,\frac{s_{n+1}^{\omega_{n+1}}}{|q_{n+1}|^2}\,, \tag{17}$$

where ω_i are the Regge trajectories and $C(q_i, q_{i-1})$ are the Reggeon-Reggeon-gluon vertices. In LLA they are

$$\omega_r = -\frac{\alpha_s N_c}{2\pi}\left(\ln\frac{|q_r^2|}{\mu^2} - \frac{1}{\epsilon}\right)\,, \quad C(q_2,q_1) = \frac{q_2\,q_1^*}{q_2^* - q_1^*}\,, \tag{18}$$

where the produced gluons are assumed to have definite helicities. The total cross-section can be calculated through squares of these amplitudes

$$\sigma_t = \sum_n \int d\Gamma_n\,|M_{2\to 2+n}|^2\,. \tag{19}$$

The production amplitudes should satisfy the Steinmann requirement of absence of simultaneous singularities in the overlapping channels

$$\Delta_{s_r}\Delta_{s_{r+1}} M_{2\to 2+n} = 0\,. \tag{20}$$

In an accordance with this property we have the following representation for the one particle production amplitude $M_{2\to 3}$ in the Regge ansatz

$$M_{2\to 3} = c_1(-s)^{j(t_2)}(-s_1)^{j(t_1)-j(t_2)} + c_2(-s)^{j(t_1)}(-s_2)^{j(t_2)-j(t_1)} + crossing\ terms\,, \tag{21}$$

where c_1 and c_2 are real quantities. In the case of two particle production the corresponding expression for the multi-Regge amplitude contains five terms. These formulas relate real and imaginary contributions to the production amplitudes. The imaginary parts can be constructed with the use of the unitarity conditions and real parts are obtained from above dispersion-type relations. In the leading logarithmic approximation the imaginary contributions to the production amplitudes $M_{m\to n}$ are small. Therefore the unitarity conditions allow to derive a closed non-linear equations for the amplitudes, which are called the bootstrap relations [5].

The knowledge of quantities $M_{2\to 2+n}$ gives a possibility to construct the total cross-section and the scattering amplitude with color singlet quantum numbers in the crossing channel. This amplitude can be expressed in terms of the Pomeron wave function satisfying the BFKL equation [5]

$$E\,\Psi(\vec{\rho}_1,\vec{\rho}_2) = H_{12}\,\Psi(\vec{\rho}_1,\vec{\rho}_2)\,,\ \sigma_t \sim s^{\Delta}\,,\ \Delta = -\frac{\alpha_s N_c}{2\pi}\,E_0\,. \tag{22}$$

The operator H_{12} is the BFKL Hamiltonian and Δ is the Pomeron intercept. In the impact parameter representation the hamiltonian can be written as follows [6]

$$H_{12} = \frac{1}{p_1 p_2^*}(\ln|\rho_{12}|^2)p_1 p_2^* + \frac{1}{p_1^* p_2}(\ln|\rho_{12}|^2)p_1^* p_2 + \ln|p_1 p_2|^2 - 4\psi(1)\,, \tag{23}$$

where

$$\rho_{12} = \rho_1 - \rho_2\,,\ \rho_r = x_r + iy_r\,,\ \Delta = 4\alpha N_c \ln 2/\pi\,. \tag{24}$$

The Schrödinger equation is invariant under the Möbius transformations [7]

$$\rho_k \longrightarrow \frac{a\rho_k + b}{c\rho_k + d} \tag{25}$$

and, as a result, the eigenfunctions and eigenvalues of H_{12} are classified by the conformal weights

$$m = \gamma + n/2\,,\ \widetilde{m} = \gamma - n/2\,, \tag{26}$$

which depend on the anomalous dimension $\gamma = 1/2 + i\nu$ of twist-2 operators and integer conformal spin n. The energies have the property of the holomorphic separability and are expressed in terms of the logarithmic derivative $\psi(x)$ of the gamma function

$$E = 2\chi(n,\gamma) = \psi(m) + \psi(1-m) + \psi(\widetilde{m}) + \psi(1-\widetilde{m}) - 4\psi(1)\,. \tag{27}$$

These properties lead to the integrability of the BFKL dynamics [8, 9].

3 BFKL Pomeron in $N = 4$ SUSY

QCD corrections to the BFKL kernel were calculated in ref. [10]. The corresponding two-loop expression for its eigenvalue has the form

$$\omega = 4\,\hat{a}\;\chi(n,\gamma) + 4\;\hat{a}^2\,\Delta(n,\gamma)\,,\;\; \hat{a} = g^2 N_c/(16\pi^2)\,, \tag{28}$$

where $\chi(n,\gamma)$ is the Born expression. This result contains non-analitic terms proportional to $\delta_{n,0}$ and $\delta_{n,2}$. But in the case of the $N = 4$ extended supersymmetric gauge theory these Kroniker symbols are canceled and the two loop result has the property of the hermitian separability [11]

$$\Delta(n,\gamma) = \phi(M) + \phi(M^*) - \frac{\rho(M) + \rho(M^*)}{2\hat{a}/\omega}\,,\; M = \gamma + \frac{|n|}{2}\,, \tag{29}$$

where

$$\rho(M) = \beta'(M) + \frac{1}{2}\zeta(2)\,,\; \beta'(z) = \frac{1}{4}\Big[\Psi'\big(\frac{z+1}{2}\big) - \Psi'\big(\frac{z}{2}\big)\Big] \tag{30}$$

and

$$\phi(M) = 3\zeta(3) + \Psi''(M) - 2\Phi(M) + 2\beta'(M)\big(\Psi(1) - \Psi(M)\big), \tag{31}$$

$$\Phi(M) = \sum_{k=0}^{\infty} \frac{(-1)^k}{k+M}\left(\Psi'(k+1)\; - \;\frac{\Psi(k+1) - \Psi(1)}{k+M}\right). \tag{32}$$

It means, in particular, that the eigenvalue in $N = 4$ has a remarkable property of the maximal transcendentality [12]. This property has allowed to calculate the anomalous dimensions of twist-two operators up to three loops in $N = 4$ SUSY [13]. It is important, that the $N = 4$ four-dimensional conformal field theory due to the Maldacena guess is equivalent to the superstrings living on the anti-de-Sitter 10-dimensional space [14, 15, 16]. Therefore the Pomeron in this model is equivalent to the reggeized graviton in the superstring theory.

On can write the kernel of the BFKL equation for N=4 SUSY in a diffusion approximation in the form

$$j = 2 - \Delta - D\,\nu^2\,,\;\; \gamma = 1 + \frac{j-2}{2} + i\nu\,, \tag{33}$$

where the parameters Δ and D depend on the gauge coupling constant. According to the energy-momentum conservation we have $D = \Delta$ and therefore the anomalous dimension of twist-2 operators is

$$\gamma = (j-2)\left(\frac{1}{2} - \frac{1/\Delta}{1 + \sqrt{1 + (j-2)/\Delta}}\right). \tag{34}$$

It turns out, that due to the AdS/CFT correspondence the diffusion approximation for the BFKL equation corresponds to the linear graviton Regge trajectory

$$j = 2 + \frac{\alpha'}{2}\,t\,,\;\; t = E^2/R^2\,,\;\; \alpha' = \frac{R^2}{2}\,\Delta\,. \tag{35}$$

Because the behavior of the anomalous dimension at large coupling constants is known

$$\gamma = -\sqrt{2\pi(j-2)}\,\hat{a}^{1/4}\,, \tag{36}$$

one can calculate the Pomeron intercept in the strong coupling limit [13, 17]

$$j = 2 - \Delta\,,\ \Delta = \frac{1}{2\pi}\,\hat{a}^{-1/2}\,. \tag{37}$$

The duality between the BFKL Pomeron and reggeized graviton means, that the Gribov Pomeron calculus could be described in the framework of the approach based on the effective action for the reggeized gravitons. It is the reason, why we consider below the high energy scattering in gravity, but initially a similar approach to the high energy processes in QCD will be reviewed.

4 Effective action for high energy processes in QCD

In this section we consider the effective field theory for reggeized gluons [18]. The effective action is local in the rapidity space for the corresponding cluster of particles

$$y = \frac{1}{2}\ln\frac{\epsilon_k + |k|}{\epsilon_k - |k|}\,,\ |y - y_0| < \eta\,,\ \eta << \ln\, s\,. \tag{38}$$

We introduce the anti-hermitian fields v_μ describing the usual gluons and the gauge invariant fields $A_\pm$ describing the production and annihilation of the reggeized gluons

$$v_\mu(x) = -iT^a v_\mu^a(x)\,,\ A_\pm(x) = -iT^a A_\pm^a(x)\,,\ \delta A_\pm(x) = 0\,. \tag{39}$$

The fields $A_\pm$ satisfy the kinematical constraints

$$\partial_- A_+ = \partial_+ A_- = 0\,, \tag{40}$$

corresponding to the fact, that the Sudakov variables of momenta of different clusters are strongly ordered in the quasi-multi-Regge kinematics.

The effective action for particles inside a cluster has the form

$$S = \int d^4x \left(L_{QCD} + Tr(V_+\partial_\mu^2 A_- + V_-\partial_\mu^2 A_+)\right)\,, \tag{41}$$

where L_{QCD} is the usual QCD action and the effective currents $V_\pm$ are given below

$$V_+ = -\frac{1}{g}\partial_+\, P\exp\left(-\frac{g}{2}\int_{-\infty}^{x^+} v_+(x')d(x')^+\right) = v_+ - gv_+\frac{1}{\partial_+}v_+ + \ldots\,. \tag{42}$$

The Feynman rules for this action are derived in ref. [19]. The effective action approach gives a possibility to construct various reggeon vertices needed to calculate NLO and NNLO corrections to the BFKL kernel.

Another application of this effective field theory is a verification of the BDS ansatz [20] for the inelastic amplitudes in the planar approximation for $N = 4$ SUSY. It was shown [21, 22] with the use of the BFKL equation for the color octet states, that in the multi-Regge kinematics, where $s, s_2 > 0$; $s_1, s_3 < 0$, the BDS amplitude $M^{2\to 4}$ is multiplied by the factor

$$M^{2\to 4} = (1 + i\Delta_{2\to 4})\, M_{2\to 4}^{BDS}\,, \tag{43}$$

where in LLA

$$\Delta_{2\to4} = \frac{a}{2}\sum_{n=-\infty}^{\infty}(-1)^n \int_{-\infty}^{\infty}\frac{d\nu}{\nu^2+\frac{n^2}{4}}\,(V^*)^{i\nu-\frac{n}{2}}\,V^{i\nu+\frac{n}{2}}\left(s_2^{\delta(\nu,n)}-1\right) \tag{44}$$

and $\delta(\nu, n)$ is the eigenvalue of the Hamiltonian. Further, the quantities V depend on transverse components of produced particle momenta k_1, k_2 and momentum transfers q_1, q_2, q_3. Using the duality transformation [9], we can present them in the simple form

$$V = \frac{q_3 k_1}{k_2 q_1} \to \frac{z_{03}z_{0'1}}{z_{0'3}z_{01}}\,. \tag{45}$$

Moreover, the correction $\Delta_{2\to4}$ turns out to be a function of 4-dimensional unharmonic ratios

$$i\Delta_{2\to4} = \frac{a^2}{4}Li_2(\chi)\ln\frac{\chi t_2 s_{13}}{s_3 t_1}\ln\frac{\chi t_2 s_{02}}{t_3 s_1} + \dots\,,\ \chi = 1 - \frac{s s_2}{s_{012}s_{123}}\,. \tag{46}$$

In the two-loop approximation this expression can be derived also from properties of analyticity and factorization [24] or directly from recently obtained exact result for $M_{2\to4}$ [25]. Recently the remainder function R_6 was obtained in the next-to-leading approximation [26] The effective action allows one to construct the Mandelstam cut contributions corresponding to the composite states of an arbitrary number n of reggeized gluons in the adjoint representation [23]. The holomorphic hamiltonian for such state has the form

$$h = \ln(Z_1^2\partial_1) - 2\psi(1) + \ln\partial_{n-1} + \sum_{k=1}^{n-2} h_{k,k+1}\,,\ p_k = Z_{k-1,k}\,. \tag{47}$$

The pair hamiltonian corresponds to the interaction of neighbors in the spin chain considered in ref. [8]

$$h_{1,2} = \ln(Z_{12}^2\partial_1) + \ln(Z_{12}^2\partial_2) - 2\ln Z_{12} - 2\psi(1)\,. \tag{48}$$

The total Hamiltonian h coincides with the hamiltonian for the integrable open spin chain [23]. The corresponding monodromy matrix has the form

$$t(u) = L_1(u)L_2(u)\dots L_{n-1}(u) = \begin{pmatrix} A(u) & B(u) \\ C(u) & D(u) \end{pmatrix}, \tag{49}$$

where L_i are L-operators constructed from the Möbius group generators. The monodromy matrix contains the integrals of motion and allows to derive the equation for the Baxter function $Q(u)$

$$[D(u), h] = 0\,,\ D(u)Q(u) = (u-i)^{n-1}Q(u-i)\,. \tag{50}$$

5 Effective action for high energy gravity

To construct the interaction of the reggeized gravitons with the usual gravitons we use again the locality in the rapidity space

$$y = \frac{1}{2}\ln\frac{\epsilon_k + |k|}{\epsilon_k - |k|}\,,\ |y - y_0| < \eta\,,\ \eta << \ln s\,. \tag{51}$$

The various effective graviton vertices and the graviton trajectory in LLA were calculated many years ago [27]. In particular, it was found that the Regge trajectory does not contain the ultraviolet divergency only in $N = 4$ SUSY [28]. The effective field theory for multi-Regge processes was suggested in ref. [29]. Here we construct the generally covariant effective action for quasi-multi-Regge processes. In the general relativity the graviton field is described by the metric tensor $g_{\mu\nu}$ [30]

$$d^2S = \sum_{\mu\nu} g_{\mu\nu}\, dx^\mu\, dx^\nu \,, \quad \delta g_{\mu\nu} = D_\mu \chi_\nu + D_\mu \chi_\nu \,, \tag{52}$$

where D_μ are covariant derivatives and χ_ν are infinitesimal parameters of general covariant transformations. The reggeon fields A^{++} and A^{--} describing production and annihilation of the reggeized gravitons are assumed to be invariant under the general covariant transformation with the parameter $\chi_\nu(x) \to 0$ at $x \to \infty$. They satisfy also the additional constraints

$$\partial_+ A^{++} = \partial_- A^{--} = 0 \tag{53}$$

as a consequence of the quasi-multi-Regge kinematics.

The effective action for the interactions of the fields local in particle rapidities is given below

$$S = \frac{1}{2\kappa} \int d^4x \left(-\sqrt{-g}\, R + j_{++} A^{++} + j_{--} A^{--} \right) , \tag{54}$$

where R is the Riemann curvature. The currents j_{++} and j_{--} are completely defined by the invariance of the action under the general coordinate transformations. They can be constructed in the perturbation theory. For example, the perturbative expansion of j_{++} has the form

$$j_{++} = h_{++} - X_{\sigma+}^2 + X_{\rho+}\, X_{\sigma+} h_{\rho\sigma} - X_{\sigma+} \frac{\partial_\sigma}{\partial_+} X_{\rho+}^2 + \ldots , \tag{55}$$

where

$$g_{\mu\nu} = \eta_{\mu\nu} + h_{\mu\nu} \,, \quad X_{\sigma+} = h_{\sigma+} - \frac{1}{2} \frac{\partial_\sigma}{\partial_+} h_{++} \tag{56}$$

and the Minkovsky metric $\eta_{\mu\nu}$ has the signature $(-, +, +, +)$. If we shall introduce the new curent $j^\mp$ according to the definition

$$j_{\pm\pm} = \partial_\pm\, j^\mp \,, \tag{57}$$

it satisfies the "fan" equation

$$g^{\rho\mp} \partial_\rho j^\mp = -g^{\mp\mp} - \frac{1}{4} g^{\rho\sigma} \left(\partial_\rho j^\mp \right) \left(\partial_\sigma j^\mp \right) . \tag{58}$$

From the above effective action one can reproduce the results of refs. [27, 28].

If we introduce the new fields $\omega^\pm$ according to the definition

$$j^\mp = -2\, x^\mp + \omega^\pm \,, \tag{59}$$

they satisfy the light front equation

$$g^{\rho\sigma}\, \partial_\rho \omega^\pm\, \partial_\sigma \omega^\pm = 0 \,. \tag{60}$$

The last equation can be obtained in an independent way if we would search the solution of the generally covariant d'Alambert equation

$$\nabla^2 \phi(x) = 0 \tag{61}$$

for the wave function of the scalar particle moving with a large momentum $p^{\mp}$ in the form

$$\phi^{\pm}(x) = \exp\left(-i|p|\,x^{\mp} - i\,\theta^{\mp}(x)\right)\,,\quad \theta^{\mp}(x) = \frac{|p|}{2}\,j^{\mp}(x)\,. \tag{62}$$

Here $\theta^{\mp}(x)$ is a rapidly changing phase and $j^{\mp}$ is the effective current in our normalization. Indeed, by neglecting the derivatives from the metric tensors in comparison with large derivatives from $\phi^{\pm}$ we obtain from the d'Alambert equation its semiclassical version

$$g^{\rho\sigma}\left(\frac{1}{2}\,\partial_\rho j^{\mp} + g^{\mp}_{\rho}\right)\left(\frac{1}{2}\,\partial_\sigma j^{\mp} + g^{\mp}_{\sigma}\right) = 0\,, \tag{63}$$

which coincides with the "fan" equation for $j^{\mp}$ derived above.

It is well known, that the solution of the light front equation

$$g^{\rho\sigma}\,\partial_\rho\omega\,\partial_\sigma\omega = 0 \tag{64}$$

can be expressed in terms of the null-geodesic trajectories of particles in the gravitational field satisfying the equation of motion

$$\frac{d^2x^\mu}{(d\tau)^2} = \Gamma^\mu_{\alpha\beta}\,\frac{dx^\alpha}{d\tau}\,\frac{dx^\beta}{d\tau}\,, \tag{65}$$

where τ is a parameter increasing along the trajectory and $\Gamma^\mu_{\alpha\beta}$ is the Christophel symbol. The geodesic equation can be presented in the form of the Hamilton equations

$$\frac{dx^\mu}{d\tau} = g^{\mu\nu}\,\omega_\nu\,,\quad \frac{d\omega_\alpha}{d\tau} = -\frac{1}{2}\,\omega_\mu\,\omega_\nu\partial_\alpha\,g^{\mu\nu}\,, \tag{66}$$

where

$$\omega_\alpha \equiv \partial_\alpha\omega = g_{\alpha\beta}\frac{dx^\beta}{dp} \tag{67}$$

plays a role of the energy - momentum vector.

For example, for the shock wave metric of Aichelburg and Sexl

$$(ds)^2 = \eta_{\mu\nu}\,dx^\mu\,dx^\nu + h_{--}(dx^-)^2\,, \tag{68}$$

where

$$h_{--} = \frac{8}{\sqrt{2}}\,G\,\mu\ln|\overrightarrow{x}|\,\delta(x^-) \tag{69}$$

and $\overrightarrow{x}$ is the transverse component of the vector x^ρ, one can construct the solution of the Hamilton and light front equations. The trajectory is given below

$$\overrightarrow{\rho} = \overrightarrow{x}\,f(z)\,,\quad z = \frac{8}{\sqrt{2}}\,G\,\mu\,\frac{x^-}{|x|^2}\,, \tag{70}$$

where

$$f(z) = \frac{1}{2}\left(1 + \sqrt{1+2z}\right) = 1 + \frac{z}{2} - \frac{z^2}{4} + \frac{z^3}{4} - \frac{5z^4}{16} + \dots . \tag{71}$$

The current j^+ in this case can be written in the form

$$j^+ = a\left(\frac{1}{4}\frac{|x|^2}{|\rho|^2}\, z - \ln \rho\right) = -a\,(\ln x + \phi(z))\,,\ a = \frac{8}{\sqrt{2}}\, G\,\mu \tag{72}$$

where

$$\phi(z) = \ln f(z) - \frac{1}{4}\frac{z}{f^2(z)} = \frac{z}{4} - \frac{z^2}{8} + \frac{5}{48}\, z^3 - \frac{7}{64}\, z^4 + \dots . \tag{73}$$

We can reproduce from the effective action also the graviton Regge trajectory [28]

$$\omega(q_\perp^2) = \frac{\kappa^2}{(2\pi)^3}\int \frac{q_\perp^2\, d^2k_\perp}{k_\perp^2 (q-k)_\perp^2}\left(\frac{(k,q-k)_\perp^2}{k_\perp^2} + \frac{(k,q-k)_\perp^2}{(q-k)_\perp^2} - q_\perp^2 + \frac{N}{2}\,(k,q-k)_\perp\right). \tag{74}$$

Here we added the contribution of N gravitinos for the N-extended supergravity [27]. Other super-partners do not give any contribution in this order.

Note, that the infrared divergency of the Regge trajectory is universal, but the logarithmic divergency at large $k_\perp$ depends on N and is absent at $N = 4$. Really the sum of the one-loop diagrams do not contain any ultraviolet divergency, because the gravity is renormalized in one loop. It means, that the integral over $k_\perp^2$ is restricted from above by the value of the order of s, which leads to the double-logarithmic asymptotics of the scattering amplitude with the graviton quantum numbers in the t-channel. In Ref. [27] the corresponding ladder diagrams in the double-logarithmic approximation were summed and the scattering amplitude was expressed in terms of the Bessel function.

Acknowledgments

I thank the Hamburg University for hospitality and my collaborators J. Bartels and A. Sabio Vera for helpful discussions. This work was supported by the grant RFFI-10-02-01338-a.

References

[1] V. N. Gribov, *Sov. Phys. JETP* **14** 478 (1962).

[2] S. Mandelstam, *Nuovo Cim.* **30**, 1148 (1963).

[3] V. N. Gribov, I. Ya. Pomeranchuk and K. A. Ter-Martirosyan, *Phys. Rev.* **B 139**, 184 (1965).

[4] V. N. Gribov, *Sov. Phys. JETP* **26**, 414 (1968).

[5] L. N. Lipatov, *Sov. J. Nucl. Phys.* **23** 338 (1976);
V. S. Fadin, E. A. Kuraev, L. N. Lipatov, *Phys. Lett.* **B 60** 50 (1975);
E. A. Kuraev, L. N. Lipatov, V. S. Fadin, Sov. Phys. JETP **44** 443 (1976).

[6] L. N. Lipatov, *Phys. Lett.* **B 309** 394 (1993).

[7] L. N. Lipatov, *Sov. Phys. JETP* **63** 904 (1986).

[8] L. N. Lipatov *High energy asymptotics of multi-colour QCD and exactly solvable lattice models*, hep-th/9311037, unpublished.

[9] L. N. Lipatov, *Nucl. Phys.* **B 548** 328 (1999).

[10] V. S. Fadin, L. N. Lipatov, *Phys. Lett.* **B 429** 127 (1998);
M. Ciafaloni and G. Camici, *Phys. Lett.* **B 430** 349 (1998).

[11] A. V. Kotikov, L. N. Lipatov, *Nucl. Phys.* **B 582** 19 (2000).

[12] A. V. Kotikov, L. N. Lipatov, *Nucl. Phys.* **B 661** 19 (2003).

[13] A. V. Kotikov, L. N. Lipatov, A. I. Onishchenko, V. N. Velizhanin, *Phys. Lett.* **B 595** 521 (2004); [Erratum-ibid. **B 632** 754 (2006)].

[14] J. M. Maldacena, *Adv. Theor. Math. Phys.* **2** 231 (1998).

[15] S. S. Gubser, I. R. Klebanov, A. M. Polyakov, *Phys. Lett.* **B 428** 105 (1998).

[16] E. Witten, *Adv. Theor. Math. Phys.* **2** 253 (1998).

[17] R. C. Brower, J. Polchinsky, M. J. Strassler, C. I. Tan, JHEP **0712** 005 (2007).

[18] L. N. Lipatov, *Nucl. Phys.* B **B 452**, 369 (1995); *Phys. Rept.* **286**, 131 (1997).

[19] E. N. Antonov, L. N. Lipatov, E. A. Kuraev, I. O. Cherednikov, *Nucl. Phys.* **B 721**, 111 (2005).

[20] Z. Bern, L. J. Dixon, V. A. Smirnov, *Phys. Rev.* **D 72**, 085001 (2005).

[21] J. Bartels, L. N. Lipatov, A. Sabio Vera, Phys.Rev.D80:045002,2009,

[22] J. Bartels, L. N. Lipatov, A. Sabio Vera, Phys.Rev.D80:045002,2009,

[23] L. N. Lipatov, *J. Phys.* **A 42**:304020 (2009).

[24] L. N. Lipatov, preprint, hep-th 1008.1015.

[25] L. N. Lipatov, A. Prygarin, preprints, hep-th 1008.1016, hep-th 1011.2673.

[26] V. S. Fadin, L. N. Lipatov, in preparation.

[27] L. N. Lipatov, *JETP*, **82** 991 (1982).

[28] L. N. Lipatov, *Phys. Lett.* **116B** 411 (1982).

[29] L. N. Lipatov, *Nucl. Phys.* **B365** 614 (1991).

[30] V. A. Fock, *The theory of space, time and gravitation*, Pergamon Press, London and Aylesbury (1969).

CHAIRMAN: L. LIPATOV

Scientific Secretaries: L. Alberte, A. Khmelnitsky

DISCUSSION

- *Galakhov:*

As I remember the Pomeron in QCD can be represented as a sum of stair gluon diagrams which seems to be a perturbative computation. You presented a perturbative computation on the gravity side although the Maldacena duality as I remember relates perturbative computation on one side of this duality to non-perturbative result on the other side. So isn't it contradictory?

- *Lipatov:*

There is no contradiction. Indeed, the AdS/CFT duality relates two different regimes for the conformal field theory in 4 dimensions and for the string theory on the Anti-de-Sitter 10-dimensional space. In particular, to find the high energy behavior of scattering amplitudes at strong coupling regime in CFT one can use the perturbation theory in the 10 dimensional gravity. My effective action in gravity was obtained by using perturbation theory. After calculating several terms of the perturbative expansion I derived the light front equation for the effective currents. According to the Maldacena guess this effective action could describe the high energy behavior of amplitudes at strong coupling regime of CFT. In principle, my study of the high energy gravity is not completed yet because I found the effective action only in the asymptotically Minkowski space assuming that the gravitational field vanishes at infinity. To use the AdS/CFT correspondence one should construct the action with the asymptotics of the metric tensor corresponding to the Anti de Sitter space. Further, it is needed to extend this approach to the case of the supergravity and include the gravitinos and dilatons. In future somebody could generalize the effective action for a full string theory on the Anti de Sitter space. Only in this case it could be used for the description of the pomeron interactions in the conformal field theory. But what is surprising, Balitsky and Kovchegov recently constructed the non-linear equation for the description of the screening corrections to the BFKL Pomeron in perturbative QCD which is similar to the light front equation for the effective currents in gravity. Analogous equations were obtained also in the framework of the color glass condensate model and dipole approach.

- *Burda:*

Could you say a few words about the object called Odderon? Could you give us a brief introduction on what is the difference between Pomeron and Odderon?

- *Lipatov:*

The notion of Pomeron and Odderon as moving singularities of the partial waves in the crossing channel appeared many years ago, even before QCD. Pomeron is related to the so-called Froissart theorem claiming, that total cross-sections should be bounded from above by the square of the logarithm of energy. On other hand there is another theorem due to I. Pomeranchuk. According to it the difference between total cross-sections for particle-particle and particle-antiparticle scattering divided by their sum should vanish at high energies. Odderon is the object responsible for the behavior of this difference of cross sections. It has the negative signature and charge parity. The hadrons lying on its trajectory, have odd spins. For the maximal Odderon the difference of the particle-particle and particle-anti-particle cross sections behaves as a logarithm of energy. In QCD Pomeron and Odderon are composite states of two and three gluons, respectively. The Pomeron wave function satisfies the BFKL equation which has the Schroedinger-like form and can be solved exactly. As you know, the three-body problem is not solvable even in classical mechanics.

However, it is remarkable, that in QCD at the leading logarithmic approximation the equations for the wave function of the composite state of three reggeized gluons are integrable. Moreover, in the t'Hooft limit of a large rank of the gauge group the Hamiltonian for the composite state of an arbitrary number of gluons coincides with the Hamiltonian of the integrable Heisenberg spin chain.

- *Burda:*

You mean for the Odderon case?

- *Lipatov:*

I mean for a general case of a composite state of several gluons. Odderon consists of three gluons and the corresponding quantum mechanical system is also integrable. From the first sight the Hamiltonian for this system looks ugly. From time to time I looked at the corresponding Schroedinger equation for about 20 years and understood only in the end that it was remarkable. To begin with, this equation is invariant under the Moebious transformations. Moreover, for the composite state of several gluons there are differential operators which commute with the Hamiltonian. As a result, the Schroedinger equation is integrable and one can calculate the wave functions and energies for an arbitrary number of gluons. The integrability is violated in the next-to-leading approximation at QCD, but presumably it remains in the N=4 supersymmetric field theory. It is important, that the reggeized gluons are gauge invariant objects and as a result, the equations for composite states of these objects can have such remarkable properties as Moebius invariance, holomorphic separability and integrability. The Hamiltonian is also symmetric under duality transformations similar to interchanging coordinates and momenta in the case of the harmonic oscillator.

- *Tapia Takaki:*

What do you think would be a good experimental measurement for detecting this Odderon?

- *Lipatov:*

In principle, the QCD predictions for Odderon were obtained in the perturbation theory, but really people attempted to find a manifestation of this object in the proton-proton and proton-antiproton scattering. In particular. Odderon should give a significant real part to the corresponding amplitudes whereas the amplitude with the Pomeron exchange in the crossing channel is almost imaginary at high energies. The experimental data obtained at FERMILAB many years ago did not observe such real part. In my opinion the Odderon effects initially should be tested in semi-hard processes, where one can calculate the amplitudes in perturbation theory due to the smallness of the coupling constant at large virtualities. In this case we can estimate the Odderon wave function and its intercept. Of course, it would be nice to find some effects of Odderon experimentally in the hadron-hadron collisions, but we don't know how to calculate these effects in perturbative QCD.

- *Tapia Takaki:*

Could you see it for instance in the production of J/ψ in p-p collisions, looking on p_t distribution?

- *Lipatov:*

Of course, one can trace some effects of the perturbative Odderon if jets with the large transverse momentum are measured. There are some theoretical suggestions how to find its effect in other reactions, but in my opinion initially the Odderon will be discovered in the deep inelastic lepton-hadron scattering.

- *Alba:*

I have several technical questions concerning the transparency 12. Is there some expression for A^{++} and A^{--} in terms of the metric or ordinary graviton field?

- *Lipatov:*

These new fields are not expressed in terms of the metric, but they have a geometric interpretation. In the similar expression for the effective action in QCD the fields A^+ and A^- describe the creation and annihilation of the reggeized gluons in the crossing channel. The corresponding currents contain the P-exponents integrated along the light paths. Thus, the fields A^+ and A^- describe a probability distribution of these paths. From the first sight the introduction of the new fields leads to different theories but it is not the case. We can rearrange the Feynmann diagram contributions in such a way that the regions of integration with the particles having approximately equal rapidities are unified and described by the effective action, whereas the interaction of particles having completely different rapidities is performed by the exchange of the reggeons. After the integration over the auxiliary fields A^+ and A^- we return to the QCD action. The high energy effective action approach is similar to the usual renormalization group approach where one separates the Feynmann integrals in such a way that the contribution of the particles with the same virtuality gives the anomalous dimensions and beta functions and the solution of the Calan-Symanzik equation with these functions allows you to find the total contribution. For example, the structure functions for the deep-inelastic ep scattering can be obtained from the solution of renormalization group equations. Completely the same situation is in the high energy approach: the contributions of the Feynmann integrals with particles having the same rapidity leads to the effective action. When we solve the BFKL-like equations, constructed with this effective action, we take into account all other contributions present in the Feynmann diagrams. In particular, the effective currents contain inversed derivatives. These nonlocal operators can be interpreted as propagators of other particles emitting gluons or gravitons in the given interval of rapidity. Such interpretation is possible because the neighboring reggeon propagator is killed by the Laplace operators present in the action. Therefore we do not obtain something different from contributions of usual Feynman diagrams.

- *Alba:*

I do not know if I understand correctly that a partial derivative with the plus or minus subscript correspond to the light-cone variables, and therefore this constraint on A^{++} and A^{--} means that the graviton is transverse?

- *Lipatov:*

This constraint on the graviton field corresponds to the quasi-multiregge kinematics of the high energy processes. In this kinematics the clusters of produced particles differ strongly in their rapidities. This is a generalization of the multiregge kinematics where there is a strong ordering of particles in their rapidities. The effective action describes the interaction of virtual and real particles belonging to the same cluster. It is helpful to parametrize the momenta of clusters by their Sudakov components with the respect to the momenta p_1 and p_2 of colliding particles. It turns out that when you go along the chain of produced clusters from particle with momentum p_1 to particle with momentum p_2, the reggeon momentum components along p_1 decrease and the reggeon components along p_2 increase rapidly. The large Sudakov components are transmitted almost completely to corresponding clusters. In the coordinate space the condition of the strong momentum ordering leads to the above kinematical constrain on the reggeon fields. This constraint is very important for the general covariance of the effective action in gravity and for its gauge invariance in QCD. The rapidity for the high energy processes plays the role of time. One obtaines a Schroedinger-like equation, in which the effective action can be considered as a Hamiltonian. Iteration of this equation corresponds to taking into account interactions of particles with completely different rapidities. I repeat, that this approach is similar to the renormalization group formalisn.

- *Fazio:*

In AdS/CFT correspondence one tries to correct the BFKL exponent at large coupling. There is a corresponding gauge theory with a huge number of colors. How the large number of colors affects the results for BFKL exponent as being computed in QCD with N_c=3?

- *Lipatov:*

Indeed, in QCD the gauge group is SU(3). But theorists believe that characteristic features of the theory are the same if the number of colors is very large. Moreover, in two first orders in perturbation theory the BFKL Hamiltonian does not contain $1/N_c$ corrections. The problem appears when you attemp to unitarize the scattering amplitudes in the s-channel to obtain total cross-sections in an agreement with the Froissart bound. Indeed, in this case the $1/N_c$ corrections are very important. For example, in the N=4 supersymmetric theory at large N_c the Pomeron intercept for large coupling constant turns out to be equal to the graviton spin 2, because here Pomeron and graviton coincide. Such value for the intercept contradicts the Froissart bound. To correct this drawback of the theory one should take into account the non-planar diagrams suppressed as $1/N_c$.

- *Csanàd:*

I would like to ask about the proton-proton differential cross-section because already such an easy thing is a big challenge for the theory to describe data. How this problem should be approached?

- *Lipatov:*

There is a group of people, for example Larry McLerran with collaborators who attemp to generalize the BFKL approach in an application to pure hadronic processes, in particular, to the proton-proton scattering. But I'm not sure that one can predict scattering amplitudes in QCD at the non-perturbative regime. The problem is that the Hamiltonian for the BFKL equation is known only in two loops. As a result, we cannot describe processes in which the characteristic transverse momenta are of the order of hadronic masses, where the coupling constant is large. But if you take into account the non-linear effects incorporated for example in the Balitsky-Kovchegov equation, the characteristic transverse momenta grow with energy. Strictly speaking the Balitsky-Kovchegov equation can be applied only to the deep inelastic scattering or to the hadron scattering off heavy nuclei. If you attempt to describe the proton-proton scattering then there is a problem related to the t-channel unitarity. Because in the Balitsky-Kovchegov equations only a subset of Feynman diagrams is taken into account, the unitarity is lost. In practice this approach could work, but I am not a specialist in fitting experimental data. Nevertheless, recently I took part in the theoretical description of the last experimental data obtained at HERA. We with collaborators incorporated in the BFKL equation some non-perturbative effects through boundary conditions at small transverse momenta and obtained a rather good fit of the experimental data.

- *Campbell:*

Where in your theory the scale at which we would expect to see the effects of the new terms in the action is set?

- *Lipatov:*

Of course I used perturbation theory. Therefore in QCD the characteristic transverse momenta should be large in comparison with the QCD scale. The spectrum of the states described by the BFKL dynamics is very rich and one can predict multiparticle amplitudes with very different scales of energies, enhanced and non-enhanced t-channel contributions.

- *Campbell:*

But where in particular do you introduce the scale?

- *Lipatov:*

For example, in the well-known Mueller-Navalet process of two jet production in proton-proton collisions the relative rapidities of the jets should be big enough with approximately equal transverse momenta. For sufficiently large transverse momenta one can use the high energy effective action in QCD.

Infrared Instability in QCD

Peter Minkowski
Albert Einstein Center for Fundamental Physics - ITP
University of Bern

Abstract

In this lecture concepts are presented which shall elucidate the set of regularity conditions for gauge fields – connection and curvature – inherent to the dynamics of all local color carrying quanta in QCD at large space - time separations. The regularity conditions do entail dynamical consequences as represented – not exclusively – through the main vacuum condensates $\langle \, \Omega \mid \frac{1}{4} \, (:) \, B^{\,t}_{\,\mu \, \nu} \, B^{\,\mu \, \nu \, t} \, (:) \mid \Omega \, \rangle$ for gauge field pairs and $\langle \, \Omega \mid (:) \, \overline{q} \, q \, (:) \mid \Omega \, \rangle$ for quark antiquark field pairs .

1-1 QCD – the two central anomalies : scale and chiral U1

1-1-a – Premises

We face the theoretical abstraction of QCD with $N_{fl} = 6$, representing strong interactions – adaptable to two or three light flavors (u , d , s) of quarks and antiquarks.

quarks: color is counted in $\pi^0 \rightarrow \gamma\gamma$

spin and flavor are clearly seen in $q\overline{q}$ and $3q, 3\overline{q}$ spectroscopy.

$$\mathcal{L} = \left[\begin{array}{c} \overline{q}\,^{\dot{c}'}_{\dot{\mathcal{A}}'\,\dot{f}} \left\{ \begin{array}{c} \frac{i}{2}\overleftrightarrow{\partial}_{\mu}\,\delta_{c'\dot{c}} \\ -\,v^{r}_{\mu}\left(\frac{1}{2}\lambda_{r}\right)_{c'\dot{c}} \end{array} \right\} \gamma^{\mu}_{\dot{\mathcal{A}}'\mathcal{A}}\, q^{c}_{\mathcal{A}\,f} \\ -\,m_{f}\,\overline{q}\,^{\dot{c}}_{\dot{\mathcal{A}}\,\dot{f}}\, q^{c}_{\mathcal{A}\,f} \end{array} \right] - \frac{1}{4g^{2}}\,B^{\mu\nu\,r}\,B^{r}_{\mu\nu} + \Delta\,\mathcal{L} \tag{1}$$

quarks : c' , $c = 1, 2, 3$ color , $f = 1, \cdots, 6$ flavor

$\mathcal{A}', \mathcal{A} = 1, \cdots, 4$ spin , m_f mass

gauge bosons :

$$B^{r}_{\mu\nu} = \partial_{\nu}\,v^{r}_{\mu} - \partial_{\mu}\,v^{r}_{\nu} - f_{rst}\,v^{s}_{\nu}\,v^{t}_{\mu}$$

$$r, s, t = 1, \cdots, dim\,(\,G = SU3_{c}\,) = 8$$

$$\longrightarrow\quad W^{r}_{\mu} \equiv -\,v^{r}_{\mu}\ ;\ B^{r}_{\mu\nu} = \partial_{\mu}\,W^{r}_{\nu} - \partial_{\nu}\,W^{r}_{\mu} + f_{rst}\,W^{s}_{\mu}\,W^{t}_{\nu}$$

Lie algebra labels, $\left[\frac{1}{2}\,\lambda^{r}\,,\,\frac{1}{2}\,\lambda^{s}\right] = i\,f_{rst}\,\frac{1}{2}\,\lambda^{t}$

perturbative rescaling :

$$v^{r}_{\mu} = g\,v^{r}_{\mu\;pert}\ ,\ B^{r}_{\mu\nu} = g\,B^{r}_{\mu\nu\;pert} \tag{2}$$

Degrees of freedom are seen in jets , in (e.g.) the energy momentum sum rule in deep inelastic scattering but not clearly in spectroscopy.
Completing $\Delta\,\mathcal{L}$ in Fermi gauges

$$\Delta\,\mathcal{L} = \left\{ \begin{array}{c} -\,\frac{1}{2\,\eta\,g^{2}}\left(\partial_{\mu}\,v^{\mu\,r}\right)^{2} \\ +\,\partial^{\mu}\,\overline{c}^{\,r}\left(D_{\mu}\,c\right)^{r} \end{array} \right\}\ ;\ \eta\ :\ \text{gauge parameter} \tag{3}$$

ghost fermion fields : c , $\overline{c}$; $(D_{\mu}\,c)^{r} = \partial_{\mu}\,c^{r} - f_{rst}\,v^{s}_{\mu}\,c^{t}$

gauge fixing constraint : $C^{r} = \partial_{\mu}\,v^{\mu\,r}$

1-1-b – Gauge boson binary bilocal and adjoint (here octet-) string operators

One goal is, to identify – not just some candidate resonance – gluonic mesons, binary and higher modes, and to relate them to the base quantities within QCD.

$$\begin{array}{l} B_{[\,\mu_1\,\nu_1\,]\,,\,[\,\mu_2\,\nu_2\,]}\,(\,x_{1}\,,\,x_{2}\,) = \\ = B^{r}_{[\,\mu_1\,\nu_1\,]}\,(\,x_{1}\,)\,U\,(\,x_{1}\,,\,r\,;\,x_{2}\,,\,s\,)\,B^{s}_{[\,\mu_2\,\nu_2\,]}\,(\,x_{2}\,) \end{array} \tag{4}$$

$r\ ,\ s\ ,\cdots = 1, \cdots, 8$; no flavor but spin $\rightarrow$

$B^{\ r}_{[\mu\nu]}(x)$ denote the local color octet of field strengths.
The quantity $U(x, r; y, s)$ in eq. (4) denotes the octet string operator, i. e. the path ordered exponential over a straight line path $\mathcal{C}$ from y to x

$$U(x, r; y, s) = P \exp\left(\int_y^x \Big|_{\mathcal{C}} dz^{\mu} \tfrac{1}{i} v^{\,t}_{\mu}(z) \mathcal{F}_t \right)\Bigg|_{rs}$$
$$= P \exp\left(- \int_y^x \Big|_{\mathcal{C}} dz^{\mu} W^{\,t}_{\mu}(z) \left(\tfrac{1}{i} \mathcal{F}_t \right) \right)\Bigg|_{rs}$$
$$(\mathcal{F}_t)_{rs} = i f_{rts} \; ; \; (ad_t)_{rs} = \tfrac{1}{i} (\mathcal{F}_t)_{rs} = f_{rts} \tag{5}$$

The path ordered exponential as a matrix function of the argument is to be performed before the matrix elements, denoted $|_{..}$ in eq. 5 , are taken. The local limit becomes

$$B_{[\mu_1\nu_1],[\mu_2\nu_2]}(x_1 = x_2 = x) = (:) B^{\ r}_{[\mu_1\nu_1]}(x)\, B^{\ r}_{[\mu_2\nu_2]}(x) (:)$$
no flavor but spin

(6)

1-1-c – $\overline{q}\, q$ bilinears and triplet-string operators

$$B^{\,q}_{[\dot{\mathcal{A}}\, \dot{f}_1, \mathcal{B}\, f_2]}(x_1, x_2) = \overline{q}^{\ \dot{c}_1}_{\dot{\mathcal{B}}\, \dot{f}_2}(x_1)\, U(x_1, c_1; x_2, \dot{c}_2)\, q^{\ c}_{\mathcal{A}\, f_1}(x_2)$$
flavor *and* spin
$$U(x, c_1; y, \dot{c}_2) = P \exp\left(\int_y^x \Big|_{\mathcal{C}} dz^{\mu} \tfrac{1}{i} v^{\,t}_{\mu}(z) \tfrac{1}{2} \lambda_t \right)\Bigg|_{c_1 \dot{c}_2}$$
$$= P \exp\left(- \int_y^x \Big|_{\mathcal{C}} dz^{\mu} W^{\,t}_{\mu}(z) \left(\tfrac{1}{i} \tfrac{1}{2} \lambda_t \right) \right)\Bigg|_{c_1 \dot{c}_2} \tag{7}$$

with the local limit

$$B^{\,q}_{[\dot{\mathcal{B}}\, \dot{f}_2, \mathcal{A}\, f_1]}(x_1 = x_2 = x) = (:)\, \overline{q}^{\ \dot{c}}_{\dot{\mathcal{B}}\, \dot{f}_2}(x)\, q^{\ c}_{\mathcal{A}\, f_1}(x)\, (:) \tag{8}$$

The symbols (:) in eqs. 6 and 8 should indicate that normal ordering of regulating the local limits is required and further that such normal ordering is *not* unique , and dependent on quark masses in the case of the $\overline{q}\, q$ bilinears .

1-1-d – Connection and curvature - forms preparing the ensuing analysis of regularity conditions

Lets begin this (sub-)section rewriting the bilocal (formally) unitary operators forming the gauge connection dependent octet- (eq.5) and triplet (eq.7) QCD strings , substituting an equivalent , matrix oriented notation

$$\begin{array}{ll} \text{octet string}: & U(x, r; y, s) = \\ & P \exp\left(- \int_y^x \Big|_C d z^{\mu} W_{\mu}^{t}(z) \left(\tfrac{1}{i} \mathcal{F}_t \right) \right) \Big|_{r s} \\ \text{triplet string}: & U(x, c_1; y, \dot{c}_2) = \\ & P \exp\left(- \int_y^x \Big|_C d z^{\mu} W_{\mu}^{t}(z) \left(\tfrac{1}{i}\tfrac{1}{2} \lambda_t \right) \right) \Big|_{c_1 \dot{c}_2} \end{array}$$

with the substitutions $\longrightarrow$

$$\left. \begin{array}{ll} \text{octet string}: & U(x, r; y, s) \rightarrow \left(U \left(x \overset{C}{\leftarrow} y \right) \right)_{rs} \\ \text{triplet string}: & U(x, c_1; y, \dot{c}_2) \rightarrow \left(U \left(x \overset{C}{\leftarrow} y \right) \right)_{c_1 \dot{c}_2} \end{array} \right\} \rightarrow$$

$$\rightarrow \quad (U (x, C, y ; \mathcal{D}))_{\alpha\beta} \in \mathcal{D} (\mathcal{G})$$

with $\mathcal{G}$ = simple compact gauge group ;

$\mathcal{D}$: general irreducible representation of $\mathcal{G}$

(9)

Here $\mathcal{G} = SU3_c$ and $\mathcal{D}$ is the adjoint- , triplet representation for the respective QCD $\mathcal{D}$ - strings.

Further let us consider matrix valued connection 1-forms , which define the bilocal matrix valued operators $(U (x, C, y ; \mathcal{D}))_{\alpha\beta} \in \mathcal{D} (\mathcal{G})$ as given in eq. 9 . To this end the form of octet and triplet strings in eq. 9 is repeated below

$$\begin{array}{ll} \text{octet string}: & U(x, r; y, s) = \\ & = P \exp\left(- \int_y^x \Big|_C d z^{\mu} W_{\mu}^{t}(z) \left(\tfrac{1}{i} \mathcal{F}_t \right) \right) \Big|_{r s} \\ \text{triplet string}: & U(x, c_1; y, \dot{c}_2) = \\ & = P \exp\left(- \int_y^x \Big|_C d z^{\mu} W_{\mu}^{t}(z) \left(\tfrac{1}{i}\tfrac{1}{2} \lambda_t \right) \right) \Big|_{c_1 \dot{c}_2} \end{array} \tag{10}$$

The two matrices in brackets to the right of the integrand expressions in eq. 10 form an antihermitian basis of the Lie algebra representation $Lie (\mathcal{D})$

for $\mathcal{D}$ = adjoint and $\mathcal{D}$ = triplet representations of $\mathcal{G} = SU3_c$ respectively

$$\begin{array}{l} d_t \equiv d_t\ (\mathcal{D}) \leftrightarrow\ (d_t)_{\alpha\beta} = \left\{ \begin{array}{ll} \left(\frac{1}{i}\ \mathcal{F}_t\right)_{rs} & Lie\ (\mathcal{D}) = \text{adjoint} \\ \left(\frac{1}{i}\frac{1}{2}\ \lambda_t\right)_{c_1 \dot{c}_2} & Lie\ (\mathcal{D}) = \text{triplet} \end{array} \right. \\ d_t = -d_t^{\dagger}\ ;\ t = 1, \cdots, dim\ \mathcal{G}\ ;\ \alpha\ ,\ \beta = 1, \cdots, dim\ \mathcal{D} \\ \text{for general } \mathcal{D} \end{array} \tag{11}$$

From eqs. 10 and 11 we construct a – hopefully – consistent notation as appropriate for matrix valued $\mathcal{D}$ connections, 1-forms and strings, as well as derived 2- and higher forms. First eq. 11 is subject to the (matrix-) commutation relations

$$\begin{array}{l} [\, d_r\ ,\ d_s\,] = f_{rst}\, d_t\ ;\ \forall\, \mathcal{D}\, (\, \mathcal{G}\,)\ ;\ r, s, t = 1, \cdots, dim\ \mathcal{G} \longrightarrow \\ (\, d_t\, (\, \mathcal{D} = \text{adjoint representation}\,)\,)_{sr} = (\, ad_t\,)_{sr} = f_{str}\ ; \\ \text{independent of } \mathcal{D} \\ f_{str}\ :\ \text{totally antisymmetric, real structure constants of } Lie\ (\, \mathcal{G}\,) \end{array} \tag{12}$$

In physics the antihermitian matrix code with respect to the representations $Lie\ (\, \mathcal{D}\,)$ is (mostly) replaced by the hermitian one [1]

$$\begin{array}{l} \left(\, d_t \equiv \frac{1}{i}\, h_t\, \right)\ (\, Lie\, (\, \mathcal{D}\,)\,)\, \big|_{\alpha\beta}\ ;\ \alpha, \beta = 1, \cdots, dim\ \mathcal{D}\, ;\ h_t = h_t^{\dagger} \\ [\, h_r\ ,\ h_s\,] = i\, f_{rst}\, h_t \alpha, \beta = 1, \cdots, dim\ \mathcal{D} \end{array} \tag{13}$$

Eq. 12 serves to define matrix valued connections built from a basis of $Lie\ (\, \mathcal{D}\,)$ representation matrices as defined in eq. 12 for general irredcible representations $\mathcal{D}\ (\, \mathcal{G}\,)$ denoted $\mathcal{W}_\mu\ (\, z\ ,\ \mathcal{D}\,)$

$$\begin{array}{l} \mathcal{W}_\mu\, (\, z\ ,\ \mathcal{D}\,)\, |_{\alpha\beta} = W_\mu^{\,r}\, (\, z\,)\ (\, d_r\,)_{\alpha\beta\ Lie\ (\, \mathcal{D}\,)} \\ \left[\begin{array}{l} r = 1, \cdots, dim\ (\, \mathcal{G}\,) \\ \alpha\ ,\ \beta = 1, \cdots, dim\ (\, \mathcal{D}\,) \end{array} \right] \\ \mathcal{W}_\mu\, (\, z\ ,\ \mathcal{D}\,)\, |_{\alpha\beta} \longrightarrow \mathcal{W}_\mu \text{ for compact matrix notation} \end{array} \tag{14}$$

In the following it is to be understood, that $\mathcal{W}_\mu$ is extended to a general collection of representations $\bigcup\, \mathcal{D}$ – thought to be carried by real and spurious spin $\frac{1}{2}$ fields – care beeing taken that asymptotic freedom in the ultraviolet is not upset.

---

[1] A (partial) collection of historical and textbook references to the topics pertaining to 'Continous transformation groups and differential geometry' is presented at the end of references and labelled by the symbols 1H , 2H $\cdots$.

From eq. 14 we define the associated matrix valued connection 1-form displayed alongside the base definition repeated from eq. 14 in eq. 15 below

$$\begin{array}{rcll} \mathcal{W}^{(1)} (z , \mathcal{D}) \mid_{\alpha\beta} & = & d z^{\mu} \mathcal{W}_{\mu} (z , \mathcal{D}) \mid_{\alpha\beta} & \longrightarrow \mathcal{W}^{(1)} \\ \mathcal{W}_{\mu} (z , \mathcal{D}) \mid_{\alpha\beta} & = & W_{\mu}^{r} (z) (d_{r})_{\alpha\beta \; Lie (\mathcal{D})} & \longrightarrow \mathcal{W}_{\mu} \end{array} \tag{15}$$

and the matrix valued field-strength tensor

$$\mathcal{W}_{\mu\nu} (z, \mathcal{D}) \mid_{\alpha\beta} = \left\{ \begin{array}{c} \partial_{\mu} \mathcal{W}_{\nu} (z, \mathcal{D}) - \partial_{\nu} \mathcal{W}_{\mu} (z, \mathcal{D}) + \\ + [\mathcal{W}_{\mu} (z, \mathcal{D}) , \mathcal{W}_{\nu} (z, \mathcal{D})] \end{array} \right\}_{\alpha\beta} \longrightarrow \mathcal{W}_{\mu\nu} \tag{16}$$

together with their associated curvature 2-form

$$\begin{array}{rcl} \mathcal{W}^{(2)} (z,\mathcal{D}) \big|_{\alpha\beta} & = & \frac{1}{2} d z^{\mu} \wedge d z^{\mu} \mathcal{W}_{\mu\nu} (z,\mathcal{D}) \big|_{\alpha\beta} \\ & & \longrightarrow \mathcal{W}^{(2)} \\ \mathcal{W}_{\mu\nu} (z,\mathcal{D}) \big|_{\alpha\beta} & = & \left\{ \begin{array}{c} \partial_{\mu} \mathcal{W}_{\nu} (z,\mathcal{D}) - \partial_{\nu} \mathcal{W}_{\mu} (z,\mathcal{D}) + \\ + [\mathcal{W}_{\mu} (z,\mathcal{D}) , \mathcal{W}_{\nu} (z, \mathcal{D})] \end{array} \right\}_{\alpha\beta} \\ & & \longrightarrow \mathcal{W}_{\mu\nu} \\ & = & W_{\mu\nu}^{r} (z) (d_{r})_{\alpha\beta \; Lie (\mathcal{D})} \end{array}$$

$$W_{\mu\nu}^{r} = \partial_{\mu} W_{\nu}^{r} - \partial_{\nu} W_{\mu}^{r} + f_{rst} W_{\mu}^{s} W_{\nu}^{t} \; ; \text{ independent of } \mathcal{D} \tag{17}$$

Two remarks are in place here

1) In order to distinguish field strength from potentials (connections) the following equivalent but different notations for the field strength shall be used

$$\mathcal{W}_{\mu\nu} \equiv \mathcal{B}_{\mu\nu} \; ; \; \mathcal{W}^{(2)} \equiv \mathcal{B}^{(2)} \; ; \; W_{\mu\nu}^{r} \equiv B_{\mu\nu}^{r} \tag{18}$$

2) From the last relation in eq. 17 it may appear redundant to extend connections and curvatures to matrix valued form with respect to a wide collection of irreducible representations $\mathcal{D} (\mathcal{G})$. This however is tantamount to neglecting nontrivial global regularity conditions in the infrared .

We end this subsection (1-1-d) displaying →

the bilocal (parallel transport-) operators defined in eq. 9 using the short-hand notation in eq. 17

$$
\begin{array}{lcl}
(U (x , C , y ; \mathcal{D}))_{\alpha\beta} & = & P \exp \left(- \int_y^x \Big|_{C} \mathcal{W}^{(1)} (z, \mathcal{D}) \right) \Bigg|_{\alpha\beta} \\
\downarrow & & \downarrow \\
U (x , C , y) & = & P \exp \left(- \int_y^x \Big|_{C} \mathcal{W}^{(1)} \right) \\
[\text{for } (\bigcup \mathcal{D})] & &
\end{array}
\tag{19}
$$

1-1-e – The U1- or singlet axial current anomaly

The U1-axial central anomaly involves the local chiral current projections from $B^{\,q}_{[\,\dot{\mathcal{B}}\,\dot{f}\,2\,,\,\mathcal{A}\,f\,1\,]}\,(x)$ in eq. 8

$$
\begin{array}{lcl}
(j^{\pm}_{\mu})_{\dot{f}_2\, f_1} (x) & = & B^{\,q}_{[\,\dot{\mathcal{B}}\,\dot{f}_2\,,\,\mathcal{A}\,f_1\,]}(x) \left(\gamma_{\mu} \frac{1}{2} (\P \pm \gamma_5) \right)_{\mathcal{B}\,\dot{\mathcal{A}}} \\
 & = & (:)\, \overline{q}^{\,\dot{c}}_{\dot{f}\,2}\, \gamma^{\pm}_{\mu}\, q^{\,c}_{f\,1} (x)\, (:) \\
\gamma_5 = \gamma_{5\,R} & = & \frac{1}{i} \gamma_0 \gamma_1 \gamma_2 \gamma_3 \; ; \; \gamma^{\pm}_{\mu} = \gamma_{\mu} \frac{1}{2} (\P \pm \gamma_5)
\end{array}
\tag{20}
$$

The equations of motion for the fermion fields are *and superficially imply* (upon $f_1 \leftrightarrow f_2$)

$$
\begin{array}{lcl}
\partial\!\!\!/ \, q^{\,c}_{f\,2} & = & \frac{1}{i} \left(v\!\!\!/^{\,c\,\dot{c}'} + \delta^{\,c\,\dot{c}'}\, m_{f\,2} \right) q^{\,c'}_{f\,2} \\
\overline{q}^{\,\dot{c}}_{\dot{f}\,1}\, \overleftarrow{\partial\!\!\!/} & = & \overline{q}^{\,\dot{c}'}_{\dot{f}\,1} \frac{1}{i} \left(- v\!\!\!/^{\,c'\,\dot{c}} - \delta^{\,c'\,\dot{c}}\, m_{f\,1} \right)
\end{array}
\; ; \quad \begin{array}{c} \text{no sums over} \\ \dot{f}_1 \, , \, f_2 \end{array} \quad \rightarrow
$$

$$
\begin{array}{lcl}
\partial^{\mu} (j^{\pm}_{\mu})_{\dot{f}_1\, f_2} & = & \frac{1}{2\,i} \left(\begin{array}{l} (m_{f\,2} - m_{f\,1})\, S_{\dot{f}\,1\,f\,2} \mp \\ (m_{f\,2} + m_{f\,1})\, P_{\dot{f}_1\, f_2} \end{array} \right) \\
S_{\dot{f}\,1\,f\,2} & = & (:)\, \overline{q}^{\,\dot{c}}_{\dot{f}\,1}\, q^{\,c}_{f\,2}\, (:) \; , \; P_{\dot{f}\,1\,f\,2} = (:)\, \overline{q}^{\,\dot{c}}_{\dot{f}\,1}\, \gamma_5\, q^{\,c}_{f\,2}\, (:)
\end{array}
\tag{21}
$$

In eq. 21 m_{f} denotes the *real* , *nonnegative* quark mass for flavor f. From eq. 21 the relations for vector and axial vector currents *superficially* follow

$$
\begin{array}{lcl}
(j_{\mu})_{\dot{f}\,1\,f\,2} & = & (j^{+}_{\mu})_{\dot{f}\,1\,f\,2} + (j^{-}_{\mu})_{\dot{f}\,1\,f\,2} \\
(j^{5}_{\mu})_{\dot{f}\,1\,f\,2} & = & (j^{+}_{\mu})_{\dot{f}\,1\,f\,2} - (j^{-}_{\mu})_{\dot{f}\,1\,f\,2} \\
\partial^{\mu} (j_{\mu})_{\dot{f}\,1\,f\,2} & = & \frac{1}{i} (m_{f\,2} - m_{f\,1})\, S_{\dot{f}\,1\,f\,2} \\
\partial^{\mu} (j^{5}_{\mu})_{\dot{f}\,1\,f\,2} & = & (m_{f\,2} + m_{f\,1})\, i\, P_{\dot{f}\,1\,f\,2}
\end{array}
\tag{22}
$$

As it follows from the original derivation by Adler and Bell and Jackiw [1-1-1969] in QED, the vecor current Ward identities in eq. 22 can be implemented also in QCD , leaving the axial current ones reduced to the flavor non-singlet case , leaving the U1 axial current divergent anomalous

$$\begin{aligned} \partial^{\mu} \, (j_{\mu})_{\dot{f}_1 f_2} &= \tfrac{1}{i} \, (m_{f_2} - m_{f_1}) \, S_{\dot{f}_1 f_2} \; \surd \\ \left\{ \begin{array}{c} j^{5}_{\mu} \\ P \end{array} \right\}^{NS}_{\dot{f}_1 f_2} &= \left\{ \begin{array}{c} j^{5}_{\mu} \\ P \end{array} \right\}_{\dot{f}_1 f_2} - \tfrac{1}{N_{fl}} \, \delta_{\dot{f}_1 f_2} \sum_f \left\{ \begin{array}{c} j^{5}_{\mu} \\ P \end{array} \right\}_{\dot{f} f} \end{aligned} \tag{23}$$

and similarly

$$\left\{ \begin{array}{c} j^{5}_{\mu} \\ P \end{array} \right\}^{S}_{\dot{f}_1 f_2} = \sum_f \left\{ \begin{array}{c} j^{5}_{\mu} \\ P \end{array} \right\}_{\dot{f} f} \tag{24}$$

1-1-f – Quark masses and splittings : m_f and $\Delta m_f = m_f - \langle m \rangle$

In the subtitle above $\langle m \rangle$ stands for the mean quark mass

$$\langle m \rangle = \tfrac{1}{N_{fl}} \sum_f m_f \tag{25}$$

The identities for vector currents in eqs. 22 and 23 can be extended separating the conributions proportional to Δm_f and $\langle m \rangle$

$$\begin{aligned} \partial^{\mu} \, (j_{\mu})_{\dot{f}_1 f_2} &= \tfrac{1}{i} \, (\Delta m_{f_2} - \Delta m_{f_1}) \, S_{\dot{f}_1 f_2} \; \surd \\ \partial^{\mu} \, (j^{5}_{\mu})^{NS}_{\dot{f}_1 f_2} &= (\Delta m_{f_2} + \Delta m_{f_1}) \, i \, P^{NS}_{\dot{f}_1 f_2} \; \surd \\ \partial^{\mu} \, (j^{5}_{\mu})^{S}_{\dot{f}_1 f_2} &= 2 \, \langle m \rangle \, i \, P^{S} \; \not{\surd} \; [\longrightarrow + \delta_5] \\ \delta_5 &= (2 N_{fl}) \, \tfrac{1}{32 \pi^2} \, B^{r}_{\mu \nu} \, \widetilde{B}^{\mu \nu r} \Big|_{\to ren.gr.inv} \\ \widetilde{B}^{r}_{\mu \nu} &= \tfrac{1}{2} \, \varepsilon_{\mu \nu \sigma \tau} \, B^{\sigma \tau r} \end{aligned} \tag{26}$$

2

We shall return to the question of how the local operator $ch_2 (B) \equiv \frac{1}{32 \pi^2} (:) \, B^{r}_{\mu \nu} \, \widetilde{B}^{\mu \nu r} (:)$ is to be normalized and rendered renormalization group invariant [1-2-1991] . Here we just assume this to have been achieved and denote the U1-axial anomaly, the first of the central two, in its general form

(eq. 26)

$$\begin{aligned} \left\{ \partial^{\mu} \, (j^{5}_{\mu})^{S} = 2 \, \langle m \rangle \, i \, P^{S} + \delta_5 \right\} (x) \\ \delta_5 = (2 N_{fl}) \, \tfrac{1}{32 \pi^2} \, (:) \, B^{r}_{\mu \nu} \, \widetilde{B}^{\mu \nu r} (:) \Big|_{\to ren.gr.inv} \end{aligned} \tag{27}$$

From here it is conceptually clear how the scale- (or trace-) anomaly arises but strictly within QCD . The renormalizability of a field theory in the limit

[2] δ_5 was – as far as I know – introduced by Murray Gell-Mann in lectures ~ 1970 in Hawaii .

of uncurved space-time gives rise to a local , symmetric and *conserved* energy momentum tensor , implying exact Poincaré invariance

$$\begin{array}{l} \{ \vartheta_{\mu \nu} = \vartheta_{\nu \mu} \} (x) \\ \partial^{\nu} \vartheta_{\mu \nu} = 0 \end{array} \tag{28}$$

In connection with the normal ordering questions it is important to admit in the precise form of the energy momentum tensor a nontrivial vacuum expected value , which
in view of exact Poincaré invariance must be of the form

$$\begin{array}{l} \langle \Omega \mid \vartheta_{\mu \nu} (x) \mid \Omega \rangle = \frac{1}{4} \eta_{\mu \nu} \tau ; \\ \left\{ \begin{array}{c} \eta_{\mu \nu} = diag (1 , -1 , -1 , -1) \\ \tau \end{array} \right\} \text{ independent of x } \longrightarrow \\ \Delta \vartheta_{\mu \nu} (x) = \\ \quad = \vartheta_{\mu \nu} (x) - \langle \Omega \mid \vartheta_{\mu \nu} (x) \mid \Omega \rangle \times \left\{ \begin{array}{l} \widehat{\P} \\ \text{or } \mid \Omega \rangle \langle \Omega \mid \end{array} \right. \\ \text{with } \partial^{\nu} \Delta \vartheta_{\mu \nu} (x) = 0 ; \langle \Omega \mid \Delta \vartheta_{\mu \nu} (x) \mid \Omega \rangle = 0 \end{array} \tag{29}$$

In eq. 29 $\widehat{\P}$ denotes the unit operator in the entire Hilbert space of states, while $P_{\Omega} = \mid \Omega \rangle \langle \Omega \mid$ stands for the projector on the ground state . Furthermore from the two local , *conserved* tensors in eq. 29 only $\Delta \vartheta_{\mu\nu} (x)$ with vanishing vacuum expected value is acceptable as representing the conserved 4 momentum *operators* in the integral form

$$\widehat{P}_{\mu} = \int_{t} d^{3} x \, \Delta \vartheta_{\mu 0} (t , \vec{x}) \tag{30}$$

All these arguments *notwithstanding* to subtract any eventual vacuum expected values of local operators , often put forward as mathematical prerequisites , it is wise *not to do so* in the presence of spontaneous parameters , the dynamical origin of spontaneous symmetry breaking, e.g. chiral symmetries in the limit or neighbourhood of some $m_{f} \rightarrow 0$.
Using the (classical) equations of motion pertaining to the Lagrangean in eqs. 1 - 3

$$\begin{array}{l} (D_{\nu} B^{\mu\nu})^{r} = j^{\mu \, r} (\overline{q} , q) ; \; B \rightarrow B_{pert} \\ (D_{\varrho} B^{\mu \nu})^{r} = \partial_{\varrho} B^{\mu \nu r} + f_{rst} W_{\varrho}^{s} B^{\mu \nu t} \\ j_{\mu}^{r} (\overline{q} , q) = g \, \overline{q}_{\dot{A} f}^{\dot{c}} (\gamma_{\mu})_{\dot{A} B} \frac{1}{2} (\lambda^{r})_{c c'} q_{A f}^{c'} \\ i (\gamma^{\mu} D_{\mu} q)_{A f}^{c} = m_{f} q_{A f}^{c} \text{ and } q \rightarrow \overline{q} \\ (D_{\mu} q)_{A f}^{c} = [\partial_{\mu} \delta_{c c'} + \frac{1}{i} W_{\mu}^{t} \frac{1}{2} (\lambda^{t})_{c c'}] q_{A f}^{c'} \\ \hline W_{\mu}^{r} \equiv - v_{\mu}^{r} = g (W_{\mu}^{r})_{pert} \equiv - g (v_{\mu}^{r})_{pert} \end{array} \tag{31}$$

the associated form of the energy momentum becomes

$$\vartheta^{(cl)}_{\mu\nu} = \left[\begin{array}{c} \frac{1}{4g^2} \left[B^t_{\mu\varrho} B^{\varrho t}_{\nu} - \frac{1}{4} \eta_{\mu\nu} B^t_{\sigma\varrho} B^{\varrho\sigma t} \right] + \\ + \frac{1}{2} \left[\overline{q}_{\dot{f}} \gamma_\mu \frac{i}{2} \overleftrightarrow{D}_\nu q_f + \mu \leftrightarrow \nu \right] \end{array} \right] \quad (32)$$

and using once more the fermion part of the equations of motion the trace of the classical energy momentum tensor becomes

$$\begin{aligned} \vartheta^{\mu}{}_{\mu}^{(cl)} &= \sum_f m_f S_{\dot{f} f} \\ S_{\dot{f}_1 f_2} &= (:) \overline{q}^{\dot{c}}_{\dot{f}_1} q^{c}_{f_2} (:) \end{aligned} \quad (33)$$

1-1-g – The scale- or trace- anomaly

From the classical soft fermionic contribution to the trace of the energy momentum tensor there is a clear conjecture, also by Murray Gell-Mann, of the anomalous contribution , which subsequently became the scale- or trace-anomaly within QCD

$$\begin{aligned} \vartheta^{\mu}{}_{\mu} &= \sum_f m_f S_{\dot{f} f} + \delta_0 \\ \delta_0 &= - \left(-2\beta(g)/g^3 \right) \left[\tfrac{1}{4} (:) B^t_{\mu\nu} B^{\mu\nu t} (:) \right]_{\rightarrow ren.gr.inv} \end{aligned} \quad (34)$$

1-1-h – The two central anomalies alongside : scale- or trace- and U1-axial anomaly

We collect the two anomalous identities in eqs. 34 and 16

$$\begin{aligned} &\left\{ \vartheta^{\mu}{}_{\mu} = \sum_f m_f S_{\dot{f} f} + \delta_0 \right\} (x) \\ &\left\{ \partial^\mu (j^5_\mu)^S = 2 \langle m \rangle i P^S + \delta_5 \right\} (x) \\ &\delta_0 = - \left(-2\beta(g)/g^3 \right) \left[\tfrac{1}{4} (:) B^t_{\mu\nu} B^{\mu\nu t} (:) \right]_{\rightarrow ren.gr.inv} \\ &\delta_5 = (2 N_{fl}) \tfrac{1}{8\pi^2} \left[\tfrac{1}{4} (:) B^t_{\mu\nu} \widetilde{B}^{\mu\nu t} (:) \right]_{\rightarrow ren.gr.inv} \\ \hline &-\beta/g^3 = \tfrac{1}{16\pi^2} b_0 + O(X) \; ; \; X = g^2/(16\pi^2 \\ &\beta(g) \; : \; \text{Callan-Symanzik rescaling function in QCD} \end{aligned} \quad (35)$$

The qualification 'central' for the anomalies in eq. 35 stands for the property that in rendering the square coupling constant and the associated ϑ – parameter in the gauge boson *renormalized* Lagrangean density x dependent

$$\begin{aligned} &\mathcal{L}_{g.b.} = - \tfrac{1}{g^2} \tfrac{1}{4} (:) B^t_{\mu\nu} B^{\mu\nu t} (:) + \vartheta \tfrac{1}{8\pi^2} \tfrac{1}{4} (:) B^t_{\mu\nu} \widetilde{B}^{\mu\nu t} \longrightarrow \\ &g^2 \rightarrow g^2(x) \; ; \; \vartheta \rightarrow \vartheta(x) \end{aligned} \quad (36)$$

maintains perturbative renormalizability and acts together with suitable boundary- – more generally – regularity conditions as external sources for the scalar and pseudoscalar local field strength bilinears

$$\tfrac{1}{4} \,(:)\, B^{\;t}_{\mu\,\nu} \, B^{\,\mu\,\nu\,t} \,(:) \;,\; \tfrac{1}{4} \,(:)\, B^{\;t}_{\mu\,\nu} \, \widetilde{B}^{\,\mu\,\nu\,t} \tag{37}$$

We will use the following definitions relative to the rescaling function β

$$\begin{aligned}
& -\beta \,/\, g \;=\; X \, B \,(\, X \,) \;;\; B \,(\, X \,) \;=\; b_{\,0} \, A \,(\, X \,) \\
& B \,(\, X \,) \;\sim\; \textstyle\sum_{\,n=0}^{\,\infty} \, b_{\,n} \, X^{\,n} \;,\; A \,(\, X \,) \;\sim\; \textstyle\sum_{\,n=0}^{\,\infty} \, a_{\,n} \, X^{\,n} \\
& \kappa \;=\; g^{\,2} \,/\, (\, 16\,\pi^{\,2} \,) \text{ generic } \;\longrightarrow\; X \,,\, Y \\
& b_{\,0} \;=\; \tfrac{1}{3} \,(\, 33 \;-\; 2\, N_{\,fl} \,) \;,\; a_{\,0} \;=\; 1 \;,\; a_{\,n} \;=\; b_{\,n} \,/\, b_{\,0} \\
& b_{\,1} \;=\; \tfrac{2}{3} \,(\, 153 \;-\; 19\, N_{\,fl} \,) \\
& b_{\,2} \;=\; \tfrac{1}{54} \left(\, 77139 \;-\; 15099\, N_{\,fl} \;+\; 325\, N^{\,2}_{\,fl} \,\right) \\
& b_{\,3} \;\sim\; 29243 \;-\; 6946.3\, N_{\,fl} \;+\; 405.089\, N^{\,2}_{\,fl} \;+\; 1.49931\, N^{\,3}_{\,fl}
\end{aligned} \tag{38}$$

References in conjunction with this section ('1-1') are presented in five (partial) collections :

1 : (R) directly related to the two central anomalies

2 : (rBsquare) establishing the one renormalization group invariant quantity of dimension $\left[\, M^{\,4} \,\right]$

3 : (r-sp-1) a recent paper by Guido Altarelli and references cited therein

4 : (r-A2x) a selection of papers and textbooks for the entire realm of QCD

5 : (r-condx) : Condensation phenomena and field theory realizations

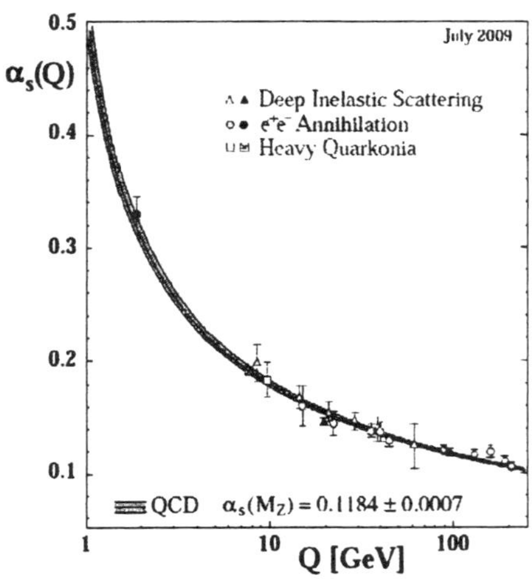

Fig A21 : $\alpha_{\,s} \,(\, Q \,) \;=\; 4\pi \; \kappa_{\;\overline{\mu}\,=\,Q}$
from ref. [A215-2009] .

1-2 – Gauge dependence of connection and curvature fields from local to large spae-time distances

We proced using the notions prepared in subsection '1-1-d' to define a connection using matrix valued form relative to $Lie\ (\mathcal{D}\ (\mathcal{G}\ )\ )$ (eqs. 11 - 15)

$$(\mathcal{W}_{\mu} (\mathcal{D}) (x))_{\alpha\beta} = W^{\,r}_{\mu} (x) (d_{r} (\mathcal{D}))_{\alpha\beta} \tag{39}$$

acting on a real or virtual [3] spin $\frac{1}{2}$ matter field , transforming according to the irredcible representation $\mathcal{D}\ (\ \mathcal{G}\ )$ denoted $f\ (\ x\ )$ understood as a column vector with the local (chargelike-) gauge group action

$$f (x) = \begin{pmatrix} f_{1} \\ \bullet \\ f_{\alpha} \\ \bullet \\ f_{dim\ \mathcal{D}} \end{pmatrix} (x) \ ; \ \Omega (x) \ : \ f (x) \rightarrow \Omega (x) f (x)$$

$$\text{with } \Omega (x) = (\Omega (x))_{\alpha\beta} \in \mathcal{D} \tag{40}$$

In eq. 40 matrix notation and multiplication is understood. The spin label pertaining to the field $f\ (\ x\ )$ is suppressed for simplicity , as it is left invariant by the local gauge transformations.
Thus the defining requirements for the connection fields take the form of characteristic $\mathcal{D}$ – covariant derivatives

$$\begin{array}{l} (D_{\mu} (\mathcal{W}_{\varrho} , \mathcal{D}))_{\alpha\beta} = \partial_{x\,\mu} \delta_{\alpha\beta} + (\mathcal{W}_{\mu} (\mathcal{D}) (x))_{\alpha\beta} \cdot f_{\beta} (x) \\ \qquad\qquad\qquad\qquad\qquad\qquad\qquad\qquad \downarrow \ \downarrow \\ \qquad\qquad\qquad\qquad\qquad\qquad\qquad\qquad D_{\mu} \cdot f \\ \text{for a suitable, maximal collection } (\bigcup \mathcal{D}) \end{array} \tag{41}$$

relating to the corresponding (locally) gauge transformed quantities

$$\begin{array}{l} \left(D_{\mu} (\mathcal{W}^{\,\Omega}_{\varrho}, \mathcal{D})\right)_{\alpha\beta} = \partial_{x\,\mu}\, \delta_{\alpha\beta} + \left(\mathcal{W}^{\,\Omega}_{\mu} (\mathcal{D}) (x)\right)_{\alpha\beta} \cdot f^{\,\Omega}_{\beta} (x) \\ \qquad\qquad\qquad\qquad\qquad\qquad\qquad\qquad \downarrow \qquad \downarrow \\ \qquad\qquad\qquad\qquad\qquad\qquad\qquad\qquad D^{\,\Omega}_{\mu} \cdot f^{\,\Omega} \\ f^{\,\Omega} (x) = \Omega (x) f (x) \longrightarrow \\ \hline D^{\,\Omega}_{\mu} f^{\,\Omega} (x) = \Omega (x) (D_{\mu} f) (x) \\ \text{for a suitable, maximal collection } (\bigcup \mathcal{D}) \end{array} \tag{42}$$

---

[3] **Here 'real' relates to quark and antiquark fields in the 3 & $\overline{3}$ color representations, while 'virtual' denotes spin $\frac{1}{2}$ fermions forming $\mathcal{D}$ & $\overline{\mathcal{D}}$ representations and removed in the limit of infinite mass from physically realized fields.**

From the last relation in eq. 42 the local gauge transfomation property follows , yet only in the environment where all field qantities are restricted to be classical

$$\mathcal{W}_{\varrho}^{\Omega} = \Omega(x) \left[\partial_{x \varrho} + \mathcal{W}_{\varrho} \right] \Omega^{-1}(x) \tag{43}$$

It is worth while to check the associative nature of the local mapping

$$\begin{aligned} \widehat{\Omega} : \mathcal{W}_{\varrho} \rightarrow \mathcal{W}_{\varrho}^{\Omega} ; \; & \widehat{\Omega}_2 \left(\widehat{\Omega}_1 \mathcal{W}_{\varrho} \right) = \widehat{\Omega}_3 \mathcal{W}_{\varrho} \\ \Omega_3(x) &= \Omega_2(x) \Omega_1(x) \\ \widehat{\Omega}_3 \mathcal{W}_{\varrho} &= \left[\begin{array}{c} \Omega_2 \left(\Omega_1 \mathcal{W}_{\varrho} \Omega_1^{-1} + \Omega_1 \partial_{\varrho} \Omega_1^{-1} \right) \Omega_2^{-1} + \\ \Omega_2 \partial_{\varrho} \Omega_2^{-1} \end{array} \right] \\ &= \begin{array}{c} (\Omega_2 \Omega_1) \mathcal{W}_{\varrho} (\Omega_2 \Omega_1)^{-1} + \\ (\Omega_2 \Omega_1) \partial_{\varrho} \left[(\Omega_2 \Omega_1)^{-1} \right] \quad \surd \end{array} \end{aligned} \tag{44}$$

Even though only first derivatives acting on *all* $\{ \Omega(x, \mathcal{D}) \}$ are required, in view of remark 2) in subsection '1-1-d' :

2) From the last relation in eq. 17 it may appear redundant to extend connections and curvatures to matrix valued form with respect to a wide collection of irreducible representations $\mathcal{D}(\mathcal{G})$. This however is tantamount to neglecting nontrivial global regularity conditions in the infrared .

– it is indeed not necessary – at this local stage of gauge dependence of the connection fields – to invoke the 'maximal collection' $\bigcup \mathcal{D}$, as shown below.

To this end we inspect the structure of both the homogeneous and inhomogeneous contribution to $\mathcal{W}_{\mu}^{\Omega}$ on the right hand side of eq. 43

$$\mathcal{W}_{\mu}^{\Omega} = W_{\mu}^{r\,\Omega} d_r = \left\{ \begin{array}{c} W_{\mu}^{r} \Omega d_r \Omega^{-1} + \\ + \Omega \partial_{\mu} \Omega^{-1} \end{array} \right\} \tag{45}$$

We trace back the unitary, irredcible representation $\mathcal{D}(\mathcal{G})$ to the – compact, simple Lie group $\mathcal{G}$ – representing group elements generically by the letters g, h, ...

$$\begin{array}{c} g \in \mathcal{G} \rightarrow \Omega(g) \in \mathcal{D} \text{ and inducing} \\ g \rightarrow g(x) \leftrightarrow \Omega \rightarrow \Omega(g(x)) \end{array} \tag{46}$$

Then independently of the local x dependence it follows

$$\begin{aligned} \Omega(g) d_r \Omega(g)^{-1} &= \left(Ad^{-1}(g) \right)_{rs} d_s \; ; \; d_t \in Lie(\mathcal{D}) \; \forall \mathcal{D} \\ \left(Ad^{-1}(g) \right)_{rs} &= \left(Ad((g)) \right)_{sr} \end{aligned} \tag{47}$$

from the defining properties of the (finite) adjoint representation $g \rightarrow Ad(g)$, which for compact Lie groups is by real, hence orthogonal matrices $Ad(g)$, as shown on the last line in eq. 47 .

Inserting the relations in eq. 47 into eq. 45 we obtain

$$\mathcal{W}_{\mu}^{\Omega} = W_{\mu}^{r\,\Omega}\, d_r = \left\{ \begin{array}{c} [\,(\,Ad\,(\,g\,)\,)_{rs}\, W_{\mu}^{s}\,]\, d_r + \\ + \Omega\,(\,g\,)\,\partial_{\mu}\,\Omega^{-1}\,(\,g\,) \end{array} \right\} \tag{48}$$

The inhomogeneous part $\Omega\,\partial_{\mu}\Omega^{-1}$ on the right hand side of eq. 48 can also be transferred back to coordinates on the gauge group $\mathcal{G}$ itself and equally involve the adjoint representation as defined on $\mathcal{G}$ factoring out the Lie algebra basis matrices d_t (Lie ($\mathcal{D}$) as only actually $\mathcal{D}$ dependent parts

$$\begin{array}{l} \Omega\,(\,g\,)\,\partial_{\mu}\,\Omega^{-1}\,(\,g\,)\,(\,x\,) = \\ \quad = \left[\,Ad\,(\,g\,)_{rs}\,\mathcal{J}_{t}^{s}\,(\,g^{-1}\,)\,\left(\,-\,\partial_{x\,\mu}\,g^{t}\,\right)\,\right]\,(\,x\,)\,d_r \\ \mathcal{J}_{t}^{s}\,(\,h\,) = \left.\dfrac{\partial\,\Pi^{r}\,(\,h\,;\,y\,)}{\partial_{y\,t}}\right|_{y\,=\,e} \\ g = g\,(\,x\,) \leftrightarrow h \longleftarrow (\,g\,(\,x\,)\,)^{-1} \end{array} \tag{49}$$

In eq. 49 we have introduced coordinates on $\mathcal{G}$, which gives rise to the coordinate transormation implementation of the group multiplication, denoted $g\,\bullet\,h$ in eq. 50 below – as references let me cite [10H-1941-1986, 11H-1951-1961]

$$\begin{array}{l} g \leftrightarrow \left(\,g^{1}\,,\,\cdots\,,\,g^{\,dim\,\mathcal{G}}\,\right) \text{ with} \\ e = \text{neutral element of } \mathcal{G} \leftrightarrow (\,0\,,\,\cdots\,,\,0\,) \\ (\,g\,\bullet\,h\,)^{r} = \Pi^{r}\,(\,g^{u}\,;\,h^{v}\,)\;;\quad {}^{r\,,\,u\,,\,v} = 1,\cdots,dim\,\mathcal{G} \end{array} \tag{50}$$

Substituting the expressions in eq. 49 in eq. 48 and factoring away the Lie ($\mathcal{D}$) basis matrices d_r yields the $\mathcal{D}$ independent, defining form of the connection. as it depends on local gauge transformations , choosing on $\mathcal{G}$ suitable coordinates, for which the adjoint representation is real , orthogonal

$$\left\{ \begin{array}{l} \Omega\,(\,x\,) = \Omega\,(\,g\,(\,x\,)\,)\;;\; Ad\,(\,g\,)^{r}_{\;s} \equiv Ad\,(\,g\,)_{rs} \\ \mathcal{J}_{t}^{s}\,(\,h\,) = \left.\dfrac{\partial\,\Pi^{r}\,(\,h\,;\,y\,)}{\partial_{y\,t}}\right|_{y\,=\,e} \end{array} \right\} \longrightarrow$$

$$\begin{array}{l} W_{\mu}^{r\,\Omega}\,(\,x\,) = \\ Ad\,(\,g\,(\,x\,)\,)^{r}_{\;s}\,\left[\,W_{\mu}^{s}\,(\,x\,) - \mathcal{J}_{t}^{s}\,(\,g^{-1}\,(\,x\,)\,)\,\left(\,\partial_{x\,\mu}\,g^{t}\,\right)\,(\,x\,)\,\right] \end{array} \tag{51}$$

Eq. 51 shows, that the defining and local regularity requirements pertaining to the connection or gauge potentials do not need any extension to include a maximal set $\bigcup \mathcal{D} (\mathcal{G})$ which can relate to 'real' and 'virtual' matter fields, as specified in eqs. 41 , 42 . This ends the general part of section '1-2', the next subsections shall extend to gauge covariant local and minimally bi-local operators , derived from the connection .

1-2-a – Curvature as local covariant field strengths and the Bianchi identity

We stay with local properties in this subsection and (yet) continue to represent the connection – as well as its associated 1-form (eqs. 14 , 15) – in matrix valued form relative to *Lie* ($\mathcal{D} (\mathcal{G})$) (eqs. 11 - 15) as defined in eq. 39 repeated below

$$\begin{array}{lcll} (\mathcal{W}_{\mu} (\mathcal{D}) (x))_{\alpha\beta} & = & W_{\mu}^{\;r} (x) (d_{r} (\mathcal{D}))_{\alpha\beta} \longrightarrow & \mathcal{W}_{\mu} \\ & & & \text{for short} \\ \mathcal{W}^{(1)} (z , \mathcal{D}) \big|_{\alpha\beta} & = & d z^{\mu} \; \mathcal{W}_{\mu} (z , \mathcal{D}) \, | _{\alpha\beta} \longrightarrow & \mathcal{W}^{(1)} \\ & & & \text{for short} \end{array}$$

$$\mathcal{W}_{\mu} (z , \mathcal{D}) \equiv \mathcal{W}_{\mu} (\mathcal{D}) (z) \tag{52}$$

We also repeat the definitions in eqs. 16 , 17 with $\mathcal{B}_{\mu\nu}$ ($\equiv \mathcal{W}_{\mu\nu}$) selected from eq. 18

$$\begin{array}{lcl} \mathcal{B}^{(2)} (z , \mathcal{D}) \, \big|_{\alpha\beta} & = & \frac{1}{2} d z^{\mu} \wedge d z^{\mu} \; \mathcal{B}_{\mu\nu} (z , \mathcal{D}) \, | _{\alpha\beta} \\ & & \longrightarrow \mathcal{B}^{(2)} \\ \mathcal{B}_{\mu\nu} (z, \mathcal{D}) |_{\alpha\beta} & = & \left\{ \begin{array}{c} \partial_{\mu} \mathcal{W}_{\nu} (z, \mathcal{D}) - \partial_{\nu} \mathcal{W}_{\mu} (z, \mathcal{D}) + \\ + [\mathcal{W}_{\mu} (z, \mathcal{D}) , \mathcal{W}_{\nu} (z, \mathcal{D})] \end{array} \right\}_{\alpha\beta} \\ & & \longrightarrow \mathcal{B}_{\mu\nu} \\ & = & B_{\mu\nu}^{\;r} (z) (d_{r})_{\alpha\beta \; Lie \, (\mathcal{D})} \end{array}$$

$$B_{\mu\nu}^{\;r} = \partial_{\mu} W_{\nu}^{\;r} - \partial_{\nu} W_{\mu}^{\;r} + f_{rst} W_{\mu}^{\;s} W_{\nu}^{\;t} \; ; \; \text{independent of } \mathcal{D} \tag{53}$$

The *Lie* $\mathcal{D}$ – valued curvature 2-form $\mathcal{B}^{(2)}$ (eq. 53) takes a concise form, whence expressed in terms of the connection 1-form and the derivative d acting according to the rules of (matrix valued extension) of de Rham cohomology [2-1-1931]

$$\begin{array}{l} \mathcal{W}^{(p)} (z , \mathcal{D}) = \frac{1}{p!} d z^{\mu_1} \wedge \cdots \wedge z^{\mu_p} \mathcal{W}_{[\mu_1 \cdots \mu_p]} (z , \mathcal{D}) \\ \text{d} : \mathcal{W}^{(p+1)} = \text{d} \mathcal{W}^{(p)} \\ \quad = d z^{\mu_0} \wedge d z^{\mu_1} \wedge \cdots \wedge z^{\mu_p} \frac{1}{p!} \partial_{z \, \mu_0} \mathcal{W}_{[\mu_1 \cdots \mu_p]} (z , \mathcal{D}) \\ \text{with } \text{d}^{2} \mathcal{W}^{(p)} = \text{d} \left(\text{d} \mathcal{W}^{(p)} \right) \equiv 0 \end{array} \tag{54}$$

The expressions $_{[\mu_1 \cdots \mu_p]}$ shall indicate the completely antisymmetric tensor structure with respect to the indices in brackets. The sought expression for $\mathcal{B}^{(2)}$ thus becomes

$$\mathcal{B}^{(2)} = \mathrm{d}\,\mathcal{W}^{(1)} + \left(\mathcal{W}^{(1)} \right)^2 \tag{55}$$

The action of a local gauge transformation $\Omega (z , \mathcal{D}))$ (eq. 45) in terms of 1-forms is

$$\mathcal{W}^{(1)\,\Omega} = \Omega\, \mathcal{W}^{(1)}\, \Omega^{-1} + \Omega\, \mathrm{d}\, \Omega^{-1} \tag{56}$$

Substituting eq. 56 in eq. 55 a series of terms forming $\mathcal{B}^{(2)\,\Omega}$ is obtained

$$\begin{aligned} &\mathcal{B}^{(2)\,\Omega} = \\ &(\mathrm{d}\,\Omega)\, \mathcal{W}^{(1)}\, \Omega^{-1} + \Omega \left(\mathrm{d}\, \mathcal{W}^{(1)} \right) \Omega^{-1} - \Omega\, \mathcal{W}^{(1)} \left(\mathrm{d}\, \Omega^{-1} \right) + \\ &\qquad + \mathrm{d} \left(\Omega\, \mathrm{d}\, \Omega^{-1} \right) \\ &\qquad + \begin{pmatrix} \Omega \left(\mathrm{d}\, \Omega^{-1} \right) + \\ + \Omega\, \mathcal{W}^{(1)}\, \Omega^{-1} \end{pmatrix} \begin{pmatrix} \Omega \left(\mathrm{d}\, \Omega^{-1} \right) \\ + \Omega\, \mathcal{W}^{(1)}\, \Omega^{-1} \end{pmatrix} \end{aligned} \tag{57}$$

We assemble terms involving equal powers (2,1,0) of $\mathcal{W}^{(1)}$ on the right hand side pf eq. 57

$$\begin{aligned} &\mathcal{B}^{(2)\,\Omega} = \\ &\qquad \Omega \left(\mathcal{W}^{(1)} \right)^2 \Omega^{-1} \\ &+ (\mathrm{d}\,\Omega)\, \mathcal{W}^{(1)}\, \Omega^{-1} + \Omega \left(\mathrm{d}\, \mathcal{W}^{(1)} \right) \Omega^{-1} - \Omega\, \mathcal{W}^{(1)} \left(\mathrm{d}\, \Omega^{-1} \right) \\ &+ \Omega \left(\mathrm{d}\, \Omega^{-1} \right) \Omega\, \mathcal{W}^{(1)}\, \Omega^{-1} + \Omega\, \mathcal{W}^{(1)}\, \Omega^{-1}\, \Omega \left(\mathrm{d}\, \Omega^{-1} \right) \\ &+ \mathrm{d} \left(\Omega\, \mathrm{d}\, \Omega^{-1} \right) + \Omega \left(\mathrm{d}\, \Omega^{-1} \right) \Omega \left(\mathrm{d}\, \Omega^{-1} \right) \end{aligned} \tag{58}$$

The last 2 (out of the 8 terms from up left to down right) in eq. 58 vanish identically. Lets rearrange the remaining 6

$$\mathcal{B}^{(2)\,\Omega} = \left[\begin{array}{c} \Omega \left(\mathrm{d}\, \mathcal{W}^{(1)} + \left(\mathcal{W}^{(1)} \right)^2 \right) \Omega^{-1} \\ + (\mathrm{d}\,\Omega)\, \mathcal{W}^{(1)}\, \Omega^{-1} - \Omega\, \mathcal{W}^{(1)} \left(\mathrm{d}\, \Omega^{-1} \right) \\ - (\mathrm{d}\,\Omega)\, \mathcal{W}^{(1)}\, \Omega^{-1} + \Omega\, \mathcal{W}^{(1)} \left(\mathrm{d}\, \Omega^{-1} \right) \end{array} \right] \tag{59}$$

of which the last 4 also cancel identically . This leaves the (well known) local gauge covariant transformation property pertaining to field strengths

$$\begin{aligned} &\mathcal{B}^{(2)\,\Omega} (z , \mathcal{D}) = \Omega (z)\, \mathcal{B}^{(2)} (z , \mathcal{D})\, \Omega^{-1} (z) \longrightarrow \\ &B_{\mu\nu}^{\;r\,\Omega} (z)\, d_r = B_{\mu\nu}^{\;r} (z) \left(\Omega (z)\, d_r\, \Omega^{-1} (z) \right) \longrightarrow \\ &B_{\mu\nu}^{\;r\,\Omega \leftarrow g} (z) = Ad \left(g \; (z) \right)_{rs} B_{\mu\nu}^{\;s} (z) \; ; \text{ independent of } \mathcal{D} \end{aligned} \tag{60}$$

The last relation in eq. 61 follows from eq. 46 and eq. 47 repeated below

$$\Omega (g) d_r \Omega (g)^{-1} = (Ad^{-1} (g))_{rs} d_s \; ; \; d_t \in Lie (\mathcal{D}) \; \forall \mathcal{D}$$
$$(Ad^{-1} (g))_{rs} = (Ad ((g))_{sr} \tag{61}$$

The Bianchi identity, only local consequence of a system of local connections and field strengths

In a system as specified in the (sub-) title above one identity follows, generalising the homogeneous Maxwell equations in QED

$$\mathcal{B}^{(3)} = \mathrm{d}\,\mathcal{B}^{(2)} + [\mathcal{W}^{(1)} , \mathcal{B}^{(2)}] \,|_{\mathcal{D}} \equiv 0$$

$$= \begin{bmatrix} \mathrm{d}\,\mathrm{d}\,\mathcal{W}^{(1)} + \\ [(\mathrm{d}\,\mathcal{W}^{(1)}) , \mathcal{W}^{(1)}] \\ + [\mathcal{W}^{(1)} , (\mathrm{d}\,\mathcal{W}^{(1)})] \\ + \left[\mathcal{W}^{(1)} , (\mathcal{W}^{(1)})^2 \right] \end{bmatrix} \tag{62}$$

$$\longleftrightarrow \mathcal{B}^{(2)} (\mathcal{D}) \rightarrow \mathcal{W}^{(2)} = \mathrm{d}\,\mathcal{W}^{(1)} + (\mathcal{W}^{(1)})^2$$

The (Bianchi -) identity in eq. 62 holds independently of whether other matter fields, e.g. q , $\overline{q}$ are included with finite masses or not . However the identity only holds if $\mathcal{B}^{(2)} (\mathcal{D})$ actually derives from a connection , and furthermore if allowed connections do satisfy regularity conditions, here those appropriate for complete such, i.e. admitting a suitable, maximal collection $(\bigcup \mathcal{D})$ (eq. 42) .
This ends subsection '1-2-a' and we turn towards nonlocal properties of complete connections in the next subsection '1-2-b'.

1-2-b – QCD $\mathcal{D} (\mathcal{G})$ – strings

building blocks of bi-local gauge covariant parallel transports $x \overset{C}{\leftarrow} y$, relative to the associated irreducible representations $\mathcal{D} \leftrightarrow x$, $\overline{\mathcal{D}} \leftrightarrow y$ of the gauge group $\mathcal{G}$

Reparametrization of straight line using a tangent vector $\underline{v} = (\underline{v}^{\mu})$

We want to use a variant of the parametrizations of the parallel transport operators $U \left(x \overset{C}{\leftarrow} y \right) = P \exp \left(- \int_C \mathcal{W}^{(1)} (\overline{x} , \mathcal{D}) \right)$ introduced in eqs. 9 , 14 and 15

$$U \left(x \stackrel{C}{\leftarrow} y \right) = P \exp \left(- \int_C \mathcal{W}^{(1)} (\overline{x}) \right)$$
$$\rightarrow \ (U (x , C , y))_{\alpha\beta} \in \mathcal{D} (\mathcal{G})$$
$$C = C \{ \overline{x} \} \ : \ \overline{x} = \overline{x} (s) \ ; \ \tau \geq s \geq 0$$
$$s \ : \ \text{path parameter} \ ; \ \overline{x} (s = \tau) = x (\tau) \ ; \ \overline{x} (s = 0) = y \tag{63}$$

The word reparametrization is used to render τ the straight line path parameter between the limit points x , y

$$x^{\mu} (\tau) = \tau \underline{v}^{\mu} + y^{\mu} \ ; \ \underline{v}^{\mu} = x^{\mu} - y^{\mu} = \text{independent of } \tau$$
$$\{ x = x (\tau = 1) ; \ y = x (\tau = 0) \} \tag{64}$$

With the definitions given in eqs. 63 and 64 a first order differential equation with respect to τ follows for

$$U \left(x (\tau) \stackrel{C}{\leftarrow} y \right) \text{ and } \widetilde{U} \left(x (\tau) \stackrel{C}{\leftarrow} y \right)$$

defined in eq. 65 below

$$\frac{d}{d\,\tau} U \left(x(\tau) \stackrel{C}{\leftarrow} y \right) = - \underline{v}^{\mu} \mathcal{W}_{\mu} \left(x(\tau), \mathcal{D} \right) U \left(x (\tau) \stackrel{C}{\leftarrow} y \right)$$
$$\frac{d}{d\,\tau} U \left(x(\tau) \stackrel{C}{\leftarrow} y \right) = \underline{v}^{\mu} \partial_{\xi\,\mu} \widetilde{U} \left(x(\tau) + \xi \right) \Big|_{\xi = 0} \longrightarrow$$
$$\longrightarrow \ \underline{v}^{\mu} \left(\partial_{x\,\mu} + \mathcal{W}_{\mu} (x) \right) \widetilde{U} (x) \Big|_{x = x(\tau)} = 0$$
$$\widetilde{U} = \widetilde{U} \left(x ; \{ \mathcal{W}^{(1)} (\mathcal{D}) \} , y \right) \ ; \ \widetilde{U} \left(y ; \{ \mathcal{W}^{(1)} (\mathcal{D}) \} , y \right) = \P \tag{65}$$

The notation of the arguments x ; $\{ \mathcal{W}^{(1)} (\mathcal{D}) \}$, y of the parallel transport operators $\widetilde{U}$ in the last line of eq. 65 shall indicate that $x \rightarrow x (\tau)$ is understood as space time variable, whereas the entire connection $\{ \mathcal{W}^{(1)} (\mathcal{D}) \}$ as well as the base point y shall be understood in a parametric sense, and specifying the initial condition for $\tau = 0$.
The relations in the last two lines of eq. 65 show the intrinsic connection between boundary - and integrability conditions pertaining to complete connections and the associated determination of an axial gauge.
The 'variable-' point x ('Aufpunkt' in german terminology) and base point y represented by the endpoints along the straight line of parallel transport through the quantity

$$\widetilde{U} \left(x ; \{ \mathcal{W}^{(1)} (\mathcal{D}) \} , y \right) \tag{66}$$

separate the adjoint representation of the local limit into the product of two distinct local representations of the gauge group , one each at x and y , transforming under relatively complex conjugate but singly arbitrary irreducible rerpresentations $\mathcal{D} \leftrightarrow x$, $\overline{\mathcal{D}} \leftrightarrow y$ of the gauge group $\mathcal{G}$ – subject to restrictions from continuity of local gauge group representations – for complete connections.
This gives rise to the (x , y) associated bi-local transformation properties under an – operator valued – local family of gauge transformations discssed in the following .
In the following we drop the $\widetilde{U} \longrightarrow U$ symbol explained in eqs. 65 - 66 for simplicity and repeat the differential 'parallel-transport'- equation (eq. 65)

$$\begin{aligned} & \underline{\mathrm{v}}^{\mu} \left(\partial_{x\,\mu} + \mathcal{W}_{\mu} (x) \right) U (x) \left.\right|_{x=x(\tau)} = 0 \\ & U (x) \longleftarrow U \left(x \, ; \, \{ \mathcal{W}^{(1)} (\mathcal{D}) \} \, , y \right) \\ & U \left(y \, ; \, \{ \mathcal{W}^{(1)} (\mathcal{D}) \} \, , y \right) = ¶ \end{aligned} \tag{67}$$

Next we consider local gauge transformations $\{ \Omega \}$, substitute the identity $\Omega^{-1} (x) \, \Omega (x) = ¶$ and use the chain rule for partial derivatives

$$\begin{aligned} & \underline{\mathrm{v}}^{\mu} \left(\partial_{x\,\mu} + \mathcal{W}_{\mu} (x) \right) \left(\Omega^{-1} (x) \, \Omega (x) \right) U (x) \left.\right|_{x=x(\tau)} = 0 \\ & \underline{\mathrm{v}}^{\mu} \, \Omega^{-1} (x) \left(\begin{array}{c} \Omega (x) \left[\begin{array}{c} \partial_{x\,\mu} + \\ \mathcal{W}_{\mu} (x) \end{array} \right] \Omega^{-1} (x) \\ + \partial_{x\,\mu} \end{array} \right) \Omega (x) \, U (x) \Bigg|_{x=x(\tau)} \\ & = 0 \end{aligned} \tag{68}$$

By eq. 43 the quantity in [ ] brackets in the second relation of eq. 68 represents the gauge transformed connection

$$\Omega (x) \left[\partial_{x\,\mu} + \mathcal{W}_{\mu} (x) \right] \Omega^{-1} (x) = \mathcal{W}_{\mu}^{\,\Omega} (x) \tag{69}$$

And so we derive the bi-locally transformed parallel transport operators, from the associated differential equation and initial conditions (eqs. 67 , 68)

$$\underline{v}^{\mu} \left(\partial_{x\,\mu} + \mathcal{W}^{\Omega}_{\mu} (x) \right) U \Big(x \, ; \, \{ \mathcal{W}^{(1)} (\mathcal{D}) \}^{\Omega} , y \Big) \Big|_{x = x(\tau)} = 0$$

$$U \left(y \, ; \left\{ \{ \mathcal{W}^{(1)} (\mathcal{D}) \}^{\Omega} \right\} , y \right) = ¶ \longrightarrow$$

$$U \left(x \, ; \, \{ \mathcal{W}^{(1)} (\mathcal{D}) \}^{\Omega} , y \right) =$$

$$\Omega (x) \, U \left(x \, ; \, \{ \mathcal{W}^{(1)} (\mathcal{D}) \} , y \right) \Omega^{-1} (y)$$

---

for a suitable, maximal collection $\bigcup \mathcal{D} (\mathcal{G})$

$$U \left(x ; \left\{ \{ \mathcal{W}^{(1)} (\mathcal{D}) \}^{\Omega} \right\} , y \right) = P \exp \left(- \int_{C} \mathcal{W}^{(1)\,\Omega} (x) \right) \tag{70}$$

We emphasize that existence and uniqueness of the operator valued families $U (\, . \,)$, necessary for the derivation of the transformation properties , displayed in eqs. 63 and 70 , constitute regularity conditions imposed on – so characterized – complete connections . A pertinent discussion of less restricted connections can be found in ref. [2-3-1991] .

1-2-c - Bare Lagrangean density, equations of motion and dim $[\, M^{4} \,]$ equations – QCD analog of QED dipole equations – for field strengths in unconstrained gauges

The bare local Lagrangean density is formed by the bare field strengths

$$\mathcal{L} (x) =$$

$$\frac{1}{4 \, T_{2} (\mathcal{D}) \, g^{2}} \, tr_{\mathcal{D}} \left[\mathcal{B}_{\mu\nu} \mathcal{B}^{\mu\nu} \right] (\mathcal{D}) (x) + \mathcal{L}_{\{q\}} (x)$$

$$\mathcal{L}_{\{q\}} (x) =$$

$$\sum_{q-fl} \overline{q}^{\,\dot{c}'} \left\{ \frac{i}{2} \gamma^{\mu} \begin{bmatrix} \left(\overrightarrow{D}_{\mu} (3) \right)_{c'\dot{c}} - \\ - \left(\overleftarrow{D}_{\mu} (\overline{3}) \right)_{\dot{c}c'} \end{bmatrix} - m_{q} \, \delta_{c'\dot{c}} \right\} q^{\,c} (x) \tag{71}$$

with the normalization conventions

$$\begin{aligned} & - tr_{\mathcal{D}} \, d^{\,p} d^{\,q} = \delta^{\,pq} \, T_{2} (\mathcal{D}) \\ & - \textstyle\sum_{r} \left(d^{\,r} \right)^{2} = C_{2} (\mathcal{D}) \, ¶_{\dim \mathcal{D} \times \dim \mathcal{D}} \longrightarrow \\ & T_{2} (\mathcal{D}) = \frac{\dim (\mathcal{D})}{\dim (\mathcal{G})} \, C_{2} (\mathcal{D}) \; ; \text{ for } \left. \begin{array}{l} \mathcal{G} = SU3 \\ \mathcal{D} = 3 \text{ or } \overline{3} \end{array} \right\} : \end{aligned} \tag{72}$$

$$\dim \mathcal{D} = 3 \, , \; \dim \mathcal{G} = 8 \; ; \; T_{2} (\mathcal{D}) = \tfrac{1}{2} \; ; \; C_{2} (\mathcal{D}) = \tfrac{4}{3}$$

The Euler equations from the lagrangean density in eq. 71 become

$$\partial_{\varrho} \left\{ \frac{\delta \mathcal{L}_{g-b}}{\delta B^{\,r}_{\mu\nu}} \frac{\delta B^{\,r}_{\mu\nu}}{\delta (\partial_{\varrho} W^{\,s}_{\sigma})} \right\} - \frac{\delta \mathcal{L}_{g-b}}{\delta B^{\,r}_{\mu\nu}} \frac{\delta B^{\,r}_{\mu\nu}}{\delta W^{\,s}_{\sigma}} = \frac{\delta \mathcal{L}_{\{q\}}}{\delta W^{\,s}_{\sigma}}$$

$$\frac{\delta \mathcal{L}_{\{q\}}}{\delta W^{\,s}_{\sigma}} = \sum_{q-fl} \overline{q}^{\,\dot{c}'} \left\{ \gamma^{\sigma} \left(\tfrac{1}{2} \lambda^{s} \right)_{c'\dot{c}} \right\} q^{\,c} = (j^{\,\sigma s})_{\{q\}} \tag{73}$$

In eq. 73 $\mathcal{L}_{g-b}$ denotes the gauge boson part of $\mathcal{L}$ and $j^{\,\sigma s}$ the color octet current due to quark and antiquark flavors .
Eq. 73 becomes

$$(D_{\varrho} (ad))_{sr} \left\{ \frac{1}{g^{2}} B^{\,\sigma\varrho\, r} \right\} = (j^{\,\sigma s})_{\{q\}}$$

$$(D_{\varrho} (ad))_{sr} = \partial_{\varrho} \delta_{sr} + W^{\,t}_{\varrho} (ad_{t})_{sr} \;;\; (ad_{t})_{sr} = f_{str}$$

$$(j^{\,\sigma s})_{\{q\}} = \sum_{q-fl} \overline{q}^{\,\dot{c}'} \left\{ \gamma^{\sigma} \left(\tfrac{1}{2} \lambda^{s} \right)_{c'\dot{c}} \right\} q^{\,c} \tag{74}$$

The equations of motion (eq. 74) as well as the Bianchi identities (eq. 62) are of engineering dimension $[M^{3}]$.
The latter enter the derivation of the dimension $[M^{4}]$ equations, for which purpose we introduce the matrix notations as adapted to the adjoint representation and convert Lorentz indices freely between covariant and contravariant ones

$$D_{\varrho} = \partial_{\varrho} + \mathcal{W}_{\varrho} \,;\, \mathcal{W}_{\varrho} = W^{\,r}_{\varrho}\, ad_{r} \,;\, D^{\varrho} \left\{ \frac{1}{g^{2}} B_{\sigma\varrho} \right\}^{s} = j^{\,s}_{\sigma} \tag{75}$$

Then we form the antisymmetric covariant (electric- and magnetic-) dipole current densities and suppress the adjoint component ^{s} and abbreviate the inverse square bare copling constant by $G = g^{-2}$

$$G = g^{-2}$$
$$D_{\tau} D_{\varrho} \{ G B^{\,\varrho}_{\sigma} \} - D_{\sigma} D_{\varrho} \{ G B^{\,\varrho}_{\tau} \} = D_{\tau} j_{\sigma} - D_{\sigma} j_{\tau} \tag{76}$$

Next we move the covariant derivatives D_{τ} , D_{σ} to the right using the identity

$$D_{\tau} D_{\varrho} = [D_{\tau} , D_{\varrho}] + D_{\varrho} D_{\tau} \tag{77}$$

Eq. 76 thus takes the form

$$\{ [D_{\tau} , D_{\varrho}] + D_{\varrho} D_{\tau} \} \{ G B^{\,\varrho}_{\sigma} \} - (\tau \leftrightarrow \sigma) = D_{\tau} j_{\sigma} - D_{\sigma} j_{\tau} \tag{78}$$

The commutators of covariant derivatives reduce to the adjoint field strength matrix valued form

$$[D_{\tau} , D_{\varrho}] = \mathcal{B}_{\tau \varrho} = B^{\,t}_{\tau \varrho} \, ad_{\,t} \; ; \; \tau \, \varrho \rightarrow \sigma \, \varrho \tag{79}$$

which yields , substituded into eq. 78

$$\begin{bmatrix} D^{\varrho} \left(D_{\tau} B_{\sigma \varrho} - D_{\sigma} B_{\tau \varrho} \right) + \\ + \mathcal{B}_{\tau \varrho} B_{\sigma}^{\;\varrho} - \mathcal{B}_{\sigma \varrho} B_{\tau}^{\;\varrho} \end{bmatrix} = g^{2} \left(D_{\tau} j_{\sigma} - D_{\sigma} j_{\tau} \right) \tag{80}$$

At this stage whence factoring out G beyond partial derivatives in eq. 80 we assume that G is not space-time dependent, and come back at a later stage to consider an arbitrary space-time dependent extension

$$\partial_{\mu} G = 0 \text{ with the extension } G \rightarrow \widetilde{G} (x) \text{ with } \lim_{x \rightarrow \infty} \widetilde{G} (x) = G \tag{81}$$

as an external source with appropriate bondary conditions for $x \rightarrow \infty$. The qantity in brackets in the first line of the left hand side of eq. 80 can be transformed , using the Bianchi identity for the covariant derivatives of field strengths

$$\begin{array}{l} D_{\tau} B_{\sigma \varrho} - D_{\sigma} B_{\tau \varrho} = D_{\tau} B_{\sigma \varrho} + D_{\sigma} B_{\varrho \tau} = - D_{\varrho} B_{\tau \sigma} \longrightarrow \\ \text{Bianchi identity : } D_{\tau} B_{\sigma \varrho} + D_{\sigma} B_{\varrho \tau} + D_{\varrho} B_{\tau \sigma} = 0 \end{array} \tag{82}$$

The dipole density second order differential equation thus can be brought to the form

$$\begin{bmatrix} D^{\varrho} D_{\varrho} B_{\sigma \tau} |^{\,s} + \\ + \mathcal{B}_{\tau \varrho} B_{\sigma}^{\;\varrho} |^{\,s} - \mathcal{B}_{\sigma \varrho} B_{\tau}^{\;\varrho} |^{\,s} \end{bmatrix} = g^{2} \left(D_{\tau} j_{\sigma} - D_{\sigma} j_{\tau} \right) |^{\,s} \tag{83}$$

We summarize the equations of motion – eqs. 73 , 75 – and the chromo-dipole-density $dim \left[M^{4} \right]$ derived equations – eq. 83 – below

$$\partial_{\varrho} \left\{ \frac{\delta \mathcal{L}}{\delta B^{r}_{\mu\nu}} \frac{\delta B^{r}_{\mu\nu}}{\delta \left(\partial_{\varrho} W^{s}_{\sigma} \right)} \right\} - \frac{\delta \mathcal{L}}{\delta B^{r}_{\mu\nu}} \frac{\delta B^{r}_{\mu\nu}}{\delta W^{s}_{\sigma}} = \frac{\delta \mathcal{L}_{\{q\}}}{\delta W^{s}_{\sigma}}$$

$$\frac{\delta \mathcal{L}_{\{q\}}}{\delta W^{s}_{\sigma}} = \sum_{q-fl} \overline{q}^{\dot{c}'} \left\{ \gamma^{\sigma} \left(\tfrac{1}{2} \lambda^{s} \right)_{c'\dot{c}} \right\} q^{c} = \left(j^{\sigma s} \right)_{\{q\}}$$

$$D_{\varrho} = \partial_{\varrho} + \mathcal{W}_{\varrho} \; ; \; \mathcal{W}_{\varrho} = W^{r}_{\varrho} \, ad_{r}$$

$$D^{\varrho} \left\{ B_{\sigma\varrho} \right\}^{s} = g^{2} j^{s}_{\sigma}$$

$$\left[\begin{array}{c} D^{\varrho} D_{\varrho} B_{\sigma\tau} + \\ + \mathcal{B}_{\tau\varrho} B^{\varrho}_{\sigma} - \mathcal{B}_{\sigma\varrho} B^{\varrho}_{\tau} \end{array} \right] = g^{2} \left(D_{\tau} j_{\sigma} - D_{\sigma} j_{\tau} \right) \Big|^{s}$$

(84)

2 – Concluding remarks , outlook

1) The present outline is an attempt to convey – in concise form – the structure and nature of 'infrared instability of QCD' , whence the restriction is imposed to colored local base fields – matter fields of $N_{fl} \leq 6$ color triplet , spin $\frac{1}{2}$, quarks and color antitriplet antiquarks and gauge fields of potentials or connection fields and color octet field strengths or curvature fields .

2) The entry point is the derivation of the two central QCD anomalies amenable to local relations among gauge invariant composite fields, covered in section '1-1' .

3) Section '1-2' is devoted to local regalurity conditions : the Bianchi identity, best known as the homogeneous Maxwell eqations, yet different and 'just gauge covariant' for nonabelian gauge fields , to bilocal $\mathcal{D}$ - strings , which decouple the representations of local gauge transformations at the two space-time endpoints with respect to the irreducible $SU3_{c}$ representations $\mathcal{D}$, $\overline{\mathcal{D}}$ from the local limit, where only the adjoint representation for field strengths prevails and also to the derivation of the chromo-dipole local and covariant field equations , a feature which is new – for me – despite previous attempts.

4) It is the nature of only partial understanding of the associated nonperturbative dynamics resulting in a 'choice' of the mentioned regularity conditions , that controversy concerning this 'choice' not only exists, but is – hopefully – a seed to increased understanding and insight.

References

[1-1-1969] S. L. Adler, 'Axial vector vertex in spinor electrodynamics', Phys.Rev.177 (1969) 2426-2438 .

J Bell and R. Jackiw, 'A PCAC puzzle: $\pi^0 \rightarrow \gamma\gamma$ in the sigma model', Nuovo Cim.A60 (1969) 47-61 .

[1-2-1991] P. Minkowski, 'Proton spin and quark spin', in Proceedings of the "Workshop on Effective Field Theories of the Standard Model", Dobogókõ, Hungary August 22 - 26, 1991 ; U. Meissner edt., World Scientific Singapore 1992 , BUTP-91-2, Aug 1991. 31pp. .

[1-3-1976] P. Minkowski, 'On the anomalous divergence of the dilatation current in gauge theories', Bern preprint 1976,
URL : http://www.mink.itp.unibe.ch .

[2-1-1931] G. de Rham, 'Sur l'analysis situs des variétés à n dimensions', J. Math. Pures Appl. (9) 10 (1931) 115-200 .

[2-2-1985] J. Manes, R. Stora and B. Zumino, 'Algebraic Study of Chiral Anomalies' LBL-19138, UCB-PTH-85/7, LAPP-TH-137, Feb 1985. 32pp., Commun.Math.Phys.102 (1985) 157 , scanned version from KEK library .

[2-3-1991] A. S. Kronfeld and U.-J. Wiese, '$SU(N)$ gauge theories with C-periodic boundary conditions – (I). Topological structure' , Nucl. Physics B357 (1991) 521-533 .

[19-1975] H. Kluberg-Stern and J. B. Zuber, 'Ward Identities and Some Clues to the Renormalization of Gauge Invariant Operators', Phys.Rev. D12 (1975) 467 ,
'Renormalization of Nonabelian Gauge Theories in a Background Field Gauge. 1. Green Functions', Phys.Rev. D12 (1975) 482 and
'Renormalization of Nonabelian Gauge Theories in a Background Field Gauge. 2. Gauge Invariant Operators', Phys.Rev.D12 (1975) 3159-3180

Historical and textbook references to 'Continous transformation groups and differential geometry'

[1H-1893] S. Lie and G. Scheffers, 'Vorlesungen über continuierliche Gruppen, mit geometrischen und anderen Anwendungen', Leipzig 1893 .

[2H-1890] W. Killing, 'Die Zusammensetzung der stetigen endlichen Transformationsgruppen', Mathematische Annalen 31 (1888) 252-290; 33 (1888) 1-48; 34 (1889) 57-122; 36 (1890) 161-189 .

[3H-1894] É. Cartan, 'Sur la structure des groupes de transformations finis et continues', Thèse, Paris 1884 , IIème édition 1933 .

[4H-1897] F. Klein and A. Sommerfeld, 'Über die Theorie des Kreisels', 4 volumes, Teubner Verlag 1897 .

[5H-1903] L. Bianchi, 'Lezioni sulla teoria dei gruppi continui finiti di trasformazioni', Pisa 1903 , IIa edizione 1918 .

[6H-1927] F. Peter und H. Weyl, 'Die Vollständigkeit der primitiven Darstellungen einer geschlossenen kontinuierlichen Gruppe' , Ann. Math. 97 (1927) 737-755 .

[7H-1933] A. Haar, 'Der Massbegriff in der Theorie der kontinuierlichen Gruppen', Ann. Math. (2) 34 (1933) 147-169 .

[8H-1933] H. Weyl, 'The structure and representation of continous groups', I and II, Princeton 1933 and 1934-35.

[9H-1933] L.P. Eisenhart, 'Continuous groups of tranformations', Princeton 1933.

[10H-1941-1986] L. S. Pontryagin, 'Topological Groups', Gordon and Breach Sci. Publ., New York/London/Paris 1986 , first edition in russian : 1941

[11H-1951-1961] G. Racah, 'Group theory and spectroscopy', CERN Yellow report 61-8, 1961, URL : http://doc.cern.ch/cernrep/1961/1961-008/1961-008.html , reprinted from lectures dlivered (by the author) at the Institute for Advanced Study in Princeton, 1951 ; and references therein quoted below.

[12H-1953] G. W. Mackey, 'Induced Representations of Locally Compact Groups II. The Frobenius Reciprocity Theorem', The Annals of Mathematics, Second Series, Vol. 58, No. 2 (Sep., 1953) 193-221 .

[13H-1962] S. Helgason, 'Differential geometry and Symmetric Spaces', Academic Press New York and London 1962 .

[14H-1963] S. Kobayashi and K. Nomizu, 'Foundations of Differential Geometry', 2 volumes, John Wiley and Sons, Inc., Hoboken, NJ 1963, Wiley Classics Library 2009 .

End of historical and textbook references to
'Continous transformation groups and differential geometry'

[1-sp-2011] G. Altarelli, 'The Early Days of QCD (as seen from Rome)', Comments: 8 pages, 2 figures. Talk in honour of Mario Greco presented at the Rencontres de la Vallee d'Aoste 2011, La Thuile, Italy, arXiv:1106.3189 [hep-ph] .

[A21-197(3)4] D. Politzer, ' Reliable Perturbative Results for Strong Interactions?', Phys. Rev. Lett. 30 (1973) 1346 ,
H. Georgi and D. Politzer, 'Electroproduction scaling in an asymptotically free theory of strong interactions', Phys. Rev. D9 (1974) 416 .

[A22-197(3)4] D.J. Gross and F. Wilczek,
'Asymptotically free gauge theories. 1', Phys.Rev.D8 (1973) 3633-3652 ,
'Asymptotically free gauge theories. 2', Phys.Rev.D9 (1974) 980-993 .

[A23-1974] H. Georgi and D. Politzer, 'Electroproduction scaling in an asymptotically free theory of strong interactions', Phys. Rev. D9 (1974) 416 and ref. [A21-197(3)4] .

[A24-1975] H. Kluberg-Stern and J. B. Zuber, 'Ward Identities and Some Clues to the Renormalization of Gauge Invariant Operators', Phys.Rev. D12 (1975) 467 ,
'Renormalization of Nonabelian Gauge Theories in a Background Field Gauge. 1. Green Functions', Phys.Rev. D12 (1975) 482 and
'Renormalization of Nonabelian Gauge Theories in a Background Field Gauge. 2. Gauge Invariant Operators', Phys.Rev.D12 (1975) 3159-3180.

[A25] C. Itzykson and J. B. Zuber, 'Quantum field theory', McGraw-Hill, New York 1980.

M. E. Peskin, D. V. Schroeder, 'An Introduction to quantum field theory', Addison-Wesley, Reading, MA 1995 .

J. Zinn-Justin, 'Quantum field theory and critical phenomena', The International Series of Monographs on Physics, Vol. 113 , Oxford University Press, Oxford 2002 .

[A26-1982] G. Altarelli, 'Partons in quantum chromodynamics', NFN-ROME-783, Jul 1981. 211pp., Phys.Rept.81 (1982) .

[A27-1975] H. Fritzsch and P. Minkowski, 'Psi Resonances, Gluons and the Zweig Rule', Caltech preprint CALT-68-492, Mar 1975. 61pp., Nuovo Cim.A30 (1975) 393 .

[A28-1970] C. G. Callan, 'Broken scale invariance in scalar field theory', Phys.Rev.D2 (1970) 1541-1547 .

[A29-1970] K. Symanzik, 'Small distance behavior in field theory and power counting', Commun.Math.Phys.18 (1970) 227-246 .

[A210-1971(2)] G. 't Hooft, 'Renormalizable Lagrangians for Massive Yang-Mills Fields', Nucl.Phys.B35 (1971) 167-188 ,
G. 't Hooft and M.J.G. Veltman, 'Regularization and Renormalization of Gauge Fields', Nucl.Phys.B44 (1972) 189-213 .

[A211-1969] I. B. Khriplovich, 'Greens functions in theories with a non-abelian gauge group' , Yad. Fiz. 10 (1969) 409 , engl. transl. Sov. J. Nucl. Phys. 10 (1970) 235 .

[A212-1972] G. 't Hooft, Remarks on the ultraviolet stable sign of the beta function at the Colloquium on renormalization of Yang-Mills fields and applications to particle physics, Marseille 1972 .

[A213-1988] S. G. Gorishnii , A. L. Kataev and S. A. Larin, 'Next-To-Leading $O(\alpha_s^3)$ QCD Correction to Σ_t ($e^+ e^- \to$ Hadrons) : Analytical Calculation and Estimation of the Parameter Lambda (MS)', JINR-E2-88-254, Apr 1988. 18pp., Phys.Lett.B212 (1988) 238-244 , also in the same proceedings as ref. [cond-11-1988] .

[A214-1997] T. van Ritbergen, J.A.M. Vermaseren and S.A. Larin, 'The Four loop beta function in quantum chromodynamics', UM-TH-97-01, NIKHEF-97-001, Jan 1997. 9pp.,
Phys.Lett.B400 (1997) 379 , hep-ph/9701390 .

[A215-2009] S. Bethke, 'The 2009 Wolrd Average of $\alpha_s(M_Z)$', arXiv:0908.1135 [hep-ph] .

[A216-1969] K. Wilson, ' Nonlagrangian models of current algebra', Phys. Rev. 179 (1969) 1499 .

[A217-1954] M. Gell-Mann and F. Low, 'Quantum electrodynamics at small distances', Phys. Rev. 95 (1954) 1300 .

[A218-1973] S. Weinberg, 'New approach to the renormalization group', Phys. Rev. D8 (1973) 3497 .

[A219-1982] O. V. Tarasov, Dubna report JINR P2-82-900 (1982) , unpublished .

[A220-1994] K. G. Chetyrkin, C. A. Dominguez, D. Pirjol and K. Schilcher, 'Mass singularities in light quark correlators : the strange quark case', Phys.Rev.D51 (1995) 5090-5100 , hep-ph/9409371 .

[A221-2006] P. A. Baikov, K. G. Chetyrkin and J. H. Kühn, Phys. Rev. Lett. 96 (2006) 012003 .

P. A. Baikov, K. G. Chetyrkin and J. H. Kühn, 'Massless propagators: Applications in QCD and QED', SFB-CPP-08-06, TTP08-03, Oct 2008, 7pp., Contributed to 8th International Symposium on Radiative Corrections (RADCOR 2007): Application of Quantum Field Theory to Phenomenology, Florence, Italy, 1-6 Oct 2007, PoS RADCOR2007 (2007) 023, arXiv:0810.4048 [hep-ph] .

[A222-2008] P. A. Baikov, K. G. Chetyrkin and J. H. Kühn, 'Hadronic Z- and tau-decays in order α_s^4 ', Phys.Rev.Lett.101 (2008) 012002, arXiv:0801.1821v2 [hep-ph] .

[A223-2008] C.A. Dominguez, N.F. Nasrallah, R.H. Röntsch and K. Schilcher, 'Up and down quark masses from Finite Energy QCD sum rules to five loops', UCT-TP-271-08, Jun 2008, 4pp., Phys.Rev.D79 (2009) 014009, arXiv:0806.0467 [hep-ph] .

[A224-1979] M. A. Shifman , A. I. Vainshtein and V. I. Zakharov , 'QCD and resonance physics. Theoretical Foundations.' , Nucl. Phys. B147 (1979) 385 , 448.

[A225-2001] J. Gasser and H. Leutwyler, Annals Phys. 158 (1984) 142, H. Leutwyler, 'Nonlattice determinations of the light quark masses', Nucl.Phys.Proc.Suppl.94 (2001) 108, hep-ph/0011049.

[A226-2008] B. Blossier et al., European Twisted Mass Collaboration, 'Light quark masses and pseudoscalar decay constants from N(f)=2 Lattice QCD with twisted mass fermions', DESY-07-148, FTUV-07-2709, IFIC-07-57, MS-TP-07-23, RM3-TH-07-11, ROM2F-2007-16, SFB-CPP-07-58, TUM-HEP-676-07, Sep 2007. 20pp., JHEP 0804 (2008) 020 , arXiv:0709.4574 [hep-lat].

[A227-2008] C. Amsler et al. (Particle Data Group), Physics Letters B667 (2008) 1.

[A228-2008] T. Becher and M. D. Schwartz, 'A Precise determination of alpha(s) from LEP thrust data using effective field theory', FERMILAB-PUB-08-048-T, Mar 2008, 34pp., JHEP 0807 (2008) 034, arXiv:0803.0342 [hep-ph] .

[A229-2011] S. Narison, 'Gluon Condensates and $m_b(m_b)$ from QCD-Exponential Moments at Higher Orders', May 2011, 4pp., arXiv:1105.5070 [hep-ph] .

[A230-2011] S. Borsanyi, G. Endrodi, Z. Fodor, A. Jakovac, S. D. Katz, S. Krieg, C. Ratti, K. K. Szabo, 'The QCD equation of state with dynamical quarks', WUB-10-12, Jul 2010. 31pp., JHEP 1011 (2010) 077, arXiv:1007.2580 [hep-lat] .

[cond-1-1993] W. H. Zurek, 'Cosmic strings in laboratory superfluids and remnants of other phase transitions', Acta Physica Polonica B7 (1993) 1301 ,
URL → http://th-www.if.uj.edu.pl/acta/vol24/pdf/v24p1301.pdf .

[cond-2-1937] P. L. Kapitza, Nature 141 (1937) 74 , JETP 11 (1941) , 1 and 581 .

[cond-3-1941] L. D. Landau, JETP 11 (1941) 592 and L. D. Landau and E. M. Lifshitz, 'Statistical Mechanics', Pergamon Press, New York 1958 .

[cond-4-1998] Complete list works of L D Landau, Physics-Uspekhi 41 (6), June, 1998.

[cond-5-1961] L. P. Pitaevskii, JETP 13 (1961) 451 .

[cond-6-195(6)7] L. D. Landau, 'The theory of a Fermi liquid', JETP 3 (1957) 920 (english translation) ,
original : JETP 30 (1956) 1058 ;
'Oscillations in a Fermi liquid', JETP 5 (1957) 101 (english translation).

[cond-7-195(6)7] L. N. Cooper, 'Bound Electron Pairs in a Degenerate Fermi Gas', Phys. Rev 104 (1956) 1189 - 1190 ,
J . Bardeen, L. N. Cooper, and J. R. Schrieffer, 'Microscopic Theory of Superconductivity', Phys. Rev. 106 (1957) 162 - 164 ,
J. Bardeen, L. N. Cooper, and J. R. Schrieffer, 'Theory of Superconductivity', Phys. Rev. 108 (1957) 1175 .

[cond-8-1947] N. N. Bogoliubov, J. Phys. (USSR) 11 (1947) 23 .

[cond-8'-1947] N. N. Bogoliubov, 'On the Theory of Superfluidity', J. Phys. XI (1947) 292-301.

[cond-9-195(4)5] N. N. Bogoliubov and D. N. Zubarew, 'The Wave Function of the Lowest State of a System of Interacting Bose Particles', Sov. Phys. JETP 1 (1955) 83-90, and references cited therein.

[cond-10-2002] J.M. Vogels,K. Xu,C. Raman,J.R. Abo-Shaeer and W. Ketterle, 'Experimental observation of the Bogolyubov transformation for a Bose-Einstein condensed gas', Phys.Rev.Lett.88 (2002) 060402, cond-mat/0109205 .

[cond-11-1988] P. Minkowski, 'Color superconducvtivity and the associated phase transition in QCD', in 'Hadron Interactions', Proceedings of the International Symposium on "Hadron Interactions – Theory and Phenomenology", Bechyne, Czechoslovakia, 26. June - 1. July, 1988, J. Fischer, P. Kolář, V. Kundrát eds., Prague 1988.

CHAIRMAN: P. MINKOWSKI

Scientific Secretaries: L. Wen, L. F. Palhares

DISCUSSION

- *Calmet*:

Can you please re-explain the meaning of the connection Omega you have introduced?

- *Minkowski:*

For the D-string, when we transform locally the connection and curvature accordingly, the transformation rule splits into a multiplication with the representative D irreducible representation on X to the left and the corresponding inverse to the right. If X and Y become local, this is just a Clebsch-Gordan transformation to the adjoint. If they are not local, this is the simplest purely gauge covariant quantity one can write. And it splits in a local way, the D representation at X from the Dbar representation at Y. The D-String acquires knowledge of the structure of the group, from its inspecting the neighborhood of X, whereas the adjoint representation, for instance, is not sensitive to the center of the group. You gain the information by the bilocal decomposition. This is the regularity condition. You may demand that certain Ds should not give rise to the representation. You can take them out, and then you weaken the regularity conditions. They have to be chosen dynamically, in the sense that one has to try out all of them. Then the physics decides what QCD chooses for the minimum of the energy.

- *Rinaldi:*

What is the punchline of your lectures? What do your infrared conditions imply on this non-perturbative region of QCD?

- *Minkowski:*

The procedure here was just to name the set of regularity conditions and where they come from. To understand its meaning one should apply it to the thermal average and compare to what is done on the lattice. If you choose the complete connections, the boundary conditions on the edges which are separated by 1/T have to be free -- they cannot be fixed -- because it breaks the extended gauge invariance. Then it means that quantities which are extended periodic are completely gauge invariant. For instance, the Polyakov loop is not. So this shows you the dynamic strings: if you assume that the Polyakov loop is invariant, and you identify the connection Omega, the boundary conditions applied in the thermal definition of the Polyakov loop will imply Omega(t, \vec{x})=Omega(t + 1/T, \vec{x}) and this breaks gauge invariance of this type. If you then choose another type of gauge invariance, a lower type -- not the complete one --, you will get a different result for the thermal functions. This illustrates the dynamical non-triviality of these regularity conditions.

- *Rinaldi:*

Did you introduce any non-perturbative effects in this formulation?

- *Minkowski:*

Yes, they are mostly non-perturbative by their very definition. Normally, we are used to perturbation theory and to applying it to many problems even if the perturbative regime does not apply. In this talk, it is shown that if you want to self-consistently check whether perturbation theory applies, when you are looking, for instance, at correlation functions between very different points x and y, you find a negative answer. The discriminant of perturbative approximations, namely the beta function, which is perfectly perturbative in the ultraviolet, does not hold in the infrared.

- *'t Hooft:*

I have a question related to that. There have been several researches about the phases of QCD: whether in the infrared it gives confinement or effectively generates a Higgs mechanism or it could generate many long-range fields, massless fields like in the Coulomb mode. Can you make similar statements? Under which conditions would you have each phase in your formulation? For instance, how many quarks do you need to make a flip-over from the confining phase to a Higgs-like phase? Or how many or what kind of representations do you need to get other kinds of behavior such as dual transformations. We could imagine transforming QCD with a number of fermions into QCD with a different number of fermions by performing dual transformations.

- *Minkowski:*

From the very first principles I cannot calculate all the phases. You probably meant thermal phases under certain conditions, but the first phase is the infinite volume phase which is necessary in every field theory; one cannot formulate any field theory without going through the finite volume theory and then taking the volume to infinity, which corresponds to zero temperature, and the usual assumption is that there is only one phase there.

-*'t Hooft:*

Normally you assume hadrons, which are confined, but in principle something else could happen. For instance the quark fields could combine in a scalar combination and they generate a Higgs mechanism. Then the hadronic spectrum looks entirely different. These two scenarios could be connected by phase transition lines.

- *Minkowski:*

Phase transitions at zero temperature, which is what you mean, are often discussed and I cannot disprove their existence. However, they usually bring problems and the easiest assumption is that these composite Higgs which have no scalar fields to drive them do not arise in the T=0 limit.

- *'t Hooft:*

You could think of modifying the theory -- having different numbers of quarks or quarks in different representations or different gauge groups -- and ask the same questions. Then the theory could flip-over to a different phase.

- *Minkowski:*

I absolutely agree. The simplest case for which I could guess the answer is the one with 17 flavors of quarks. However, this theory is not well-defined: it is not asymptotic and not infrared stable. Another case is N=4 Super Yang-Mills theory. This theory is conformal, so it has no trace anomaly, and has the usual chiral anomaly. In this sense, this theory is in principle well-defined, but it is not clear how to formulate it in terms of regularity conditions.

- *Csanàd:*

Can and should the Lattice QCD community use your formulation in calculating phases or quark masses?

- *Minkowski:*

Of course, yes. But they did not do it yet. Perhaps my prescription is not precise enough, but I am trying to interact with them.

- *Palhares:*

You said that the presence of gluon and chiral condensates in the vacuum stabilize the infrared regime of QCD. Could you manage to stabilize the theory in the infrared with only one of these condensates or do you need both of them?

- *Minkowski:*

One can show that these condensates become related in a theory with modified quark content. Consider for instance six quark flavors, going up to the top quark. The anomalies are flavor-number-dependent but mass-independent, so they are not affected if one considers unrealistic quark masses that are not so large. If all quarks in the theory are at most as heavy as the strange quark, we can take any subgroup out of the six flavors and take them out of the world, in an adiabatic way, for instance, as is often done to discuss the S matrix. In this limit, (- m_quark qbar q) is a well-defined multiple of F^2, so that the two condensates are dependent. For realistic quark masses, however, their dependence is one of the most difficult problems. Since the gluon condensate, F^2, is a dimension 4 object, its relation to the 3-dimensional object (qbar q) requires an extra mass scale, indicating an anomalous dimension. Even though one cannot compute it explicitly, the suspicion is that, even in this case, these condensates are dependent.

- *Palhares:*

Then you believe that whenever you have a gluon condensate you should have a chiral condensate as well?

- *Minkowski:*

Yes, and this can be derived indirectly from these regularity conditions.

June 24, 2012

Notes on Strings and Higher Spins

A. Sagnotti

Scuola Normale Superiore and INFN
Piazza dei Cavalieri, 7
I-56126 Pisa ITALY
e-mail: sagnotti@sns.it

Abstract

These notes are devoted to the intriguing and still largely unexplored links between String Theory and Higher Spins, the types of excitations that lie behind its most cherished properties. A closer look at higher–spin fields provides some further clues that String Theory describes a broken phase of a Higher–Spin Gauge Theory. Conversely, string amplitudes contain a wealth of information on higher–spin interactions that can clarify long–standing issues related to their infrared behavior.

Based on the lectures presented at the International School for Subnuclear Physics "Searching for the Unexpected at LHC and Status of Our Knowledge" (Erice, June 24–July 3 2011), and on the talks presented at "Strings, Branes and Supergravity" (Istanbul, July 31–Aug 5 2011), at "Quantum Theory and Symmetries (QTS7)" (Prague, Aug. 7–13 2011) and at "Fundamental Fields and Particles (FFP12)" (Udine, Nov 21–23 2011).

1 INTRODUCTION

More than forty years ago, Veneziano [1] gave birth to String Theory [2] relating Euler's Beta–function $B(u, v)$ to tree–level S–matrix amplitudes. Together with remarkable subsequent developments, this linked its two arguments u and v to α', a parameter reflecting a string tension, and to the Mandelstam variables s and t of four–particle amplitudes, in what eventually became tachyon amplitudes of the 26–dimensional bosonic string. This model, however, also contains infinitely many massive excitations, and among them infinitely many higher–spin (HS) ones [3] that correspond to multi–symmetric tensors of the type $\varphi_{\mu_1 \ldots \mu_{s_1}; \nu_1 \ldots \nu_{s_2}; \ldots}$. These massive HS excitations are crucial in order to guarantee "planar duality", the manifest symmetry of $B(u, v)$ under the interchange of its two arguments, despite their different origins in diagrammatic constructions. For instance, the amplitude built exhibiting *only* s–channel poles also possesses t–channel poles that can be recovered by resumming polynomial residues, and clearly their infinite number is instrumental to this end.

One can argue, in similar naive terms, that other key properties of String Theory, such as modular invariance or open–closed duality, rest on the presence of infinitely many HS modes. Yet, HS amplitudes have not attracted much attention over the years, possibly because they appear somewhat remote from applications, but certainly because they are rather unwieldy in general. Perhaps this has some bearing on the present, not fully satisfactory, state of String Theory. No neat geometrical principles underlying it have emerged over the years, and this incomplete grasp of the foundations goes along with a lack of universal agreement on the actual lessons that it has in store. Could it be, then, that the massive excitations hold the key for penetrating more deeply into the whole setup? And if this were the case, could the key for understanding the massive excitations lie in the breaking of HS symmetries? This picture, in fact, has long been in the minds of many authors, at least since the 1980's, when String Field Theory [4] displayed Stueckelberg–like kinetic terms for massive string modes [1]. Stueckelberg symmetries are indeed deformations of ordinary massless gauge symmetries induced by "debris" that is brought about by interactions in the presence of symmetry breaking. As we shall see in the final part of these notes, a closer look at string amplitudes provides further evidence to this effect, although the order parameter, related to α', lacks a dynamical origin in the current formulation of String Theory.

In mentioning HS fields, I have stressed that they are very complicated in general. As we shall see in the next Section, they predated String Theory by a few decades, since their history started in the 1930's [8, 9]. Yet, they have long been lagging somewhat behind, so much so that basic ingredients like their free Lagrangian formulations could still receive new inputs during the last decade, while both the nature of their interactions and their actual meaning, let alone their subtle behavior when infra–red cutoffs (masses and/or a cosmological constant) are removed, remain largely mysterious. On the other hand, this framework can exhibit a signpost, the Vasiliev system [10], which sits in some respects slightly ahead of those laid by String Theory. Lack of space forces me to leave it out

[1]The arguments of [5], inspired by the AdS/CFT correspondence [6] and by the important observations of [7], are a notable example of this body of evidence.

of these notes, so that I can only mention that it describes the mutual interactions of infinitely many symmetric tensors in the presence of a parameter that cuts them off in the infra–red. The actual meaning of this infra–red cutoff is quite interesting in its own right, as stressed in [11]. The Vasiliev system has the looks of an effective description for the first Regge trajectory of the open bosonic string, in a regime where it should dominate the low–energy dynamics and in a language that extends the frame–like formulation of gravity.

The two main sections that follow are devoted almost exclusively to free and interacting massless HS fields around flat space, the starting point for building massive counterparts in the presence of symmetry breaking. They focus on an extension of the metric–like formulation of gravity that allows a direct comparison with String Theory and begin with some cursory historical remarks that are tailored to the ensuing discussions. Sections 2.1 and 2.2 discuss free irreducible HS fields and their reducible counterparts that emerge from String Theory in the $\alpha' \to \infty$ limit, while Section 3.1 discusses free fields in (A)dS backgrounds, stressing role and meaning of their mass–like terms. Section 3.2 discusses external currents and the van Dam–Veltman–Zakharov discontinuity, while Section 3.3 illustrates some definite lessons that String Theory already provides for HS interactions. Section 3.4, strictly speaking, would not belong to the Erice course, but it was part of the subsequent talks and I took the freedom to include it since it contains, in my opinion, very interesting clues on the behavior of HS interactions around flat space.

2 Free Higher Spins

This Section is devoted to some properties of free symmetric HS tensors, fields of the form $\varphi_{\mu_1 \ldots \mu_s}$. Lack of space forces me to confine my attention to Bose fields and to leave out altogether HS fields of mixed symmetry [12, 13, 14, 15], although they constitute the vast majority of string excitations. The material is meant to prepare the grounds for the discussion of string interactions of Section 3, so that I review Fronsdal's constrained formulation of irreducible HS fields [16] and its minimal unconstrained extension [17], their links with the geometry of free HS fields and the reducible "triplets" [18] that emerge from String Theory in the $\alpha' \to \infty$ limit.

2.1 Free Symmetric Higher Spins

It is commonly stated that the theory of HS fields originates from the works of Dirac, Fierz and Pauli [9]. Their key results were the elegant DFP physical state conditions

$$\begin{aligned} \left(\Box + m^2\right) \varphi_{\mu_1 \ldots \mu_s} &= 0 \, , \\ \partial^{\mu_1} \varphi_{\mu_1 \mu_2 \ldots \mu_s} &= 0 \, , \\ \eta^{\mu_1 \mu_2} \varphi_{\mu_1 \mu_2 \ldots \mu_s} &= 0 \end{aligned} \qquad (2.1)$$

for symmetric tensors $\varphi_{\mu_1 \dots \mu_s}$ of arbitrary rank and their counterparts for symmetric tensor–spinors $\psi_{\mu_1 \dots \mu_s}$ of arbitrary rank

$$
\begin{aligned}
(\not{\partial} - m)\, \psi_{\mu_1 \dots \mu_s} &= 0 \,, \\
\partial^{\mu_1}\, \psi_{\mu_1 \mu_2 \dots \mu_s} &= 0 \,, \\
\gamma^{\mu_1}\, \psi_{\mu_1 \dots \mu_s} &= 0 \,,
\end{aligned}
\tag{2.2}
$$

whose first members are the Klein–Gordon and Dirac equations, and the first evidence that a naive coupling to Electromagnetism faces unexpected problems. For any given s, the first member of the sets (2.1) and (2.2) defines the mass–shell, the second eliminates unwanted "time" components, and finally the last confines the available excitations to irreducible multiplets. Equations of this type lie at the heart of Quantum Field Theory and are key building blocks for most current and past constructions. The logic behind their prominence surfaced long ago in Wigner's work [19]: eqs. (2.1) and (2.2), while based on *finite–dimensional non–unitary* representations of the Lorentz group, lead upon quantization to one–particle states filling *unitary* irreducible representations of the Poincaré group.

The emergence of higher spins, however, dates back to an older and long neglected paper of Majorana [8]. Although this work lies somewhat outside the scope of these notes, I would like to comment briefly on its striking content, since after all the Erice School is rightfully named after this great Italian scientist. Majorana tried to formulate a Dirac–like equation free of the then disturbing negative–energy solutions, and in doing so was led to consider *infinite–dimensional unitary* representations of the Lorentz group SO(3,1). Let me stress that in 1932 this subject was largely unfamiliar in Mathematics, let alone in Theoretical Physics. Majorana readily found out that, in the massive case, his equation was not describing a single type of particle, but rather a whole "trajectory" of states with

$$
M(s) \sim \frac{1}{s + \frac{1}{2}} \,. \tag{2.3}
$$

He was thus anticipating, to some extent, both Regge's idea [20] and its eventual realization in the Veneziano amplitude [1], and thus in String Theory altogether [2], by over thirty years! On the mathematical side, he had come across an oscillator realization of the $so(3,1)$ algebra, well before Mathematics touched upon a tool that only in the 1970's became relatively familiar in Theoretical Physics [21]. The paper is striking, since its techniques are a sublimation of elementary facts of angular momentum theory that were then common tools in Atomic Physics. His analysis rests on oscillator generalizations of the Dirac γ–matrices, but it was perhaps too early even for him to appreciate that he was actually building, at no extra cost, unitary representations of the larger $so(3,2)$ algebra. The extension reflects, in modern terms, a familiar fact: if one combines γ^{μ}'s and $\gamma^{\mu\nu}$'s, the resulting commutators generate indeed $so(3,2)$. In fact, without being fully aware Majorana had stumbled upon Dirac's "singletons" [22], which lie at the heart of beautiful sequels by Flato, Fronsdal and others [23], and more recently of both the AdS/CFT correspondence [6] and Vasiliev's theory [10]. I will end here these parenthetic remarks, referring the interested reader to [8] and to some interesting recent papers [24] on Majorana's 1932 contribution, where the subsequent history of the subject is also addressed.

The discovery of the Rarita–Schwinger theory [25], a development that played a key role in Theoretical Physics toward the end of the last Century as a key building block of supergravity [26], followed rather closely the Schrödinger–like DFP conditions (2.1) and (2.2), but World War II forced the scientific community to leave aside these themes for a while. As a result, a systematic search for action principles, the next natural task along this line, had to await about four decades, since in the 1950's and 1960's, when a sizable activity on fundamental questions resumed, the focus was rather on the S–matrix, with an eye on important issues in current algebra.

Rather than spurring progress in a constructive sense, key developments on HS that took place in the 1960's actually spotted further unexpected difficulties. In retrospect, these results were unveiling important information on the nature of HS interactions, but for a while their net effect was to force most of the community away from these themes. I am referring, in particular, to Weinberg's 1964 argument on soft HS particles and the S matrix [27], to the Coleman–Mandula theorem [28] and to the Velo–Zwanziger violation of causality in the presence of electromagnetic interactions [29]. These were then followed, after a few years, by the Aragone–Deser field–theory argument [30] on the conflict between "minimal", two–derivative, gravitational couplings and HS fields around flat space, and by the Weinberg–Witten theorem [31, 32], a result that points in the same direction but is again inspired by S–matrix techniques. I will return to these issues in Section 3.

Some major developments that took place in Field Theory during the 1970's were certainly not foreign to the renewed interest that ultimately led to action principles for massive symmetric Bose and Fermi HS fields [33]. These results were the starting point for the subsequent work by Fronsdal and Fang and Fronsdal [16], where gauge–invariant actions were finally recovered in the massless limit.

Fronsdal arrived at a natural generalization of the Maxwell and linearized Einstein equations. He extended the familiar $s = 1, 2$ cases

$$\begin{aligned} &\Box A_\mu \,-\, \partial_\mu \partial \cdot A \,=\, 0 \ , \\ &\Box \varphi_{\mu\nu} \,-\, (\,\partial_\mu \partial \cdot \varphi_\nu \,+\, \partial_\nu \partial \cdot \varphi_\mu\,) \,+\, \partial_\mu \partial_\nu \varphi^{\,\prime} \,=\, 0 \ , \end{aligned} \tag{2.4}$$

where the "prime" denotes a trace, that are invariant under the gauge transformations

$$\begin{aligned} \delta A_\mu \,&=\, \partial_\mu \Lambda \ , \\ \delta \varphi_{\mu\nu} \,&=\, \partial_\mu \Lambda_\nu \,+\, \partial_\nu \Lambda_\mu \ , \end{aligned} \tag{2.5}$$

to symmetric tensors $\varphi_{\mu_1 \ldots \mu_s}$ of arbitrary rank, arriving at the second–order equations

$$\mathcal{F}_{\mu_1 \ldots \mu_s} \,\equiv\, \Box\, \varphi_{\mu_1 \ldots \mu_s} \,-\, (\,\partial_{\mu_1} \partial \cdot \varphi_{\mu_2 \ldots \mu_s} \,+\, \ldots) \,+\, \left(\partial_{\mu_1} \partial_{\mu_2} \varphi^{\prime}_{\mu_3 \ldots \mu_s} \,+\, \ldots\right) \,=\, 0 \ , \tag{2.6}$$

where the combinations $\mathcal{F}_{\mu_1 \ldots \mu_s}$ are often referred to as Fronsdal's tensors. As these are natural generalizations of eqs. (2.4), one would expect that eqs. (2.5) leave way to the natural spin–s gauge transformations

$$\delta \varphi_{\mu_1 \ldots \mu_s} \,=\, \partial_{\mu_1} \Lambda_{\mu_2 \ldots \mu_s} \,+\, \ldots \ , \tag{2.7}$$

but as we shall see shortly this is not quite the case.

In eqs. (2.6) and (2.7), a number of terms needed to complete the symmetrizations are inevitably left implicit. In order to streamline the ensuing discussion, it is very convenient to resort to a notation introduced in [17] that may be regarded as a counterpart, for these symmetric tensors, of the language of forms. Lorentz labels are thus eliminated altogether, so that a generic spin–s field is simply denoted by φ, and in this fashion eqs. (2.6) and (2.7) take the simpler forms

$$\mathcal{F} \equiv \Box\varphi - \partial\partial\cdot\varphi + \partial^2\varphi' = 0 , \tag{2.8}$$

and

$$\delta\varphi = \partial\Lambda . \tag{2.9}$$

Let me stress that this choice is not merely a shorthand. Not only does it simplify greatly all expressions at hand, in fact, but it also embodies the key properties that are needed to manipulate them at will. To this end, one only needs a handful of rules, that are discussed in some detail, for instance, in [11]:

$$\begin{aligned}
(\partial^p\varphi)' &= \Box\,\partial^{p-2}\varphi + 2\,\partial^{p-1}\partial\cdot\varphi + \partial^p\varphi' , \\
\partial^p\,\partial^q &= \binom{p+q}{p}\ \partial^{p+q} , \\
\partial\cdot(\partial^p\,\varphi) &= \Box\,\partial^{p-1}\,\varphi + \partial^p\,\partial\cdot\varphi , \\
\partial\cdot(\eta^k\varphi) &= \partial\,\eta^{k-1}\varphi + \eta^k\,(\partial\cdot\varphi) , \\
(\eta^k\varphi)' &= [D + 2\,(s+k-1)]\,\eta^{k-1}\varphi + \eta^k\varphi' .
\end{aligned} \tag{2.10}$$

Suffice it to say, here, that in this notation every implicit symmetrization is meant to involve the *least* number of terms that are needed to effect it, and that the overall coefficients are always one. One can thus understand, for instance, why $\partial\,\partial = 2\,\partial^2$, the counterpart of the vanishing of the squared exterior derivative: since derivatives commute, a naive symmetrization would overcount the terms actually needed by a factor two.

Returning to the Fronsdal equations (2.6), their structure is clearly quite natural in view of eqs. (2.4), and yet their behavior under gauge transformations entails a little surprise, since

$$\delta\,\mathcal{F} = 3\,\partial^3\Lambda' . \tag{2.11}$$

In other words, gauge invariance demands that $\Lambda' = 0$, *i.e.* that the parameters be *traceless.* This condition first shows up for $s = 3$ and is often referred to as *Fronsdal's first constraint.* A second, subtler constraint, is also needed in order that the natural counterpart of the free Einstein Lagrangian, that up to partial integrations can be presented in the form

$$\mathcal{L} = \varphi\left(\mathcal{F} - \frac{1}{2}\,\eta\,\mathcal{F}'\right) , \tag{2.12}$$

be gauge invariant, even with traceless parameters. As we have seen, in this case $\mathcal{F}$ is gauge invariant, while up to total derivatives the variation of the "naked" φ yields

$$\delta\,\mathcal{L} = -\,s\,\Lambda\left(\partial\cdot\mathcal{F} - \frac{1}{2}\,\partial\mathcal{F}'\right) , \tag{2.13}$$

an expression that would vanish identically if a HS counterpart of the familiar Noether or Bianchi identity for the linearized version of Einstein's gravity,

$$\partial^\mu \mathcal{R}_{\mu\nu} - \frac{1}{2}\,\partial_\nu \mathcal{R} = 0\,, \tag{2.14}$$

were to hold. As it turns out, however, this is not the case, since in general

$$\partial \cdot \mathcal{F} - \frac{1}{2}\,\partial \mathcal{F}' = -\frac{3}{2}\,\partial^3 \varphi''\,, \tag{2.15}$$

so that for $s \geq 4$ gauge invariant action principles obtain only if the gauge fields are subject to *Fronsdal's second constraint*, $\varphi'' = 0$, the condition that φ be *doubly traceless.*

The need for these constraints is a sign that Fronsdal's formulation is incomplete, since eqs. (2.6) and (2.13) do not accommodate the natural gauge symmetry (2.7), despite the fact that algebraic constraints certainly do not raise doubts on the nature of the actual propagating degrees of freedom. This problem was first bypassed in the 1990's by the Dubna group [34], adapting to free higher spins the BRST construction [35] of free String Field Theory [4], but at the price of introducing $\mathcal{O}(s)$ different fields for the case of a rank–s tensor $\varphi_{\mu_1\ldots\mu_s}$. A "minimal" solution was then obtained in [17, 11]. Its "compensator equations"

$$\mathcal{A} \equiv \mathcal{F} - 3\,\partial^3 \alpha = 0 \tag{2.16}$$

involve, for $s \geq 3$, only one additional field, a spin–$(s-3)$ compensator $\alpha_{\mu_1\ldots\mu_{s-3}}$, and the resulting extension $\mathcal{A}$ of Fronsdal's $\mathcal{F}$ tensor is gauge invariant provided

$$\delta\alpha = \Lambda'\,. \tag{2.17}$$

Actually, for $s=3$ the compensator made an early appearance in the work of Schwinger [36], as I was told some time ago by G. Savvidy. Schwinger's books contain indeed a discussion, for $s=3$, of this additional field, which he also introduced in order to bypass Fronsdal's first constraint but readily abandoned since it brings along higher derivatives. In retrospect, these higher derivatives are harmless since they affect non–propagating components, and we now understand how to trade them for a few more fields, whose number does not grow with the spin, as in [37, 38].

An additional field, a Lagrange multiplier $\beta_{\mu_1\ldots\mu_{s-4}}$ such that

$$\delta\beta = \partial\cdot\partial\cdot\partial\cdot\Lambda\,, \tag{2.18}$$

is also present for $s \geq 4$ in the complete Lagrangians [17, 11]

$$\mathcal{L} = \varphi\left(\mathcal{A} - \frac{1}{2}\,\eta\mathcal{A}'\right) - \frac{3}{4}\binom{s}{3}\,\alpha\,\partial\cdot\mathcal{A}' + 3\binom{s}{4}\,\beta\,\mathcal{C}\,, \tag{2.19}$$

where

$$\mathcal{C} = \varphi'' - 4\,\partial\cdot\alpha - \partial\,\alpha' \tag{2.20}$$

is a gauge–invariant completion of the double trace of φ. The resulting Lagrangian equations for φ are

$$\begin{aligned} \mathcal{A} - \frac{1}{2}\eta\mathcal{A}' + \eta^2\mathcal{B} &= 0\,, \\ \varphi'' - 4\,\partial\cdot\alpha - \partial\alpha' &= 0\,, \end{aligned} \tag{2.21}$$

where η is the Minkowski metric and

$$\mathcal{B} = \beta - \frac{1}{2}(\partial\cdot\partial\cdot\varphi' - 2\,\Box\,\partial\cdot\alpha - \partial\partial\cdot\partial\cdot\alpha) \tag{2.22}$$

is a gauge invariant completion of the Lagrange multiplier β. Let me stress that eqs. (2.21) generalize the linearized Lagrangian equation for gravity,

$$\mathcal{R}_{\mu\nu} - \frac{1}{2}\eta_{\mu\nu}\mathcal{R} = 0\,, \tag{2.23}$$

and are thus more complicated than eq. (2.8) that is rather the counterpart for these systems of the simpler linearized non–Lagrangian equation

$$\mathcal{R}_{\mu\nu} = 0\,. \tag{2.24}$$

A recursive argument [17, 11] shows that eqs. (2.21) can turned into

$$\mathcal{B} = 0 \qquad \text{and} \qquad \mathcal{A} = 0 \tag{2.25}$$

once they are combined with their traces.

Removing Fronsdal's constraints via the compensator α (and the Lagrange multiplier β) is clearly eliminating an asymmetry between the first two cases, $s = 1, 2$, and the HS ones. Interestingly, this also opens a small and yet instructive window on the geometrical nature of HS constructions. After all, eqs. (2.4) admit a different but equally familiar presentation in terms of the Maxwell curvature $F_{\mu\nu}$ and, as we have already anticipated, of its counterpart $\mathcal{R}_{\mu\nu}$ for the linearized Einstein theory. There is a key difference between these objects, however: for $s = 2$ one arrives at the linearized Riemann curvature tensor $\mathcal{R}_{\mu\nu\rho\sigma}$ via an intermediate object, the Christoffel connection, that in the linearized case reads

$$\Gamma^{\mu}{}_{\nu\rho} = \frac{1}{2}\left(\partial_\nu\,\varphi^{\mu}{}_{\rho} + \partial_\rho\,\varphi^{\mu}{}_{\nu} - \partial^{\mu}\,\varphi_{\nu\rho}\right)\,. \tag{2.26}$$

As a result, while the Maxwell curvature contains one derivative of A_μ, its $s = 2$ counterpart is bound to contain two derivatives of $\varphi_{\mu\nu}$.

It is convenient to focus here on a variant of the usual Riemann curvature that is *symmetric* under interchanges within its two groups of indices, although it is merely a combination of conventional Riemann tensors on account of the cyclic identity. Yet, when proceeding to arbitrary values of s, this choice has the virtue of suggesting a recursive construction where the linearized $\Gamma_{\mu,\,\nu\rho}$ leaves way to a tower of $s-1$ Christoffel–like connections, $\Gamma_{\mu_1,\,\nu_1\ldots\nu_s}$, $\ldots$, $\Gamma_{\mu_1\ldots\mu_{s-1},\,\nu_1\ldots\nu_s}$. Each of these quantities is obtained from combinations of derivatives of the previous member of the list in such a way that, in the gauge transformations, all μ indices fall on the gauge parameter. These expressions carry, by construction, 1, ..., s-1 derivatives of the original field, and the next member of the sequence is a spin–s analogue of the linearized $s = 2$ curvature in its doubly–symmetric incarnation, a tensor $\mathcal{R}_{\mu_1 \ldots \mu_s, \nu_1 \ldots \nu_s}$ built from combinations of s derivatives of the original $\varphi_{\mu_1 \ldots \mu_s}$ that is *gauge invariant* under transformations like (2.7), albeit with unconstrained parameters. One can also show that these curvature tensors possess the additional symmetry

$$\mathcal{R}_{\mu_1 \ldots \mu_s, \nu_1 \ldots \nu_s} = (-1)^s \, \mathcal{R}_{\nu_1 \ldots \nu_s, \mu_1 \ldots \mu_s} , \tag{2.27}$$

which is already visible in the $s = 1, 2$ cases.

The beautiful construction that I just sketched, due to de Wit and Freedman [39], marks the real entry point of HS geometry into the game. Since, as I have stressed, the relevant HS curvatures are by construction invariant under unconstrained gauge transformations of the type (2.7), in this fashion the authors came short of bypassing Fronsdal's constraints. They did not make this further step, however, since they insisted on the apparently natural condition that Fermi or Bose equations contain, respectively, one or two derivatives. As a result, while they could recognize that the Fronsdal tensor $\mathcal{F}$ is related to the second Christoffel–like connection according to

$$\mathcal{F}_{\nu_1 \ldots \nu_s} \sim \eta^{\mu_1 \mu_2} \, \Gamma_{\mu_1 \mu_2, \nu_1 \ldots \nu_s} , \tag{2.28}$$

all in all their HS geometry appeared for a while somewhat remote from the actual free dynamics.

The compensator equations do better in this respect, as they should since they embody an unconstrained gauge symmetry. Francia and I obtained these equations directly [17], before formulating the compensator theory, but it is instructive to retrace the argument in this fashion, starting from the compensator equations. Let us do it in some detail for the simplest case, $s = 3$, in which the non–Lagrangian equations (2.25) reduce to

$$\mathcal{F}_{\mu\nu\rho} = 3 \, \partial_\mu \partial_\nu \partial_\rho \, \alpha \, . \tag{2.29}$$

One can formally solve this equation for $\partial_\mu \, \alpha$, obtaining

$$\partial_\mu \, \alpha = \frac{1}{3} \, \frac{1}{\Box} \, \mathcal{F}'_\mu , \tag{2.30}$$

or directly for α, obtaining

$$\alpha = \frac{1}{3} \, \frac{1}{\Box^2} \, \partial \cdot \mathcal{F}' , \tag{2.31}$$

where I have resorted again to the shorthand "prime" notation to indicate traces. This is tantamount to turning the compensator equation into a pair of distinct *non–local* equations for $\varphi_{\mu\nu\rho}$ alone. By construction, the end result is invariant under the unconstrained gauge transformations (2.7), but it is clearly *not* unique, since eq. (2.30) leads to

$$\mathcal{F}^{(1)}{}_{\mu\nu\rho} \equiv \mathcal{F}_{\mu\nu\rho} - \frac{1}{3\Box} \left(\partial_\mu \, \partial_\nu \, \mathcal{F}'_\rho + \partial_\nu \, \partial_\rho \, \mathcal{F}'_\mu + \partial_\rho \, \partial_\mu \, \mathcal{F}'_\nu \right) = 0 \, , \tag{2.32}$$

while eq. (2.31) leads to the alternative gauge invariant non–Lagrangian equation

$$\widetilde{\mathcal{F}}^{(1)}_{\mu\nu\rho} \equiv \mathcal{F}_{\mu\nu\rho} - \frac{1}{\Box^2}\, \partial_\mu\, \partial_\nu\, \partial_\rho\, \partial \cdot \mathcal{F}' = 0 \,. \tag{2.33}$$

Having more than one option might seem problematic, but as we shall see it is actually instrumental for the very consistency on the non–local formulation, and to begin with the reader can verify that eqs. (2.32) and (2.33) can be turned into one another, so that they are nicely equivalent in the absence of external sources. Interestingly, factoring out the inverse d'Alembertian present in eq. (2.32) according to

$$\mathcal{F}^{(1)}{}_{\mu\nu\rho} \equiv \frac{1}{\Box}\left[\Box\, \mathcal{F}_{\mu\nu\rho} - \frac{1}{3}\left(\partial_\mu\, \partial_\nu\, \mathcal{F}'{}_\rho + \partial_\nu\, \partial_\rho\, \mathcal{F}'{}_\mu + \partial_\rho\, \partial_\mu\, \mathcal{F}'{}_\nu\right)\right] = 0 \,, \tag{2.34}$$

leaves room for a total of four derivatives within the square brackets, as many as the divergence of the spin–3 curvature would contain, according to the preceding discussion. All in all, one is thus led to the conclusion that the spin–3 non–local equations of motion, and thus a fortiori their local compensator counterparts, are equivalent to

$$\frac{1}{\Box}\, \eta^{\mu_1\mu_2}\, \partial^{\mu_3}\, \mathcal{R}_{\mu_1\mu_2\mu_3,\nu_1\nu_2\nu_3} = 0 \,, \tag{2.35}$$

with $\mathcal{R}$ the spin–3 de Wit–Freedman curvature, symmetric in its two sets of Lorentz labels and antisymmetric under an overall interchange of them. This can be simply verified, and clearly eq. (2.35) is a very natural and satisfactory spin–3 counterpart of the Maxwell equation, albeit *not* a Lagrangian equation.

This pattern extends to arbitrary spin, as can be appreciated via an inductive argument that rests on the sequence of integro–differential operators

$$\mathcal{F}^{(n+1)} = \mathcal{F}^{(n)} + \frac{1}{(n+1)(2n+1)}\, \frac{\partial^2}{\Box}\, \mathcal{F}'^{(n)} - \frac{1}{n+1}\, \frac{\partial}{\Box}\, \partial \cdot \mathcal{F}^{(n)} \,, \tag{2.36}$$

whose Bianchi identities take the form

$$\partial \cdot \mathcal{F}^{(n)} - \frac{1}{2n}\, \partial\, \mathcal{F}^{(n)\,\prime} = -\left(1 + \frac{1}{2n}\right) \frac{\partial^{\,2n+1}}{\Box^{\,n-1}}\, \varphi^{[n+1]} \,, \tag{2.37}$$

where $\varphi^{[n+1]}$ denotes the $(n+1)$–st trace of the HS field φ. For these combinations, the original compensator equation (2.8) translates into the sequence

$$\mathcal{F}^{(k)} = (2\,k + 1)\, \frac{\partial^{2k+1}}{\Box^{k-1}}\, \alpha^{[k-1]} \,, \tag{2.38}$$

where $\alpha^{[k-1]}$ denotes the $(k-1)$–fold trace of α, whose members have thus the virtue of involving successive traces of the compensator. For any given s, these traces are simply not available for k sufficiently large, so that fully gauge invariant equations obtain after completing $\left[\frac{s}{2}\right]$ iterations. Taking into account for any s the first of these equations, one is finally led to an infinite family of fully gauge invariant *non–local* counterparts of the Maxwell and Einstein equations,

$$\begin{aligned} s = 2n+1: &\quad \frac{1}{\Box^n}\,\partial_\mu \mathcal{R}^{\mu[n];\nu_1\dots\nu_s} = 0\,, \\ s = 2n: &\quad \frac{1}{\Box^{n-1}}\,\mathcal{R}^{[n];\nu_1\dots\nu_s} = 0\,. \end{aligned} \tag{2.39}$$

While these non–Lagrangian equations involve natural geometric structures for these systems in the simplest combinations that one could a priori conceive, the corresponding Lagrangian equations rest not only on traces of the first gauge–invariant $\mathcal{F}^{(k)}$ but also on higher members of the set, obtained via further iterations of eq. (2.36). We shall return shortly to this point, after introducing some additional tools.

Before closing this section, let me stress briefly that what I have said extends to Fermi fields carrying a set of totally symmetric vector labels, up to standard complications brought about by γ–matrices [16, 34, 17, 11, 37]. Tensors and spinor–tensors of mixed symmetry, on the other hand, rest on a richer theory that originates from the work of Labastida and others [12, 13] and would require a detailed discussion. I will content myself with stating that one can extend to this case the non–local equations (2.39) [14] and, up to some subtleties that are discussed in detail in [15], the compensator formulation as well. The subject is very interesting and is crucial in order to come eventually to terms with String Theory, but lack of space forces me to leave it out here.

2.2 TRIPLETS AND FREE STRING FIELD THEORY

A key question that these notes are meant to address is what these notions about higher spins are teaching us about the massive string excitations.

At the free level, the link between String Theory and HS is relatively simple, and yet it is interesting and quite instructive. In short, String Theory favors reducible representations, so that it drops somehow the last DFP condition of eq. (2.1). For symmetric Bose fields this entails the replacement of the Fronsdal Lagrangian or its compensator extensions with an interesting system of three fields that is often called a "triplet" [18, 34, 17, 40, 41] in the literature. The three fields are a rank–s tensor $\varphi_{\mu_1\dots\mu_s}$, the dynamical field in this context, and two additional tensors of ranks $s-1$ and $s-2$, $C_{\mu_1\dots\mu_{s-1}}$ and $D_{\mu_1\dots\mu_{s-2}}$ that disappear on shell but are nonetheless crucial to attain an *unconstrained* gauge symmetry. The $s=1$ case is degenerate, and the system reduces to the description of an electromagnetic potential in the Nakanishi–Lautrup presentation.

In the index–free notation, the triplet equations read

$$\begin{aligned} &\Box\,\varphi \;=\; \partial\, C\,, \\ &\partial\cdot\varphi \;-\; \partial\, D \;=\; C\,, \\ &\Box\, D \;=\; \partial\cdot C\,, \end{aligned} \tag{2.40}$$

and are invariant, as anticipated, under the *unconstrained* gauge transformations

$$\begin{aligned} \delta\,\varphi &= \partial\Lambda\,, \\ \delta\,C &= \Box\,\Lambda\,, \\ \delta\,D &= \partial\cdot\Lambda\,. \end{aligned} \tag{2.41}$$

If one is ready to take for granted that C and D disappear on shell, which can be shown in a few steps, the triplet equations (2.40) clearly boil down to the first two DFP conditions (2.1).

The triplet equations (2.40) follow from the elegant BRST [35] formulation of String Field Theory [4]. Its field equation,

$$\mathcal{Q}\,|\Psi\rangle\,=\,0\,, \tag{2.42}$$

rests on the BRST operator $\mathcal{Q}$ of world–sheet reparametrizations, whose nilpotency in the critical dimension brings about the unconstrained gauge symmetry

$$\delta\,|\Psi\rangle\,=\,\mathcal{Q}\,|\Lambda\rangle\,, \tag{2.43}$$

together with gauge–for–gauge counterparts. However, while the string BRST operator is only nilpotent in the critical dimension, its contracted version that emerges in the formal limit $\alpha' \to \infty$ is *identically* nilpotent [34], and as a result the triplet system (2.40) is indeed gauge invariant in any number of dimensions. An alternative presentation of eqs. (2.40) obtains if C is eliminated algebraically via the second of them, reducing the first to

$$\mathcal{F}\,=\,\partial^2\,(\varphi'\,-\,2\,D)\;. \tag{2.44}$$

One can simply select the spin–s excitations of this system [17], adding to eqs. (2.40) the further condition

$$\varphi' - 2\,D\,=\,\partial\,\alpha \tag{2.45}$$

that enforces on shell, in a gauge invariant fashion, the last DFP condition of eq. (2.1). This step brings about a compensator α with the gauge transformation of eq. (2.17), while eq. (2.44) turns readily into eq. (2.8). Following [38], one can also perform this reduction at the Lagrangian level, at the price of a few additional fields whose number, however, does not grow with the spin.

Another interesting question concerns the non–local formulation of the triplet systems, based on φ alone. One can build it systematically [41] and the end results,

$$\mathcal{L}\,\sim\,\mathcal{R}^{\mu_1\dots\mu_s\,;\,\nu_1\dots\nu_s}\,\frac{1}{\Box^{s-1}}\,\mathcal{R}_{\mu_1\dots\mu_s\,;\,\nu_1\dots\nu_s}\,, \tag{2.46}$$

are strikingly natural non–local counterparts of the Maxwell Lagrangian.

3 Interacting Higher Spins

We can now turn to HS interactions, and to begin with I would like to comment on some classic results of the 1960's and 1970's that revealed unexpected difficulties.

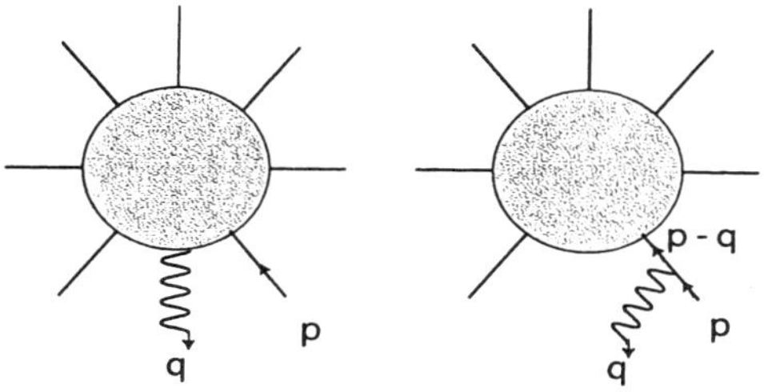

Figure 1: Factorization of an amplitude containing a "soft" emission.

The first and most striking of these results was obtained by S. Weinberg in 1964 [27]. It concerns the behavior that S–matrix amplitudes for a single *massless* spin–s particle and a number of scalar particles would exhibit in the presence of long–range HS interactions. Referring to fig. 1, the "soft" limit $q \to 0$ of one such amplitude, here denoted by A_s, would expose long–range HS effects and would be dominated by poles proportional to $p_i \cdot q$ originating from scalar propagators close to their mass shells. Moreover, Lorentz–invariant vertices for two scalars and a spin–s particle giving rise to long–range HS effects would lead, in the "soft" limit, to s powers of p_i contracting the HS polarization tensor. For $q \to 0$, an N–point A_s amplitude would thus behave asymptotically as

$$A_s(1,\ldots,N) \;\sim\; A(1,\ldots,N-1) \times \sum_{i=1}^{N-1} \frac{p_{i,\mu_1}\cdots p_{i,\mu_s}\,\varphi^{\mu_1\ldots\mu_s}(q)}{2\,p_i\cdot q}\,, \qquad (3.1)$$

where $A(1,\ldots,N-1)$ denotes a more conventional $(N-1)$–point scalar amplitude, the residue of the pole.

Weinberg's key observation was that, although an *on–shell* Abelian gauge transformation

$$\varphi(q) \to \varphi(q) \;+\; i\,q\,\Lambda(q) \qquad (3.2)$$

of the spin–s polarization tensor would seem to cause to no problems due the "soft" nature of the momentum q, the pole would make the end result *finite* in the limit. Consequently, HS amplitudes might well be *not* invariant under these shifts, with the disastrous consequence that unphysical polarizations might *not* decouple. Unless, of course, some conditions are met, and a clear way around does exists, but only for $s = 1$ or 2. In the former case, the limiting contribution (3.1) would be indeed proportional to the total electric charge of the scattered particles, which ought to vanish in S–matrix amplitudes, while in the latter it would be proportional to the sum of the scalar momenta weighted by the corresponding "gravitational charges". Hence, with gravity coupling universally, for $s = 2$ consistency would translate into momentum conservation and therefore would be automatically guaranteed. Conversely, the requirement of consistency for long–range gravitational interactions emerges, in this fashion, as a reason for their universality. On the other hand, for $s \geq 3$ eq. (3.1) would involve polynomials in the momenta, that clearly would not add up to zero in general. All in all, generic amplitudes with a single *massless* HS particle would thus seem inconsistent, but there are possible loopholes. First, as noticed in [27], if more powers of q were present in the vertices, they could compensate the pole, making the terms in eq. (3.1) vanish individually in the "soft" limit. Vertices of this type, however, would not mediate long–range HS effects, and actually can often be eliminated altogether by field redefinitions. Moreover, a recent analysis [42] that takes into account the explicit form of the cubic HS vertices of [43, 44] revealed that the problem presents itself *only* when the exchanged particles have spins *lower* than the external ones. We shall return to this important point in Section 3.4. However, the reader should already appreciate that, with these amplitudes alone, the possible long–range effects of HS particles would be confined to their own world, a scenario that was nicely foreseen by Porrati in [32].

The next classic result, the Coleman–Mandula theorem [28], is a formal S–matrix argument that has had a wide impact over the years. One of its key assumptions, however, the presence of finitely many HS excitations below any given mass level, makes it not directly applicable to the HS constructions one is after, whose gauge algebras bring about infinitely many massless gauge fields. The theorem does provide a rationale for the failure of naive attempts involving finitely many HS gauge fields, but I cannot help finding it less palatable and instructive than Weinberg's analysis.

The Velo–Zwanziger argument [29] can be regarded as a refinement of the old work of Fierz and Pauli [9]. It concerns the loss of causality that HS fields tend to manifest in the presence of minimal couplings to uniform electromagnetic fields, and in this context String Theory has proved strikingly instructive. Indeed, Argyres and Nappi [45] showed long ago that for the massive $s = 2$ mode of the open string, that in some respects behaves like a HS field, the actual coupling to uniform electromagnetic fields, which rests on an exact conformal field theory and can thus be thoroughly analyzed, is free from such pathologies. The crux of the matter lie in the peculiar highly non–linear contact terms that are present in String Theory, and a recent extension of their analysis shows that all symmetric tensors of the first Regge trajectory work exactly in the same way, without any loss of causality [46]. However, the minimal couplings of these states to Electromagnetism fade out when the string tension or the HS masses tend to zero, so that HS interactions loose their infrared cutoff. The result resonates again with Weinberg's argument, and reinforces the feeling that HS fields do not like to be involved in *long–range* interactions mediated by low–spin particles.

The last classic results that I would like to describe point again in the same direction, albeit in different ways. They both indicate that HS fields *do not* interact with gravity around flat space with two–derivative minimal couplings as their low–spin counterparts.

The Weinberg–Witten argument [31] compares two ways of determining the effect of rotations on matrix elements of the energy–momentum tensor $\langle f|T_{\mu\nu}(q)|i\rangle$ between two spin–s states in above three dimensions. If $T_{\mu\nu}$ is assumed to behave as a Lorentz tensor, a property that it possesses when it is gauge invariant, and to involve only two derivatives, the argument shows that the matrix element vanishes for any type of *massless* HS particles, be they fundamental or composite, and the authors present a similar argument for spin–one currents. The assumption is hardly a weak one, since already for $s = 2$ the energy–momentum tensor $T_{\mu\nu}$ is *not* gauge invariant, while already for $s = 1$ spin–one currents are also not gauge invariant. Still, a recent refinement of [31] took into account the subtle behavior of $T_{\mu\nu}$ but reached nonetheless the same conclusion [32]. The problem has to do, once more, with the number of derivatives in HS couplings: if more derivatives were present, the matrix elements would vanish as $q \to 0$ and the arguments of [31, 32] would not apply.

Finally, the Aragone–Deser argument [30] rests on an explicit attempt to construct a standard gravitational coupling for a spin–$\frac{5}{2}$ field. The attempt fails, since the computation generates Weyl tensors, which cannot be compensated by the Einstein–Hilbert terms and thus constitute a genuine obstruction to the gauge symmetry. Similar results obtain for $s > 5/2$ Fermi fields or for $s > 2$ Bose fields. This, however, is only true around flat space, and remarkably in the 1980's Fradkin and Vasiliev [47] could show that the obstruction disappears in the presence of a cosmological term λ. In other words, in the presence of a cosmological term, and up to the cubic order, HS fields do allow conventional–looking gravitational couplings, albeit with an important proviso: they are inevitably accompanied by higher–derivative partners that, for dimensional reasons, bring along negative powers of λ. As a result, the solution to the Aragone–Deser problem found by Fradkin and Vasiliev has a singular behavior in the limit of a flat background.

The Aragone–Deser argument has had a profound influence on subsequent developments, which have been strongly biased toward the search for "minimal–like" HS interactions in the presence of a cosmological term. The main outcome of this research, the Vasiliev system [10], is remarkable and confusing at the same time. It is remarkable as a mathematical realization of free–differential algebras with non–polynomial scalar couplings and as an example of classically consistent HS interactions, and not surprisingly starts to play an important role in the AdS/CFT correspondence. It is confusing, however, since it points to intriguing dynamical effects. Vasiliev's fields are in fact a scalar, a vector and infinitely many symmetric tensors, one for each rank $s \geq 2$, all valued in (anti)symmetric representations of a Chan–Paton group [48]. Hence, they are in one–to–one correspondence with the first Regge trajectory of the *open* bosonic string. These fields have *minimal*–like interactions, albeit around (A)dS backgrounds, and moreover the spin–2 singlet is naturally associated with gravity, while in String Theory gravity has to do with *closed*, rather than open, strings. This circumstance would seem to suggest that a Cabibbo–like mixing takes place between closed and open fields, when they become degenerate in mass in the tensionless limit $\alpha' \to \infty$. This point was raised in [11] and clearly deserves further attention. For all these reasons, and mainly to stress the key role of non–minimal interactions, both in the lectures that I have presented at Erice and in subsequent talks, I have kept a (slightly embellished) slide of my Strings 2009 presentation, where I had elaborated on the apparent tension between the Fradkin–Vasiliev minimal–like couplings and two recent results. One of these is due to Metsaev [49], who completed the analysis that in the 1980's opened our first window on HS interactions [50], and concerns the light–cone classification of cubic HS vertices in flat space, while the other is the subsequent scaling analysis of Boulanger, Leclercq and Sundell [51]. Briefly stated, Metsaev's analysis showed that cubic light–cone vertices exist around flat space in general, albeit with *lower bounds* on the number of derivatives that depend on the collection of fields involved. For example, the $s = 2$ couplings of a spin–s field involve at least $2s - 2$

derivatives. Boulanger, Leclercq and Sundell proposed a scaling argument that pulls out of the Fradkin–Vasiliev couplings the dominant contributions (what I called "seeds" in [52]), and in so doing recovered explicit examples of covariant (as opposed to light–cone, as in Metsaev's analysis) cubic flat–space vertices. The existence of the scaling limit indicates that terms containing lower numbers of derivatives are really dressings due to the (A)dS background, no more no less than the mass–like terms that I shall revisit shortly. A later section will illustrate how String Theory sheds light on the issue. In the low–tension limit of its amplitudes, in fact, one can identify whole chains of covariant terms that fill Metsaev's list, whose high–derivative couplings naturally forego the previous restrictions.

Although this section is mostly devoted to the behavior around flat space, it actually begins with a brief discussion of how mass–like terms emerge in (A)dS backgrounds, starting from the free equations of Section 2. I then turn to external currents and the van Dam–Veltman–Zakharov discontinuity [53], whose origin can now be related to special types of HS gauge symmetries that are allowed in dS backgrounds. External currents are also a first step toward more general Lagrangian couplings: the HS currents of free scalar fields, for instance, embody key information on their possible HS interactions and on their behavior at high energies [54]. Section 3.3 is then devoted to disk amplitudes for the open bosonic string and their neat lessons for the cubic interactions of symmetric Bose fields, and actually also of symmetric tensor–spinors [43]. The end result of this analysis provides further clues that massive string spectra draw their origin from broken HS symmetries. The section ends with a brief review of some recent results of Taronna [42] that address explicitly the subtle behavior of the higher–point functions for massless HS fields in flat space.

3.1 FREE HS FIELDS IN (A)DS BACKGROUNDS

As we have seen, the classic Aragone–Deser problem [30] is signalled by terms involving the Weyl tensor, which disappear in conformally flat backgrounds. This is an important exception that I cannot refrain from touching upon, since it underlies HS dynamics in the presence of a cosmological constant. The corresponding maximally symmetric backgrounds are (A)dS, rather than Minkowski, space times, where ordinary derivatives leave way to covariant derivatives such that

$$[\nabla_\mu , \nabla_\nu]\, V_\rho \;=\; \frac{1}{L^2}\,(g_{\mu\rho}V_\nu \;-\; g_{\nu\rho}V_\mu)\;. \tag{3.3}$$

For definiteness, here the sign is tailored to the dS case, g is the background metric and L, the dS radius, is related to the cosmological constant in D dimensions according to

$$\lambda \;=\; \frac{(D-1)(D-2)}{2\,L^2}\;. \tag{3.4}$$

Corresponding results for AdS obtain replacing L^2 with $-L^2$.

The commutator (3.3) brings about modifications of both $\mathcal{F}$ and its unconstrained completion $\mathcal{A}$ that make them compatible with the deformed gauge transformations

$$\delta\,\varphi \;=\; \nabla\,\Lambda\;. \tag{3.5}$$

These modifications involve the replacement of $\mathcal{F}$ with

$$\mathcal{F}_L = \mathcal{F} + \frac{1}{L^2}\left\{\left[(3-D-s)(2-s)-s\right]\varphi + 2g\varphi'\right\}, \tag{3.6}$$

where D denotes the number of space-time dimensions and

$$\mathcal{F} = \Box\varphi - \nabla\nabla\cdot\varphi + \nabla^2\varphi' \tag{3.7}$$

is now built in terms of the covariant derivatives (3.3). Notice that the deformed non–Lagrangian equations of motion, that are now $\mathcal{F}_L = 0$, involve *mass–like terms*. Mass–like terms, by no means ordinary mass terms, since they are precisely as needed for gauge invariance. At the same time the gauge transformations of α and β become

$$\begin{aligned} \delta\alpha &= \Lambda', \\ \delta\beta &= \nabla\cdot\nabla\cdot\nabla\cdot\Lambda. \end{aligned} \tag{3.8}$$

Consequently, the AdS deformations of $\mathcal{A}$ and $\mathcal{C}$ read

$$\begin{aligned} \mathcal{A}_L &= \mathcal{F}_L - 3\nabla^3\alpha - \frac{4}{L^2}g\nabla\alpha, \\ \mathcal{C}_L &= \varphi'' - 4\nabla\cdot\alpha - \nabla\alpha', \end{aligned} \tag{3.9}$$

and are rather simple, while

$$\begin{aligned} \mathcal{B}_L = \beta - &\left\{\frac{1}{2}\nabla\cdot\nabla\cdot\varphi' - \Box\nabla\cdot\alpha - \frac{1}{2}\nabla\nabla\cdot\nabla\cdot\alpha \right. \\ &\left. + \frac{1}{L^2}\left(2\nabla\alpha' + 2g\nabla\cdot\alpha' + \left[(s-3)(5-s-D)\right]\nabla\cdot\alpha\right)\right\} \end{aligned} \tag{3.10}$$

is more involved.

Let me stress that the actual form of the mass–like terms, which are mere curvature–induced dressings of the flat–space kinetic terms, is *not* unique, but depends on the ordering chosen for the covariant derivatives. This is strikingly clear in the $s = 1$ case, since the Maxwell equation admits the two naively different presentations

$$\begin{aligned} &\Box A_\mu - \nabla_\nu\nabla^\mu A^\nu = 0, \\ &\Box A_\mu - \nabla_\mu\nabla\cdot A - \frac{D-1}{L^2}A_\mu = 0, \end{aligned} \tag{3.11}$$

which are actually equivalent in view of the commutator (3.3).

3.2 External Currents and the vDVZ Discontinuity

External currents are responsible for the simplest interactions in Field Theory, but much can be learned from them notwithstanding. Static sources give rise in this fashion to Coulomb's law or its HS generalizations, while generic sources encode important information on the number of propagating degrees of freedom.

One can start from the general Lagrangian field equations for symmetric tensors and extend them allowing for the presence of an external current $\mathcal{J}$, so that they become

$$\begin{aligned} &\mathcal{A} - \frac{1}{2}\eta\,\mathcal{A}' + \eta^2\,\mathcal{B} = \mathcal{J}\ , \\ &\partial\cdot\mathcal{A}' - (2\,\partial + \eta\,\partial\cdot)\,\mathcal{B} = 0\ , \\ &\varphi'' - 4\,\partial\cdot\alpha - \partial\,\alpha' = 0\ , \end{aligned} \tag{3.12}$$

where the unconstrained gauge symmetry demands that $\mathcal{J}$ be conserved. In the familiar $s = 1$ case, in momentum space these equations reduce to

$$-\,p^2\,A_\mu + p_\mu\,p\cdot A = \mathcal{J}_\mu\ , \tag{3.13}$$

and the scalar product of this expression with the conserved current $\mathcal{J}_\mu$,

$$\mathcal{E}_0(D,1) \equiv -\,p^2\,\mathcal{J}^\mu\,A_\mu = \mathcal{J}^\mu\,\mathcal{J}_\mu\ , \tag{3.14}$$

makes it possible, so to speak, to invert the kinetic operator without the need for gauge fixing, so that the "current–exchange amplitude" (3.14) identifies the residue of the pole. With an eye to the ensuing discussion, the subscript of $\mathcal{E}$ is meant to emphasize that the amplitude is computed in flat space, while its two arguments reflect the space–time dimension and the spin of the exchanged field. Notice that the square of the conserved current in (3.14) rests on $D-2$ independent contributions, as many as the independent polarizations of a massless spin–1 field in D dimensions.

Before discussing the actual solution of the system (3.12), let us elaborate briefly on the important lessons that this analysis has in store for the non–local formulation. In principle, the non–local formulation could be derived directly, integrating out both the compensator α and the Lagrange multiplier β along the lines of [41], but current exchange amplitudes provide an alternative path that is conceptually important. The issue is selecting a non–local Lagrangian equation that yields the same physical effects, and thus the same current exchange amplitudes, as the local system (3.12). Complete details can be found in [11], but here we can give some clues that a proper non–local Lagrangian does indeed exist and is actually unique by taking a closer look at the $s = 3$ case. In Section 2.1 we identified two alternatives for the $s = 3$ field, and now we can see explicitly how external currents can distinguish the pseudo–differential operators $\mathcal{F}^{(1)}$ and $\widetilde{\mathcal{F}}^{(1)}$ of eqs. (2.32) and (2.33). Their differences reflect themselves in their Bianchi identities, that in index–free notation read

$$\begin{aligned} &\partial\cdot\mathcal{F}^{(1)} - \frac{1}{4}\,\partial\,\mathcal{F}^{(1)\,\prime} = 0\ , \\ &\partial\cdot\widetilde{\mathcal{F}}^{(1)} - \frac{1}{2}\,\partial\,\widetilde{\mathcal{F}}^{(1)\,\prime} = 0\ , \end{aligned} \tag{3.15}$$

where the first is a special case of eq. (2.37) while the second can be computed directly and coincides with the Bianchi identity for the Fronsdal operator $\mathcal{F}$. As a result, in the two cases one is led to define the divergence–free Einstein–like tensors

$$\mathcal{G}^{(1)} = \mathcal{F}^{(1)} - \frac{1}{4}\,\eta\,\mathcal{F}^{(1)\,\prime} \qquad \text{and} \qquad \widetilde{\mathcal{G}}^{(1)} = \widetilde{\mathcal{F}}^{(1)} - \frac{1}{2}\,\eta\,\widetilde{\mathcal{F}}^{(1)\,\prime}\ , \tag{3.16}$$

and the corresponding Lagrangian equations, where they are sourced by conserved external currents, then yield the current exchange amplitudes

$$\mathcal{J} \cdot \mathcal{F}^{(1)} \equiv \mathcal{J} \cdot \Box\varphi = \mathcal{J} \cdot \mathcal{J} - \frac{3}{D-2}\, \mathcal{J}' \cdot \mathcal{J}' \tag{3.17}$$

and

$$\mathcal{E}_0(D,3) \equiv \mathcal{J} \cdot \widetilde{\mathcal{F}}^{(1)} \equiv \mathcal{J} \cdot \Box\varphi = \mathcal{J} \cdot \mathcal{J} - \frac{3}{D}\, \mathcal{J}' \cdot \mathcal{J}' \,. \tag{3.18}$$

$\widetilde{\mathcal{G}}^{(1)}$ is the correct Einstein–like tensor in this case, since one can verify that only the last expression identifies a traceless and transverse $s = 3$ current.

The detailed analysis in [11] provides a clear explanation of how the Bianchi identities for the $\mathcal{F}^{(n)}$ operators of Section 2.1 determine the proper non–local Einstein–like tensors. For any given s, for n large enough these operators become gauge invariant, while their Bianchi identities loose memory of the term proportional to φ'' and simply relate divergences to gradients of their traces. Therefore, combining traces of the first gauge–invariant $\mathcal{F}^{(n)}$ with corresponding powers of $\frac{\partial^2}{\Box}$ one ought to arrive at a non–local analogue of the gauge–invariant $\mathcal{A}$ tensor of Section 2.1. As shown in [11], this is indeed true, and moreover there is a unique choice to this effect. Consequently, the non–local Lagrangians make full use of the independent gauge–invariant operators that are built by eq. (2.36), as we had stated in Section 2.1, and the end results yield the same current exchange amplitudes as the local theories. Here we can content ourselves with a brief discussion of the final result,

$$\mathcal{E}_0(D,s) \equiv \sum_{n=0}^{N} \rho_n(D-2,s)\, \frac{s!}{n!\,(s-2n)!\,2^n}\, \mathcal{J}^{[n]} \cdot \mathcal{J}^{[n]} \,, \tag{3.19}$$

$$\rho_{n+1}(D,s) = -\, \frac{\rho_n(D,s)}{D+2(s-n-2)} \,, \qquad \rho_0(D,s) = 1 \,, \tag{3.20}$$

which can be justified noticing that the recursive definition (3.20) of the ρ_n makes the sum (3.19) transverse and traceless, as in the preceding $s = 3$ example. The first non–trivial case corresponds to $s = 2$, where eq. (3.19) reduces to

$$\mathcal{E}_0(D,2) = T^{\mu\nu}\, T_{\mu\nu} - \frac{1}{D-2}\, (T^{\mu}{}_{\mu})^2 \tag{3.21}$$

where, abiding to standard conventions, we have called $T_{\mu\nu}$ the corresponding current.

Let me stress that the amplitude $\mathcal{E}_0(D,s)$ depends on the spin s of the HS field determining the exchange and on the number D of space–time dimensions. If one insists on currents that are *conserved* in D dimensions, it is possible to adapt the analysis to *massive* spin–s fields. A harmonic dependence on an internal circle coordinate suffices in fact to introduce masses *à la* Stueckelberg, so that massive exchanges driven by currents that are still *conserved* in D dimensions and lack internal components obtain replacing D with $D+1$ in eq. (3.19). The difference between the exchanges for any given value of s computed in $D+1$ and D dimensions,

$$\Delta_0(D,s) = \sum_{n=0}^{N} \left[\rho_n(D-1,s) - \rho_n(D-2,s)\right] \frac{s!}{n!\,(s-2n)!\,2^n}\, \mathcal{J}^{[n]} \cdot \mathcal{J}^{[n]} \,, \tag{3.22}$$

thus encodes the generalization of the discontinuity originally found by van Dam, Veltmar and Zakharov [53] for $s = 2$, that can be computed starting from eq. (3.21).

This phenomenon has a very instructive counterpart in the presence of a cosmological constant λ that deforms Minkowski space to (A)dS backgrounds depending on its (negative)positive sign. Starting from the deformed equations of Section 3.1 and focussing on the dS case, for $s = 2$ the resulting massive exchange for a generic mass M, computed some time ago in [55], reads[2]

$$\mathcal{E}_\lambda(D,2) \;=\; T^{\mu\nu}\,T_{\mu\nu} \;-\; \frac{1}{D-1}\,\frac{(ML)^2-(D-1)}{(ML)^2-(D-2)}\,\left(T^{\mu}{}_{\mu}\right)^2 \;, \tag{3.23}$$

where the relation between L and the cosmological constant is given in eq. (3.4). Notice that this exchange is a rational function of ML with a simple pole determined by the condition

$$(ML)^2 = D-2 \;. \tag{3.24}$$

The pole in eq. (3.23) is a manifestation of an interesting phenomenon, a shortening that occurs in dS representations precisely for the value (3.24) of $(ML)^2$ [56]. Indeed, as first noticed in [57], when eq. (3.24) holds the theory acquires the "partial gauge symmetry"

$$\delta\varphi_{\mu\nu} = \nabla_\mu\nabla_\nu\,\zeta \,+\, \frac{M^2}{D-2}\,g_{\mu\nu}\,\zeta \;, \tag{3.25}$$

with ζ a scalar parameter.

There is a marked difference between eq. (3.25) and the standard massless gauge symmetry

$$\delta\varphi_{\mu\nu} = \nabla_\mu\,\xi_\nu \,+\, \nabla_\nu\,\xi_\mu \;, \tag{3.26}$$

with ξ_μ a vector parameter, since the coupling to a generic conserved energy–momentum tensor is *not* compatible with the second term in eq. (3.25) unless the trace of $T_{\mu\nu}$ vanishes identically. This fact also brings about the vDVZ discontinuity, since the limiting values of the rational function in eq. (3.23) for $ML \to 0$, that identifies the flat massless case, and for $ML \to \infty$, that identifies the flat massive case, differ in compliance with Liouville's theorem. In other words, the vDVZ discontinuity in flat space is somehow a reflection of the partial gauge symmetry that emerges, for real fine–tuned values of $(ML)^2$, in dS space.

More details can be found in [11], where eq. (3.23) was extended to symmetric tensors of arbitrary rank. The first terms, drawn from [11] but adapted to the index–free notation

[2]This mass deformation should not be confused with the special mass parameters of Section 3.1, which guarantee the massless gauge symmetry in (A)dS backgrounds.

of Section 2.1, read

$$\begin{aligned}
\mathcal{E}_\lambda(D,s) &= \mathcal{J}_s \,+\, \frac{g\,\mathcal{J}_s{}'}{2(\frac{5}{2}-\zeta)}\,\frac{(ML)^2+2(\frac{5}{2}-\zeta)}{(ML)^2-2(\zeta-3)} \\
&+ \frac{g^2\,\mathcal{J}_s{}^{[2]}}{8}\,\frac{(ML)^4+8(ML)^2(\frac{7}{2}-\zeta)+12\left(\frac{5}{2}-\zeta\right)_2}{\left(\frac{5}{2}-\zeta\right)_2[(ML)^2-2(\zeta-3)][(ML)^2-6(\zeta-4)]} \\
&+ \frac{g^3\,\mathcal{J}_s{}^{[3]}}{48}\,\frac{(ML)^6-(ML)^4(18\zeta-77)+92(ML)^2\left(\frac{7}{2}-\zeta\right)_2+120\left(\frac{5}{2}-\zeta\right)_3}{\left(\frac{5}{2}-\zeta\right)_3[(ML)^2-2(\zeta-3)][(ML)^2-6(\zeta-4)][(ML)^2-10(\zeta-5)]} \\
&+ \dots ,
\end{aligned} \tag{3.27}$$

where $\mathcal{J}_s{}^{[2,3]}$ are higher traces of $\mathcal{J}_s$, g is the background dS metric, $\zeta = \frac{D}{2} + s$ and

$$(a)_n = a(a+1)\dots(a+n-1) \tag{3.28}$$

are Pochhammer symbols.

In the dS case the poles lie precisely at real values of ML where partially massless gauge transformations involving terms like the second in eq. (3.25), not protected by gradients and thus incompatible with generic conserved currents, emerge. For a rank–s tensor there are in fact $s-1$ partially massless points, for which

$$(ML)^2 = (s-1-r)(D+s+r-4)\,, \tag{3.29}$$

where $r = 0,\dots,s-2$, while $r = s-1$ corresponds to the more familiar massless point. For all $s \geq 2$ a first pole thus appears at $r = s-2$, and for $s > 3$ others lie at alternate values of r below it. For instance, for $s = 3$ the relevant value is $r = 1$, since for $r = 0$ all terms present in the partially massless gauge transformation contain gradients of the parameter and are thus manifestly compatible with conserved currents.

One can derive eq. (3.27) starting from a $D+1$–dimensional flat Minkowski space and performing a radial reduction [58], a generalization of the polar decomposition of Euclidean space. Among the technical complications discussed in [11], I would like to mention here that the end result was recast in a form that depends manifestly on $(ML)^2$ via interesting identities for the generalized hypergeometric functions ${}_3F_2$, more complicated cousins of the familiar ${}_2F_1$ [59].

Let me stress that a close analogy exists between the singular behavior of dS exchanges at partially massless points and the massless limit of the more familiar flat–space exchanges in the presence of non–conserved currents. For instance, in the Proca theory a non–conserved current would lead to the exchange

$$\left(p^2 + m^2\right) J \cdot A \,=\, J \cdot J - \frac{1}{m^2}\,(p \cdot J)^2 \ , \tag{3.30}$$

whose massless limit is singular precisely because a current that is not conserved conflicts with the emerging gauge symmetry. Following [53], we insisted on conserved currents also in the massive case, and this allowed a direct comparison between massive and massless exchange amplitudes.

A further step along these lines was made by Bekaert, Joung and Mourad in [54]. The key idea of their work was to apply the current–exchange formula (3.19) to the *dynamical* Noether HS currents of a complex scalar field Φ. These can be defined, as in [60], via the Wigner function

$$J(x,v) = \bar{\varphi}(x+iv)\,\varphi(x-iv)\,, \tag{3.31}$$

where the spin–s term is combined with s powers of the vector v. At the same time, the strength of the interactions with generic HS symmetric tensors can be characterized via a single dimensionful coupling, that by analogy with String Theory I will call α' in the following, and via an infinite number of dimensionless couplings a_k, or rather via a *coupling function*

$$a(z) = \sum_r \frac{z^r}{r!}\, a_r\,, \tag{3.32}$$

that in String Theory would be dominated by an exponential. Classic results in the theory of orthogonal polynomials allow one to combine the current–exchange amplitude (3.19), a pair of HS currents as in eq. (3.31) and the coupling function (3.32) into a compact formula for the interaction between complex scalars and infinitely many massless HS fields. In four dimensions the result is strikingly neat, and for the $\varphi+\varphi \rightarrow \varphi+\varphi$ amplitude reads

$$\begin{aligned}\mathcal{A}^{(s)} = -\frac{1}{\alpha' s}\left[a\left(\frac{\alpha'}{4}(u-t)+\frac{\alpha'}{2}\sqrt{-ut}\right)+a\left(\frac{\alpha'}{4}(u-t)-\frac{\alpha'}{2}\sqrt{-ut}\right)-a_0\right]\times \\ \varphi_1(p_1)\,\varphi_2(p_2)\,\varphi_3(p_3)\,\varphi_4(p_4)\,,\end{aligned} \tag{3.33}$$

where s, t and u are the familiar Mandelstam variables, since the sum contributing to the current exchange can be related to Chebyshev polynomials [59].

This beautiful expression has a number of interesting lessons in store. For one matter, it is a consistent four–scalar amplitude involving the exchange of infinitely many *massless* HS particles. Moreover, the detailed discussion in [54] shows that, in principle, a soft behavior at high energies can be attained working only with (infinitely many) symmetric fields, provided the coupling function tends to zero for large negative real values of its argument. In String Theory the essential singularity of $a(z)$ may be held ultimately responsible for the presence of lower Regge trajectories, since a soft behavior for the conjugate amplitude $\varphi+\bar{\varphi} \rightarrow \varphi+\bar{\varphi}$ would also demand that $a(z)$ tend to zero for large positive real arguments. Therefore, as stressed in [54], in the present setting a soft behavior for all conjugate amplitudes would require that the coupling function $a(z)$ tend to zero at infinity in the complex plane. This is a subtle condition, since Liouville's theorem would then require the presence of singularities in the finite plane, which in their turn signal in general an extended nature for the objects involved. There is clearly more to be understood here, and other intriguing properties will show up in the ensuing discussion.

3.3 String lessons for cubic HS couplings

I can now turn to the scattering amplitudes of the open bosonic string involving its massive HS excitations. As we shall see, they contain a wealth of information on HS couplings [43], and this analysis provides some further evidence for the long–held, but never fully quantified, expectation that String Theory describes a broken phase for HS gauge symmetries. This brings about a natural link between cubic couplings and Noether currents, with the potential of clarifying also the nature of the latter. The situation is particularly manageable for the symmetric tensors of the first Regge trajectory, whose cubic disk amplitudes can be simply computed from the path integral for the free action

$$S_P[X,\gamma] \;=\; -\,\frac{1}{4\pi\alpha'}\int_M d^2\xi\,\eta^{ab}\,\partial_a X^\mu\,\partial_b X_\mu \tag{3.34}$$

in the presence of special external currents. Eq. (3.34) is in fact the remnant of the bosonic string action after a complete gauge fixing in the critical dimension, while the vertex operators of the first Regge trajectory, combinations of an exponential with powers of ∂X^μ, can all be recovered as Taylor coefficients in the "symbols" ξ^μ of

$$\mathcal{V}(\sigma,p,\xi) \;=\; \exp\left[ip\cdot X(\sigma) \,+\, \xi\cdot\partial X(\sigma)\right]\,. \tag{3.35}$$

Up to a measure factor that can be ascribed to ghost fields, tree–level correlation functions of these types of vertices on the disk are then tantamount to a gaussian path integral in the presence of the boundary currents

$$J(\sigma,p,\xi) \;=\; \sum_{i=1}^{N}\left(p_i\,\delta^2(\sigma-\sigma_i) \;-\; \frac{\xi_i}{\sqrt{2\alpha'}}\,\partial_\sigma\,\delta^2(\sigma-\sigma_i)\right)\,. \tag{3.36}$$

All these amplitudes, and three–point functions in particular, can be extracted from

$$\begin{aligned} S^{open}_{j_1\cdots j_n} = \int_{\mathbb{R}^{n-3}} & dy_4\cdots dy_n\;|y_{12}y_{13}y_{23}|\;\times \\ & \langle\,\mathcal{V}_{j_1}(\hat{y}_1)\mathcal{V}_{j_2}(\hat{y}_2)\mathcal{V}_{j_3}(\hat{y}_3)\cdots\mathcal{V}_{j_n}(y_n)\,\rangle Tr(\Lambda^{a_1}\cdots\Lambda^{a_n})\,, \end{aligned} \tag{3.37}$$

where the trace is a Chan–Paton factor [48] and the integrals are along the real axis, so that the interior of the disk corresponds to the upper half–plane. For brevity, we have left implicit momenta and ξ–variables for the n vertex operators in (3.37). Standard field theory techniques then imply that correlation functions of the vertex operators, determined by the two–dimensional Green function, take the form

$$\begin{aligned} &\langle\,\mathcal{V}_{j_1}(\hat{y}_1)\cdots\mathcal{V}_{j_n}(y_n)\,\rangle = \\ &\qquad \exp\left[\sum_{i\neq j}^{N}\left(\alpha' p_i\cdot p_j\,\ln|y_{ij}| + \sqrt{2\alpha'}\,\frac{\xi_i\cdot p_j}{y_{ij}} + \frac{1}{2}\,\frac{\xi_i\cdot\xi_j}{y_{ij}^2}\right)\right]\,, \end{aligned} \tag{3.38}$$

where the y_i denote the locations of the punctures along the real axis and $y_{ij} = y_i - y_j$.

Eq. (3.38) would seem to violate basic tenets of String Theory, since the dependence on the location of the punctures ought to disappear from three–point functions, up to a measure factor compensating exactly the one introduced by ghosts in eq. (3.37). Of course, this property ought to hold solely for on–shell external states satisfying the Virasoro conditions, so that a closer look into the matter is necessary before coming to definite conclusions. The independent Virasoro conditions,

$$\begin{aligned} (L_0 - 1)\,|\Psi\rangle &= 0\,, \\ L_1\,|\Psi\rangle &= 0\,, \\ L_2\,|\Psi\rangle &= 0 \end{aligned} \tag{3.39}$$

translate directly into the DFP conditions (2.1) of Section 2, as the reader can simply verify, and considering three external states such that

$$-p_1^2 = \frac{n_1 - 1}{\alpha'}\,, \qquad -p_2^2 = \frac{n_2 - 1}{\alpha'}\,, \qquad -p_3^2 = \frac{n_3 - 1}{\alpha'}\,, \tag{3.40}$$

the three–point correlation functions (3.38) take the form

$$\left|\frac{y_{12}y_{13}}{y_{23}}\right|^{n_1} \left|\frac{y_{12}y_{23}}{y_{13}}\right|^{n_2} \left|\frac{y_{13}y_{23}}{y_{12}}\right|^{n_3} \times \exp\left[\sum_{i\neq j}^{3}\left(\frac{1}{2}\frac{\xi_i\cdot\xi_j}{y_{ij}^2} + \sqrt{2\alpha'}\,\frac{\xi_i\cdot p_j}{y_{ij}}\right)\right]\,. \tag{3.41}$$

Apparently, this expression still depends on the location of the punctures. However, on account of the Virasoro conditions, the spins of the external states should be correlated to their mass levels, so that in eq. (3.41) one should only retain terms containing n_1 powers of ξ_1, n_2 powers of ξ_2 and n_3 powers of ξ_3. It is then pleasing to verify how, for this class of terms, the dependence on the y_i disappears altogether, so that one is finally left with

$$\begin{aligned} \exp\Bigg\{\sqrt{\frac{\alpha'}{2}}\,\bigg[\xi_1\cdot p_{23}\left\langle\frac{y_{23}}{y_{12}y_{13}}\right\rangle + \xi_2\cdot p_{31}\left\langle\frac{y_{13}}{y_{12}y_{23}}\right\rangle \\ + \xi_3\cdot p_{12}\left\langle\frac{y_{12}}{y_{13}y_{23}}\right\rangle\bigg] + \big[\xi_1\cdot\xi_2 + \xi_1\cdot\xi_3 + \xi_2\cdot\xi_3\big]\Bigg\}\,. \end{aligned} \tag{3.42}$$

In this expression the brackets are merely signs: they reflect the flip symmetry [48] of string amplitudes, which is implemented by projective–disk amplitudes and plays a crucial role whenever external states of odd spin are present.

Eq. (3.42) should be translated into the language of Field Theory, taking into account that s powers of the ξ_i are meant to correspond to a spin–s field of the first Regge trajectory, a rank–s symmetric tensor. Alternatively, one would like to associate to the external fields generic polarization tensors rather than simply powers of the ξ_i's, and this is effected by a $\star$ – product, a convenient procedure to multiply pairs of series. Thus, if

$$A(\xi) = \sum_{k=0}^{\infty} A_{\mu_1\ldots\mu_k}\,\frac{\xi^{\mu_1}\ldots\xi^{\mu_k}}{k!} \tag{3.43}$$

and

$$B(\xi) = \sum_{k=0}^{\infty} B_{\mu_1\ldots\mu_k}\,\frac{\xi^{\mu_1}\ldots\xi^{\mu_k}}{k!}\,, \tag{3.44}$$

then

$$A \star B = \sum_{k=0}^{\infty} \frac{1}{k!} A^{\mu_1 \dots \mu_k} B_{\mu_1 \dots \mu_k} , \qquad (3.45)$$

which is indeed reconstructing generic polarization tensors from the special low–rank tensors determined by the generating function (3.38). The standard presentation of this $\star$ – product is

$$A \star B = \exp\left(\frac{\partial}{\partial\xi} \cdot \frac{\partial}{\partial\eta}\right) A(\xi)\, B(\eta)\Big|_{\xi=\eta=0} , \qquad (3.46)$$

while a second presentation,

$$A \star B = \int \frac{d^d U}{(2\pi)^{d/2}} \tilde{A}(U)\, B(i\,U) , \qquad (3.47)$$

where $\tilde{A}$ denotes the Fourier transform of S with respect to the symbols, follows inserting into the first the definition of A via an inverse Fourier transform. More details on how these results lead directly to the general cubic couplings

$$\varphi_1\left(p_1, \partial_\xi \pm \sqrt{\frac{\alpha'}{2}}\, p_{31}\right) \varphi_2\left(p_2, \xi + \partial_\xi \pm \sqrt{\frac{\alpha'}{2}}\, p_{23}\right) \varphi_3\left(p_3, \xi \pm \sqrt{\frac{\alpha'}{2}}\, p_{12}\right)\Bigg|_{\xi=0} \qquad (3.48)$$

can be found in [43]. This expression is to be computed at $\xi = 0$, $p_{ij} = p_i - p_j$ and the notation is as in eq. (3.43), so that for example the first factor is

$$\sum_{k=0}^{\infty} \varphi_{\mu_1 \dots \mu_k}(p_1)\, \frac{\left(\partial_\xi \pm \sqrt{\frac{\alpha'}{2}}\, p_{31}\right)^{\mu_1} \dots \left(\partial_\xi \pm \sqrt{\frac{\alpha'}{2}}\, p_{31}\right)^{\mu_k}}{k!} , \qquad (3.49)$$

where derivatives act to the right. For traceless φ fields there is no ordering ambiguity.

It is instructive to expand eq. (3.48) for the first few cases,

$$\begin{aligned}
\mathcal{A}^{\pm}_{0-0-s} &= \left(\pm\sqrt{\frac{\alpha'}{2}}\right)^{s} \varphi_1\, \varphi_2\, \varphi_3 \cdot p_{12}^{s} , \\
\mathcal{A}^{\pm}_{1-1-s} &= \left(\pm\sqrt{\frac{\alpha'}{2}}\right)^{s-2} s(s-1)\, A_{1\,\mu}\, A_{2\,\nu}\, \varphi^{\mu\nu\dots}\, p_{12}^{s-2} \\
&+ \left(\pm\sqrt{\frac{\alpha'}{2}}\right)^{s} \Big[A_1 \cdot A_2\, \varphi \cdot p_{12}^{s} + s\, A_1 \cdot p_{23}\, A_{2\,\nu}\, \varphi^{\nu\dots}\, p_{12}^{s-1} \\
&+ s\, A_2 \cdot p_{31}\, A_{1\,\nu}\, \varphi^{\nu\dots}\, p_{12}^{s-1}\Big] \\
&+ \left(\pm\sqrt{\frac{\alpha'}{2}}\right)^{s+2} A_1 \cdot p_{23}\, A_2 \cdot p_{31}\, \varphi \cdot p_{12}^{s} ,
\end{aligned} \qquad (3.50)$$

the first of which corresponds to the Wigner currents (3.31) used in [54]. Notice that the $0-0-s$ amplitude in (3.50) does not retain any memory, so to speak, of the fact that for $s > 1$ the φ fields are *massive* in the open bosonic string. Indeed, if all momenta were those of massless fields it would be automatically invariant under the gauge transformation (2.9) up to Virasoro or DFP conditions, which would imply $p_i \cdot p_j = 0$ for all i and j. On the other hand, even with all massless momenta the first line of the $1-1-s$ amplitude, which contains its lowest–derivative terms, would be incompatible with the spin–s gauge symmetry.

This feature of cubic vertices is along the lines of what we have stressed in preceding sections: in the free string spectrum, mass generation *à la* Stueckelberg rests precisely on terms that interactions feed into the quadratic level, consistently with the fact that a conventional gauge symmetry is recovered if these are left aside. This is what happens in the limit $\alpha' \to \infty$, *i.e.* in the limit of vanishing string tension. Hence, if a breaking phenomenon were really at work in String Theory, contributions from higher orders ought to pop up amidst cubic couplings. This phenomenon is strikingly visible at the cubic level: once the non–invariant terms, that carry along higher powers of the order parameter $\frac{1}{\sqrt{\alpha'}}$ for the breaking, are taken apart, they leave behind a conventional gauge symmetry. Unfortunately, α' lacks a dynamical origin in our present formulation of String Theory, while a minor subtlety is that all this occurs up to divergences and traces in these S–matrix amplitudes. Still, any gauge variations of the cubic vertices that are proportional to p_i^2 signal the mere need for deformed gauge transformations, up to trivial redefinitions, since any non–linear gauge variations of the kinetic terms that could in principle compensate them would be inevitably proportional to the free field equations.

Let us therefore remove from the cubic vertices all terms like the first in eq. (3.50) and focus, for all values of s_1, s_2 and s_3, on contributions that become gauge invariant on shell when the masses of the external states are removed, The end result, rather rewarding and strikingly simple, is captured by the expression [43] [3]

$$\mathcal{A}^{\pm} = e^{\pm\Gamma}\, \varphi_1(p_1, \xi_1)\, \varphi_2(p_2, \xi_2)\, \varphi_3(p_3, \xi_3)\Big|_{\xi_i=0}\,, \tag{3.51}$$

so that it rests on the operator

$$\begin{aligned}\Gamma = \sqrt{\frac{\alpha'}{2}}\Big\{ &[1+(\partial_{\xi_1}\cdot\partial_{\xi_2})]\,(\partial_{\xi_3}\cdot p_{12}) + [1+(\partial_{\xi_2}\cdot\partial_{\xi_3})]\,(\partial_{\xi_1}\cdot p_{23}) \\ &+ [1+(\partial_{\xi_3}\cdot\partial_{\xi_1})]\,(\partial_{\xi_2}\cdot p_{31})\Big\}\,,\end{aligned} \tag{3.52}$$

which commutes with gauge transformations (in this notation, with terms of the form $p_i \cdot \xi_i$), up to Virasoro or DFP conditions. There is therefore a unique term containing $s_1 + s_2 + s_3$ powers of momenta, proportional to

$$p_{23}^{s_1} \cdot \varphi_1(p_1)\; p_{31}^{s_2} \cdot \varphi_2(p_2)\; p_{12}^{s_3} \cdot \varphi_3(p_3)\,, \tag{3.53}$$

[3]Let me stress that retaining all these gauge–invariant terms is a weaker condition than simply letting $\alpha' \to \infty$, which would only leave the leading terms with $s_1 + s_2 + s_3$ derivatives that I shall describe shortly. The leading terms, incidentally, were first identified long ago by Gross and Mende [61].

together with other groups of terms containing lower numbers of derivatives, whose structure is best captured by an equivalent expression for $\mathcal{A}$,

$$\begin{aligned}\mathcal{A}^{\pm} &= e^{\pm \mathcal{G}}\, \phi_1\left(p_1\,, \xi_1 \pm \sqrt{\frac{\alpha'}{2}}\, p_{23}\right) \times \\ & \phi_2\left(p_2\,, \xi_2 \pm \sqrt{\frac{\alpha'}{2}}\, p_{31}\right)\, \phi_3\left(p_3\,, \xi_3 \pm \sqrt{\frac{\alpha'}{2}}\, p_{12}\right)\Bigg|_{\xi_i=0}\,,\end{aligned} \tag{3.54}$$

where $p_{ij} = p_i - p_j$, the expression is to be evaluated at $\xi_i = 0$ and

$$\mathcal{G} = \sqrt{\frac{\alpha'}{2}}\left[(\partial_{\xi_1}\cdot\partial_{\xi_2})(\partial_{\xi_3}\cdot p_{12}) + (\partial_{\xi_2}\cdot\partial_{\xi_3})(\partial_{\xi_1}\cdot p_{23}) + (\partial_{\xi_3}\cdot\partial_{\xi_1})(\partial_{\xi_2}\cdot p_{31})\right]. \tag{3.55}$$

Notice that $\mathcal{G}$ reduces indeed by two units the total number of momenta and removes at the same time three ξ_i's, one from each of the fields. Therefore if s_3, say, is the smallest of the three spins, the operation can be repeated at most s_3 times, leading precisely to the removal of $2\, s_3$ of the original momenta. In particular, for three spin–1 external states eqs. (3.51) and (3.54) yield both the three–derivative vertex displayed in Schwarz's 1982 review [48] and the cubic Yang–Mills vertex. Similarly, for three spin–2 external states they yield three groups of terms with six, four and two derivatives, the last of which is the standard cubic Einstein vertex. As a result, eqs. (3.51) and (3.54) involve terms with overall powers of momenta ranging between $s_1 + s_2 + s_3 - 2\min\{s_i\}$ and $s_1 + s_2 + s_3$, precisely as in Metsaev's light–cone classification of cubic HS interactions [49].[4]

Infinitely many gauge–invariant three point functions are tantamount to infinitely many conserved currents that generalize the Wigner function of eq. (3.31),

$$\begin{aligned}\mathcal{J}^{\pm}(x\,;\,\xi) &= \exp\left(\mp i\sqrt{\frac{\alpha'}{2}}\,\xi_\alpha\left[\partial_{\zeta_1}\cdot\partial_{\zeta_2}\,\partial_{12}^{\alpha} - 2\,\partial_{\zeta_1}^{\alpha}\,\partial_{\zeta_2}\cdot\partial_1 + 2\,\partial_{\zeta_2}^{\alpha}\,\partial_{\zeta_1}\cdot\partial_2\right]\right) \times \\ & \varphi_1\left(x_1 \mp i\sqrt{\frac{\alpha'}{2}}\,\xi, \zeta_1 \mp i\sqrt{2\alpha'}\partial_2\right)\, \varphi_2\left(x_2 \pm i\sqrt{\frac{\alpha'}{2}}\xi, \zeta_2 \pm i\sqrt{2\alpha'}\partial_1\right)\Bigg|_{\zeta_i=0}\,,\end{aligned} \tag{3.56}$$

where ∂_1, ∂_2 and ∂_{12} are shorthands for ∂_{x_1}, ∂_{x_2} and for their difference, and where at the end of the computation one is to set x_1 and x_2 equal to x and $\zeta_i = 0$. In [43], where the reader can find more details, we also presented an educated guess for the counterpart of eq. (3.51) for the superstring, where (3.56) leaves way to bosonic and fermionic currents. Notice that the spin–2 currents built from a pair of spin–s fields contain at least $2s$-2 derivatives, and therefore *are not* the energy–momentum tensors of the spin–s fields implied by the Lagrangians of Section 2.

At this stage we have really left String Theory to return to a field theory framework, albeit in an apparently incomplete fashion, since the starting point was provided by cubic

[4] I cannot fail to mention that a very recent (A)dS extension [62] shows in a concise and elegant fashion how the covariant derivatives (3.3) generate Fradkin–Vasiliev minimal–like dressings, the counterparts for cubic vertices of the mass–like terms of Section 3.1, although this goes well beyond the Erice School.

on–shell string amplitudes. Complete cubic vertices that are gauge invariant up to the Fronsdal equations (2.8) or their compensator analogs (2.16) can be recovered completing eq. (3.51), as in [43, 44], with two types of terms that we ignored so far. These involve divergences and traces of the three external fields, and actually fall into the combinations

$$\mathcal{H}_{ij} = \left(1 + \partial_{\xi_i} \cdot \partial_{\xi_j}\right) i\,\mathcal{D}_j - \frac{1}{2}\, p_j \cdot \partial_{\xi_i}\, \partial_{\xi_j} \cdot \partial_{\xi_j} \,, \tag{3.57}$$

with

$$i\,\mathcal{D}_i = p_i \cdot \partial_{\xi_i} - \frac{1}{2}\, p_i \cdot \xi_i\, \partial_{\xi_i} \cdot \partial_{\xi_i} \tag{3.58}$$

the de Donder operator. In terms of the $\mathcal{H}_{ij}$ the complete cubic vertex in Fronsdal's constrained form reads [5]

$$\begin{aligned} \mathcal{A}^{\pm} = e^{\pm\Gamma} \Bigg[1 + \left(\frac{\alpha'}{2}\right) \left(\mathcal{H}_{12}\,\mathcal{H}_{13} + \mathcal{H}_{21}\,\mathcal{H}_{23} + \mathcal{H}_{31}\,\mathcal{H}_{32}\right) \pm \left(\frac{\alpha'}{2}\right)^{\frac{3}{2}} \times \\ \left(: \mathcal{H}_{21}\,\mathcal{H}_{32}\,\mathcal{H}_{13} : - \ : \mathcal{H}_{12}\,\mathcal{H}_{31}\,\mathcal{H}_{23} : \right) \Bigg] \varphi_1\left(p_1, \xi_1\right)\, \varphi_2\left(p_2, \xi_2\right)\, \varphi_3\left(p_3, \xi_3\right) \,, \end{aligned} \tag{3.59}$$

where the normal ordering moves de Donder operators to the right in products with ∂_{ξ_i}'s. This is the point where the construction of [43] meets the independent work of Manvelyan, Mkrtchyan and Ruhl [44], who obtained the same cubic vertices from field theory considerations. Strictly speaking, however, in [43] we were more precise with the Chan–Paton symmetry [48], which makes full use in general of the sign option in eqs. (3.51) or (3.54).

3.4 Four–point functions and beyond

In [42] Taronna brought the discussion forward by some stimulating steps that can potentially lead to further progress. He moved from the observation that the operator of eq. (3.52), whose exponential generates the whole sequence of cubic vertices, comprises two different groups of familiar terms. The first,

$$(\partial_{\xi_1} \cdot \partial_{\xi_2})(\partial_{\xi_3} \cdot p_{12}) + (\partial_{\xi_2} \cdot \partial_{\xi_3})(\partial_{\xi_1} \cdot p_{23}) + (\partial_{\xi_3} \cdot \partial_{\xi_1})(\partial_{\xi_2} \cdot p_{31}) \,, \tag{3.60}$$

in the usual tensor notation would become the familiar cubic Yang–Mills vertex

$$(\mathrm{YM3})_{\mu\nu\rho} = \eta_{\mu\nu}\, p_{12,\,\rho} + \eta_{\nu\rho}\, p_{23,\,\mu} + \eta_{\rho\mu}\, p_{31,\,\nu} \,, \tag{3.61}$$

while the rest is a combination of scalar–scalar–vector vertices. Let me stress that the key property of Γ is that it commutes with gauge transformations, up to Virasoro or DFP conditions. As we have seen, this makes it possible, for an action combining the quadratic terms of Section 2.1 with the vertex (3.51), to be gauge invariant under transformations that are generally deformed. While String Theory uses the exponential of Γ, it would seem possible to consider more general functions of this operator in HS constructions,

[5] Appendix A of [43] explains how to pass to the compensator formulation of Section 2.1.

altering the relative weights of terms involving different numbers of derivatives. This is another point that deserves further investigation.

A key question addressed in [42] is what should be the counterparts, for four and higher–point functions, of the exponent Γ that builds cubic HS vertices. For brevity, let me leave aside the scalar–scalar–vector terms present in Γ to concentrate on the four–point functions built starting solely from (3.61). To begin with, these are four–point Yang–Mills S–matrix amplitudes, but interestingly they suffice to define a whole class of HS amplitudes. In order to display them, let us first recast the Yang–Mills amplitudes in terms of *independent* group–theory factors, or independent combinations of Chan–Paton factors, that a reasoning along the lines of [48] would guarantee to also allow the factorization of S–matrix amplitudes involving generic HS states. The end result combines exchange amplitudes in pairs of channels and quartic Yang–Mills interactions and thus, amusingly, has *planar duality*, albeit in a very simple form where the poles are manifest in pairs of channels. It is uniquely determined once, making use of the Jacobi identity, one sorts out the contributions to the Yang–Mills amplitude that carry along independent products of structure constants. One of these contributions takes the form

$$A_\mu(p_1)\, A_\nu(p_2)\, A_\rho(p_3)\, A_\sigma(p_4)\; \mathcal{A}^{\mu\nu\rho\sigma}\; \mathrm{tr}\left([T_1,T_2]\,[T_3,T_4]\right)\ , \tag{3.62}$$

where the dynamical amplitude has been combined with its Chan–Paton factors and, schematically,

$$\mathcal{A}^{\mu\nu\rho\sigma}\ =\ \left(\mathrm{YM3}\ \frac{1}{s}\ \mathrm{YM3}\ +\ \mathrm{YM4}\right)^{\mu\nu\rho\sigma}\ +\ \left(\mathrm{YM3}\ \frac{1}{u}\ \mathrm{YM3}\ +\ \mathrm{YM4}\right)^{\mu\sigma\nu\rho}\ . \tag{3.63}$$

Here s and u are Mandelstam variables, while the YM4's are the usual quartic Yang–Mills couplings. Notice that the decoupling of longitudinal polarizations from (3.63) translates into an invariance of (3.62) under on–shell Abelian gauge transformations of the type

$$A_i(p_i)\ \longrightarrow\ A_i(p_i)\ +\ i\,p_i\,\Lambda(p_i)\ . \tag{3.64}$$

This key property of Yang–Mills tree amplitudes must hold individually for the two groups of terms, since they carry along independent group theory factors, and in the complete amplitude the (s,u) contribution (3.63) is accompanied by a similar (s,t) contribution,

$$A_\mu(p_1)\, A_\nu(p_2)\, A_\rho(p_3)\, A_\sigma(p_4)\; \mathcal{A}^{\mu\rho\sigma\nu}\; \mathrm{tr}\left([T_1,T_3]\,[T_4,T_2]\right)\ . \tag{3.65}$$

The spin–s amplitudes proposed in [42] include

$$\varphi_{\mu_1\ldots\mu_s}(p_1)\, \varphi_{\nu_1\ldots\nu_s}(p_2)\, \varphi_{\rho_1\ldots\rho_s}(p_3)\, \varphi_{\sigma_1\ldots\sigma_s}(p_4)\ \, (su)^{s-1} \prod_{k=1}^{s} \mathcal{A}^{\mu_k\nu_k\rho_k\sigma_k}\ , \tag{3.66}$$

that should be properly dressed with the same Chan–Paton factors as in eq. (3.62) and then combined with corresponding (s,t) contributions. Notice that these amplitudes are invariant by construction under the HS counterparts of eq. (3.64), Abelian on–shell transformations that in the index–free notation of Section 2 would read

$$\varphi(p_i) \longrightarrow \varphi(p_i) + i\,p_i\,\Lambda(p_i)\ . \tag{3.67}$$

Moreover, the crucial power of su has a two–fold effect: it guarantees that only single poles in s and u be present and raises the spin of the exchanged particles to an *odd* value, $2s - 1$, which is higher than the spin of the external states for $s > 1$. Notice that the end result possesses the same symmetries as the Chan–Paton factor of eq. (3.62), which it should carry in a construction along the lines of [48], where the flip symmetries would be directly correlated to the spins of the massless external states.

Tensoring the independent contributions to four–point Yang–Mills amplitudes has thus led to an infinite family of HS amplitudes that by construction do not suffer from Weinberg's factorization problem. How can this be the case? In order to answer this question, Weinberg's argument was extended in [42] to generic soft amplitudes, taking into account the explicit form of the vertex (3.51). The conclusion is that *no problems are encountered if the intermediate spins are not lower than the external ones*, precisely as was the case for eq. (3.66). The very existence of the amplitudes (3.66), however, raises another important question. Namely, once the cubic vertices are given, what prevents one from building with them "bad" amplitudes, with internal spins that are too low to comply with Weinberg's factorization? The answer proposed in [42] is very interesting and controversial at the same time. It makes use of non–local quartic vertices, in fact the minimal set of them needed to barely remove the "bad" exchange amplitudes built from any given pair of vertices (3.51). In this fashion, one would be left with a subset of non–vanishing amplitudes, all invariant by construction under the Abelian gauge transformations that signal the decoupling of unphysical polarizations and thus complying, a fortiori, with the extension of Weinberg's soft emission argument. As we have already mentioned, however, in this fashion external HS particles would interact only via exchanges of other HS particles with spins not lower than the external ones! This state of affairs brings to one's mind the peculiar superstring states considered in [63], and it would be interesting to elaborate further on the possible links between the two situations. Unfortunately, as noticed in [42], with finitely many spins the removal of the unwanted exchanges is apparently in conflict with unitarity, which is somehow the Coleman–Mandula argument haunting back again. A proper understanding of the singular behavior of infinitely many HS fields once all infrared cutoffs, be they masses or a cosmological term, are removed, will thus require more effort. Clearly, further progress along these lines will be highly instructive.

Acknowledgments

This work was supported in part by Scuola Normale Superiore, by INFN, by the MIUR-PRIN contract 2009-KHZKRX and by the ERC Senior Grant n. 226455, "Supersymmetry, Quantum Gravity and Gauge Fields" (SUPERFIELDS). I am very grateful to Mirian Tsulaia and Jihad Mourad for pleasant and instructive collaborations on some of the topics reviewed in these notes, and to Dario Francia, Andrea Campoleoni and Massimo Taronna for long and fruitful collaborations and also for a number of useful suggestions on the manuscript, whose final form also benefitted from some input from Karapet Mkrtchyan. Finally, I would like to thank Prof. Noriaki Kitazawa of Tokyo Metropolitan University for a careful "non–expert reading" that helped me greatly to streamline the presentation.

References

[1] G. Veneziano, Nuovo Cim. **A57** (1968) 190-197.

[2] See, for instance: M. B. Green, J. H. Schwarz and E. Witten, "Superstring Theory", 2 vols. (Cambridge Univ. Press, Cambridge, UK, 1987); J. Polchinski, "String theory", 2 vols. (Cambridge Univ. Press, Cambridge, UK, 1998); B. Zwiebach, "A first course in string theory", (Cambridge Univ. Press, Cambridge, UK, 2004); K. Becker, M. Becker and J. H. Schwarz, "String theory and M-theory: A modern introduction", (Cambridge Univ. Press, Cambridge, UK, 2007); E. Kiritsis, "String theory in a nutshell", (Princeton Univ. Press, Princeton, NJ, USA, 2007).

[3] "Higher-Spin Gauge Theories", Proceedings of the First Solvay Workshop, held in Brussels on May 12-14, 2004, eds. R. Argurio, G. Barnich, G. Bonelli and M. Grigoriev (Int. Solvay Institutes, 2006). This collection contains some contributions closely related, in spirit, to the present work, including: M. Bianchi and V. Didenko, arXiv:hep-th/0502220; D. Francia and C. M. Hull, arXiv:hep-th/0501236; N. Bouatta, G. Compere and A. Sagnotti, arXiv:hep-th/0409068; X. Bekaert, S. Cnockaert, C. Iazeolla and M. A. Vasiliev, arXiv:hep-th/0503128; A. Sagnotti, E. Sezgin and P. Sundell, arXiv:hep-th/0501156. Other recent reviews include: D. Sorokin, AIP Conf. Proc. **767** (2005) 172 [hep-th/0405069]; D. Francia and A. Sagnotti, J. Phys. Conf. Ser. **33** (2006) 57 [arXiv:hep-th/0601199]; A. Fotopoulos and M. Tsulaia, arXiv:hep-th/0805.1346; A. Campoleoni, Riv. Nuovo Cim. **033** (2010) 123-253. [arXiv:0910.3155 [hep-th]]; A. Sagnotti, arXiv:hep-th/1002.3388; X. Bekaert, N. Boulanger and P. Sundell, arXiv:1007.0435 [hep-th].

[4] W. Siegel, Nucl. Phys. **B263** (1986) 93; W. Siegel and B. Zwiebach, Nucl. Phys. **B263** (1986) 105; T. Banks and M. E. Peskin, Nucl. Phys. **B264** (1986) 513; M. Kato and K. Ogawa, Nucl. Phys. **B212** (1983) 443; N. Ohta, Phys. Rev. **D33** (1986) 1681, Phys. Lett. **B179** (1986) 347, Phys. Rev. Lett. **56** (1986) 440 [Erratum-ibid. **56** (1986) 1316]; A. Neveu, H. Nicolai and P. C. West, Nucl. Phys. **B264** (1986) 573; A. Neveu and P. C. West, Nucl. Phys. **B268** (1986) 125; E. Witten, Nucl. Phys. **B268** (1986) 253.

[5] M. Bianchi, J. F. Morales and H. Samtleben, JHEP **0307** (2003) 062 [arXiv:hep-th/0305052 [hep-th]]; N. Beisert, M. Bianchi, J. F. Morales and H. Samtleben, JHEP **0402** (2004) 001 [hep-th/0310292], JHEP **0407** (2004) 058 [hep-th/0405057].

[6] J. M. Maldacena, Adv. Theor. Math. Phys. **2** (1998) 231-252. [hep-th/9711200]; E. Witten, Adv. Theor. Math. Phys. **2** (1998) 253-291. [hep-th/9802150]; S. S. Gubser, I. R. Klebanov, A. M. Polyakov, Phys. Lett. **B428** (1998) 105-114. [hep-th/9802109]. For a review see: O. Aharony, S. S. Gubser, J. M. Maldacena, H. Ooguri, Y. Oz, Phys. Rept. **323** (2000) 183-386. [hep-th/9905111].

[7] E. Sezgin and P. Sundell, Nucl. Phys. B **644** (2002) 303 [Erratum-ibid. B **660** (2003) 403] [hep-th/0205131].

[8] E. Majorana, Nuovo Cim. **9** (1932) 335.

[9] P. A. M. Dirac, Proc. Roy. Soc. Lond. **155A** (1936) 447; M. Fierz, Helv. Phys. Acta **12** (1939) 3-37; M. Fierz and W. Pauli, Proc. Roy. Soc. Lond. **A173** (1939) 211.

[10] M. A. Vasiliev, Annals Phys. **190** (1989) 59, Phys. Lett. **B238** (1990) 305, Phys. Lett. **B243** (1990) 378, Class. Quant. Grav. **8** (1991) 1387, Phys. Lett. **B257** (1991) 111, Phys. Lett. **B285** (1992) 225, Fortsch. Phys. **52** (2004) 702 [arXiv:hep-th/0401177]; For a review see also: M. A. Vasiliev, Int. J. Mod. Phys. **D5** (1996) 763 [arXiv:hep-th/9611024], arXiv:hep-th/9910096.

[11] D. Francia, J. Mourad and A. Sagnotti, Nucl. Phys. **B773** (2007) 203 [arXiv:hep-th/0701163]; Nucl. Phys. **B804** (2008) 383-420. [arXiv:0803.3832 [hep-th]].

[12] T. Curtright, Phys. Lett. **B165** (1985) 304; C. S. Aulakh, I. G. Koh and S. Ouvry, Phys. Lett. **B173** (1986) 284; C. S. Aulakh, I. G. Koh and S. Ouvry, Phys. Lett. **B173** (1986) 284; S. Ouvry and J. Stern, Phys. Lett. **B177** (1986) 335; W. Siegel and B. Zwiebach, Nucl. Phys. **B282** (1987) 125; W. Siegel, Nucl. Phys. **B284** (1987) 632;

[13] J. M. F. Labastida and T. R. Morris, Phys. Lett. **B180** (1986) 101; J. M. F. Labastida, Phys. Rev. Lett. **58** (1987) 531; J. M. F. Labastida, Nucl. Phys. **B322** (1989) 185.

[14] X. Bekaert and N. Boulanger, Commun. Math. Phys. **245** (2004) 27 [arXiv:hep-th/0208058], Phys. Lett. **B561** (2003) 183 [arXiv:hep-th/0301243]; P. de Medeiros and C. Hull, JHEP **0305** (2003) 019 [arXiv:hep-th/0303036].

[15] A. Campoleoni, D. Francia, J. Mourad and A. Sagnotti, Nucl. Phys. **B815** (2009) 289 [arXiv:hep-th/0810.4350], Nucl. Phys. **B828** (2010) 405 [arXiv:hep-th/0904.4447].

[16] C. Fronsdal, Phys. Rev. **D18** (1978) 3624; J. Fang and C. Fronsdal, Phys. Rev. **D18** (1978) 3630.

[17] D. Francia and A. Sagnotti, Phys. Lett. **B543** (2002) 303 [arXiv:hep-th/0207002], Class. Quant. Grav. **20** (2003) S473 [arXiv:hep-th/0212185]; Phys. Lett. **B624** (2005) 93 [arXiv:hep-th/0507144].

[18] A. K. H. Bengtsson, Phys. Lett. **B182** (1986) 321; M. Henneaux and C. Teitelboim, in "Quantum Mechanics of Fundamental Systems, 2", eds. C. Teitelboim and J. Zanelli (Plenum Press, New York, 1988), p. 113; G. Bonelli, Nucl. Phys. **B669** (2003) 159-172. [hep-th/0305155]; A. K. H. Bengtsson, Class. Quant. Grav. **5** (1988) 437; A. Fotopoulos and M. Tsulaia, Int. J. Mod. Phys. **A24** (2009) 1 [arXiv:0805.1346 [hep-th]].

[19] E. P. Wigner, Annals Math. **40** (1939) 149-204.

[20] T. Regge, Nuovo Cim. **14** (1959) 951.

[21] For a review see: M. Gunaydin, Lect. Notes Phys. **180** (1983) 192-213.

[22] P. A. M. Dirac, J. Math. Phys. **4** (1963) 901-909.

[23] C. Fronsdal, Rev. Mod. Phys. **37** (1965) 221-224; M. Flato, C. Fronsdal, Lett. Math. Phys. **2** (1978) 421-426; E. Angelopoulos, M. Flato, C. Fronsdal, D. Sternheimer, Phys. Rev. **D23** (1981) 1278.

[24] M. S. Plyushchay, Electron. J. Theor. Phys. **3N10** (2006) 17 [math-ph/0604022]; R. Casalbuoni, PoS **EMC2006** (2006) 004. [arXiv:hep-th/0610252 [hep-th]]; X. Bekaert, M. R. de Traubenberg, M. Valenzuela, JHEP **0905** (2009) 118. [arXiv:0904.2533 [hep-th]]; S. Esposito, [arXiv:1110.6878 [physics.hist-ph]].

[25] W. Rarita, J. Schwinger, Phys. Rev. **60** (1941) 61.

[26] D. Z. Freedman, P. van Nieuwenhuizen and S. Ferrara, Phys. Rev. **D13** (1976) 3214; S. Deser and B. Zumino, Phys. Lett. **B62** (1976) 335.

[27] S. Weinberg, Phys. Rev. **135** (1964) B1049-B1056.

[28] S. R. Coleman, J. Mandula, Phys. Rev. **159** (1967) 1251-1256.

[29] G. Velo and D. Zwanziger, Phys. Rev. **186** (1969) 1337, Phys. Rev. **188** (1969) 2218; G. Velo, Nucl. Phys. **B43** (1972) 389.

[30] C. Aragone and S. Deser, Phys. Lett. **B86** (1979) 161, Nuovo Cim. **B57** (1980) 33.

[31] S. Weinberg, E. Witten, Phys. Lett. **B96** (1980) 5.

[32] M. Porrati, Phys. Rev. **D78** (2008) 065016 [arXiv:0804.4672 [hep-th]].

[33] L. P. S. Singh and C. R. Hagen, Phys. Rev. **D9** (1974) 910, Phys. Rev. **D9** (1974) 898.

[34] A. Pashnev and M. M. Tsulaia, Mod. Phys. Lett. **A12** (1997) 861 [arXiv:hep-th/9703010]; Mod. Phys. Lett. **A13** (1998) 1853 [arXiv:hep-th/9803207], C. Burdik, A. Pashnev and M. Tsulaia, Nucl. Phys. Proc. Suppl. **102** (2001) 285 [arXiv:hep-th/0103143]; I. L. Buchbinder, A. Pashnev and M. Tsulaia, Phys. Lett. **B523** (2001) 338 [arXiv:hep-th/0109067] arXiv:hep-th/0206026; X. Bekaert, I. L. Buchbinder, A. Pashnev and M. Tsulaia, Class. Quant. Grav. **21** (2004) S1457 [arXiv:hep-th/0312252]; I. L. Buchbinder, V. A. Krykhtin and A. Pashnev, Nucl. Phys. **B711** (2005) 367 [arXiv:hep-th/0410215].

[35] C. Becchi, A. Rouet and R. Stora, Commun. Math. Phys. **42** (1975) 127, Annals Phys. **98** (1976) 287; I.V. Tyutin, preprint FIAN n. 39 (1975); E. S. Fradkin and G. A. Vilkovisky, Phys. Lett. **B55** (1975) 224; I. A. Batalin and G. A. Vilkovisky, Phys. Lett. **B69** (1977) 309; M. Henneaux, Phys. Rept. **126** (1985) 1.

[36] J. S. Schwinger, Reading, USA: Addison-Wesley (1989) 306 p. (Advanced book classics series).

[37] D. Francia, Nucl. Phys. **B796** (2008) 77-122. [arXiv:0710.5378 [hep-th]], Fortsch. Phys. **56** (2008) 800-808. [arXiv:0804.2857 [hep-th]].

[38] I. L. Buchbinder, A. V. Galajinsky, V. A. Krykhtin, Nucl. Phys. **B779** (2007) 155-177. [hep-th/0702161]; I. L. Buchbinder, V. A. Krykhtin, A. A. Reshetnyak, Nucl. Phys. **B787** (2007) 211-240. [hep-th/0703049]; I. L. Buchbinder, A. V. Galajinsky, JHEP **0811** (2008) 081. [arXiv:0810.2852 [hep-th]].

[39] B. de Wit and D. Z. Freedman, Phys. Rev. **D21** (1980) 358.

[40] A. Sagnotti and M. Tsulaia, Nucl. Phys. **B682** (2004) 83 [arXiv:hep-th/0311257].

[41] D. Francia, J. Phys. Conf. Ser. **222** (2010) 012002. [arXiv:1001.3854 [hep-th]]; Phys. Lett. **B690** (2010) 90 [arXiv:1001.5003 [hep-th]].

[42] M. Taronna, "Higher-Spin Interactions: four-point functions and beyond," [arXiv:1107.5843 [hep-th]].

[43] M. Taronna, "Higher Spins and String Interactions", Master Thesis, arXiv:1005.3061 [hep-th]; A. Sagnotti, M. Taronna, Nucl. Phys. **B842** (2011) 299-361. [arXiv:1006.5242 [hep-th]].

[44] R. Manvelyan, K. Mkrtchyan, W. Ruhl, Nucl. Phys. **B836** (2010) 204-221. [arXiv:1003.2877 [hep-th]]; Phys. Lett. **B696** (2011) 410 [arXiv:1009.1054 [hep-th]]; R. Manvelyan, K. Mkrtchyan, W. Ruhl, M. Tovmasyan, Phys. Lett. **B699** (2011) 187-191. [arXiv:1102.0306 [hep-th]]. For a review, see: K. Mkrtchyan, arXiv:1101.5643 [hep-th].

[45] P. C. Argyres, C. R. Nappi, Phys. Lett. **B224** (1989) 89.

[46] M. Porrati, R. Rahman, A. Sagnotti, Nucl. Phys. **B846** (2011) 250-282. [arXiv:1011.6411 [hep-th]].

[47] E. S. Fradkin and M. A. Vasiliev, Nucl. Phys. **B291** (1987) 141, Phys. Lett. **B189** (1987) 89.

[48] J. E. Paton and H. M. Chan, Nucl. Phys. **B10** (1969) 516; J. H. Schwarz, "Gauge Groups For Type I Superstrings", CALT-68-906-REV *Presented at 6th Johns Hopkins Workshop on Current Problems in High-Energy Particle Theory, Florence, Italy, Jun 2-4, 1982*; N. Marcus and A. Sagnotti, Phys. Lett. **B119** (1982) 97, Phys. Lett. **B188** (1987) 58. For reviews see: J. H. Schwarz, Phys. Rept. **89** (1982) 223-322; C. Angelantonj, A. Sagnotti, Phys. Rept. **371** (2002) 1-150. [hep-th/0204089].

[49] R. R. Metsaev, arXiv:hep-th/0712.3526, Nucl. Phys. **B759** (2006) 147 [arXiv:hep-th/0512342].

[50] A. K. H. Bengtsson, I. Bengtsson and L. Brink, Nucl. Phys. **B227** (1983) 31, 41.

[51] N. Boulanger, S. Leclercq and P. Sundell, JHEP **0808** (2008) 056 [arXiv:hep-th/0805.2764].

[52] A. Sagnotti, talk at Strings 09, http://strings2009.roma2.infn.it/program.html.

[53] H. van Dam and M. J. G. Veltman, Nucl. Phys. **B22**, 397 (1970); V. I. Zakharov, JETP Lett. **12** (1970) 312 [Pisma Zh. Eksp. Teor. Fiz. **12** (1970) 447].

[54] X. Bekaert, E. Joung and J. Mourad, JHEP **0905** (2009) 126 [arXiv:hep-th/0903.3338]. JHEP **1102** (2011) 048 [arXiv:1012.2103 [hep-th]].

[55] A. Higuchi, Nucl. Phys. **B282** (1987) 397, J. Math. Phys. **28** (1987) 1553 [Erratum-ibid. **43** (2002) 6385]; M. Porrati, Phys. Lett. **B498**, 92 (2001) [arXiv:hep-th/0011152]; I. I. Kogan, S. Mouslopoulos and A. Papazoglou, Phys. Lett. **B503**, 173 (2001) [arXiv:hep-th/0011138].

[56] J. Dixmier, "Représentations intégrables du groupe de De Sitter", Bulletin de la Société Mathématique de France, 89 (1961), p. 9-41. Earlier works on the representations of SO(1,4) include: L. Thomas, "On unitary representations of the group of the De Sitter group," Ann. of Math. **42** (1941) 113; T. Newton, "A note on the representation of the De Sitter group," Ann. of Math. **52** (1950) 730; A. W. Knapp and E. M. Stein, "Interwining operators for semisimple groups," Ann. of Math. (2) **93** (1971) 489; E. Thieleker, "The unitary representations of the generalized Lorentz groups," Transactions of the American mathematical Society **199** (1974) 327; N. Ja. Vilenkin and A. U. Kilmyk, "Representation of Lie groups and special fucntions," Kluwer Academic Publishers (1993). The two-dimensional case was considered in V. Bargmann, Annals Math. **48** (1947) 568.

[57] S. Deser and R. I. Nepomechie, Annals Phys. **154** (1984) 396; S. Deser and A. Waldron, Phys. Rev. Lett. **87** (2001) 031601 [arXiv:hep-th/0102166], Nucl. Phys. **B607** (2001) 577 [arXiv:hep-th/0103198], Phys. Lett. **B501**, 134 (2001) [arXiv:hep-th/0012014];

[58] C. Fronsdal, Phys. Rev. **D20** (1979) 848. J. Fang and C. Fronsdal, Phys. Rev. **D22** (1980) 1361; T. Biswas and W. Siegel, JHEP **0207** (2002) 005 [arXiv:hep-th/0203115]; K. Hallowell and A. Waldron, Nucl. Phys. **B724** (2005) 453 [arXiv:hep-th/0505255]; X. Bekaert, I. L. Buchbinder, A. Pashnev and M. Tsulaia, Class. Quant. Grav. **21** (2004) S1457 [arXiv:hep-th/0312252].

[59] See, for instance, E.S. Whittaker and G.N. Watson, *A Course of Modern Analysis* (Cambridge Univ. Press, Cambridge, 1973).

[60] F. A. Berends, G. J. H. Burgers and H. van Dam, Nucl. Phys. **B271** (1986) 429, Z. Phys. **C24** (1984) 247, Nucl. Phys. **B260** (1985) 295.

[61] D. J. Gross and P. F. Mende, Phys. Lett. **B197** (1987) 129, Nucl. Phys. **B303** (1988) 407.

[62] E. Joung and M. Taronna, arXiv:1110.5918 [hep-th].

[63] D. Polyakov, Phys. Rev. **D82** (2010) 066005 [arXiv:0910.5338 [hep-th]]; Int. J. Mod. Phys. **A25** (2010) 4623 [arXiv:1005.5512 [hep-th]].

CHAIRMAN: A. SAGNOTTI

Scientific Secretaries: Q. Dorosti, A. Campoleoni

DISCUSSION

- *Biassouni:*

Are these higher-spin objects just a tool of String Theory or do they describe some physical objects?

- *Sagnotti:*

Higher-spin states are ubiquitous among the massive excitations of String Theory, but their appearance in this context is only one of the motivations to look at them, albeit an important one. They are instrumental in granting key properties of String Theory, and yet their interactions appear to experience difficulties already at the classical level: this should be taken as a warning that we only control a small corner of Classical Field Theory.

- *Alba:*

What was the main problem that forbade to solve the free theory?

What is the geometrical meaning of the Labastida constraints?

- *Sagnotti:*

The main problem was actually the incompleteness of the free theory. The first approaches to it tried to follow the path apparently set out by the Maxwell and Einstein equations, that are two-derivative equations. It was thus seen that the free propagation of higher spins can be described via two-derivative equations, but this requires some algebric constraints, that as such are a sign of incompleteness. Incomplete descriptions can be quite effective, of course, but they usually hide something. For instance, people have been able to perform important computations in light-cone gauge formulations of gauge or string models, but these hide the covariance of these models at intermediate stages, and thus their very structure. In a similar fashion, it was shown that one can reformulate the free equations in terms of geometrical objects that generalize the Maxwell field strength and the Riemann curvature. The field equations are non-local and contain higher-derivative terms, but they work without any constraint on the gauge parameters. And, I should add, their differences with respect to the original constrained form entails only terms that are pure gauge, so that physical quantities like current exchanges coincide in the two formulations. This extension is important because the final goal is the understanding of what type of generalization of space-time geometry drives higher-spin interactions. For instance, as I have stressed in the lectures, the approach of Vasiliev, that is the most complete example of higher-spin interactions available at present, can be regarded as an extension of the frame formulation of gravity. While effective, this type of setup is driven by the tangent-space algebra and, as a result, it is far from clear what takes the role of general covariant transformations in this more complicated higher-spin setting. The geometrical free field equations are just a first step, but nonetheless they rest precisely on the geometrical objects that one expects to see in a geometrical formulation of the full interacting theory. And, actually, if one looks closely at String Theory or at the Vasiliev theory alike, in the proper setup the constraint-free extension of the free theory does show up in an instructive and interesting fashion.

I'm not aware of any geometrical meaning of the Labastida constraints, just as for the trace constraints of Fronsdal. Rather, the lesson should be that they hide the underlying geometry. This is true, in a similar fashion, for both symmetric and mixed-symmetry fields, despite the additional technical complications entailed by the latter systems. I wanted to stress that the Labastida constraints select, in a rather elegant fashion, only a subset of the available Young-tableaux for the leftover traces. As a result, the Labastida Lagrangians are rather complicated, and despite the constraints still contain non-vanishing higher traces with increasing numbers of rows in the corresponding Young tableaux. On the other hand, for symmetric fields the double trace vanishes and, as a result, the Lagrangian only contains the Fronsdal operator and its trace.

- *Galakhov:*

Can the generalization of the spin connection in the case higher-spin theories can be treated in a similar way?

- *Sagnotti:*

Yes, but the main problem of the Vasiliev formulation is that it hides geometry as we discussed before.

- *Csanad:*

Do you see the possibility of observing higher-spin elementary particles?

- *Sagnotti:*

We have seen plenty of higher-spin composite objects, hadronic resonances, that as such massive and have a non-vanishing size, and a key issue for the theory is how important these two properties, mass and size, are for higher-spin particles. Truly enough, we have not seen yet any higher-spin particles, but after all we have not observed yet even the graviton. And, of course, the same is true for all particles proposed by any model beyond the Standard Model, but surely this has not prevented motivated searches for extensions. Here the search, as I have stressed, is motivated by some difficulties with or understanding of String Theory, on the one hand, and of difficulties with the formulation of higher-spin systems on the other.

- *Azimov:*

Familiar descriptions of higher-spin fields reveal effects violating causality. How does this problem look like in your approach?

- *Sagnotti:*

This point will be discussed tomorrow, but I can anticipate that these problems originated from the natural, but with hindsight naive and incorrect assumption, of insisting with minimal couplings. They can be overcome, and are overcome in String Theory, via the introduction of contact terms in the interactions. The key lesson of these investigations is that higher spins dislike minimal couplings.

- *Dunin-Barakowski:*

Can you say something about recent works stating the duality between free field theory on the boundary of AdS space and higher-spin theory in the bulk?

- *Sagnotti:*

Typically the AdS/CFT correspondence is used to connect a boundary gauge theory at strong coupling to classical supergravity in the bulk. However, one can consider a different setting, with weak coupling on the boundary and thus strong coupling in the bulk. In this regime, part of the string excitations can be described by the Vasiliev equations, and recently there were various non-trivial checks of this correspondence.

- *Mironov:*

Could you give some physical arguments why vDVZ discontinuities should appear in a theory?

- *Sagnotti:*

In my lecture I have tried to stress that the exchanges reflect the available degrees of freedom. In flat space, massive and massless particles describe different number of degrees of freedom, and the latter can be related to massless particles in one more dimension. This is the origin of the discontinuities in the limits of zero mass, up to a small but important technical assumption, that is instrumental in allowing the comparison: both types of exchanges, massless and massive, are computed for identical conserved external currents.

The Mystery of Neutrino Mixings

Guido Altarelli
Dipartimento di Fisica, Universita' di Roma Tre
Rome, Italy
and
CERN, Department of Physics, Theory Division
CH-1211 Genève 23, Switzerland
E-mail: guido.altarelli@cern.ch

ABSTRACT

In the last years we have learnt a lot about neutrino masses and mixings. Neutrinos are not all massless but their masses are very small. Probably masses are small because neutrinos are Majorana particles with masses inversely proportional to the large scale M of lepton number (L) violation, which turns out to be compatible with the GUT scale. We have understood that there is no contradiction between large neutrino mixings and small quark mixings, even in the context of GUTs and that neutrino masses fit well in the SUSY GUT picture. Out of equilibrium decays with CP and L violation of heavy RH neutrinos can produce a B-L asymmetry, then converted near the weak scale by instantons into an amount of B asymmetry compatible with observations (baryogenesis via leptogenesis). It appears that active neutrinos are not a significant component of Dark Matter in the Universe. A long list of models have been formulated over the years to understand neutrino masses and mixings. With the continuous improvement of the data most of the models have been discarded by experiment. The surviving models still span a wide range going from a maximum of symmetry, with discrete non-abelian flavour groups, to the opposite extreme of anarchy.

1. Introduction

Experiments on neutrino oscillations, which measure differences of squared masses as well as mixing angles [1) 2)] have established that neutrinos have a mass. Two distinct oscillation frequencies have been first measured in solar and atmospheric neutrino oscillations and later confirmed by experiments on earth, like KamLAND, K2K and MINOS. Two well separated differences need at least three different neutrino mass eigenstates involved in oscillations so that the three known neutrino species can be sufficient. Then at least two ν's must be massive while, in principle, the third one could still be massless. A signal corresponding to a third mass difference was claimed by the LSND experiment (with antineutrinos) but not confirmed by KARMEN. More recently MiniBooNE [3)] has reported some possible supporting evidence for the LSND effect in their antineutrino run while no oscillation is observed in the neutrino run. The existence of a third oscillation frequency would imply the need for additional sterile neutrinos (i.e. with no weak interactions, as any new light active neutrino was excluded by LEP) or CPT violation (as, in this case, the masses of neutrinos and antineutrinos can be different).

The main recent developments on the experimental side were the results on θ_{13} from T2K [4)] and MINOS [5)] (very recently also DOUBLE CHOOZ [6)]) and the coming back of sterile neutrinos. As well known, the T2K run was suddenly interrupted by the devastating earthquake that hit Japan on March 11, just minutes away from the scheduled presentation of the first T2K data. Later T2K released the first publication on their data [4)], reporting a 2.5σ signal for $\sin^2 2\theta_{13}$ that indicates a value of θ_{13} close to the previous upper bound, of the order of the Cabibbo angle θ_C.

On the evidence for sterile neutrinos a number of hints have been recently reported. They do not make yet a clear evidence but certainly pose an experimental problem that needs clarification. First, there is the MiniBooNE experiment [5)] that in the antineutrino channel reports an excess of events supporting the LSND oscillation signal (originally observed with antineutrinos). More recently an update of the MiniBooNE data in the antineutrino channel shows less supporting evidence [7)]. In the neutrino channel MiniBooNE did not observe a signal in the LSND domain. However, in these data there is a unexplained excess at low energy over the (reliably?) estimated background. In the neutrino data sample, for the search of a LSND-like signal, only the events with neutrino energy above a threshold value E_{th} were used, leaving the issue of an explanation of the low energy excess unanswered. In the antineutrino channel most of the support to the LSND signal appears to arise from an excess above E_{th} but quite close to it, so that there is, in my opinion, some room for perplexity. Then there is the reactor anomaly: a reevaluation of the reactor flux [8)] produced an apparent gap between the theoretical expectations and the data taken at small distances from the reactor ($\lesssim$ 100 m). The discrepancy is of the same order of the quoted systematic error whose estimate, detailed in the paper, should perhaps be reconsidered. Similarly the Gallium anomaly [9)] depends on the assumed cross-section which could be questioned. The reactor anomaly and the Gallium anomaly do not really agree on the oscillation parameters that they point to: the Δm^2 values are compatible but the central values of $\sin^2 2\theta$ differ by about an order of magnitude, with Gallium favouring the larger angle. Cosmological data allow the existence of one sterile neutrino, while the most stringent bounds arising form nucleosynthesis disfavour two or more sterile neutrinos [10)]. Over all, only a small leakage from active to sterile neutrinos is allowed by present neutrino oscillation data, as discussed in refs. [11)] If all the indications listed above were confirmed (it looks unlikely) then one sterile neutrino would not be enough and at least two would be needed with sub-eV masses. Establishing the existence of sterile neutrinos would be a great discovery. In fact a sterile neutrino is an exotic particle not predicted by the most popular models of new physics. A sterile neutrino is not a 4th generation neutrino: the latter is coupled to the weak interactions (it is active) and heavier than half the Z mass. A sterile neutrino would probably be a remnant of some hidden sector. The issue is very important so that new and better experimental data are badly needed.

As already mentioned, in neutrino oscillations the leakage from the three active species towards the sterile neutrinos is any case small and, in fact, the best established oscillation phenomena are well described in terms of 3-neutrino models. In this domain the main recent developments have been the T2K and MINOS results on θ_{13}. The T2K result [4], based on the observation of 6 electron events when 1.5 ± 0.3 are expected for $\theta_{13} = 0$, is converted into a confidence interval $0.03(0.04) \leq \sin^2 2\theta_{13} \leq 0.28(0.34)$ at 90% c.l. for $\sin^2 2\theta_{23} = 1$, $|\Delta m^2| = 2.4\ 10^{-3} eV^2$, $\delta_{CP} = 0$ and for normal (inverted) neutrino mass hierarchy. Also the MINOS Collaboration released [5] their corresponding 90% c.l. range as $0(0) \leq \sin^2 2\theta_{13} \leq 0.12(0.19)$, which is displaced towards smaller values with respect to that of T2K. Finally DOUBLE CHOOZ [6] finds (with only the far detector in operation): $\sin^2 2\theta_{13} = 0.085 \pm 0.051$ at 1σ.

2. Neutrino Masses and Lepton Number Violation

Neutrino oscillations imply non vanishing neutrino masses which in turn demand either the existence of right-handed (RH) neutrinos (Dirac masses) or lepton number L violation (Majorana masses) or both. Given that neutrino masses are certainly extremely small, it is really difficult from the theory point of view to avoid the conclusion that L conservation must be violated. In fact, in terms of lepton number violation the smallness of neutrino masses can be explained as inversely proportional to the very large scale where L is violated, of order M_{GUT} or even M_{Pl}.

If L conservation is violated neutrinos can be Majorana fermions. For a Majorana neutrino each mass eigenstate with given helicity coincides with its own antiparticle with the same helicity. As well known, for a charged massive fermion there are four states differing by their charge and helicity (the four components of a Dirac spinor) as required by Lorentz and CPT invariance. For a massive Majorana neutrino, neutrinos and antineutrinos can be identified and only two components are needed to satisfy the Lorentz and CPT invariance constraints. Neutrinos can be Majorana fermions because, among the fundamental fermions (i.e. quarks and leptons), they are the only electrically neutral ones. If, and only if, the lepton number L is not conserved, i.e. it is not a good quantum number, then neutrinos and antineutrinos can be identified. For Majorana neutrinos both Dirac mass terms, that conserve L ($\nu \rightarrow \nu$), and Majorana mass terms, that violate L by two units ($\nu \rightarrow \bar{\nu}$), are in principle possible. Of course the restrictions from gauge invariance must be respected. So for neutrinos the Dirac mass terms ($\bar{\nu}_R \nu_L$ +h.c.) arise from the couplings with the Higgs field, as for all quarks and leptons. For Majorana masses, a $\nu_L^T \nu_L$ mass term has weak isospin 1 and needs two Higgs fields to make an invariant. On the contrary a $\nu_R^T \nu_R$ mass term is a gauge singlet and needs no Higgs. As a consequence, the right-handed neutrino Majorana mass M_R is not bound to be of the order of the electroweak symmetry breaking (induced by the Higgs vacuum expectation value) and can be very large (see below).

Some notation: the charge conjugated of ν is ν^c, given by $\nu^c = C(\bar{\nu})^T$, where $C = i\gamma_2\gamma_0$ is the charge conjugation matrix acting on the spinor indices (in the following, when dealing with the flavour structure of couplings, the C matrix will be often omitted but understood). In particular $(\nu^c)_L = C(\bar{\nu_R})^T$, so that, instead of using ν_L and ν_R, we can refer to ν_L and $(\nu^c)_L$, or simply ν and ν^c.

Once we accept L non-conservation we gain an elegant explanation for the smallness of neutrino masses. If L is not conserved, even in the absence of heavy RH neutrinos, Majorana masses for neutrinos can be generated by dimension five operators [12] of the form

$$O_5 = \frac{(Hl)_i^T \lambda_{ij} (Hl)_j}{\Lambda} \quad , \tag{1}$$

with H being the ordinary Higgs doublet, l_i the SU(2) lepton doublets, λ a matrix in flavour space, Λ a large scale of mass, of order M_{GUT} or M_{Pl} and a charge conjugation matrix C between the lepton fields is understood. Neutrino masses generated by O_5 are of the order $m_\nu \approx v^2/\Lambda$ for $\lambda_{ij} \approx$ O(1), where $v \sim$ O(100 GeV) is the vacuum expectation value of the ordinary Higgs.

We consider that the existence of RH neutrinos ν^c is quite plausible because most GUT groups larger than SU(5) require them. In particular the fact that ν^c completes the representation 16 of SO(10): 16=$\bar{5}$+10+1, so that all fermions of each family are contained in a single representation of the unifying group, is too impressive not to be significant. At least as a classification group SO(10) must be of some relevance in a more fundamental layer of the theory! Thus in the following we both assume that ν^c exist and L is not conserved. With these assumptions the see-saw mechanism [13] is possible. We recall, also to fix notations, that in its simplest form it arises as follows. Consider the SU(3) $\times$ SU(2) $\times$ U(1) invariant Lagrangian giving rise to Dirac and ν^c Majorana masses (for the time being we consider the ν (versus ν^c) Majorana mass terms as comparatively negligible):

$$\mathcal{L} = -\nu^{cT} y_\nu (Hl) + \frac{1}{2}\nu^{cT} M \nu^c + \ h.c. \tag{2}$$

The Dirac mass matrix $m_D \equiv y_\nu v/\sqrt{2}$, originating from electroweak symmetry breaking, is, in general, non-hermitian and non-symmetric, while the Majorana mass matrix M is symmetric, $M = M^T$. We expect the eigenvalues of M to be of order M_{GUT} or more because ν^c Majorana masses are SU(3)$\times$ SU(2)$\times$ U(1) invariant, hence unprotected and naturally of the order of the cutoff of the low-energy theory. Since all ν^c are very heavy we can integrate them away. For this purpose we write down the equations of motion for ν^c in the static limit, *i.e.* neglecting their kinetic terms:

$$-\frac{\partial \mathcal{L}}{\partial \nu^c} = y_\nu (Hl) - M\nu^c = 0 \quad . \tag{3}$$

From this, by solving for ν^c, we obtain:

$$\nu^c = M^{-1} y_\nu (Hl) \quad . \tag{4}$$

We now replace in the lagrangian, eq. (2), this expression for ν^c and we get the operator O_5 of eq. (1) with

$$\frac{2\lambda}{\Lambda} = -y_\nu^T M^{-1} y_\nu \quad , \tag{5}$$

and the resulting neutrino mass matrix reads:

$$m_\nu = m_D^T M^{-1} m_D \quad . \tag{6}$$

This is the well known see-saw mechanism result [13)]: the light neutrino masses are quadratic in the Dirac masses and inversely proportional to the large Majorana mass. If some ν^c are massless or light they would not be integrated away but simply added to the light neutrinos. Notice that the above results hold true for any number n of heavy neutral fermions R coupled to the 3 known neutrinos. In this more general case M is an n by n symmetric matrix and the coupling between heavy and light fields is described by the rectangular n by 3 matrix m_D. Note that for $m_\nu \approx \sqrt{\Delta m^2_{atm}} \approx 0.05$ eV (see Table(1)) and $m_\nu \approx m_D^2/M$ with $m_D \approx v \approx 200\ GeV$ we find $M \approx 10^{15}\ GeV$ which indeed is an impressive indication for M_{GUT}.

If additional non-renormalizable contributions to O_5, eq. (1), are comparatively non-negligible, they should simply be added. For instance in SO(10) or in left-right extensions of the SM, an $SU(2)_L$ triplet can couple to two lepton doublets and to two Higgs and may induce a sizeable contribution to neutrino masses. At the level of the low-energy effective theory, such contribution is still described by the operator O_5 of eq. (1), obtained by integrating out the heavy $SU(2)_L$ triplet. This contribution is called type II to be distinguished from that obtained by the exchange of RH neutrinos (type I). One can also have the exchange of a fermionic $SU(2)_L$ triplet coupled to a lepton doublet and a Higgs (type III). After elimination of the heavy fields, at the level of the effective low-energy theory, the three types of see-saw terms are equivalent. In particular they have identical transformation properties under a chiral change of basis in flavour space. The difference is, however, that in type I see-saw mechanism, the Dirac matrix m_D is presumably related to ordinary fermion masses because they are both generated by the Higgs mechanism and both must obey GUT-induced constraints. Thus more constraints are implied if one assumes the see-saw mechanism in its simplest type I version.

3. Basic Formulae and Data for Three-Neutrino Mixing

We assume in the following that there are only two distinct neutrino oscillation frequencies, the atmospheric and the solar frequencies. These two can be reproduced with the known three light neutrino species (for more than three neutrinos see, for example, ref.[11]).

Neutrino oscillations are due to a misalignment between the flavour basis, $\nu' \equiv (\nu_e, \nu_\mu, \nu_\tau)$, where ν_e is the partner of the mass and flavour eigenstate e^- in a left-handed (LH) weak isospin SU(2) doublet (similarly for ν_μ and ν_τ) and the mass eigenstates $\nu \equiv (\nu_1, \nu_2, \nu_3)$ [14,15]:

$$\nu' = U\nu \quad , \tag{7}$$

where U is the unitary 3 by 3 mixing matrix. Given the definition of U and the transformation properties of the effective light neutrino mass matrix m_ν in eq. (1):

$$\begin{aligned} \nu'^T m_\nu \nu' &= \nu^T U^T m_\nu U \nu \\ U^T m_\nu U &= \mathrm{Diag}\,(m_1, m_2, m_3) \equiv m_{diag} \quad , \end{aligned} \tag{8}$$

we obtain the general form of m_ν (i.e. of the light ν mass matrix in the basis where the charged lepton mass is a diagonal matrix):

$$m_\nu = U^* m_{diag} U^\dagger \quad . \tag{9}$$

The matrix U can be parameterized in terms of three mixing angles θ_{12}, θ_{23} and θ_{13} ($0 \le \theta_{ij} \le \pi/2$) and one phase φ ($0 \le \varphi \le 2\pi$) [16], exactly as for the quark mixing matrix V_{CKM}. The following definition of mixing angles can be adopted:

$$U = \begin{pmatrix} 1 & 0 & 0 \\ 0 & c_{23} & s_{23} \\ 0 & -s_{23} & c_{23} \end{pmatrix} \begin{pmatrix} c_{13} & 0 & s_{13}e^{i\varphi} \\ 0 & 1 & 0 \\ -s_{13}e^{-i\varphi} & 0 & c_{13} \end{pmatrix} \begin{pmatrix} c_{12} & s_{12} & 0 \\ -s_{12} & c_{12} & 0 \\ 0 & 0 & 1 \end{pmatrix} \tag{10}$$

where $s_{ij} \equiv \sin\theta_{ij}$, $c_{ij} \equiv \cos\theta_{ij}$. In addition, if ν are Majorana particles, we have the relative phases among the Majorana masses m_1, m_2 and m_3. If we choose m_3 real and positive, these phases are carried by $m_{1,2} \equiv |m_{1,2}|e^{i\phi_{1,2}}$ [17]. Thus, in general, 9 parameters are added to the SM when non-vanishing neutrino masses are included: 3 eigenvalues, 3 mixing angles and 3 CP violating phases.

In our notation the two frequencies, $\Delta m_I^2/4E$ (I=sun,atm), are parametrized in terms of the ν mass eigenvalues by

$$\Delta m^2_{sun} \equiv |\Delta m^2_{12}|, \qquad \Delta m^2_{atm} \equiv |\Delta m^2_{23}| \quad . \tag{11}$$

where $\Delta m^2_{12} = |m_2|^2 - |m_1|^2 > 0$ and $\Delta m^2_{23} = m_3^2 - |m_2|^2$. The numbering 1,2,3 corresponds to our definition of the frequencies and in principle may not coincide with the ordering from the lightest to the heaviest state. In fact, the sign of Δm^2_{23} is not known and its determination is one of the existing experimental challenges. A positive (negative) sign corresponds to normal (inverse) hierarchy.

With the above definitions the present data are summarised in Table(1) [18], [19].

Quantity	Fogli et al [18]	Schwetz et al [19]
Δm^2_{sun} $(10^{-5}$ eV$^2)$	$7.58^{+0.22}_{-0.26}$	$7.59^{+0.20}_{-0.18}$
Δm^2_{atm} $(10^{-3}$ eV$^2)$	$2.35^{+0.12}_{-0.09}$	$2.50^{+0.09}_{-0.16}$
$\sin^2\theta_{12}$	$0.312^{+0.017}_{-0.016}$	$0.312^{+0.017}_{-0.015}$
$\sin^2\theta_{23}$	$0.42^{+0.08}_{-0.03}$	$0.52^{+0.06}_{-0.07}$
$\sin^2\theta_{13}$	0.025 ± 0.007	$0.013^{+0.007}_{-0.005}$

Table 1:

Fits to neutrino oscillation data. The results correspond to the new reactor fluxes. The fit of Schwetz et al [19] refers to the normal hierarchy case (in the inverse hierarchy case the main difference is that $\sin^2\theta_{13} = 0.016 + 0.008 - 0.006$)

Oscillation experiments do not provide information about the absolute neutrino mass scale. Limits on that are obtained [1] from the endpoint of the tritium beta decay spectrum, from cosmology and from neutrinoless double beta decay ($0\nu\beta\beta$). From tritium we have an absolute upper limit of 2.2 eV (at 95% C.L.) [20] on the antineutrino mass eigenvalues involved in beta decay, which, combined with the observed oscillation frequencies under the assumption of three CPT-invariant light neutrinos, also amounts to an upper bound on the masses of the other active neutrinos. Complementary information on the sum of neutrino masses is also provided by the galaxy power spectrum combined with measurements of the cosmic microwave background anisotropies. According to recent analyses of the most reliable data [21] one obtains $\sum_i |m_i| < 0.60 \div 0.75$ eV (at 95% C.L.), depending on the retained data and the cosmological model priors assumed. These numbers for the sum have to be divided by 3 in order to obtain a limit on the mass of each light neutrino. The discovery of $0\nu\beta\beta$ decay would be very important because it would establish lepton number violation and the Majorana nature of ν's, and provide direct information on the absolute scale of neutrino masses. The present limit from $0\nu\beta\beta$ (with large ambiguities from nuclear matrix elements) is about $|m_{ee}| < (0.3 \div 0.8)$ eV [22] (see eq. (13)).

By now, after KamLAND, SNO and the upper limits on the absolute value of neutrino masses, not too much hierarchy in the spectrum of neutrinos is indicated by experiments [18], [19]:

$$r = \Delta m^2_{sol}/\Delta m^2_{atm} = 0.032 \pm 0.002 \sim 1/30. \tag{12}$$

Thus, for a hierarchical spectrum, $m_2/m_3 \sim \sqrt{r} \sim 0.2$, which is comparable to $\lambda_C \sim 0.226$ (thoughout this article $\lambda_C = \sin\theta_C$, with θ_C being the Cabibbo angle) or $\sqrt{m_\mu/m_\tau} \sim 0.24$. This suggests that the same hierarchy parameter (raised to powers with O(1) exponents) may apply for quark, charged lepton and neutrino mass matrices. In fact, $m_\mu/m_\tau \sim 0.06 \sim \lambda_C^2$ and $m_e/m_\mu \sim 0.005 \sim \lambda_C^{3-4}$).

For the near future the most important experimental challenges on neutrino oscillation experiments are more precise measurements of the absolute scale of neutrino mass (KATRIN, MARE), the accurate determination of θ_{13} (from MINOS, T2K and the reactor experiments DOUBLE CHOOZ, Daya Bay and RENO) and of the shift from maximal of θ_{23}, the fixing of the sign of Δm^2_{23} (normal or inverse hierarchy) (e.g. NOνA), the detection of CP violation in ν oscillations. Related to neutrino physics is the issue of the non conservation of the separate e, μ and τ lepton numbers. The recent new limit $\mathrm{Br}(\mu \to e\gamma) \lesssim 2.4 \cdot 10^{-12}$ obtained by the MEG experiment [23] is largely satisfied in the SM but it imposes a strong constraint on SUSY-GUT models.

4. Importance of Neutrinoless Double Beta Decay

The detection of neutrino-less double beta decay [24] would provide direct evidence of L non conservation and of the Majorana nature of neutrinos. It would also offer a way to possibly disentangle the 3 cases of degenerate, normal or inverse hierachy neutrino spectrum. The quantity which is bound by experiments on $0\nu\beta\beta$ is the 11 entry of the ν mass matrix, which in general, from $m_\nu = U^* m_{diag} U^\dagger$, is given by :

$$|m_{ee}| = |(1 - s_{13}^2)\,(m_1 c_{12}^2 + m_2 s_{12}^2) + m_3 e^{2i\phi} s_{13}^2| \qquad (13)$$

where $m_{1,2}$ are complex masses (including Majorana phases) while m_3 can be taken as real and positive and ϕ is the U_{PMNS} phase measurable from CP violation in oscillation experiments. Starting from this general formula it is simple to derive the bounds for degenerate, inverse hierarchy or normal hierarchy mass patterns shown in Fig.1 [25].

In the next few years a new generation of experiments (CUORE, GERDA, ...) will reach a larger sensitivity on $0\nu\beta\beta$ by about an order of magnitude. Assuming the standard mechanism through mediation of a light massive Majorana neutrino, if these experiments will observe a signal this would indicate that the inverse hierarchy is realized, if not, then the normal hierarchy case remains a possibility.

5. Baryogenesis via Leptogenesis from Heavy ν^c Decay

In the Universe we observe an apparent excess of baryons over antibaryons. It is appealing that one can explain the observed baryon asymmetry by dynamical evolution (baryogenesis) starting from an initial state of the Universe with zero baryon number. For baryogenesis one needs the three famous Sakharov conditions: B violation, CP violation and no thermal equilibrium. In the history of the Universe these necessary requirements have possibly occurred at different epochs. Note however that

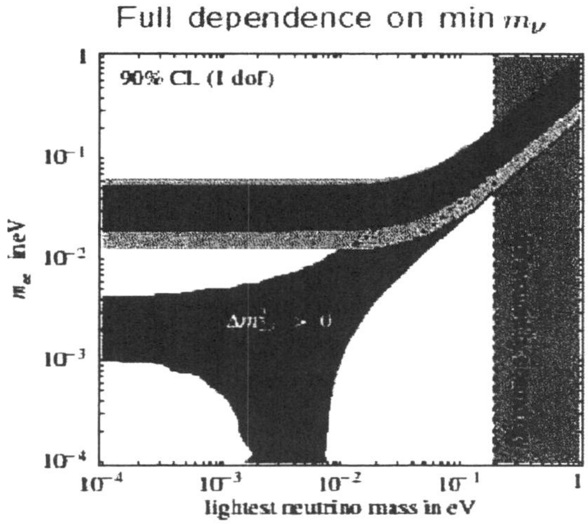

Figure 1: A plot [25] of m_{ee} in eV, the quantity measured in neutrino-less double beta decay, given in eq.(13), versus the lightest neutrino mass m_1, also in eV. The upper (lower) band is for inverse (normal) hierarchy.

the asymmetry generated during one such epoch could be erased in following epochs if not protected by some dynamical reason. In principle these conditions could be fulfilled in the SM at the electroweak phase transition. In fact, when kT is of the order of a few TeV, B conservation is violated by instantons (but B-L is conserved), CP symmetry is violated by the CKM phase and sufficiently marked out-of- equilibrium conditions could be realized during the electroweak phase transition. So the conditions for baryogenesis at the weak scale in the SM superficially appear to be present. However, a more quantitative analysis [26] shows that baryogenesis is not possible in the SM because there is not enough CP violation and the phase transition is not sufficiently strong first order, unless the Higgs mass is below a bound which by now is excluded by LEP. In SUSY extensions of the SM, in particular in the MSSM, there are additional sources of CP violation and the bound on m_H is modified but also this possibility has by now become at best marginal after the results from LEP2.

If baryogenesis at the weak scale is excluded by the data it can occur at or just below the GUT scale, after inflation. But only that part with $|\mathrm{B}-\mathrm{L}| > 0$ would survive and not be erased at the weak scale by instanton effects. Thus baryogenesis at $kT \sim 10^{10} - 10^{15}$ GeV needs B-L violation and this is also needed to allow m_ν if neutrinos are Majorana particles. The two effects could be related if baryogenesis arises from leptogenesis then converted into baryogenesis by instantons [27]. The decays of heavy Majorana neutrinos (the heavy eigenstates of the see-saw mechanism)

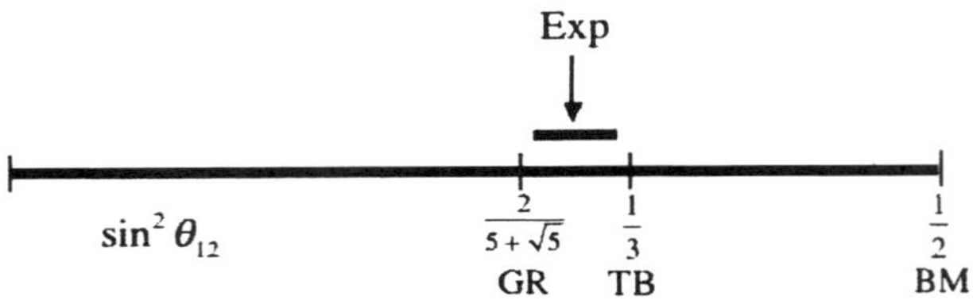

Figure 2: The values of $\sin^2\theta_{12}$ for TB or GR or BM mixing are compared with the data

happen with violation of lepton number L, hence also of B-L and can well involve a sufficient amount of CP violation. Recent results on neutrino masses are compatible with this elegant possibility. Thus the case of baryogenesis through leptogenesis has been boosted by the recent results on neutrinos.

6. Models of Neutrino Mixing

Neutrino mixing is important because it could in principle provide new clues for the understanding of the flavour problem. Even more so since neutrino mixing angles show a pattern that is completely different than that of quark mixing: for quarks all mixing angles are small, for neutrinos two angles are large (one is even compatible with the maximal value) and only the third one is small. We first consider the case of models based on discrete flavour groups that have received a lot of attention in recent years [28]. There are a number of special mixing patterns that have been studied in this context. These mixing matrices all have $\sin^2\theta_{23} = 1/2$, $\sin^2\theta_{13} = 0$ and differ by the value of $\sin^2\theta_{12}$ (see Fig. 2). The corresponding mass matrices are 2-3 symmetric , i.e. $\mu - \tau$ symmetric (see, as examples, the early work in ref.[29] and the recent paper ref.[30]). The observed value of $\sin^2\theta_{12}$ [18], [19] the best measured mixing angle, is very close, from below, to the so called Tri-Bimaximal (TB) value [31] which is $\sin^2\theta_{12} = 1/3$. Alternatively it is also very close, from above, to the Golden Ratio (GR) value [32], [33], [34] which is $\sin^2\theta_{12} = \frac{1}{\sqrt{5}\phi} = \frac{2}{5+\sqrt{5}} \sim 0.276$, where $\phi = (1+\sqrt{5})/2$ is the GR (for a different connection to the GR in this context, see [35], [36]). Thus, a possibility is that one or the other of these coincidences is taken seriously and this leads to models where either TB or GR mixing is naturally predicted as a good first approximation. Here I will mainly refer to TB mixing which is the simplest and most studied first approximation to the data.

The TB mixing matrix (in a particular phase convention) is given by:

$$U_{TB} = \begin{pmatrix} \sqrt{\frac{2}{3}} & \frac{1}{\sqrt{3}} & 0 \\ -\frac{1}{\sqrt{6}} & \frac{1}{\sqrt{3}} & -\frac{1}{\sqrt{2}} \\ -\frac{1}{\sqrt{6}} & \frac{1}{\sqrt{3}} & \frac{1}{\sqrt{2}} \end{pmatrix} \quad . \tag{14}$$

Note that the mixing angles are independent of mass ratios (while, for quark mixings, relations like $\lambda_C^2 \sim m_d/m_s$ are typical). In the basis where charged lepton masses are diagonal, the effective neutrino mass matrix in the TB case is given by $m_\nu = U_{TB}\mathrm{diag(m_1, m_2, m_3)U_{TB}^T}$:

$$m_\nu = \left[\frac{m_3}{2}M_3 + \frac{m_2}{3}M_2 + \frac{m_1}{6}M_1\right] \quad . \tag{15}$$

where:

$$M_3 = \begin{pmatrix} 0 & 0 & 0 \\ 0 & 1 & -1 \\ 0 & -1 & 1 \end{pmatrix}, \quad M_2 = \begin{pmatrix} 1 & 1 & 1 \\ 1 & 1 & 1 \\ 1 & 1 & 1 \end{pmatrix}, \quad M_1 = \begin{pmatrix} 4 & -2 & -2 \\ -2 & 1 & 1 \\ -2 & 1 & 1 \end{pmatrix}. \tag{16}$$

The eigenvalues of m_ν are m_1, m_2, m_3 with eigenvectors $(-2,1,1)/\sqrt{6}$, $(1,1,1)/\sqrt{3}$ and $(0,1,-1)/\sqrt{2}$, respectively. The expression in eq.(15) can be reproduced in models with sequential dominance or with form dominance, discussed by S. King and collaborators [37].

As we see the most general neutrino mass matrix corresponding to TB mixing, in the basis of diagonal charged leptons, is of the form:

$$m = \begin{pmatrix} x & y & y \\ y & x+v & y-v \\ y & y-v & x+v \end{pmatrix}, \tag{17}$$

This is a symmetric, 2-3 symmetric matrix with $a_{11} + a_{12} = a_{22} + a_{23}$.

We now discuss models that naturally produce TB mixing in first approximation. Discrete non-abelian groups naturally emerge as suitable flavour symmetries [28]. In fact the TB mixing matrix immediately suggests rotations by fixed, discrete angles. In a series of papers, started by [38] (for a rather complete list of references see [28], some recent papers are in ref.[39]) it has been pointed out that a broken flavour symmetry based on the discrete group A_4 appears to provide a simplest realization of this specific mixing pattern in Leading Order (LO). We recall that A_n is the group of even permutations of n objects (n!/2 elements). In the case of GR mixing the simplest choice is the group A_5 [34]. Other solutions for TB mixing based on alternative discrete or continuous flavour groups have also been considered [28], [40] but the A_4 models have a very economical and attractive structure, e.g. in terms of group representations and of field content.

We recall that A_4 can be generated by the two elements S and T obeying the relations (a "presentation" of the group):

$$S^2 = (ST)^3 = T^3 = 1 \quad . \tag{18}$$

The 12 elements of A_4 are obtained as: 1, S, T, ST, TS, T^2, ST^2, STS, TST, T^2S, TST^2, T^2ST. The inequivalent irreducible representations of A_4 are 1, 1', 1" and 3. Note that the squares of the dimensions of all these representations add up to 12, the dimension of A_4. It is immediate to see that one-dimensional unitary representations are given by:

$$\begin{array}{lll} 1 & S=1 & T=1 \\ 1' & S=1 & T=e^{i4\pi/3} \equiv \omega^2 \\ 1'' & S=1 & T=e^{i2\pi/3} \equiv \omega \end{array} \tag{19}$$

The three-dimensional unitary representation, in a basis where the element T is diagonal, is given by:

$$T = \begin{pmatrix} 1 & 0 & 0 \\ 0 & \omega^2 & 0 \\ 0 & 0 & \omega \end{pmatrix}, \qquad S = \frac{1}{3}\begin{pmatrix} -1 & 2 & 2 \\ 2 & -1 & 2 \\ 2 & 2 & -1 \end{pmatrix} \quad . \tag{20}$$

Note that the generic mass matrix for TB mixing in eq.(17) can be specified as the most general matrix that is invariant under $\mu - \tau$ symmetry, implemented by the unitary matrix $A_{\mu\tau}$:

$$A_{\mu\tau} = \begin{pmatrix} 1 & 0 & 0 \\ 0 & 0 & 1 \\ 0 & 1 & 0 \end{pmatrix} \tag{21}$$

and under the S transformation:

$$m = SmS, \qquad m = A_{\mu\tau} m A_{\mu\tau} \tag{22}$$

where S is given in eq.(20).

The m mass matrix of eq.(17) is derived in the basis where charged leptons are diagonal. It is useful to consider the product $m^2 = m_e^\dagger m_e$, where m_e is the charged lepton mass matrix (defined as $\overline{\psi}_R m_e \psi_L$), because this product transforms as $m'^2 = U_e^\dagger m^2 U_e$, with U_e the unitary matrix that rotates the left-handed (LH) charged lepton fields. The most general diagonal m^2 is invariant under a diagonal phase matrix with 3 different phase factors:

$$m_e^\dagger m_e = T^\dagger m_e^\dagger m_e T \tag{23}$$

and conversely a matrix $m_e^\dagger m_e$ satisfying the above requirement is diagonal. If $T^n = 1$ the matrix T generates a cyclic group Z_n. The simplest case is $n = 3$, which corresponds to Z_3 (but $n > 3$ is equally possible) and to the T matrix in eq.(20).

We can now see why A_4 works for TB mixing. It works because S and T are matrices of A_4 (in fact they satisfy eqs.(18)). One could object that the matrix A_{23} is not an element of A_4 (because the 2-3 exchange is an odd permutation). But it can be shown that in A_4 models the 2-3 symmetry is maintained by imposing that there are no flavons transforming as $1'$ or $1''$ that break A_4 with two different VEV's: in particular one can assume that there are no flavons in the model transforming as $1'$ or $1''$ [41].

The group A_4 has two obvious subgroups: G_S, which is a reflection subgroup generated by S and G_T, which is the group generated by T, which is isomorphic to Z_3. If the flavour symmetry associated to A_4 is broken by the VEV of a triplet $\varphi = (\varphi_1, \varphi_2, \varphi_3)$ of scalar fields, there are two interesting breaking pattern. The VEV

$$\langle\varphi\rangle = (v_S, v_S, v_S) \tag{24}$$

breaks A_4 down to G_S, while

$$\langle\varphi\rangle = (v_T, 0, 0) \tag{25}$$

breaks A_4 down to G_T. We have seen that G_S and G_T are the relevant low-energy symmetries of the neutrino and the charged-lepton sectors, respectively. Indeed we have shown that the TB mass matrix is invariant under G_S and, for charged leptons, a diagonal $m_e^\dagger m_e$ is invariant under G_T. A crucial part of all serious A4 models is the dynamical generation of this alignment in a natural way. In most of the models A_4 is accompanied by additional flavour symmetries, either discrete like Z_N or continuous like U(1), which are necessary to eliminate unwanted couplings, to ensure the needed vacuum alignment (obtained from the minimization of the most general potential compatible with the assumed symmetries) and to reproduce the observed mass hierarchies. Explicit realizations of models for TB mixing based on A_4 can be found, for example, in [41,42,43,44,45]. The possible origin of A_4 from a deeper level of the theory has been discussed in the context of extra dimensions and orbifolding [43], [46] or as related to the fact that A_4 is a subgroup of the modular group [42], which plays a role in string theory. In passing, we note that we are mainly interested here in the possibility that the flavour symmetry is valid at the GUT scale and is broken at lower scales.

7. Applying A_4 to lepton masses and mixings

In the lepton sector a typical A_4 model works as follows [42]. One assigns leptons to the four inequivalent representations of A_4: LH lepton doublets l transform as a triplet 3, while the RH charged leptons e^c, μ^c and τ^c transform as 1, $1''$ and $1'$, respectively. Here we consider a see-saw realization, so we also introduce conjugate neutrino fields ν^c transforming as a triplet of A_4. The fact that LH lepton doublets l and, in the see-saw case, also the RH neutrinos ν^c, transform as triplets is crucial to realize the fixed ratios of mass matrix elements needed to obtain TB mixing. A drawback is that for the ratio r, defined in eq.(12), one would expect $\sqrt{r} \approx \mathcal{O}(1)$ to be compared with the experimental value is $\sqrt{r} \approx 0.2$, which implies a moderate fine tuning.

One adopts a supersymmetric (SUSY) context also to make contact with Grand Unification (flavor symmetries are supposed to act near the GUT scale). In fact, as well known, SUSY is important in GUT's for offering a solution to the hierarchy problem, for improving coupling unification and for making the theory compatible with bounds on proton decay. But, in models of lepton mixing, SUSY also helps for obtaining the vacuum alignment, because the SUSY constraints are very strong and limit the form of the superpotential very much. Thus SUSY is not necessary but it is a plausible and useful ingredient. The flavor symmetry is broken by two triplets φ_S and φ_T (with the vacuum alignment in eqs.(24, 25)) and by one or more singlets ξ. All these fields are invariant under the SM gauge symmetry. Two Higgs doublets $h_{u,d}$, invariant under A_4, are also introduced. One can obtain the observed hierarchy among m_e, m_μ and m_τ by introducing an additional $\mathrm{U}(1)_{FN}$ flavor symmetry [47] under which only the RH lepton sector is charged (recently some models were proposed with a different VEV alignment such that the charged lepton hierarchies are obtained without introducing a $U(1)$ symmetry [48,45]). We recall that $U(1)_{FN}$ is a simplest flavor symmetry where particles in different generations are assigned (in general) different values of an Abelian charge. Also Higgs fields may get a non zero charge. When the symmetry is spontaneously broken the entries of mass matrices are suppressed if there is a charge mismatch and more so if the corresponding mismatch is larger. We assign FN-charges 0, q and $2q$ to τ^c, μ^c and e^c, respectively. There is some freedom in the choice of q. Here we take $q = 2$. By assuming that a flavon θ, carrying a negative unit of FN charge, acquires a VEV $\langle\theta\rangle/\Lambda \equiv \lambda < 1$, the Yukawa couplings become field dependent quantities $y_{e,\mu,\tau} = y_{e,\mu,\tau}(\theta)$ and we have

$$y_\tau \approx \mathcal{O}(1) \quad , \qquad y_\mu \approx O(\lambda^2) \quad , \qquad y_e \approx O(\lambda^4) \quad . \tag{26}$$

Had we chosen $q = 1$, we would have needed $\langle\theta\rangle/\Lambda$ of order λ^2, to reproduce the above result. The superpotential term for lepton masses, w_l is given by:

$$w_l = y_e e^c(\varphi_T l) + y_\mu \mu^c(\varphi_T l)' + y_\tau \tau^c(\varphi_T l)'' + y(\nu^c l) + (x_A\xi + \tilde{x}_A\tilde{\xi})(\nu^c\nu^c) + x_B(\varphi_S\nu^c\nu^c) + \ldots \tag{27}$$

with dots denoting higher dimensional operators that lead to corrections to the LO approximation. In our notation, the product of 2 triplets (33) transforms as 1, $(33)'$ transforms as $1'$ and $(33)''$ transforms as $1''$. To keep our formulae compact, we omit to write the Higgs and flavon fields $h_{u,d}$, θ and the cut-off scale Λ. For instance $y_e e^c(\varphi_T l)$ stands for $y_e e^c(\varphi_T l) h_d \theta^4/\Lambda^5$. The parameters of the superpotential w_l are complex, in particular those responsible for the heavy neutrino Majorana masses, $x_{A,B}$. Some terms allowed by the A_4 symmetry, such as the terms obtained by the exchange $\varphi_T \leftrightarrow \varphi_S$, (or the term $(\nu^c\nu^c)$) are missing in w_l. Their absence is crucial and, in each version of A_4 models, is motivated by additional symmetries.

As for the neutrino spectrum both normal and inverted hierarchies can be realized. It is interesting that A_4 models with the see-saw mechanism typically lead to a light neutrino spectrum which satisfies the sum rule (among complex masses):

$$\frac{1}{m_3} = \frac{1}{m_1} - \frac{2}{m_2} \quad . \tag{28}$$

A detailed discussion of a spectrum of this type can be found in refs. [42,45]. The above sum rule gives rise to bounds on the lightest neutrino mass. As a consequence, for example, the possible values of $|m_{ee}|$ are restricted. For normal hierarchy we have

$$|m_{ee}| \approx \frac{4}{3\sqrt{3}} \Delta m^2_{sun} \approx 0.007 \text{ eV} \quad . \tag{29}$$

while for inverted hierarchy

$$|m_{ee}| \geq \sqrt{\frac{\Delta m^2_{atm}}{8}} \approx 0.017 \text{ eV} \quad . \tag{30}$$

In a completely general framework, without the restrictions imposed by the flavor symmetry, $|m_{ee}|$ could vanish in the case of normal hierarchy. In this model $|m_{ee}|$ is always different from zero, though its value for normal hierarchy is probably too small to be detected in the next generation of $0\nu\beta\beta$ experiments.

In the leading approximation A_4 models lead to exact TB mixing. In these models TB mixing is implied by the symmetry at the leading order approximation which is corrected by non leading effects. Given the set of flavour symmetries and having specified the field content, the non leading corrections to TB mixing, arising from higher dimensional effective operators, can be evaluated in a well defined expansion. In the absence of specific dynamical tricks, in a generic model, all the three mixing angles receive corrections of the same order of magnitude. Since the experimentally allowed departures of θ_{12} from the TB value, $\sin^2\theta_{12} = 1/3$, are small, numerically not larger than $\mathcal{O}(\lambda_C^2)$, it follows that both θ_{13} and the deviation of θ_{23} from the maximal value are expected to also be typically of the same general size. The central values $\sin\theta_{13} \sim 0.16 - 0.11$ that can be derived from the experimental results in the two columns of Table(1), respectively, are in between $\mathcal{O}(\lambda_C^2) \sim \mathcal{O}(0.05)$ and $\mathcal{O}(\lambda_C) \sim \mathcal{O}(0.23)$. Thus models based on TB or GR mixing are still viable with preference for the lower side of the experimental range for θ_{13}. It is also to be noted that one can introduce some additional theoretical input to enhance the value of θ_{13}. In the case of A_4, examples are provided by the model of ref.[49], formulated before the T2K and MINOS results were known and the modified A_4 model of ref.[50] (see also [51]).

8. A_4, quarks and GUT's

Much attention has been devoted to the question whether models with TB mixing in the neutrino sector can be suitably extended to also successfully describe the observed pattern of quark mixings and masses and whether this more complete framework can be made compatible with (supersymmetric) SU(5) or SO(10) Grand Unification.

The simplest attempts of directly extending models based on A_4 to quarks have not been satisfactory. At first sight the most appealing possibility is to adopt for quarks the same classification scheme under A_4 that one has used for leptons (see, for example, ref.[42]). Thus one tentatively assumes that LH quark doublets Q transform as a triplet 3, while the antiquarks (u^c, d^c), (c^c, s^c) and (t^c, b^c) transform as 1, $1''$ and $1'$, respectively. This leads to $V_u = V_d$ and to the identity matrix for $V_{CKM} = V_u^\dagger V_d$ in the lowest approximation. This at first appears as very promising: a LO approximation where neutrino mixing is TB and $V_{CKM} = 1$ is a very good starting point. But there are some problems. First, the corrections to $V_{CKM} = 1$ turn out to be strongly constrained by the leptonic sector, because lepton mixing angles are very close to the TB values, and, in the simplest models, this constraint leads to a too small V_{us} (i.e. the Cabibbo angle is rather large in comparison to the allowed shifts from the TB mixing angles). Also in these models, the quark classification which leads to $V_{CKM} = 1$ is not compatible with A_4 commuting with SU(5). An additional consequence of the above assignment is that the top quark mass arises from a non-renormalizable dimension-5 operator. In that case, to reproduce the top mass, we need to compensate the cutoff suppression by some extra dynamical mechanism. Alternatively, we have to introduce a separate symmetry breaking parameter for the quark sector, sufficiently close to the cutoff scale.

Due to this, larger discrete groups have been considered for the description of quarks. A particularly appealing set of models is based on the discrete group T', the double covering group of A_4 [52], [53], [54]. The representations of T' are those of A_4 plus three independent doublets 2, $2'$ and $2''$. The doublets are interesting for the classification of the first two generations of quarks [55]. For example, in ref.[53] a viable description was obtained, i.e. in the leptonic sector the predictions of the A_4 model are maintained, while the T' symmetry plays an essential role for reproducing the pattern of quark mixing. But, again, the classification adopted in this model is not compatible with Grand Unification.

As a result, the group A_4 was considered by many authors to be too limited to also describe quarks and to lead to a grand unified description. But it has been shown [44] that this negative attitude is not justified and that it is actually possible to construct a viable model based on A_4 which leads to a grand unified theory (GUT) of quarks and leptons with TB mixing for leptons and with quark (and charged lepton) masses and mixings compatible with experiment. At the same time this model offers an example of an extra dimensional SU(5) GUT in which a description of all fermion masses and mixings is accomplished. The formulation of SU(5) in extra dimensions has the usual advantages of avoiding large Higgs representations to break SU(5) and of solving the doublet-triplet splitting problem. The choice of the transformation properties of the two Higgses H_5 and $H_{\bar{5}}$ has a special role in this model. They are chosen to transform as two different A_4 singlets 1 and 1′. As a consequence, mass terms for the Higgs colour triplets are not directly allowed and their masses are introduced by orbifolding, à la Kawamura [56]. In this model, proton decay is dominated by gauge vector boson exchange giving rise to dimension-6 operators, while the usual contribution of dimension-5 operators is forbidden by the selection rules of the model. Given the large M_{GUT} scale of SUSY models and the relatively huge theoretical uncertainties, the decay rate is within the present experimental limits. A see-saw realization in terms of an A_4 triplet of RH neutrinos ν^c ensures the correct ratio of light neutrino masses with respect to the GUT scale. In this model extra dimensional effects directly contribute to determine the flavour pattern, in that the two lightest tenplets T_1 and T_2 are in the bulk (with a doubling T_i and T'_i, $i = 1, 2$ to ensure the correct zero mode spectrum), whereas the pentaplets F and T_3 are on the brane. The hierarchy of quark and charged lepton masses and of quark mixings is determined by a combination of extra dimensional suppression factors and of $U(1)_{FN}$ charges, both of which only apply to the first two generations, while the neutrino mixing angles derive from A_4 in the usual way. If the extra dimensional suppression factors and the $U(1)_{FN}$ charges are switched off, only the third generation masses of quarks and charged leptons survive. Thus the charged fermion mass matrices are nearly empty in this limit (not much of A_4 effects remain) and the quark mixing angles are determined by the small corrections induced by the above effects. The model is natural, since most of the small parameters in the observed pattern of masses and mixings as well as the necessary vacuum alignment are justified by the symmetries of the model. However, in this case, like in all models based on $U(1)_{FN}$, the number of $\mathcal{O}(1)$ parameters is larger than the number of measurable quantities, so that in the quark sector the model can only account for the orders of magnitude (measured in terms of powers of an expansion parameter) and not for the exact values of mass ratios and mixing angles. A moderate fine tuning is only needed to enhance the Cabibbo mixing angle between the first two generations, which would generically be of $\mathcal{O}(\lambda_C^2)$.

The problem of constructing GUT models based on $SU(5) \otimes G_f$ or $SO(10) \otimes G_f$ with approximate TB mixing in the leptonic sector has also been considered by many authors. Examples are: for $G_f = A_4$ ref.[57], for T' ref.[54], for S_4 ref.[58]. As for the models based on $SO(10) \otimes G_f$ recent examples were discussed with $G_f = S_4$ [59] and $G_f = PSL_2(7)$ [60]. Clearly the case of $SO(10)$ is even more difficult than that of $SU(5)$ because the neutrino sector is tightly related to that of quarks and charged leptons as all belong to the 16 of $SO(10)$. For a discussion of $SO(10) \otimes A_4$ models, see [61]. More in general see refs.[62]. In our opinion most of the models are incomplete (for example, the crucial issue of VEV alignment is not really treated in depth as it should) and/or involve a number of unjustified steps and ad-hoc fine tuning of parameters.

While A_4 is the minimal flavor group leading to TB mixing, alternative flavor groups have been studied in the literature and can lead to interesting variants with some specific features. Actually, in ref.[63], the claim was made that, in order to obtain the TB mixing "without fine tuning", the finite group must be S_4 or a larger group containing S_4. For us this claim is not well grounded being based on an abstract mathematical criterium for a natural model (see also [64]). For us a physical field theory model is natural if the interesting results are obtained from the most general lagrangian compatible with the stated symmetry and the specified representation content for the flavons. For example, we obtain from A_4 (which is a subgroup of S_4) a natural (in our sense) model for the TB mixing by simply not including symmetry breaking flavons transforming like the $1'$ and the $1''$ representations of A_4. This limitation on the transformation properties of the flavons is not allowed by the rules specified in ref.[63] which demand that the symmetry breaking is induced by all possible kinds of flavons (note that, according to this criterium, the SM of electroweak interactions would not be natural because only Higgs doublets are introduced!). Rather, for naturalness we also require that additional physical properties like the VEV alignment or the hierarchy of charged lepton masses also follow from the assumed symmetry and are not obtained by fine tuning parameters: for this actually A_4 can be more effective than S_4 because it possesses three different singlet representations 1, $1'$ and $1''$.

Models of neutrino mixing based on S_4 have in fact been studied [65]. The group of the permutations of 4 objects S_4 has 24 elements and 5 equivalence classes that correspond to 5 inequivalent irreducible representations, two singlets, one doublet, two triplets: 1_1, 1_2, 2, 3_1 and 3_2. Note that the squares of the dimensions of all these representations add up to 24.

9. Bimaximal Mixing and S4

The new results showing that probably θ_{13} is not far from its former upper bound could alternatively be interpreted as an indication that the agreement with the TB or GR mixing is accidental. Then a scheme where instead the Bimaximal (BM) mixing is the correct first approximation modified by terms of $\mathcal{O}(\lambda_C)$ could be relevant. In BM mixing θ_{12} and θ_{23} are both maximal while $\theta_{13} = 0$ (see. Fig. 2). This is in line with the well known empirical observation that $\theta_{12} + \theta_C \sim \pi/4$, a relation known as quark-lepton complementarity [66], or similarly $\theta_{12} + \sqrt{m_\mu/m_\tau} \sim \pi/4$. No compelling model leading, without parameter fixing, to the exact complementarity relation has been produced so far. Probably the exact complementarity relation becomes more plausible if replaced with $\theta_{12} + \mathcal{O}(\theta_C) \sim \pi/4$ or $\theta_{12} + \mathcal{O}(m_\mu/m_\tau) \sim \pi/4$ (which we could call "weak" complementarity). One can think of models where, because of a suitable symmetry, BM mixing holds in the neutrino sector at leading order and the necessary, rather large, corrective terms for θ_{12} arise from the diagonalization of charged lepton masses [66]. These terms of order $\mathcal{O}(\lambda_C)$ from the charged lepton sector would then generically also affect θ_{13} and the resulting value could well be compatible with the present experimental values of θ_{13}. A word of caution must be kept in mind: in the presence of these relatively large off diagonal terms in the charged lepton diagonalizing matrix one must arrange that not too large contributions to the decays $\mu \to e\gamma$ or $\tau \to \mu\gamma$ are generated [28].

The BM mixing matrix is given by:

$$U_{BM} = \begin{pmatrix} \frac{1}{\sqrt{2}} & -\frac{1}{\sqrt{2}} & 0 \\ \frac{1}{2} & \frac{1}{2} & -\frac{1}{\sqrt{2}} \\ \frac{1}{2} & \frac{1}{2} & \frac{1}{\sqrt{2}} \end{pmatrix} . \tag{31}$$

Along this line of thought, we have used the expertise acquired with non Abelian finite flavour groups to construct a model [67] based on the permutation group S_4 which naturally leads to the BM mixing at leading order. We have adopted a supersymmetric formulation of the model in 4 space-time dimensions. The complete flavour group is $S_4 \times Z_4 \times U(1)_{FN}$. In leading order, the charged leptons are diagonal and hierarchical and the light neutrino mass matrix, after see-saw, leads to the exact BM mixing. The model is built in such a way that the dominant corrections to the BM mixing, from higher dimensional operators in the superpotential, only arise from the charged lepton sector at next-to-the-leading-order and naturally inherit λ_C (which fixes the charged lepton mass hierarchies) as the relevant expansion parameter. As a result the mixing angles deviate from the BM values by terms of $\mathcal{O}(\lambda_C)$ (at most), and weak complementarity holds. A crucial feature of the model is that only θ_{12} and θ_{13} are corrected by terms of $\mathcal{O}(\lambda_C)$ while θ_{23} is unchanged at this order (which is essential for a better agreement of the model with the present data). Recently the model was extended to include quarks in a $SU(5)$ Grand Unified version [68] or in a Pati-Salam framework [69]. An $SO(10)$ model is discussed in ref.[70].

10. Anarchy versus Symmetry

We now briefly turn to models that do not take seriously any of the coincidences described above (i.e. the proximity of the data to the TB or GR patterns or the quark-lepton complementarity: these patterns cannot all be true and it is possible that none of them is true) and are therefore based on a less restrictive flavour symmetry. It is clear that the T2K hint that θ_{13} may be large is great news for the most extreme position of this type, which is "anarchy" [71]: no symmetry at all in the lepton sector, only chance. This view predicts generic mixing angles, so the largest angle, θ_{23}, should somewhat deviate from maximal and the smallest angle, θ_{13}, should be as large as possible within the experimental bounds. Anarchy can be formulated in a $SU(5) \otimes U(1)$ context by taking different Froggatt-Nielsen charges [47] only for the

$SU(5)$ tenplets (for example 10: (3,2,0) where 3 is the charge of the first generation, 2 of the second, zero of the third) while no charge differences appear in the $\bar{5}$: $\bar{5}$: (0,0,0). This assignment is in agreement with the empirical fact that the mass hierarchies are more pronounced for up quarks in comparison with down quarks and charged leptons. In a non see-saw model, with neutrino masses dominated by the contribution of the dimension-5 Weinberg operator in eq.(1), the $\bar{5}$ vanishing charges directly lead to random neutrino mass and mixing matrices. In anarchical see-saw models also the charges of the SU(5) singlet RH neutrinos must be undifferentiated among the 3 generations: 1: (0,0,0). Anarchy can be mitigated by assuming that it only holds in the 2-3 sector: e.g $\bar{5}$: (2,0,0) with the advantage that the first generation masses and the angle θ_{13} are then naturally small (see ref.[72)] for a recent discussion of this model). In models with see-saw one can alternatively play with the charges for the RH SU(5) singlet neutrinos. If, for example, we take 1: (1, -1, 0), together with $\bar{5}$: (2,0,0), it is possible to get a normal hierarchy model with θ_{13} small and also with $r = \Delta m^2_{solar}/\Delta m^2_{atm}$ naturally small (see, for example, ref.[73)]). In summary anarchy and its variants, all based on chance, offer a rather economical class of models that are among those that are compatible with the recent θ_{13} results, with preference with the upper side of the experimentally allowed range.

11. Conclusion

In the last decade we have learnt a lot about neutrino masses and mixings. A list of important conclusions have been reached. Neutrinos are not all massless but their masses are very small. Probably masses are small because neutrinos are Majorana particles with masses inversely proportional to the large scale M of lepton number violation. It is quite remarkable that M is empirically not far from M_{GUT}, so that neutrino masses fit well in the SUSY GUT picture. Also out of equilibrium decays with CP and L violation of heavy RH neutrinos can produce a B-L asymmetry, then converted near the weak scale by instantons into an amount of B asymmetry compatible with observations (baryogenesis via leptogenesis) [27)]. It has been established that most probably active neutrinos are not a significant component of dark matter in the Universe. We have also understood there there is no contradiction between large neutrino mixings and small quark mixings, even in the context of GUTs.

This is a very impressive list of achievements. Coming to a detailed analysis of neutrino masses and mixings a long collection of models have been formulated over the years. With continuous improvements of the data and more precise values of the mixing angles most of the models have been discarded by experiment. Still the surviving models span a wide range going from a maximum of symmetry, with discrete non-abelian flavour groups, to the opposite extreme of anarchy. By now, besides the detailed knowledge of the entries of the V_{CKM} matrix we also have a reasonable determination of the neutrino mixing matrix U_{PMNS}. The data appear to suggest some special patterns (recall Fig. 2) like TB or GR or BM mixing to be valid in some leading approximation, corrected by small non leading terms. If one takes these "coincidences" seriously, then non-abelian discrete flavour groups emerge as the main road to an understanding of this mixing pattern. Indeed the entries of e.g. TB mixing matrix are clearly suggestive of "rotations" by simple, very specific angles. It is remarkable that neutrino and quark mixings have such a different qualitative pattern. An obvious question is whether some additional indication for discrete flavour groups can be obtained by considering the extension of the models to the quark sector, perhaps in a Grand Unified context. The answer appears to be that, while the quark masses and mixings can indeed be reproduced in models where TB or BM mixing is realized in the leptonic sector through the action of discrete groups, there are no specific additional hints in favour of discrete groups that come from the quark sector. Further important input could come from $\mu \to e\gamma$ and in general from lepton flavour violating processes, from $b \to s\gamma$ and from LHC physics. In fact, new physics at the weak scale could have important feedback on the physics of neutrino masses and mixing.

In conclusion, one could have imagined that neutrinos would bring a decisive boost towards the formulation of a comprehensive understanding of fermion masses and mixings. In reality it is frustrating that no real illumination was sparked on the problem of flavour. We can reproduce in many different ways the observations, in a wide range that goes from anarchy to discrete flavour symmetries) but we have not yet been able to single out a unique and convincing baseline for the understanding of fermion masses and mixings. In spite of many interesting ideas and the formulation of many elegant models the mysteries of the flavour structure of the three generations of fermions have not been much unveiled.

I imagine that by the next edition of this by now classic School, we will know the value of θ_{13} with a better accuracy, from the continuation of T2K, MINOS and DOUBLE CHOOZ and from the start of Daya Bay and RENO. Some existing models will be eliminated and the surviving ones will be updated to become more quantitative in order to cope with a precisely known mixing matrix. A definitely non vanishing θ_{13} value will encourage the planning of long baseline experiments for the detection of CP violation in neutrino oscillations. Along the way the important issue of the existence of sterile neutrinos must be clarified. The on going or in preparation experiments on the absolute value of neutrino masses, on $0\nu\beta\beta$, on $\mu \to e\gamma$, on the search for dark matter etc can also lead to extremely important developments in the near future. So this field is very promising and there are all reasons to expect an exciting time ahead of us.

12. Acknowledgments

I warmly thank Prof. Antonino Zichichi for his kind invitation and for the impressive hospitality received in Erice, as usual. I also thank the Staff of the E. Majorana School, in particular Mrs. Fiorella Ruggiu for their assistance in organizing and managing my stay at the School. I am indebted with Luca Merlo for his critical reading of a preliminary version of this work. I am also glad to acknowledge interesting discussions on this subject with him and with Ferruccio Feruglio and Davide Meloni. Supported in part by PRIN 2008 and by the EU network LHCPHENONET.

13. References

1) G. Altarelli and F. Feruglio, New J. Phys. **6** (2004) 106 [ArXiv:hep-ph/0405048]; G. Altarelli, [ArXiv:1011.5342].
2) R. N. Mohapatra and A. Y. Smirnov, Ann. Rev. Nucl. Part. Sci. 56, 569 (2006) [ArXiv:hep-ph/0603118]; W. Grimus, PoS P2GC:001,2006. [ArXiv:hep-ph/0612311]; M. C. Gonzalez-Garcia and M. Maltoni, Phys. Rept. **460** (2008) 1 [ArXiv:0704.1800].
3) The MiniBooNE Collaboration Phys.Rev.Lett. **105** 181801 (2010), [ArXiv:1007.1150].
4) T2K Collaboration: K.Abe et al, Phys. Rev. Lett. 107 (2011) 041801, [ArXiv:1106.2822].
5) MINOS Collaboration, P. Adamson et al, [ArXiv:1108.0015].
6) DOUBLE CHOOZ Collaboration, H. de Kerret, Talk at LowNu, Seoul, November 2011.
7) Z. Djurcic, Talk at NuFact2011, Geneva, August 2011.
8) G. Mention et al, Phys.Rev.D83:073006,2011, [ArXiv:1101.2755]; Th. A. Mueller et al, Phys.Rev.C83:054615,2011, [ArXiv:1101.2663]; P. Huber, Phys.Rev.C84:024617,2011, [ArXiv:1106.0687] .
9) P. Anselmann et al. (GALLEX), Phys. Lett. B342, 440 (1995); W. Hampel et al. (GALLEX), Phys. Lett. B420, 114 (1998); F. Kaether et al. Phys. Lett. B685, 47 (2010); J. N. Abdurashitov et al. (SAGE), Phys. Rev. Lett. 77, 4708 (1996); J. N. Abdurashitov et al. (SAGE), Phys. Rev. C59, 2246 (1999); J. N. Abdurashitov et al., Phys. Rev. C73, 045805 (2006); J. N. Abdurashitov et al. (SAGE), Phys. Rev. C80, 015807 (2009).
10) E. Giusarma et al, [ArXiv:1102.4774]
11) C. Giunti, [ArXiv:1110.3914] and references therein.
12) S. Weinberg, Phys.Rev.Lett. **43** 1566 (1979).
13) P. Minkowski, Phys. Letters B67 (1977)421; T. Yanagida, in *Proc. of the Workshop on Unified Theory and Baryon Number in the Universe*, KEK, March 1979; S. L. Glashow, in "Quarks and Leptons", Cargèse, ed. M. Lévy et al., Plenum, 1980 New York, p. 707; M. Gell-Mann, P. Ramond and R. Slansky, in *Supergravity*, Stony Brook, Sept 1979; R. N. Mohapatra and G. Senjanovic, Phys. Rev. Lett. **44** (1980) 912.

14) B. Pontecorvo, Sov. Phys. JETP **6**, 429 (1957) [Zh. Eksp. Teor. Fiz. **33**, 549 (1957)]; Z. Maki, M. Nakagawa and S. Sakata, Prog. Theor. Phys. **28** (1962) 870; B. Pontecorvo, Sov. Phys. JETP **26**, 984 (1968) [Zh. Eksp. Teor. Fiz. **53**, 1717 (1968)]; V. N. Gribov and B. Pontecorvo, Phys. Lett. B **28**, 493 (1969).
15) B. W. Lee, S. Pakvasa, R. Shrock, and H. Sugawara, Phys. Rev. Lett. 38, 937 (1977); B. W. Lee and R. Shrock, Phys. Rev. D16, 1444 (1977).
16) N. Cabibbo, Phys. Lett. B **72**, 333 (1978).
17) S. M. Bilenky, J. Hosek and S. T. Petcov, Phys. Lett. B **94** (1980) 495; J. Schechter and J. W. F. Valle, Phys. Rev. D **22** (1980) 2227; M. Doi, T. Kotani, H. Nishiura, K. Okuda and E. Takasugi, Phys. Lett. B **102** (1981) 323. Mass matrices with a general dependence on ϕ, $\phi_{1,2}$ have been analyzed in: M. Frigerio and A. Y. Smirnov, Nucl. Phys. B **640** (2002) 233 and Phys. Rev. D **67** (2003) 013007.
18) G. L. Fogli et al, [ArXiv:1106.6028].
19) T. Schwetz, M. Tortola and J. W. F. Valle, [ArXiv:1108.1376].
20) Ch. Weinheimer et al, Phys. Lett. B460 (1999) 219.
21) G. L. Fogli et al, [ArXiv:0805.2517], E. Komatsu, et al., Astrophys.J.Suppl.192:18,2011, [ArXiv:1001.4538], M. C. Gonzalez-Garcia, M. Maltoni and J. Salvado, JHEP 1008:117,2010, [ArXiv:1006.3795].
22) W. Rodejohann, Int.J.Mod.Phys. E20, 1833(2011), [ArXiv:1106.1334].
23) T. Mori, Talk at the EPS Conference, Grenoble, July 2011.
24) , A. S. Barabash, Phys. Atom. Nucl. 73 (2010) 162.
25) F. Feruglio, A. Strumia and F Vissani, Nucl.Phys.B637:345-377,2002., Addendum-ibid.B659:359-362, 2003. [ArXiv:hep-ph/0201291].
26) For a review see, for example,: M. Trodden, Rev. Mod. Phys. **71**, 1463 (1999).
27) For a review see for example: W. Buchmuller, R.D. Peccei and T. Yanagida, Ann.Rev.Nucl.Part.Sci.55:311-355,2005. [ArXiv:hep-ph/0502169].
28) G. Altarelli and F. Feruglio, Rev.Mod.Phys. 82 (2010) 2701, [ArXiv:1002.0211].
29) T. Fukuyama and H. Nishiura, [ArXiv:hep-ph/9702253].
30) H.-J. He and F.-R. Yin, Phys.Rev. D84 (2011) 033009, [ArXiv:1104.2654].
31) P. F. Harrison, D. H. Perkins and W. G. Scott, Phys. Lett. B **530** (2002) 167, [ArXiv:hep-ph/0202074]; P. F. Harrison and W. G. Scott, Phys. Lett. B **535** (2002) 163, [ArXiv:hep-ph/0203209]; Phys. Lett. B **547** (2002) 219, [ArXiv:hep-ph/0210197]; Phys. Lett. B **557** (2003) 76, [ArXiv:hep-ph/0302025]; Z. z. Xing, Phys. Lett. B **533** (2002) 85, [ArXiv:hep-ph/0204049].
32) Y. Kajiyama, M. Raidal and A. Strumia, Phys. Rev. D **76** (2007) 117301 [ArXiv:0705.4559].
33) L. L. Everett and A. J. Stuart, Phys.Rev. D79:085005, 2009, [ArXiv:0812.1057], G.-J. Ding, L. L. Everett and A. J. Stuart, [ArXiv:1110.1688].

34) F. Feruglio and A. Paris, JHEP 1103 (2011) 101, ArXiv:1101.0393.
35) **W. Rodejohann, Phys. Lett. B671 (2009) 267, [ArXiv:0810.5239 [hep-ph]].**
36) A. Adulpravitchai, A. Blum and W. Rodejohann, New J. Phys. **11** (2009) 063026 [ArXiv:0903.0531].
37) S. F. King, [ArXiv:1106.4239] and references therein.
38) E. Ma and G. Rajasekaran, Phys. Rev. D **64** (2001) 113012 [ArXiv:hep-ph/0106291]; E. Ma, Mod. Phys. Lett. A **17** (2002) 627 [ArXiv:hep-ph/0203238]. K. S. Babu, E. Ma and J. W. F. Valle, Phys. Lett. B **552** (2003) 207 [ArXiv:hep-ph/0206292]; M. Hirsch, J. C. Romao, S. Skadhauge, J. W. F. Valle and A. Villanova del Moral, [ArXiv:hep-ph/0312244], Phys. Rev. D **69** (2004) 093006 [ArXiv:hep-ph/0312265]; E. Ma, Phys. Rev. D **70** (2004) 031901; Phys. Rev. D **70** (2004) 031901 [ArXiv:hep-ph/0404199]; New J. Phys. **6** (2004) 104 [ArXiv:hep-ph/0405152]; ArXiv:hep-ph/0409075; S. L. Chen, M. Frigerio and E. Ma, Nucl. Phys. B **724** (2005) 423 [ArXiv:hep-ph/0504181]; E. Ma, Phys. Rev. D **72** (2005) 037301 [ArXiv:hep-ph/0505209].
39) F. del Aguila, A. Carmona and J. Santiago, [ArXiv:1001.515]; A. Kadosh and E. Pallante, [ArXiv:1004.0321]; D. Meloni, S. Morisi and E. Peinado, [ArXiv:1011.1371; E. Peinado, [ArXiv:1010.2614]; F. Feruglio and A. Paris, Nucl.Phys. **B840** (2010) 405, [ArXiv:1005.5526].
40) H. Ishimori et al, Prog.Theor.Phys.Suppl.183:1, 2010; [ArXiv:1003.3552]; P. O. Ludl, J.Phys.A A43 (2010) 395204, [ArXiv:1006.1479]; W. Grimus and P.O. Ludl, J.Phys.A43 (2010) 445209 [ArXiv:1006.0098]; K. M. Parattu and A. Wingerter, Phys.Rev.D84:013011,2011, [ArXiv:1012.2842].
41) G. Altarelli and F. Feruglio, Nucl. Phys. B **720** (2005) 64 [ArXiv:hep-ph/0504165].
42) G. Altarelli and F. Feruglio, Nucl. Phys. B **741** (2006) 215 [ArXiv:hep-ph/0512103].
43) G. Altarelli, F. Feruglio and Y. Lin, Nucl. Phys. B **775** (2007) 31 [ArXiv:hep-ph/0610165].
44) G. Altarelli, F. Feruglio and C. Hagedorn, JHEP **0803** (2008) 052 [ArXiv:0802.0090].
45) G. Altarelli and D. Meloni, J.Phys.G G36 (2009) 085005, [ArXiv:0905.0620].
46) A. Adulpravitchai, A. Blum and M. Lindner, JHEP **07** (2009) 053 [ArXiv:0906.0468]; A. Adulpravitchai, M.A. Schmidt, [ArXiv:1001.3172]; T.J. Burrows, S.F. King, Nucl.Phys. **B842** (2011) 107, [ArXiv:1007.2310].
47) C. D. Froggatt and H. B. Nielsen, Nucl. Phys. **B147** (1979) 277.
48) Y. Lin, Nucl. Phys. B **813**, 91 (2009) [ArXiv:0804.2867]; [ArXiv:0903.0831];
49) Y. Lin, Nucl.Phys. B824 (2010) 95, [ArXiv:0905.3534].
50) E. Ma and D. Wegman, [ArXiv:1106.4269].
51) I. de Medeiros Varzielas and L. Merlo, JHEP 1102 (2011) 062,

[ArXiv:1011.6662].
52) P.H. Frampton and T.W. Kephart, Int. J. Mod. Phys. **A10** (1995) 4689, [ArXiv:hep-ph/9409330]; JHEP **09** (2007) 110, [ArXiv:0706.1186]; A. Aranda, C.D. Carone and R.F. Lebed, Phys. Lett. **B474** (2000) 170, [ArXiv:hep-ph/9910392]; Phys. Rev. **D62** (2000) 016009, [ArXiv:hep-ph/0002044]; P.D. Carr and P.H. Frampton, [ArXiv:hep-ph/0701034]; A. Aranda, Phys. Rev. **D76** (2007) 111301, [ArXiv:0707.3661]; P.H. Frampton and S. Matsuzaki, Phys. Lett. **B679** (2009) 347, [ArXiv:0902.1140]; G.J. Ding, Phys. Rev. **D78** (2008) 036011, [ArXiv:0803.2278];
53) F. Feruglio *et al*, Nucl. Phys. **B775** (2007) 120, [ArXiv:hep-ph/0702194].
54) M.C. Chen and K.T. Mahanthappa, Phys. Lett. **B652** (2007) 34, ArXiv:0705.0714.
55) A. Pomarol and D. Tommasini, Nucl. Phys. **B466** (1996) 3, [ArXiv:hep-ph/9507462]; R. Barbieri, G.R. Dvali and L. Hall, Phys. Lett. **B377** (1996) 76, [ArXiv:hep-ph/9512388] R. Barbieri *et al*, Nucl. Phys. **B493** (1997) 3, [ArXiv:hep-ph/9610449]; R. Barbieri, L. Hall and A. Romanino, Phys. Lett. **B401** (1997) 47, [ArXiv:hep-ph/9702315].
56) Y. Kawamura, Prog. Theor. Phys. **105** (2001) 999, [ArXiv:hep-ph/0012125].
57) E. Ma, Mod. Phys. Lett. A **20** (2005) 2767, [ArXiv:hep-ph/0506036]; E. Ma, H. Sawanaka and M. Tanimoto, Phys. Lett. B **641** (2006) 301 [ArXiv:hep-ph/0606103]; E. Ma, Mod. Phys. Lett. A **21** (2006) 2931 [ArXiv:hep-ph/0607190]; S. Morisi, M. Picariello and E. Torrente-Lujan, Phys. Rev. D **75** (2007) 075015 [ArXiv:hep-ph/0702034]; W. Grimus and H. Kuhbock, [ArXiv:0710.1585]; P. Ciafaloni *et al*, Phys. Rev. **D79** (2009) 116010, [ArXiv:0901.2236]; F. Bazzocchi *et al*, J. Phys. **G36** (2009) 015002, [ArXiv:0802.1693]; S. Antusch, S.F. King and M. Spinrath, [ArXiv:1005.0708].
58) C. Hagedorn, S.F. King and C. Luhn, [ArXiv:1003.4249]; H. Ishimori*et al*, [ArXiv:1004.5004].
59) B. Dutta, Y. Mimura and R.N. Mohapatra, Phys. Rev. **D80** (2009) 095021, [ArXiv:0910.1043]; JHEP **05** (2010) 034, [ArXiv:0911.2242];
60) S. F. King and C. Luhn, [ArXiv:0912.1344].
61) F. Bazzocchi, F. Frigerio and S. Morisi, Phys. Rev. **D78** (2008)116018, [ArXiv:0809.3573].
62) G. Altarelli and G. Blankenburg, JHEP 1103 (2011) 133, [ArXiv:1012.2697]; A.S. Joshipura, K.M. Patel, JHEP09(2011)137, [ArXiv:1105.5943]; Phys.Rev.D83:095002,2011, [ArXiv:1102.5148]; P. S. Bhupal Dev, R. N. Mohapatra and M. Severson, [ArXiv:1107.2378].
63) C. S. Lam, Phys.Rev.Lett.101:121602,2008, [ArXiv:0804.2622]; Phys. Rev. **D 78** (2008) 073015, [ArXiv:0809.1185].
64) W. Grimus, L. Lavoura and P. O. Ludl, J.Phys. **G36** (2009) 15007, [ArXiv:0906.2689].

65) R.N. Mohapatra, M.K. Parida and G. Rajasekaran, Phys. Rev. **D69** (2004) 053007, [ArXiv:hep-ph/0301234]; C. Hagedorn, M. Lindner and R.N. Mohapatra, JHEP **06** (2006) 042, [ArXiv:hep-ph/0602244]; Y. Cai and H.B Yu, Phys. Rev. **D74** (2006) 115005, [ArXiv:hep-ph/0608022]; E. Ma, Phys. Lett. **B632** (2006) 352; [ArXiv:hep-ph/0508231]; F. Bazzocchi and S. Morisi, Phys. Rev. **D80** (2009) 096005; [ArXiv:0811.0345]; H. Ishimori, Y. Shimizu and M. Tanimoto, Prog. Theor. Phys. **121** (2009) 769, [ArXiv:0812.5031]; F. Bazzocchi, L. Merlo and S. Morisi, Nucl. Phys **B816** (2009) 204, [ArXiv:0901.2086]; Phys. Rev. **D80** (2009) 053003, [ArXiv:0902.2849]; D. Meloni, J. Phys. **G37** (2010) 055201,[ArXiv: 0911.3591]. G.J. Ding, Nucl. Phys. **B827** (2010) 82, [ArXiv:0909.2210]; S. Morisi and E. Peinado, Phys. Rev. **D81** (2010) 085015, [ArXiv:1001.2265];
66) M. Raidal, Phys. Rev. Lett. **93** (2004) 16180, [ArXiv:hep-ph/0404046]; H. Minakata and A. Y. Smirnov, Phys. Rev. D **70** (2004) 073009, [ArXiv:hep-ph/0405088]; H. Minakata, ArXiv:hep-ph/0505262; P. H. Frampton and R. N. Mohapatra, JHEP **0501**, 025 (2005), ArXiv:hep-ph/0407139;J. Ferrandis and S. Pakvasa, Phys. Rev. D **71** (2005) 033004, ArXiv:hep-ph/0412038; S. K. Kang, C. S. Kim and J. Lee, [ArXiv:hep-ph/0501029]; G. Altarelli, F. Feruglio and I. Masina, Nucl. Phys. B **689** (2004) 157 [ArXiv:hep-ph/0402155]; N. Li and B. Q. Ma, [ArXiv:hep-ph/0501226]; K. Cheung, et al, [ArXiv:hep-ph/0503122]; Z. z. Xing, [ArXiv:hep-ph/0503200]; A. Datta, L. Everett and P. Ramond, [ArXiv:hep-ph/0503222]; T. Ohlsson, [ArXiv:hep-ph/0506094]; S. Antusch, S. F. King and R. N. Mohapatra, [ArXiv:hep-ph/0504007]; M. Lindner, M. A. Schmidt and A. Y. Smirnov, [ArXiv:hep-ph/0505067]; S. F. King, JHEP **0508** (2005) 105 [ArXiv:hep-ph/0506297]; A. Dighe, S. Goswami, and P. Roy, Phys.Rev.**D73** (2006) 07130, [ArXiv:hep-ph/0602062]; B. C. Chauhanet al, Eur.Phys. **J.C50**(2007) 573, [ArXiv:hep-ph/0605032] M. A. Schmidt and A. Yu. Smirnov, Phys.Rev. **D74**(2006)113003, [ArXiv:hep-ph/0607232]; K. A Hochmuth and W. Rodejohann, Phys.Rev. **D75**(2007) 073001, [ArXiv:hep-ph/0607103]; F. Plentinger, G. Seidl and W. Winter, Nucl.Phys. **B791** (2008) 60, [ArXiv: hep-ph/0612169]; Phys.Rev. **D76** (2007)113003, [ArXiv:0707.2379]; Y. H. Ahn, H-Y. Cheng and S. Oh, Phys.Rev.D83:076012,2011, [ArXiv:1102.0879].
67) G. Altarelli, F. Feruglio and L. Merlo, JHEP 0905 (2009) 020, [ArXiv:0903.1940].
68) D. Meloni, JHEP 1110 (2011) 010, [ArXiv:1107.0221].
69) R. de Adelhart Toorop., F. Bazzocchi and L. Merlo, JHEP 1008 (2010) 001, [ArXiv:1003.4502].
70) K. M. Patel, Phys.Lett.B695:225, 2011 [ArXiv:1008.5061].
71) L. J. Hall, H. Murayama and N. Weiner, Phys.Rev.Lett. 84 (2000) 2572, [ArXiv:hep-ph/9911341].

72) W. Buchmuller, V. Domcke, and K. Schmitz, [ArXiv:1111.3872].
73) G. Altarelli, F. Feruglio and I. Masina, JHEP 0301 (2003) 035, ArXiv:hep-ph/0210342.

CHAIRMAN: G. ALTARELLI

Scientific Secretaries: A. Agócs, Á. Lukács

DISCUSSION I

- Burda:

I always thought that we cannot measure the Dirac/Majorana nature of neutrinos in experiments involving real neutrinos, for example, in the final state, so the only experimental evidence is neutrinoless double beta decay?

- Altarelli:

Well, first of all, there is no experimental evidence, since nobody so far has seen neutrinoless double beta decay. In principle, one could observe the decay of those very heavy neutrinos that enter in the see-saw mechanism. These particles decay with manifest violation of lepton number, and then one could see the violation of lepton number in a process with real neutrino states.

- Burda:

But I think that such very heavy neutrinos already contradict experimental observation, don't they?

- Altarelli:

No, I do not see why. When I discussed leptogenesis, I have argued that leptogenesis may happen just because these particles decay with lepton number and CP violation. However, I don't know of any process that one can observe in practice in the lab. involving real neutrinos, that could lead to the same conclusion.

- Martinez:

Could you explain in more detail, why is it necessary to introduce sterile neutrinos, and how will it be possible to detect them, at the LHC, for example?

- Altarelli:

First of all, at present, there is no definitive evidence for sterile neutrinos. But eventually they might be imposed on us by some experiment. I have indicated a number of possible hints for the existence of these particles. For example, if the LSND signal is confirmed (at present, the experimental situation with LSND and MiniBoone is still somewhat messy), then we would have a third oscillation frequency with Δm2 of the order of 1 eV2, in addition to the atmospheric and the solar frequencies. With the three active neutrinos that we know, one can only form two independent Δm2 differences, and there is no space for a third one, so one should invoke extra neutrinos. Remember that LEP has measured the number of neutrinos which are coupled to the Z and this number is 3, so if there is another neutrino coupled to the Z, it must be heavier than 45 GeV. This would be a fourth generation neutrino, while the sterile neutrinos have very small masses (in order to be coupled to the active ones in the oscillations). If one wants to do a fit for all the existing hints one needs at least 2 sterile neutrinos, and the differences of masses with active neutrinos should be below 1 eV with small mixings. I do not think that sterile neutrinos can produce sizable effects detectable at the LHC which is not the right machine for this search.

- *Wen:*

I have two related questions. First question: in the global fits, are neutrinos treated as Dirac or Majorana particles?

- *Altarelli:*

You mean the fit to oscillations?

- *Wen:*

Yes.

- *Altarelli:*

There, the values of the masses and the mixing angles are fitted independent of this question.

- *Wen:*

Is the mixing matrix related to the question whether neutrinos are Dirac or Majorana particles?

- *Altarelli:*

There are differences, but these differences do not play a role in neutrino oscillations. For example, for quarks – and quarks are Dirac particles – there are three mixing angles and one phase. For Majorana neutrinos, there are three mixing angles and three phases, but the two extra phases are not observable in oscillations. So from the point of view of neutrino oscillations, there is no difference between the two descriptions.

- *Wen:*

My second question: currently the global fit indicates a not-so-small θ_{13} . So, from a theoretical point of view, what is the current explanation of this not-so-small θ_{13} ?

- *Altarelli:*

The theory of the neutrino mixing matrix is the subject of tomorrow's lecture, but I can already tell you the following: before T2K we had the following picture from experiment: the preferred central value for θ_{13} was different from zero, although with a significance still below 3σ . How does one get this result from the global fit? θ_{13} establishes a connection between the solar oscillation sector and the atmospheric oscillation sector. The solar is 12, and the atmospheric is 23, and 13 connects the two. If θ_{13} is zero, one can do a two-neutrino fit for solar oscillations, and another two-neutrino fit for the atmospheric ones. If in so doing one gets a tension between the two fits, then one gets a hint that the third parameter is needed to better match the two sectors. The nonzero central value for θ_{13} was obtained from such a tension between doing a two-neutrino fit for either atmospheric and solar neutrino oscillations. As a matter of principle, there is no reason, why θ_{13} should be zero. Establishing the precise value of θ_{13} would be a very important result.

- *Wen:*

My last question: What is the fundamental connection between neutrino mixing and quark mixing?

- *Altarelli:*

This is a good question and tomorrow this is going to be discussed. The main relation is that mixing is the same kind of phenomenon for quarks and for leptons: in both cases the mixing arises because of the mismatch between the basis of mass eigenstates and the basis of flavour eigenstates. However, quarks and leptons are very different, for example, quark masses are very large with respect to neutrino masses. We interpret the smallness of neutrino masses as a consequence of the Majorana nature of neutrinos. It is possible that the differences in the mixing angles are also due to the same reason. However, as I said in the beginning, there is no standard model for either quark or neutrino mixing, everything is still to be proved.

- *Miyazaki:*

I think that the case of the Majorana neutrino is very attractive. However, I would like to ask you what are the future prospects of the Dirac neutrino, and especially how to determine the precise masses of Dirac neutrinos.

- *Altarelli:*

For example, the measurement of the neutrino masses from the endpoint of the spectrum of Tritium beta decay is independent of this question, so in principle one can measure the mass of the neutrino without entering this question. If neutrinos were Dirac particles their masses would be very puzzling because they are so extremely different from quark and charged lepton masses. One would need an enormous difference between the Yukawa couplings of the Higgs to the neutrinos with respect to the Higgs to the quarks, for example. So the advantage of the Majorana mass idea is that using this mechanism one gets a very natural explanation for this gap. Because of the fact that right-handed neutrinos are gauge singlets, they can have a huge Majorana mass without the need for electroweak symmetry breaking. Thus we can see that the smallness of the neutrino masses is related to the very large energy scale where the lepton number is violated. In order to exclude Majorana masses completely, one would have to introduce it as a dogma that lepton number is absolutely conserved, but, for example, we are strongly attracted to the picture of grand unification, and in grand unification theories lepton number is not conserved. Assuming that lepton number is conserved it would be a very weak hypothesis from a modern point of view. Once one assumes that lepton number is violated and that the Majorana masses are large, from the observed oscillation frequencies and the typical Dirac masses generated by the electroweak symmetry breaking, one finds that the scale of the Majorana masses is close to that of grand unification. So, one gets an internal consistency, that the scale of the masses is exactly the right order to be generated by lepton number violation in grand unification. This is very consistent, very plausible, and therefore very attractive.

- *Miyazaki:*

So, you think that there is no possibility for Dirac neutrinos?

- *Altarelli:*

In principle, until lepton number violation is not observed in neutrinoless double beta decay, one cannot say that neutrinos are Majorana fermions. So I am not sure, but, as we have seen, it is very plausible that lepton number is violated and that the neutrino masses have a Majorana component.

- *Csanád:*

I am wondering if, in terms of baryon asymmetry, the new mixing results will help to decide between baryogenesis and leptogenesis scenarios, or neutrinoless double beta decay measurements will be needed for that.

- *Altarelli:*

The really essential condition for CP violation is that θ_{13} is different from zero. The differences in the value of θ_{13} we are discussing here in connection with the MINOS or the T2K ranges for θ_{13} are not crucial for the production of a sufficient baryon asymmetry, at least within the accuracy allowed by the numerous uncertainties that we have in this problem. I don't see a reason for a change of perspective in this respect,

- *Biassoni:*

May the possible presence of a heavy right-handed Majorana neutrino change the predicted branching fraction for processes like tau -> mu gamma or similar decay modes?

- *Altarelli:*

Yes, the fact that neutrino mixing angles are large, two of them are large, especially when put together with supersymmetry in a grand unified context, certainly this has dramatic implications on (separate e, mu, tau) lepton number violation in channels like $\mu \to e+\gamma$, $\tau \to \mu+\gamma$, $\tau \to e+\gamma$, in fact, in many realizations of such models -- I repeat, SUSY grand unified models -- there are strong constraints, that come from the present bounds. There is this beautiful experiment, MEG, at PSI near Zürich, which is trying to improve the limit on $\mu \to e+\gamma$. There is well motivated hope to see a signal and to find a striking evidence for physics beyond the standard model, like is the case for SUSY grand unified models. In fact in the SM you cannot observe $\mu \to e+\gamma$, because it is suppressed by the GIM mechanism involving the exceedingly small ratio of neutrino to W masses.

- *Biassoni:*

Actually, you can use results from this measurement to put lower or upper limits on the Majorana heavy neutrino mass.

- *Altarelli:*

Well, in these grand unified models you have many more parameters. But the $\mu \to e+\gamma$ limit is already important to put constraints in these models.

- *Inguglia:*

There is something that looks interesting to me in neutrino mixing: the sum of θ_{12} and θ_{Cabibbo} is close to 45°, a very nice value. Could there be another matrix above these two which behaves as the CKM matrix in the case of quarks and as the PMNS matrix in the case of neutrinos?

- *Altarelli:*

You are referring to the fact that if you take the sum of the solar angle for neutrino oscillations and the Cabbibo angle, this sum is empirically very close to 45°. This is called lepton-quark complementarity. If this relation is not accidental then this would be very important for the construction of a theory of mixing, taking both quarks and leptons into account. Tomorrow, we will discuss this, I'll present models where this sort of relation is realized. Actually in models one can more easily get the result that the correction is of the order of the Cabbibo angle rather than precisely the Cabbibo angle.

- *Inguglia:*

Do you include CP-violation in the τ sector in the model you are going to present?

- *Altarelli:*

Yes, but I don't compute or predict it precisely. Namely, when in the mixing matrix the phase multiplying θ_{13} is different from zero then CP-violation occurs. I would say that the lepton quark complementarity relation has more to do with $\mu \rightarrow e+\gamma$ and $\tau \rightarrow \mu+\gamma$, rather than with CP-violation in the lepton sector.

- *Badziak:*

Suppose we have a simple supersymmetric GUT, where the GUT symmetry is broken directly to the SM and we assume minimal flavour violation at the GUT scale. Can we set a lower bound for the right-handed neutrino mass from the $\mu \rightarrow e+\gamma$ decay, so for example if the mass is of order $10^{13}\,\mathrm{GeV}/c^2$, is it still possible or is it excluded by the effects of right-handed neutrinos on the running of the diagonal elements of this lepton mixing matrix?

- *Altarelli:*

I don't think you can derive a very well defined connnection between the present limits and, for example, the mass of the right-handed neutrino, as you said. Because lepton mixing is a very complicated phenomenon. It involves many matrices: it involves the matrix that diagonalises the charged lepton mass, it involves the Dirac neutrino mass, it involves the Majorana neutrino mass. In different models there can be different structures for these various ingredients, that lead to a fit of observed masses and mixings. So, the relation between $\mu \rightarrow e+\gamma$ $\mu \rightarrow e+\gamma$ and the fundamental parameters of the model would be different in these different versions. There is not a unique framework with a unique relation, to the best of my knowledge.

- *Badziak:*

Suppose, that θ_{13} is indeed larger than the Cabbibo angle. Could you comment on the implications of this on the theories of flavour?

- *Altarelli:*

Yes, tomorrow I will discuss these in detail. Suppose that θ_{13} is really on the large side of this 90% confidence level T2K range and, for example, it is larger than the Cabbibo angle. Then whole classes of models are eliminated because there are models that predict a small θ_{13} , typically of the order of at least the square of the Cabbibo angle. There are models that instead prefer a relatively large . For example, in many models the corrections you get to each mixing angle, starting from a given leading order prediction, tend to be all of the same order. For example, there is a correction to θ_{12} that could arise from the diagonalisation of charged leptons. Then if this has to be large, of the order of the Cabbibo angle, then you would expect also θ_{13} to get a correction from zero of the same order. Models that have this type of relation, tend to predict θ_{13} of the order of the Cabbibo angle, so they would be enhanced by the T2K result. There are other models where the leading order $\sin^2(\theta_{12}) = 1/3$. The experimental result is approximately 0.30, which requires only a small correction to go from 1/3 to the experimental result. Then in these models θ_{13} is also shifted from 0 by an amount which is equally small. These types of models are disfavoured by the T2K upper range of values. Either way, if you find that θ_{13} is large or small, both will kill a population of models and also will determine the future of experiments in neutrino oscillations because a large θ_{13} means you can probably detect CP-violations in neutrino oscillations in a not too far future.

- *Agócs:*

There are detailed calculations of neutrino oscillations which take into account, for example, entaglement with the other lepton from the original decay or the propagation in a medium with matter. Will it be necessary at some point to include these details in our calculations? Can there be experiments that can distinguish between the "simple" picture and the detailed picture?

- *Altarelli:*

In the analysis of neutrino oscillations many complications are present, for example, matter effects. In many experiments neutrinos are crossing vast amounts of matter, for example solar neutrinos are produced in the center of the Sun. When neutrinos are sent from Fermilab to Soudan, they go underground, so they cross matter. Then for the interpretation of neutrino oscillations you have to know their interactions with matter. In the SM you have a precise description of these interactions. However, if there is new physics, exotic interactions between neutrinos and either quarks or charged leptons, these interactions would induce distortions in the probability of neutrino oscillations. Since matter is made of matter and not antimatter, that can create differences between neutrino and antineutrino properties when they go through matter. These distortions could simulate CP-violation effects. In taking into account these effects, you have to assume that the interactions between neutrinos and matter are according to the SM. If there is a new exotic interaction you would make a mistake in relating neutrinos and antineutrinos and you can even simulate an effect of CPT violation, which is not there in reality.

- *Dorosti Hasankiadeh:*

Do the different neutrino mass limits really furnish us any hint about other domains of physics?

- *Altarelli:*

Yes, the determination of the absolute scale of neutrino masses is extremely important both for particle physics and cosmology. The scale of the neutrino mass from the observed atmospheric oscillations is 0.05 eV. The limit from beta-decay is $m<2.2$ eV. The cosmological limit is about 0.25 eV. This is 5 times the value derived from atmospheric oscillations. At least one neutrino must be heavier than 0.05 eV. KATRIN will try to go to this level of sensitivity with an experiment in the laboratory which is completely under control.

CHAIRMAN: G. ALTARELLI

Scientific Secretaries: D.Salvatore, L.Wen

DISCUSSION II

- *Badziak:*

Are the flavor symmetries gauged? Are these models automatically anomaly free?

- *Altarelli*

No, discrete symmetry groups are not gauged. U(1) can be gauged. The question of cancellation of anomalies can be delayed to more fundamental layers of theory. Anomalies are an ultraviolet problem. One can hope there will be anomaly cancellation at the gravity level. Now we are only trying to see what the phenomenology of masses and mixing is telling us about the charges. We are still far from a really complete theory.

- *de With:*

If neutrinos are Majorana particles, two extra phases appear. Is it possible to measure all three phases by experiments now or in the future?

- *Altarelli:*

Although these phases do not affect neutrino oscillations, they affect CP-violating processes. For example, neutrinoless double beta decay and leptogenesis are effected, so they are in principle observable, but actually we are far from a practical measurement.

- *Aidala:*

Before the discovery that neutrinos have finite mass, were there specific motivations to expect that they should not have mass?

- *Altarelli:*

When some quantity is really small, the simplest assumption is that it is zero. It is often easier to justify zero rather than a specific finite number: the same assumption was done for the cosmological constant. At the beginning of the Standard Model, the zero mass neutrino was thought to correspond to the non existence of the right-handed neutrinos (which do not have any gauge interaction) and to the conservation of lepton number. These two principles prevent both Dirac and Majorana neutrino masses. But in GUT's the conservation of lepton number is violated. Indeed, since you want baryogenesis, you do not want conservation of lepton and baryon numbers. So there are no good reasons to expect a zero mass. It's very interesting that the value of the scale of the Majorana mass is so close to the GUT scale. I don't believe that the origin of neutrino mass is at the TeV scale, because neutrino masses seem to confirm the idea that we need a completion of the SM near to the electro-weak scale and then a desert up to much higher energies like in GUT's.

- *'t Hooft:*

The mass terms for neutrinos require an extra field in the SM. So discovering a non-zero neutrino mass is a bit like discovering a new particle. As long as you haven't seen it, you assume it isn't there. On the other hand, it was realized from the '70s that it would be natural for neutrino to have mass, because quarks and other leptons have mass. The first indication came when the solar neutrinos appeared to be reduced by a factor 3, which could be explained by mixing. Then the SuperKamiokande measurements came with undeniable details.

- *Blankenburg:*

In which sense a Froggatt-Nielsen U(1) can be gauged?

- *Altarelli:*

Since the grand unified group SU(5) is gauged, if you want to enlarge it also including flavor, it's natural to also gauge flavor symmetries. Many attempts were done in this direction, but no one was successful: so we don't have an established flavor symmetry yet. U(1) is the simplest possible continuous flavor symmetry, so many people took it into account since the Froggatt-Nielsen proposal in 1979, both in gauged and ungauged versions, especially at the phenomenological level. These models are toy-models: they do not appear to make a complete theory. Many U(1) models have been considered: in particular the phenomenology of the associated gauge bosons have been described.

- *Badziak:*

I assume this symmetry has to be spontaneously broken and if it would not be gauged there would be some massless Goldstone boson. That would be a phenomenological problem, right?

- *Altarelli:*

Yes, it's true. But we are not using the gauge structure of the theory, so this problem is not really discussed.

- *Martinez*

In the See-Saw formula the neutrino mass is a combination of Dirac and Majorana masses. You said that the Majorana mass can explain the larger mixing of neutrinos with respect to quark mixing. Can you explain a bit more?

- *Altarelli:*

If neutrinos are Majorana particles and have a Majorana mass, this ingredient is not shared by quarks, so in the neutrino mass matrix we really have a new special contribution. Experimentally there is a sharp difference between mixing of neutrinos and mixing of quarks. The new ingredient is a possible reason for this difference. Of course the statement that the neutrino large mixing appears in the Majorana mass matrix and not in the Dirac mass matrix, depends on the basis you choose. By a basis rotation, you can always diagonalize the right handed Majorana mass matrix. In this case, the mixing would be shifted to somewhere else. But, independent of the basis chosen, the fact remains that the Majorana mass is a very important physical difference between the case of neutrinos and that of quarks.

- *Dunin-Barkowski:*

Could you please comment once more on the mechanism that explains the extreme smallness of neutrino masses? See-saw or some other one?

- *Altarelli:*

The crucial statement is that neutrino masses are small because they are inversely proportional to very large scale where lepton number non conserving interactions appear. This is a more general statement than See-Saw. See-Saw is a particular realization of a more general mechanism. The most general form for the neutrino mass is in terms of the dimension five Weinberg operator. This operator violates lepton number and it's suppressed by a large parameter with the dimension of mass. Masses of neutrinos have to be small because they are inversely proportional to the scale of lepton number non conservation.

- *Rinaldi*

Are there models of neutrino mixing that include extra dimensions? In other words, is there any problem including the discrete flavor groups into the models of extra dimensions?

- *Altarelli:*

For example, discrete groups have been proposed for the description of flavor. Where could these discrete groups come from? They could come as a consequence of going from a larger number of dimensions down into 4 dimensions. If you start from 6 dimensions, instead of 4, then you compactify the 2 extra dimensions. Since you have two more dimensions, you have a torus. The corresponding orbifolding has four fixed points on this torus. The invariance under permutations of the 4 fixed points could be the remnant of the original 6-dimensional Poincare symmetry.

- *Wen:*

It's very nice to see a summary of the theory models for neutrino mixing. Are there other experiments that one can do to investigate the symmetry properties of neutrino mixing?

- *Altarelli:*

I have discussed some models that are at present viable. Additional confirmation or constraints for the symmetries could be found in lepton flavor violating processes like $\mu \rightarrow e+\gamma$, $\tau \rightarrow \mu+\gamma$, especially in a grand unified symmetry context. For example, in a SUSY GUT model, there may be relations between lepton and s-lepton mixings. For example, $\mu \rightarrow e+\gamma$ can go through s-leptons very efficiently, because the s-leptons are heavy and the GIM suppression does not work as it does for leptons. So the rates for lepton flavor violating processes in these models are very dangerous for such theories, given the experimental bounds. The experiment MEG is now taking data searching for $\mu \rightarrow e+\gamma$. It is already at the level of the present bound which, for the branching ratio of $\mu \rightarrow e+\gamma$ is 10^{-11} and they plan to go down to 10^{-13}. A lot of models could be ruled out by this measurement.

- *Preghenella:*

If the neutrino mass is of Dirac nature, could that still imply that neutrinos are different from charged fermions?

- *Altarelli:*

Suppose there are only Dirac mass terms for neutrinos, this does not necessarily imply that neutrino mixings are very similar to those of quarks. For example, if you have a SU(5) grand unification, right handed neutrinos are classified as singlets. So their properties are completely unrelated to those of charged leptons and quarks (which belong to tenplets and pentuplets). In SO(10) the situation is different. Actually in minimal SO(10), the constraints on the mass matrices would be definitely too strict. When the most general mass matrices are allowed in SO(10) GUT's the relation between charged fermions and neutrinos for Dirac neutrinos is somewhat relaxed.

S. C.C. TING

"The Alpha Magnetic Spectrometer (AMS) Experiment, in the honors of the 50 anniversary of the Erice School"

The Slides of the Lecture can be found at:

http://www.ccsem.infn.it/issp2011/

CHAIRMAN: S.C.C. TING

Scientific Secretaries: M. deWith, R. Schoefbeck

DISCUSSION

- *Csanàd:*

Can the AMS detector be rotated towards a source if needed?

- *Ting:*

As the ISS is the size of three football fields, this is not possible.

- *Csanàd:*

Can small repairs be done in case of for example high voltage trips or something like that?

- *Ting:*

Not really, the design is highly redundant to make sure that the AMS could operate during the whole lifetime of the ISS.

- *Rinaldi:*

Does the efficiency of the detector change as the energy of the detected particle changes?

- *Ting:*

No, the efficiency does not change and moreover, the resolution to distinguishing particles does not change.

- *Rinaldi*:

How are the data taken by the AMS transmitted back to Earth?

- *Ting:*

The data are collected by the onboard computer, sent to satellites and back to the NASA centre. After that, the data are sent to CERN for analysis.

- *Inguglia:*

Is it possible to identify the source of the high energy events that you are collecting.

- *Ting*:

No.

- *Wen:*

If there is an enhancement on the positron fraction spectrum, it is an indication of dark matter. With the very good energy resolution, we may see the fine structure of the enhancement. Are there any studies about the structure? What expectations do we have about which kind of structure corresponds to which kind of dark matter?

- *Ting*:

The very important thing is, we have to make sure the data is right. There are theory models, but the first thing is to make sure the experimental result is correct. There will be no such experiments in the future, we have to do it right.

- *Wen*:

Is dark matter uniformly distributed around our Earth? At what scale is it condensed, at what scale is it uniform?

- *Ting*:

I don't know, this is a question for theorists.

- *Campbell:*

You search for gamma ray sources, such as pulsars. The Fermi gamma ray telescope is also doing this. Can you compare your experiment to theirs?

- *Ting*:

I know that Fermi has less background rejection at higher energies, but it has a higher acceptance.

- *Campbell*:

To clarify, does this mean that you will just ignore the Fermi point source catalogue and independently create your own?

- *Ting:*

I will not ignore it.

- *Tapia Takaki:*

How does the rejection of random cosmic rays work?

- *Ting:*

It makes use of the angular distribution and rejects those events outside the acceptance.

- *Biassoni*:

Is the PAMELA $e^+/(e^+ + e^-)$ excess significant? Can this originate from known physics?

- *Ting*:

The AMS-01 and PAMELA $e^+/(e^+ + e^-)$ distribution shows a discrepancy with respect to expectation from cosmic ray collisions. Such an expectation is based on theoretical calculations. We should measure the cosmic ray spectrum for all the nuclei first. The neutralino signal might be identified if there is a decrease after the raise in the spectrum. If we see just a raise in the spectrum, the interpretation is difficult.

- *Prewitt*:

Do you have any triggers besides cosmic ray trajectory?

- *Ting*:

Yes, we trigger on electron and photon pulse height, calorimeter energy and heavy ions. We have seven triggers in total.

- *Prewitt*:

Do you keep all of it?

- *Ting:*

Yes.

- *Dunin-Barkowski:*
What is the highest energy of the particles you have seen to the moment?

- *Ting*:
Much higher than 100 GeV. This is a spectroscopy type experiment, the detector can take particles up to 2 TeV.

- *Vlasov*:
Do the shown points in the e^+ plot prove the existence of dark matter?

- *Ting*:
No, there are not enough statistics for this.

- *Dorosti:*
How long would it take to measure the antihelium versus helium ratio with AMS? And can we get a feeling about the matter to antimatter ratio within a few months of data taking?

- *Ting:*
It is very important that next to the antihelium versus helium ratio we are able to also measure anticarbon over carbon and that would take a longer time, and has to be analyzed carefully.

- *Manoni*:
Given the different conditions between your data-taking environment and test beams, do you rely on test beam only to calibrate your detector, or do you have any other handle?

- *Ting*:
We have many calibrations with cosmic rays directly from the Space Station.

The QGCW Project - Technological Challenges to Study the New World

Horst Wenninger, CERN, Geneva, CH

Status Report

The ISSP 2011 Status Report is a continuation of previous lectures on the QGCW (Quark Gluon Coloured World) Project and can therefore be short, only outlining developments during the last year.

At the International School of Subnuclear Physics 2010 (1) (ISSP 2010), I reported about the Quark Gluon Coloured World (QGCW) Project, initiated by Prof. A. Zichichi during his lecture at the 2006 Erice School (2) (ISSP 2006).

Professor A. Zichichi presented the **QGCW Project** to study the properties of - what he called - the " **new world** " which should be a source of totally unexpected phenomena, produced in a collision between heavy nuclei ($_{208}Pb^{82+}$) at extreme energy available with the LHC heavy ion collisions at CERN.

In the extremely high energy collisions between heavy ions, many QCD open-colour-states should be produced, far higher in number than the baryons and mesons known so-far, since these baryons and mesons have to obey the condition of being QCD-colourless. In conclusion: one must prepare a most advanced accelerator and detector technology for the discovery of totally unexpected events.

Since last year's ISSP 2010 report, the LHC has started operation and we can compare now and in future years RHIC data with new LHC data. Steve Vigdor (3) described the dynamical stages of a RHIC collision (see Figure 1), which are also valid for the collisions at LHC energies.

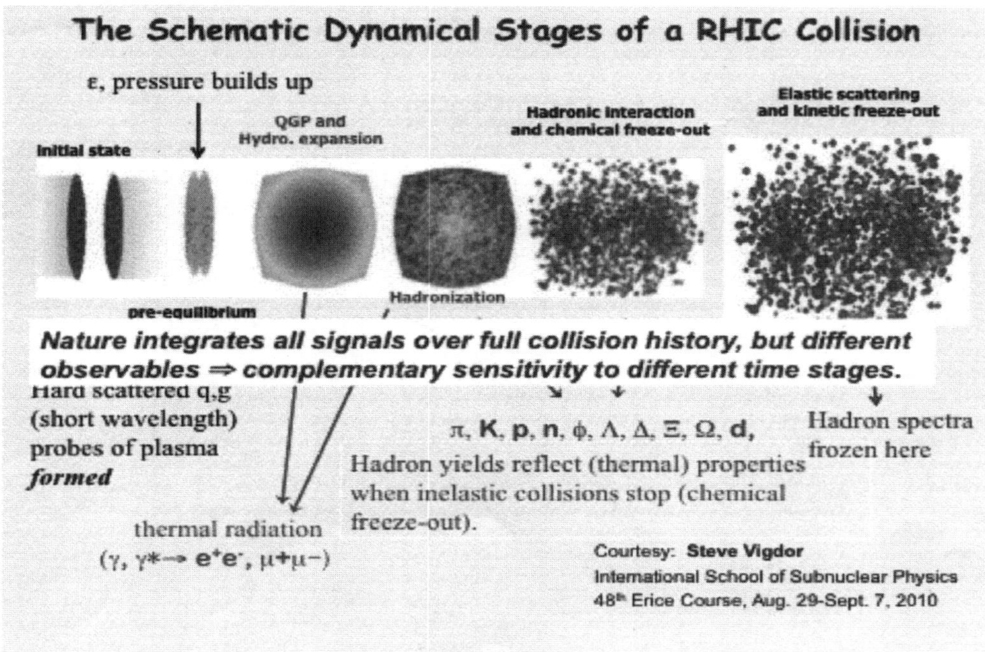

Figure 1: Schematic Dynamical Stages of a RHIC collision

At LHC energies there are, however, changes of some characteristic parameters, e.g.:

Characteristics of Central Pb-Pb Collisions at 2.76 TeV – LHC:

Freeze-out volume ~ 300 fm^3
Twice the volume measured at RHIC (AuAu@200 GeV)
Lifetime until freeze-out ~ 10 fm/c
Density reached is 3 times that of RHIC ($\varepsilon \approx 15$ GeV/fm^3)

These values will not influence the need for development work - as proposed last year - to prepare the future of the QGCW project:

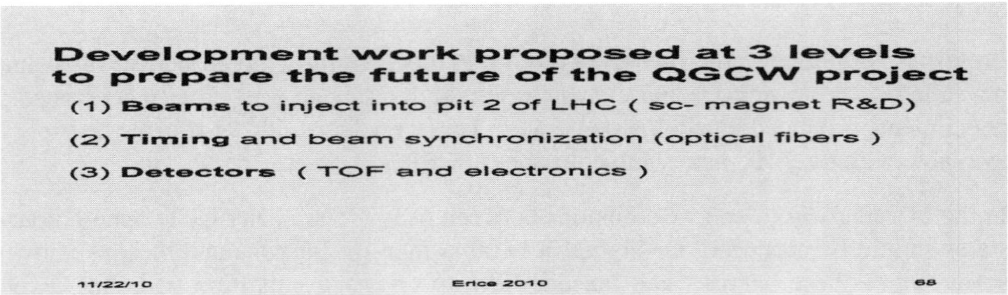

For the beams to inject particle probes into LHC pit 2 of the ALICE experiment, the ongoing R&D on superconducting dipoles – as described last year – continues with funding support from the EU. Of importance for the QGCW project are also the current plans around LHC, presented by S Bertolucci at the ISSP 2011[(4)]. The LHC injector update programs related to the LHC Time-line determine the progress on R&D on superconducting magnets and the ways beams will be eventually injected into pit 2 of LHC.

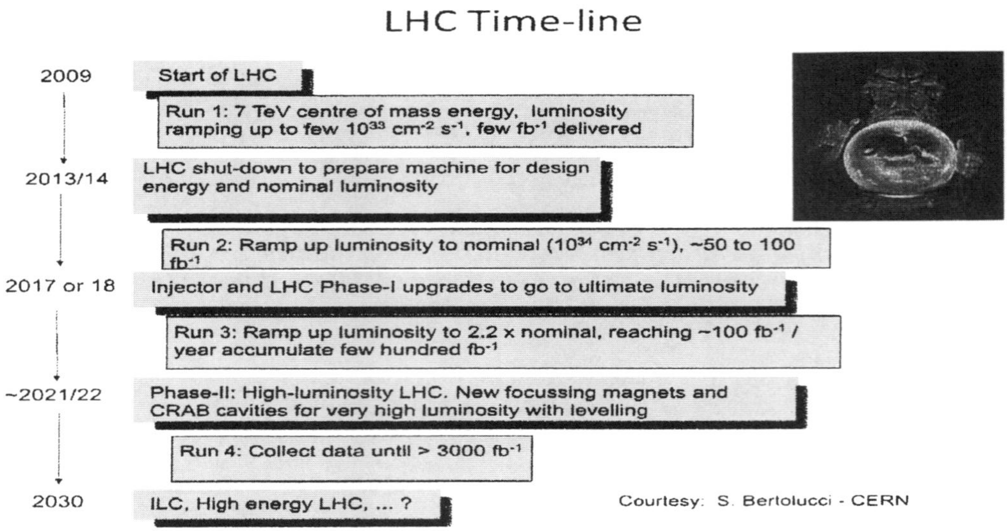

The central problem with any probes (*beams of known particles (p, n, π, K, μ, e, γ, ν) bombarding the QGCW and a special set of detectors measuring the properties of the out-coming particles*) **is that they must be synchronized with the system they probe.** Considering the time scales and evolution of the QCD matter produced in heavy ion collisions that is of order

$$\boxed{10^{-22} - 10^{-24}\ seconds}$$

our present Status Report concentrates on the issue:

Timing and beam synchronization

We need timing and synchronization between:

the colliding heavy ions, forming in collision the quark gluon colored world,
the bombarding particle, simultaneously injected to probe the QGCW, and
the detectors, triggering on the emerging particles

Outlined in Figures 2 and 3 are the most interesting actual examples of precise timing and synchronization activities, as presented by L. Serrano (CERN) [(5)] together with a summary on the present status of the field, presented by Floria Loehl [(5)] (Cornell University) at the Particle Accelerator Conference PAC11.

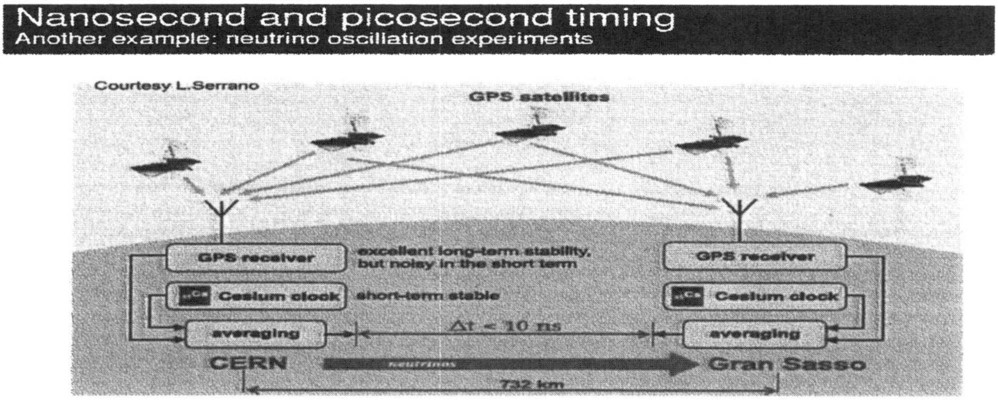

Figure 2: Nanosecond timing (CERN – Gran Sasso)

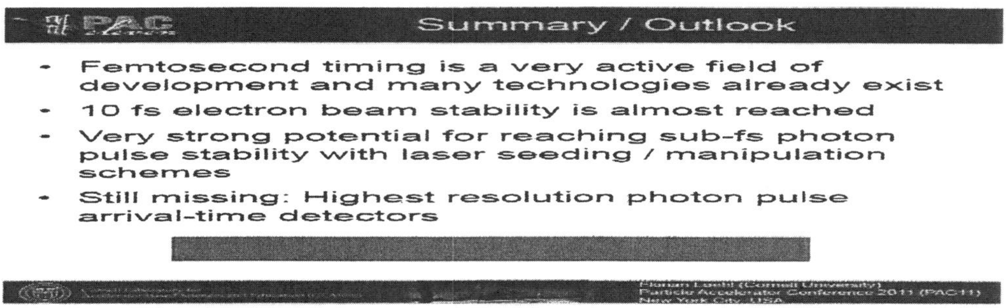

Figure 3: Timing developments

Work also continues on high speed radiation resistant optical fiber transmission at LHC (CERN) and FAIR (GSI).

The detector R&D is related to the LAA MRPC development for the TOF of the ALICE experiment and continues with improvements of the successful NINO frontend electronic chip allowing already to measure timing with 20 ps resolution.

Many R&D activities are going on, but also in future we need more focused development. A successful example of such focused R&D support is the LAA project implemented at CERN by Prof. Zichichi.

References:

(1) Horst Wenninger, Lecture at Erice School ISSP 2010
(2) Antonino Zichichi, Lecture at Erice School ISSP 2006
(3) Steve Vigdor, Lecture at Erice School ISSP 2010
(4) Sergio Bertolucci, Lecture at Erice ISSP 2011
(5) L. Serrano and F. Loehl Particle Accelerator Conference (New York) PAC 2011
(5) TIPP 2011 2nd International Conference on Technology and Instrumentation in Particle Physics

A detailed outline of the QGCW Project by A. Zichichi can be found in the document:
(Nanyang Executive Centre, Nanyang Technological University, Signapore 24-26 February 2010)
CONFERENCE IN HONOUR OF MURRAY GELL-MANN'S 80th BIRTHDAY CELEBRATION.

CHAIRMAN: H. WENNINGER

Scientific Secretaries: V. Alba, S. Mironov

DISCUSSION

- *Wenninger:*

I don't expect as many questions as others lecturers get. The message I wanted to give this morning was that following discussions on RHIC results at Erice a new project (QGCW) was proposed, showing that these discussions at this school will have an impact on future experiments (heavy ion collisions). The LHC accelerator complex and the experiments discuss upgrades and also the RHIC will be upgraded. With R&D we have to prepare these next steps.

To start with the QGCW project R&D we have chosen some very specific topics together with a group of experts who had experience as they were involved in the construction of LEP and LHC for the last 20 years. We will use the experience of the people to develop ideas for a new generation experiment at LHC.

To bombard with probes of different kind the QGCW in LHC heavy ion collisions we plan to bring a beam to the Alice pit. It would also be easy to bring a laser to this area. The basic idea behind the QGCW project is to promote already now R&D for detectors for electronics and machine physics. A very special topic concerns triggering and synchronization aspects between two beams and the third one which must be focused onto a very small volume.

My model is the LAA project: thanks to Prof. Zichichi, CERN got a great chance in LEP construction times to have many positions opened for engineers, physicists and technicians and filled to start intensive R&D development at CERN.This LAA project allowed to bring microelectronics into CERN. We could develop microelectronics, silicone strip detectors, and later pixel detectors. This has now applications in others fields. It was also Prof. Zichichi's idea to trigger developments of experimental technologies for future brilliant experiments at LHC.

- *Biassoni:*

What do you expect to come out from the collision of probe beam with the high-temperature deconfined QGP?

- *Wenninger:*

We don't know, we think about getting new independent measurements to understand what is going on in the QGP. We have no prediction of the outcome, but we think to the next generation QGP experiment. The main concern now is to trigger R&D.

- *Csanad:*

How do you see the search for the critical point and the detailed investigation of the QCD phase diagram? Will this be in focus at FAIR? How do you see the future of this research?

- *Wenninger:*

FAIR will do this research, PHENIX does it. We are planning to investigate these questions in three or four experiments, including Sharkov's group from ITEP, Moscow. Due to Prof. Zichichi's pioneer view on physics, we had been always thinking about a new experiment for LHC even in former times while we had to construct 27km of tunnel for LEP.

- *Prewitt:*

You said you want students more involved in R&D. Any ideas how to do that when professors push students and post-docs to focus on analysis?

- *Wenninger:*

This is a trend right now but students should also be interested in hardware. This is our opportunity to recruit students for technology and R&D and not compete with LHC analysis. The ALICE TOF was also used to get young pupils from Italian schools involved in building detectors within the Extreme Energy Events program, initiated by Prof. Zichichi, and they were excited. We have also a lot of opened positions for FAIR experiment, but we do not have students. We still try to get students interested early. We certainly need more funding for R&D.

- *Preghenella:*

You said there are already some smart ideas to achieve the synchronization precision, could you comment on that?

- *Wenninger:*

It is a question, which we consider concerning FAIR's future and CERN's future. And we try to combine all studies by also asking the people at DESY. There are lots of clever people, who would like to participate and they try to set up the work on equipment components. We can't go to the level of 10^{-23}-10^{-24} sec, but at high energies and with technique coming up, we must be more precise in triggering and synchronization as today. The people who do the laser research are also interested in this extreme timing resolutions. Unfortunately now we can't yet tell you exactly how to do it.

- *Tannenbaum:*

If the LHC discovers neither SM Higgs nor other new particles, do you have plans or a goal to go for a much higher energy machine?

- *Wenninger:*

First of all, we all believe that LHC will provide interesting results. We hope that the results will tell us where to go and what to do next. The particle physics should not end with LHC. Some time ago people told that LEP provided nothing new - but LEP gave us a lot of precision information on QED. And clearly we all hope that LHC will provide us with a lot of interesting information beyond the Standard Model.

- *Bertolucci:*

When I was a young student and didn't know whether to remain theoretician or to become experimental particle physicist, I went to my professor and asked him what should I do. He said that we had understood everything about QED and everything about hadron physics (there are Regge trajectories). I liked his explanation so much that I immediately went to particle physics. Then started a great revolution in the field, so everybody forgot about the death of particle physics. I have seen the same situation several times since then. The stuff becomes more and more expensive so I suppose that we should move smarter, not only more expensive. And the point which was mentioned to us by our young colleagues is that they are forced just to build if we are building something and just to make analysis if we have lots of data. We should give them the right to decide what to do, because otherwise we will grow unbalanced. It will be completely bad for them and for the field.

- *Wenninger:*

You are right. While preparing the LHC we got lots of complaints from the universities about PhD students and their theses. However it is not necessary to do data analysis in order to write PhD thesis. It is also possible to do experiment development. And it is clear that we always should keep a balance between those good people who design new experiments and those who do data analysis. I'm happy that we have new good experiments running now and it is really a good time both for young people and for us elderly people. Thanks to Prof. Ting we now have also the control center of AMS at CERN, and it is very good because there is connection between accelerator and space experimental physics.

S. BERTOLUCCI

"Start up and First LHC Results"

The Slides of the Lecture can be found at:

http://www.ccsem.infn.it/issp2011/

CHAIRMAN: S. BERTOLUCCI

Scientific Secretaries: N. L. Martinez, G. Snidero

DISCUSSION I

- *Rothkopf:*

Is there a possibility to study the nature of beam "UFOs" in a machine as LHC?

- *Bertolucci:*

In all accelerators the presence of sort of "dust" particles is found but this effect has never been seen so hugely as at the LHC. It is known that the number of these particles grows with the number of bunches in the beam. Also, they are not spread uniformly along the ring, in fact it seems more likely to hit one of them where insulators and conductors are located close to each other (e.g. kickers) or nearby ceramic materials.

- *Csanàd:*

Can you give some information about the plans to use of special optics like the roman pots for TOTEM experiment?

- *Bertolucci:*

In order to measure cross sections, as TOTEM aims to do, there is no need of a very high luminosity but rather of very long optics, allowing the measurement of very low transverse momentum transfers. Running a machine with a very long beta function, that for the case of TOTEM is more than 1km, is a special situation (as comparison now LHC is running with beta* = 1.5 m and it will go as low as 30 cm). The fact that in this case we are not using the strong focusing and that the dimension of the beam is quite large (beam size =sqrt(beta*emittance)), requires a full understanding of your machine. During the last machine development runs, one month ago, we performed some tests with un-squeezed beams and TOTEM is expected to begin with beta*=90 m to be able to measure the proton-proton cross section with an uncertainty of around 2-3%.

- *Preghenella:*

Could the Chamonix meeting decision to stop LHC at the end of 2012 be influenced by theoretical expectations as the X boson described by Prof. Fritzsch?

- *Bertolucci:*

There is a constant effort to keep the LHC running plans as flexible as possible. For instance, the recent decision to run till end of 2012 was based on the impressive increase of the machine performances, which opened the possibility to reach important physics results. These type of decisions have to be carefully evaluated in their pros and cons, and are taken after discussion with the management of all the experiments involved.

- *Schofbeck:*

What are the prospects of increasing single bunch instantaneous luminosity?

- *Bertolucci:*

We have already produced collisions with about 1.7 times the nominal number of particles per bunch at 450 GeV and all went fine. This is equal to an increase of 3 times of the instantaneous luminosity. But this will triple the in time pile up vertices from 20 to 60 and will be much more demanding for the experiments.

- *Schofbeck:*

When will we go to 25 ns bunch crossing?

- *Bertolucci:*

We have not taken a final decision but it is unlikely that we will do it during this run. There will be for sure periods dedicated to machine studies on the 25 ns bunch pattern, but it is unlikely that we will adopt it for the normal running.

- *Nguyenber:*

What are the prospects for p-Pb collisions?

- *Bertolucci:*

We will try in 2012, it has been decided to do it, although LHC was not originally planned for such collisions.

- *Mager:*

In p-Pb collisions, which direction will the proton take (clock-wise or counter clock-wise with respect the LHC ring)?

- *Bertolucci:*

In both directions, in order to try to cancel all possible systematics that could arise.

- *Mager:*

And are there plans for collisions between deuteron and atoms?

- *Bertolucci:*

No, there are not such plans for the time being.

- *Prewitt:*

What material are radio frequency "fingers" made of? And was their failure mechanically or thermally originated?

- *Bertolucci:*

They are made of Beryllium and Copper. The source of the failure was definitely mechanical, in fact this happened as a consequence of a poor design.

- *Rinaldi:*

What are the difficulties in going from protons to ions collisions?

- *Bertolucci:*

They are:

- To change the injection system (different LINAC to inject Pb ions)
- Create and pre accelerate ions beam of sufficient intensity and quality
- Tune the collimation system to protect the machine against the increased beam losses due to ion fragmentation.

Despite the difficulties, we have been able to go from proton to ions in a couple of days and to run the ion program very smoothly.

- *Takari:*

Would electron cloud effect stop us from going to 25 ns bunch crossing?

- *Bertolucci:*

It will be more challenging, but II believe that it will not. We have already done some tests on this.

- *Lorenzo Martinez:*

Problems seem to appear when increasing the luminosity. Do you think that is a good idea to increase more and more luminosity rather than staying in a lower regime but with higher efficiency in data taking?

- *Bertolucci:*

What it counts at the end is the integrated amount of *usable* data recorded, which is a function of machine availability, data taking efficiency (very high at the moment, > 95%), and data quality (low background, etc). So far, we have seen that the increase of luminosity has not significantly affected these parameters and therefore we have kept increasing the machine performances.

- *Wen:*

What is a spurious quench?

- *Bertolucci:*

Sometimes our machine protection system over-reacts, detecting a signal of magnet quench and triggering a beam dump, without an actual magnet quench. This behavior, if not too frequent, is acceptable, since we prefer definitively to avoid any damage to the machine and to the detectors caused by out of control beams.

- *Schlofbeck:*

What is the physics of the beam dumps system?

- *Bertolucci:*

You have to steer the beam into an absorbing block within the time it takes a particle to make a single turn.

- *Newman:*

Do you foresee to set the goal to record several fb-1 of data and therefore discover the Higgs and/or a possible light sector of SUSY already this year?

- *Bertolucci:*

For the Higgs we should have first hints and a vastly improved exclusion range by the end of the year. For physics beyond the Standard Model, the discovery range is increasing with the increased integrated luminosity.

- *Snidero:*

What gives origin to the electron cloud effect? And what are the scrubbing runs?

- *Bertolucci:*

Due to the presence of high electric fields, electrons are accelerated and hit the beam pipe walls, extracting secondary electrons, with the development of an avalanche. This phenomenon eventually saturates (the electron cloud), but this cloud will adversely affect the circulating beam in several ways. The first adverse effect is the extraction of trapped gas (desorption), mostly hydrogen, trapped in the metal crystal lattice, which will degrade the vacuum. There are various strategies to mitigate this effect, like choice of the beam pipe material, fabrication techniques, baking, coating, etc. After all these measures have been applied, the final cleaning is made using the beams ("beam scrubbing"), by using the beam to force electron cloud and desorption, while pumping out the produced gas.

CHAIRMAN: S. BERTOLUCCI

Scientific Secretaries: M. Brida, M. Perani

DISCUSSION II

- *Wen*:

Could you highlight the critical specifications of CMS and ATLAS detectors performance for finding Higgs? If some detector is not as good as expected what's the impact on the measurements?

- *Bertolucci:*

Currently, the detectors performances are at least as good as expected. Differences in the design of the two general purpose experiments add to their complementarity in discovery potential. As you have seen from the Higgs significance plot, the global sensitivitiy of the experiments are quite similar, which is a good thing for cross-checks and measurement combination.

- *Wen:*

Do you observe any Higgs candidate?

- *Bertolucci:*

Yes and no, depending on what it is the meaning of your question. Finding the Higgs doesn't mean of course to observe a single event, it is finding a statistically significant signal over a background. So, in a sense, both experiments have observed events whose topology is consistent with an Higgs candidate, but their statistical significance is still not enough to claim discovery. Both experiments are keeping analyzing the collected data, each with many subgroups working on different decay channels, with a strict internal peer-review process on the analysis results.

- *Salvatore:*

Provided the good status of both ATLAS and CMS and their common sensitivity to detect th Higgs boson, do we expect the discovery at the same time or not?

- *Bertolucci:*

Difficult to say. My guess is yes, because the experiments have a very similar sensitivity, but of course I cannot exclude that one of them might come up first. Anyway, before communicating the results world wide, the two experiments (and Cern) have subscribed an agreement to share the information with the other and to allow a certain period of time to confirm the observation. We have already tested this procedure with satisfactory results.

- *Csanàd:*

Going back to the old physics, $\sigma^{pp}{}_{elastic}$ will be measured by the TOTEM experiment and these data are already a big challenge to QCD+QED models. How do you see prospects on this measurement?

- *Bertolucci:*

These measurements need special optics ($\beta^*=90°$), it will be done soon because it has to be done for 7 TeV before 2012, because we will not go back to this energy later.

- *Inguglia:*

What do you think about the leakage of information?

- *Bertolucci:*

It's very difficult just to avoid it completely in such a big community. Nevertheless, I am against any form of censorship. One has to enforce a collaboration-wide code of conduct, and count on the fact that in our field reputation is the most precious value .

- *Inguglia:*

Can you combine ATLAS and CMS data soon?

- *Bertolucci:*

The experiments might start combining their data after this summer conferences. A suitable group of experts from both the experiments has been formed, because combining the data it is a delicate procedure, which requires a careful handling of the correlated systematics.

- *Inguglia:*

What about LHCb luminosity?

- *Bertolucci:*

LHCb will stay between a luminosity of 3 and 4 10^{32} $cm^{-2}s^{-1}$ until the end of next year, running at three time their design luminosity. This means that by the end of this year they might get 1 fb^{-1}. There are still 16 weeks to go in proton physics. Their future plans are very ambitious, with a substantial upgrade program, which includes a they 40 MHz readout (now at 1 MHz) to further increase their capability to use the LHC luminosity.

- *Kaminska:*

How do you get limits on W' and Z' masses? What processes are analyzed?

- *Bertolucci:*

Experiments are looking to high momentum leptons plus missing energy or high mass lepton pairs, looking for resonances.

- *Mager:*

There are rumors that LHCb wants to run with heavy ions. Could you comment?

- *Bertolucci:*

This is correct, it is a rumor,for the time being! I undestand that they are considering to take data when we will be running protons against ions.

- *Inguglia:*

Do you think that a sinergy between SuperB and LHCb will be possible?

- *Bertolucci:*

Yes, if SuperB comes up at an appropriate time. If SuperKEK-B and SuperB will be ready and running in 5-6 years, they will need a considerable integrated luminosity before they can match LHCb capabilities on many decay channels of the b meson. Of course the e+e- machines will have always a reserved ground on the b decays containing neutrinos and (to a lesser extent) neutral pions and photons.

- *Rinaldi:*

Are there any plans to study glueballs at LHC?

- *Bertolucci:*

Glueballs are predicted to have a low mass. The design of LHC is definitely not optimal for the study of these exotic states (if they exist). It might be possible for ALICE in ion runs or for LHCb in the p-Pb run, but it it might difficult (or even impossible) to trigger on these states.

- *Aidala:*

You described the LHC detectors as a "new generation" of collider detectors because they were built with a higher level of redundancy than previous detectors. How systematic and quantitative a part of the original design process was this new level of redundancy?

- *Bertolucci:*

By merging new ideas, new technological developments, a lot of prototyping into a physics driven global design. These experiments have also enjoyed a much closer collaboration between physicists and engineers than previous projects.

- *Dunin-Barkowski:*

What can you say about result on four quark states of molecule type and what can LHCb do in this directions?

- *Bertolucci:*

This is an old story, which is difficult to solve because of the lack of precise predictions from the theory, which cannot rely on a perturbative approach. I don't think that LHCb is the right instrument to shed more light on the subject. My idea is that one should probably look in more details at older data, taken by several previous experiments at low energy.

D. DENISOV

"From the Discovery of Top to the Higgs Boson Search at FERMILAB"

The Slides of the Lecture can be found at:

http://www.ccsem.infn.it/issp2011/

CHAIRMAN: D. DENISOV

Scientific Secretaries: A. Aristova, M. Prewitt

DISCUSSION

- Inguglia:

You mentioned a strange bump which CDF observed at about 145 GeV. And you say that you don't know yet if it is a problem with the energy scale of a jet or if it is a problem of misunderstanding the background, or, maybe, something physical. I would like to know, how do you plan to proceed with your analysis and, moreover, how did you proceed with analysis last year when CDF presented a measurement of anomalous CP-violation with a 3.2 sigma significance and it was also not confirmed by D0? Could it mean that something's going wrong with CDF?

- Denisov:

Let's start with the last question about anomalous dimuon asymmetry. First, last year3.2 sigma result was presented by D0, not by CDF. And CDF actually didn't do similar measurement. D0 has a special ability to change the polarity of it's magnets which reduces the systematic error and produces a very precise result. An updated result with improved techniques and more data will be the subject of a Fermilab Wine & Cheese today (30 Jun 2011). Going back to the dijet feature at around 145 GeV - 150 GeV, I wish I could answer what's going on here, because one experiment sees a peak while the other does not. What we see here is that backgrounds are relatively high, and most of them are coming from W+ jets. So, experimentalists have to subtract large background from small excess in data, which is tricky. CDF and D0 have convened a group to look into the matter. It could be a mismeasurement of the jet energy scale or real physics. We will have to wait for the outcome of the joint working group study.

- Inguglia:

Is it true that adding more data you reached 4.8 sigma significance?

- Denisov:

With more data CDF, not D0, reported an approximate 4.1 sigma significance of the "bump" in di-jet mass spectra for W+jets events.

- Alba:

Fermilab last year announced that they are preparing muon collider. Can you tell us the current status of the muon collider?

- Denisov:

The Tevatron will be closing at the end of the year. Serious discussion is underway about future projects at the Tevatron. The muon collider project has many challenges to overcome, so an exact timetable is not available.

- Alba:

Do you discuss tau-leptons collider?

- Denisov:

No, tau decays even faster than muon and their masses are higher. At this moment, there is no plan for the tau-collider. But the muon collider originally proposed by Budker at Novosibirsk is potentially a very interesting idea.

- *Burda:*

In the slide showing data collection at D0, what are the dips labeled 1, 2, 3, and 4?

- *Denisov:*

Good question. The red curve is the luminosity. When experimentalists collect data, we collect them in so-called "runs". During a run, the conditions are changing minimally so we can use triggers to optimize data collection. In the beginning of the store when the luminosity is high, we turn on triggers for Higgs searches and top physics. As the luminosity drops we increase the number of triggers for b-quark physics and other “soft” processes. The jumps labeled by numbers are the run changes. This is for internal D0 use and does not affect the Tevatron performance.

- *Burda:*

And my second question is about the baryons with heavy quarks, with b-quark in particular. Are baryons with more than one b-quark observed?

- *Denisov:*

No. There are baryons with more than one b-quark expected, but they are not yet observed because of the low cross sections and short decay times.

- *Burda:*

Is it possible to compare b-baryons experimental data with lattice calculations?

- *Denisov:*

The masses we find match the models with pretty good accuracy. We can predict masses of baryons with more than one b-quark with high accuracy. The problem is the lifetime is short and the cross section is low. It is similar to the antimatter situation. After the antiproton was discovered it took some time for anti-helium to be found.

- *Schroeder:*

I have a very technical question: could you comment how the jet energy scale corrections are determined at the Tevatron experiments? Is it based on the Monte-Carlo simulation or entirely data-driven? If so, what processes are used for the scale?

- *Denisov:*

Several methods are used, both MC simulation and data driven. For the data driven methods, P_T balance in the photon + jet or Z + jets events are used to determine energy scale.

- *Miyazaki:*

Did CDF estimate the background in "bump" events with any data-driven methods?

- *Denisov:*

Small backgrounds, such as QCDs, were estimated with data. But the shape of W+jets background can be estimated by Monte-Carlo only.

- *Csanad:*

How do you measure total & differential elastic cross-sections? How do these compare yet to the QCD calculations?

- D. Denisov:

We have "Roman pots" technique to measure differential elastic cross-sections. Nothing surprising has been found, while measurements in large kinematic region have been performed.

- Badziak:

How is the "expected" line on the Higgs mass exclusion limit plot is obtained?

- Denisov:

The "expected" line and its fluctuation are based on simulations using Standard Model without Higgs. The indication of the existence of the Higgs boson would be the statistically significant difference between the observed data and the result of Monte-Carlo simulation without the Higgs boson.

- Rinaldi:

Quarks are confined within hadrons and are not observed as physical particles. How do you define the mass of a top quark? Does that mass depend on the model used to make predictions?

- Denisov:

The mass we quote is the analog of the parameter used in the Pythia simulations. We compare the predictions coming from simulations with the experiments and find what mass gives compatible results. We also did a completely independent measurement using the total cross section measured comparing it with QCD predictions as a function of the top mass.

- Martinez:

When is the Tevatron supposed to stop, and do you think that you will have enough data to do an exclusion of the Higgs boson at low masses?

- Denisov:

The Tevatron will turn off on September 30th, 2011 at 14:00 CST. We expect to have enough data to do a CDF and D0 combined exclusion of a low mass Higgs.

- Biassoni:

The significance of the CDF bump in W+2 jets is estimated by generating events according to the background distribution and fitting with signal + background. The significance thus depends on the signal parameters. I think that the most correct procedure to use in such cases is to compute the data/MC background disagreement with no signal in the fit. The CDF paper report a χ^2 value of 26/20 which will actually result in a lower significance.

- Denisov:

The statistical part of CDF analysis is performed based on commonly used methods and takes into account the background shape in the absence of the signal, so the significance they quote is reasonable.

- de With:

How long will it take to analyze all Tevatron data?

- Denisov:

The full reconstruction of the data will be finished later this year. After 6 – 12 months 'hot topics' like the Higgs search will be finalized. Complex topics for which deep understanding of the detector and data are needed, like W mass, will take longer. Results will continue after that, especially on analysis that can only be done at a ppbar collider.

- *Schóefbeck:*

With respect to the W+2 jets bump, did you look at the dependence on the value of the 3rd jet?

- *Denisov:*

Yes, we looked at and no correlations were found.

- *Mager:*

Why do you record events at only 200 Hz?

- *Denisov:*

This is a technical limit of using equipment from 1990s. We decided not to have upgraded since majority of interesting events are being recorded.

- *Dunin-Barkovskiy:*

You said that we now know the Higgs mass with 20% accuracy. What about the high mass region above ~180 GeV?

- *Denisov:*

It is true that in that region we do not know the Higgs mass accurately, but it is up to the LHC to discover if the Higgs has high mass.

P. GIUBELLINO

"Highlights from ALICE"

The Slides of the Lecture can be found at:

http://www.ccsem.infn.it/issp2011/

CHAIRMAN: P. GIUBELLINO

Scientific Secretaries: D. Tapia Takaki, M. Mager

DISCUSSION

- *Inguglia:*

How can you keep the temperature of electronic devices in the ITS under control?

- *Giubellino:*

Although the temperature of the medium produced in heavy-ion collisions is very high, it has no influence on the detectors since these events are at the microscopic scale. However, there are other issues in the context of radiation damage, which depend on the particle flux and are dominant in high luminosity proton-proton collisions.

- *Biassoni:*

1) How do you simulate quark-gluon plasma effects in your Monte Carlo?
2) Why do you publish on three different journals and not only one?

- *Giubellino:*

1) There are a number of models to simulate heavy-ion collisions. They describe what happens after the deconfined phase but are not able to reproduce all phenomena that are observed. One typically uses a model that starts from parton-parton interactions and adds modifications to mimic the phase transition. To incorporate further effects (e.g. interference in HBT) one runs a so-called "after-burner", which alters the Monte Carlo generator.

2) People from the collaboration have different tastes and feelings and you have to satisfy balances. What we decide at the beginning is whether a paper becomes a letter or a long paper. Then a complex negotiation starts between the 115 institutes.There are also more general considerations: nowadays there is a very strong move in high-energy physics towards open-access journals. This causes a wrestling with the publishers.

- *Wen:*

1) What technique does ALICE use for vertexing (and tracking alignment)?
2) If there is a big energy difference between the two particles in a heavy-ion collision, does it affect the properties of the quark-gluon plasma?

- *Giubellino:*

1) The vertexing is done with silicon detectors, which allow the highest spacial resolution. This resolution depends on three main factors: a) the distance of the detector to the interaction, b) the intrinsic resolution of the detector, and c) the material between the collision and the detector. ALICE put lots of effort in optimizing these even though this causes severe practical difficulties. For example, its silicon pixels are located practically on the beam pipe. To access them all other detectors around have to be moved, which is a major undertaking.

2) Due to the higher collision energy at LHC with respect to RHIC, the system at LHC gets bigger and will have a longer lifetime. This will make it easier to access its properties. But once you have created a perfect fluid it cannot become more perfect. In that sense it has the same properties.

- *Rinaldi:*

1) How can ALICE determine the deconfinement temperature?

2) Can you comment a bit more on J/ψ suppression?

- *Giubellino:*

1) In heavy-ion collision you generally study the temperature of the re-confinement transition. This can be accessed for example by the chemical composition, i.e. by particle ratios. A remarkable result is that this temperature obtained keeps the same beginning from SPS beam energies since it tells that it is independent of the initial temperature of the system. However, this is only one of the models and it has some difficulties to describe all effects observed at LHC.

2) There are two processes, suppression and regeneration, that can explain this phenomena. The interplay of these two processes is not yet clear at LHC energies. It is too early to get conclusive results, but more data (five times more statistics are expected in 2011), and p-Pb collisions will provide a better understanding.

- *Csanàd:*

At the new ALICE R_{AA} measurement there is a significant rise with increasing p_t which was not seen at RHIC in the same interval. Do you know about any calculation that explains this?

-*Giubellino:*

It is probably due to shadowing. In the next two years we will have a short run with p-Pb to address this issue.

- *Golovanov:*

ALICE studies of underlying events are mainly based on minimum bias triggers. Are you going to study underlying events at high p_t?

- *Giubellino:*

Yes, we are studying these events. However, CMS and ATLAS are better suited for such measurements.

- *Lorenzo Martinez:*

You use the photon conversion to measure the material budget. In ATLAS there is the same method but the problem is that we do not know the initial flux of photons. How do you deal with this problem?

-*Giubellino:*

There are regions for which we know the material budget very well (e.g. the TPC gas). They are used to calibrate the absolute scale of the measurement. The Monte Carlo is finally tuned following an iterative procedure.

- *Preghenella:*

1) You said ALICE will upgrade its beam pipe. Is this going to happen during the long LHC shutdown?

2) What are the other upgrades foreseen during this period?

-*Giubellino:*

1) At present we are proposing this upgrade for 2017.

2) There is a large list of projects (e.g. more electromagnetic calorimeters at forward rapidity and an improved inner tracker system) that are being considered. Final decisions are yet to be taken by the ALICE management.

Highlights from BNL-RHIC

M. J. Tannenbaum *
Physics Department, 510c,
Brookhaven National Laboratory,
Upton, NY 11973-5000, USA
mjt@bnl.gov

November 8, 2011

1 Introduction

High energy nucleus-nucleus collisions provide the means of creating nuclear matter in conditions of extreme temperature and density [1, 2, 3]. The kinetic energy of the incident projectiles would be dissipated in the large volume of nuclear matter involved in the reaction. At large energy or baryon densities, a phase transition is expected from a state of nucleons containing confined quarks and gluons to a state of "deconfined" (from their individual nucleons) quarks and gluons, in chemical and thermal equilibrium, covering a volume that is many units of the confining length scale. This state of nuclear matter was originally given the name Quark Gluon Plasma (QGP) [4], a plasma being an ionized gas.

A typical proposed phase diagram of nuclear matter [5] is shown in Fig. 1 together with the presumed trajectories of the evolution of the medium for collisions at RHIC and LHC c.m. energies, where the axes are the temperature T vs. the baryon chemical potential μ_B. The temperature for the transition from the Quark Gluon Plasma (QGP) to a hadron gas is taken as 170 MeV for $\mu_B = 0$ and the phase boundary is predicted to be a smooth crossover down to a critical point below which the phase boundary becomes a first order phase transition. Also shown are idealized trajectories for the RHIC c.m. energy scan and future experiments at FAIR which are being performed in order to find the QCD critical point.

*Research supported by U.S. Department of Energy, DE-AC02-98CH10886.

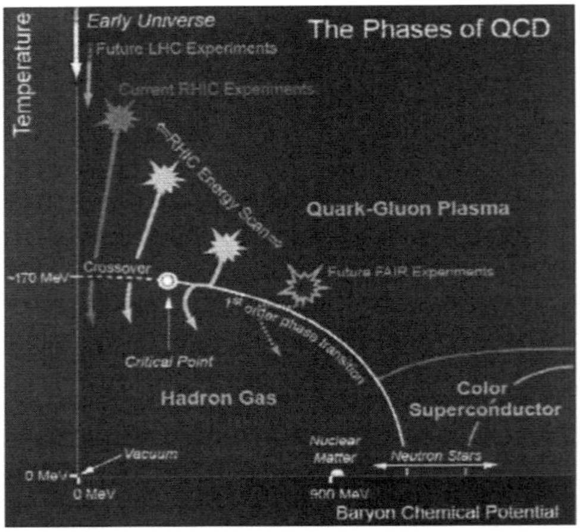

Figure 1: (left) A proposed phase diagram for nuclear matter [5]: Temperature, T, vs Baryon Chemical Potential, μ_B.

2 Discovery of the QGP

The QGP was discovered at RHIC, and announced on April 19, 2005. However the results at RHIC [2] indicated that instead of behaving like a gas of free quarks and gluons, the matter created in heavy ion collisions at nucleon-nucleon c.m. energy $\sqrt{s_{NN}} = 200$ GeV appears to be more like a *liquid*. This matter interacts much more strongly than originally expected, as elaborated in peer reviewed articles by the 4 RHIC experiments [6, 7, 8, 9], which inspired the theorists [10] to give it the new name "sQGP" (strongly interacting QGP). These properties were quite different from the "new state of matter" claimed in a press-conference [11] by the CERN fixed target heavy ion program on February 10, 2000, which was neither peer-reviewed nor published.

In spite of not being published, the CERN press-release had a major effect on the press in the United States resulting in an article on the front page of the New York Times [12]. Ironically, on this very same front page was an article announcing that the true version of the famous Italian sausage, Mortadella, would, for the first time, be allowed to be imported into the United States. A photograph of the iconic Bologna sausages appeared right next to the article about the CERN "qgp". Unfortunately, the first European Baloney to arrive in the U. S. was the CERN announcement. [1]

[1] It is important for the reader of these proceedings to be aware that a high official of CERN was in the audience during this talk and made no objection to this comment. Furthermore, the author has a long, positive and productive relationship with this great

3 What's new at BNL and the RHIC machine this year.

RHIC (Fig. 2) is one of only two Heavy Ion Colliders in the world, the other being the CERN LHC. RHIC is composed of two independent rings of su-

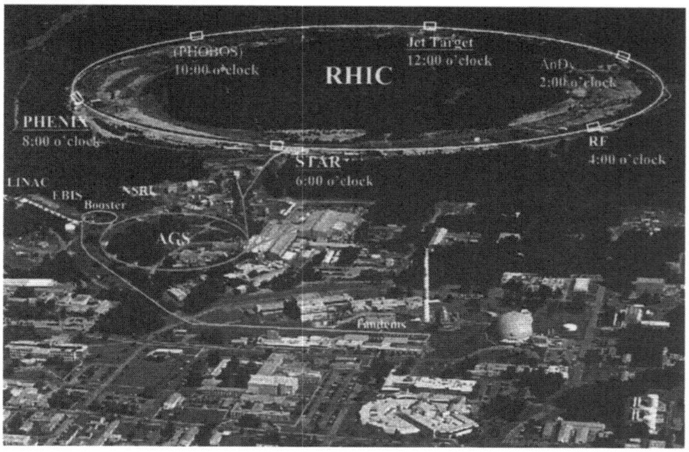

Figure 2: Aerial view of RHIC/AGS facility. The two principal experiments still running are PHENIX and STAR. The LINAC is the injector for polarized protons. The TANDEM injector for Ions will be replaced by the Electon Beam Ion Source (EBIS) in fall 2011. The dome is the decommissioned High Flux Beam Reactor while the enclosed ring on the lower right center is the National Synchrotron Light Source (NSLS).

perconducting magnets, 3.8 km in diameter (see Fig. 4a, below). RHIC can collide any species with any other species and so far has provided Au+Au, d+Au, Cu+Cu collisions at 12 different values of nucleon-nucleon c.m. energy $\sqrt{s_{NN}}$. For the past two years, an Au+Au energy scan has been performed with Energy/beam 3.85, 5.75, 9.8, 13.5, 19.5, 31.2, 100 GeV/nucleon. Also RHIC is the world's first and only polarized proton collider. The performance history of RHIC with A+A and polarized p-p collisions is shown in Fig. 3. At present RHIC operates at 15 times design luminosity for Au+Au and has shown a factor 2 progress in integrated luminosity $\mathcal{L}_{\rm int}$ per week from Run-4 (2004) to Run-7 to Run-10 to Run-11. This year, 3 dimensional stochastic cooling was introduced for the Au+Au collisions which will

laboratory and has praised its many successes. Thus he feels justified in commenting on one of their rare misjudgments. See reference [3] for a detailed scientific discussion.

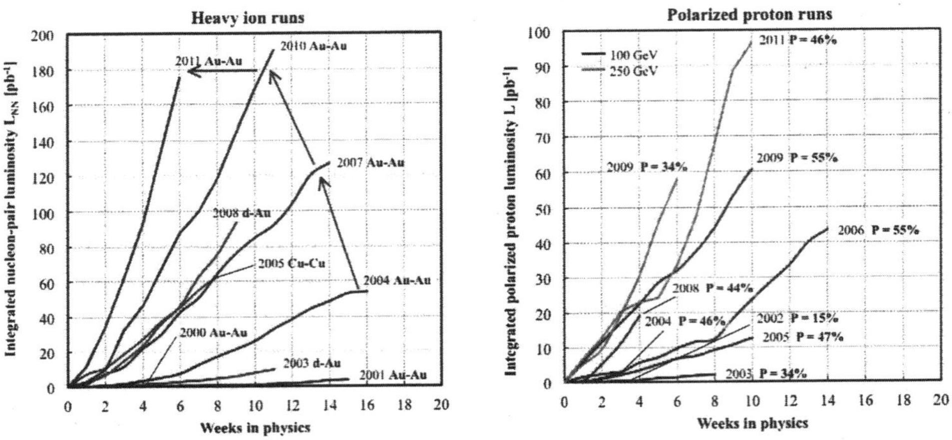

Figure 3: a)(left) Au+Au performance. b) (right) polarized p-p performance. Courtesy Wolfram Fischer.

improve the storage lifetime. Future improvements in the longitudinal profile, i.e. smaller diamond size without bucket migration, will be made with increased longitudinal focusing from new 56 MHz storage r.f. now under construction, with commissioning planned for Run-14.

A significant improvement for the upcoming Run-12 will be the replacement of the 40 year old Tandem Van de Graaff injector with an Electron Beam Ion Source (EBIS). A 10 A electron beam creates the desired charge state(s) in a trap within a 5 T superconducting solenoid. This is then accelerated through the RFQ and linac and injected into the AGS Booster (Fig. 2). All ion species including noble gases, uranium and polarized ^{3}He are available. A uranium cathode has been received in preparation for a pilot U+U run in 2012. Commissioning of the EBIS started during early 2011 and it has already operated and supplied He^+, He^{2+}, Ne^{5+}, Ne^{8+}, Ar^{11+}, Ti^{18+} and Fe^{20+} for the NASA Space Radiation Research Laboratory (NSRL) at BNL [13].

3.1 BNL's Superconducting Magnet Division

All of the upgrades to the RHIC machine mentioned above involved superconducting magnets built or developed in BNL's Superconducting Magnet Division. The Magnet Division also developed the RHIC machine superconducting magnets (Palmer magnet [14]) which are the basis for the other post-Tevatron machines such as HERA and the LHC. The RHIC dipole

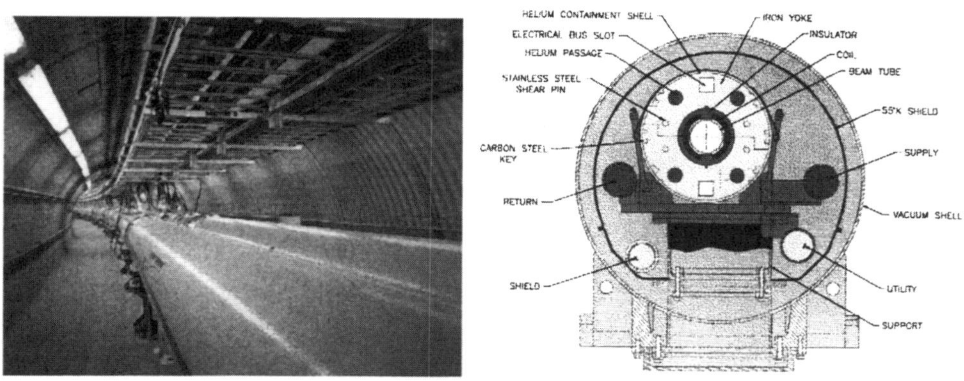

Figure 4: a)(left) Photo of the RHIC machine composed of two independent rings, with a total of 1740 superconducting dipole, quadrupole and corrector magnets. b) (right) Cross section of RHIC dipole.

design (Fig. 4b) is based on a relatively large bore (80 mm inner diameter), single-layer "cosine theta" coil, wound from a (partially) keystoned, kapton-insulated, 30-strand Rutherford-type cable, arranged in coil blocks with intervening copper wedges, in order to meet the stringent field quality specifications, and mechanically supported by a laminated, cold steel yoke encased in a stainless steel shell. The shell contains the helium and is also a load bearing part of the assembly.

It is amusing to note that in order to build the RHIC magnets for the purpose of making the QGP and studying its phase diagram, it is important to understand another phase-diagram, that of Fe+C, i.e. magnet steel (Fig. 5). The Fe+C phase diagram is quite complicated with many phases,

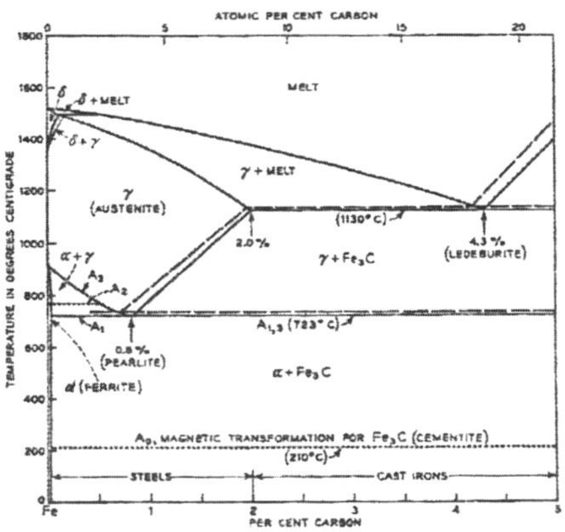

Figure 5: Phase diagram of iron-carbon alloys [15].

but is well known; while the proposed phase diagram of nuclear matter (Fig. 1) seems much simpler, probably because it is largely unknown.

3.2 Latest results from the Magnet Division

BNL's superconducting magnet division is an international resource and actively participates in many projects, several in the news this year. The CERN Courier of March 2011 featured on its cover (Fig. 6a) the beautiful magnet used to trap anti-hydrogen for 1000 seconds in the ALPHA experiment [16] in the Antiproton Decelerator at CERN. The octupole magnet

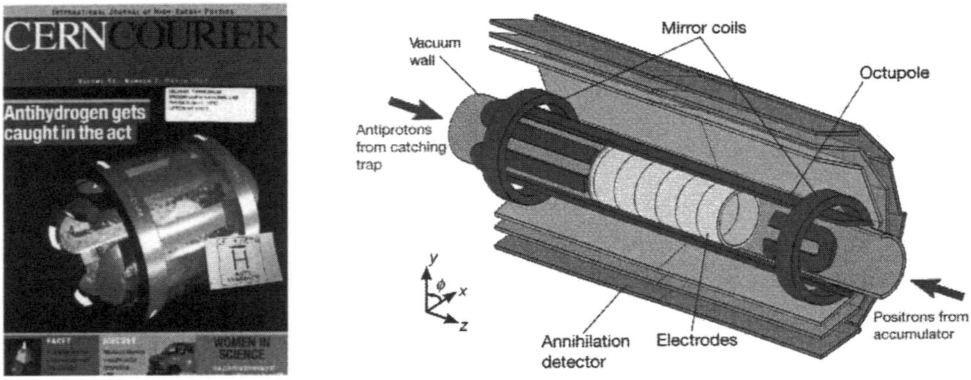

Figure 6: a)(left) CERN Courier cover March 2011. b) (right) Antihydrogen synthesis and trapping region of the ALPHA apparatus [16]. The atom-trap magnets, the modular annihilation detector and some of the Penning trap electrodes are shown (not to scale).

(Fig. 6b) was built by the BNL Magnet Division and provides a very pure octupole field which is zero on the axis and rises sharply near the vacuum wall to keep the anti-hydrogen confined radially.

Another recent press release featuring BNL magnets was the "Indication of Electron Neutrino Appearance" at the T2K (Tokai to Kamioka) Experiment in Japan on June 15, 2011 [18]. BNL provided 5 superconducting dipole corrector magnets in the proton beam at the Japan Proton Accelerator Research Complex (J-PARC) in Tokai which produced the μ-neutrinos that were detected in the Super-Kamiokande detector after transforming to e-neutrinos over the 295 km flight path.

Other work of the Magnet Division includes: NbTi magnets for the RHIC Electron Lens upgrade; NbTi final focus quadrupole for the ILC; Nb_3Sn 11.5

Tesla strand-test-barrel magnet and two coils for the LHC luminosity upgrade; High Temperature Superconductor (HTS) Quadrupole for the Facility for Rare Isotope Beams at Michigan State University; Spare NbTi dipole for LHC; HTS solenoid R&D for muon collider and Energy Storage; Nb_3Sn Open Midplane Dipole for the Muon Collider.

3.3 National Synchrotron Light Source-II

In addition to accelerators for high energy particle and nuclear physics, Brookhaven has been innovative in synchrotron radiation light sources. The National Synchrotron Light Source (NSLS) which started operations in 1982 was the first to use the Chasman-Green double-bend achromat lattice [17], which is now the standard lattice at the major synchrotron light sources worldwide. A new third generation light source, NSLS-II, 4.66 times larger in circumference than NSLS, is now under construction at BNL, with unique design features of high brightness, small source size and long beam lines which will replace NSLS in 2014. NSLS-II is designed to deliver photons with high average spectral brightness in the 2 keV to 10 keV energy range exceeding 10^{21} ph/s/0.1%BW/mm^2/mrad^2. The spectral flux density should exceed 10^{15} ph/s/0.1%BW in all spectral ranges. This cutting-edge performance requires the storage ring to support a very high-current electron beam (I = 500 mA) with sub-nm-rad horizontal emittance (down to 0.5 nm-rad) and diffraction-limited vertical emittance at a wavelength of 1 Å(vertical emittance of 8 pm-rad) [19].

4 QGP Physics—Highlights from RHIC

Given that I already said that the QGP was discovered at RHIC, what further studies are important?

The QGP is the only place in the universe where we can in principle and in practice study Quantum Chromo-Dynamics (QCD) for color-charged quarks in a color-charged medium. For instance, how long will it take before we understand the passage of a quark through a QGP medium as well as we understand the passage of a muon through Copper in QED (Fig. 7)?

Of course, in addition to understanding the behavior of QCD in a medium, the central goal of our field is a quantitative study of the phases of nuclear matter. This requires a broad, quantitative study of the fundamental properties of the QGP including the extraction of the transport coefficients of the medium such as critical temperature, T_c, speed of sound, c_s, the ratio of shear viscosity to entropy density, η/s, etc. To help understand how we

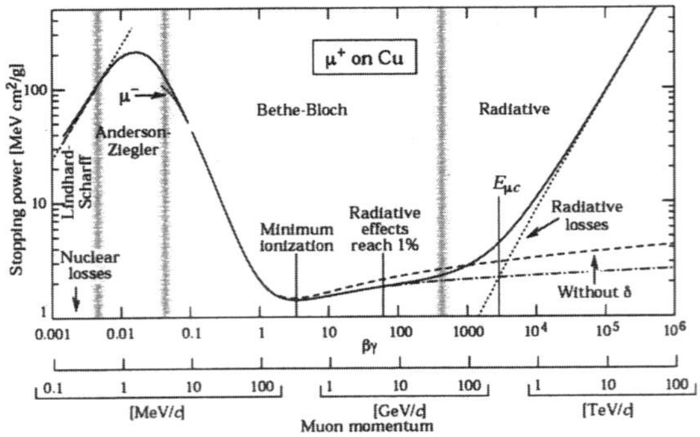

Figure 7: dE/dx of a μ^+ in Copper as a function of muon momentum [20].

shall proceed to address these issues, it is important to understand how we got to this point.

This is the 11th year of RHIC operation, and the two major detectors PHENIX and STAR which study the QGP at RHIC are basically first round detectors with a few incremental upgrades. Thus, the design of these detectors was heavily influenced by the c. 1990 view of the signatures of the QGP, which as noted above were quite different from what was discovered.

4.1 J/Ψ suppression—the original "gold-plated" QGP signature

Since 1986, the 'gold-plated' signature of deconfinement was thought to be J/Ψ suppression. Matsui and Satz [21] proposed that J/Ψ production in A+A collisions would be suppressed by Debye screening of the quark color charge in the QGP. The J/Ψ is produced when two gluons interact to produce a $c, \bar{c}$ pair which then resonates to form the J/Ψ. In the plasma the $c, \bar{c}$ interaction is screened so that the $c, \bar{c}$ go their separate ways and eventually pick up other quarks at the periphery to become *open charm*.

"Anomalous suppression" of J/Ψ was found in $\sqrt{s_{NN}} = 17.2$ GeV Pb+Pb collisions at the CERN SpS [22] (Fig. 8a). This is the CERN fixed target heavy ion program's main claim to fame: but the situation has always been complicated because the J/Ψ is suppressed in p+A collisions. For example, in $\sqrt{s_{NN}} = 38.8$ GeV p+A collisions [23] (Fig. 8b) the Drell-Yan $\bar{q}q \rightarrow \mu^+\mu^-$ cross-section per nucleon is constant as a function of mass number, A, which indicates the expected absence of shadowing in a nucleus for point-like production processes; while the J/Ψ and Υ cross sections per nucleon are suppressed by an amount A^α with $\alpha = 0.920 \pm 0.008$ for both J/Ψ and Ψ' and $\alpha = 0.96 \pm 0.01$ for both the Υ_{1s} and Υ_{2s+3s}. This is called

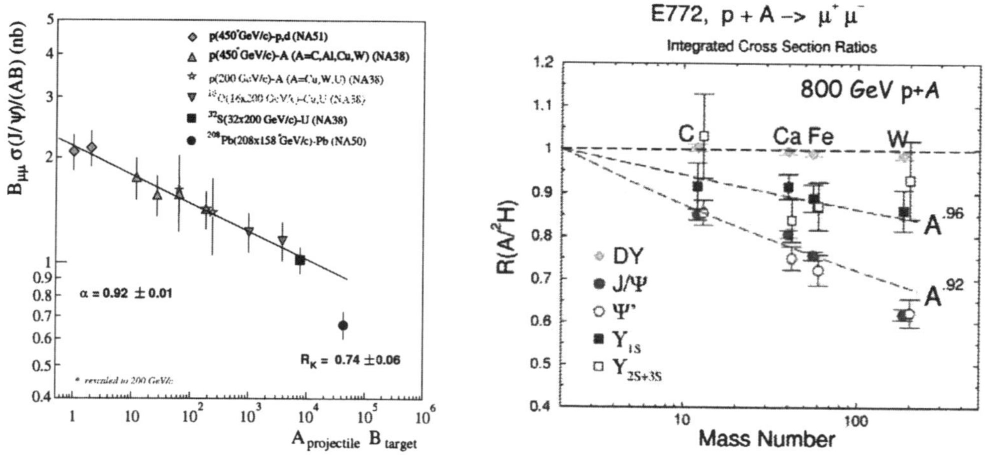

Figure 8: a) (left) Total cross section for J/Ψ production divided by AB in A+B collisions at 158–200A GeV [22] (note the *). b) (right) A dependence of charmonium and Drell-Yan pair production in 800 GeV p+A collisions [23] expressed as the ratio of heavy nucleus to deuterium cross sections per nucleon. The dashed lines are fits to A^α for the CNM effect, with the values of α indicated.

a Cold Nuclear Matter or CNM effect and is shown as the line with $\alpha = 0.92$ on Fig. 8a. The "Anomalous suppression" is the difference between the data point at $AB = 208^2$ and the line, provided that the CNM effect is the same at $\sqrt{s_{NN}} = 17.2$ and 38.8 GeV.

The search for J/Ψ suppression and thermal photon/dilepton radiation from the QGP drove the design of the RHIC experiments.

4.2 Detector issues in A+A compared to p-p collisions

A main concern of experimental design in RHI collisions is the huge multiplicity in A+A central collisions compared to p-p collisions. A schematic drawing of a collision of two relativistic Au nuclei is shown in Fig. 9a. In the center of mass system of the nucleus-nucleus collision, the two Lorentz-contracted nuclei of radius R approach each other with impact parameter b. In the region of overlap, the "participating" nucleons interact with each other, while in the non-overlap region, the "spectator" nucleons simply continue on their original trajectories and can be measured in Zero Degree Calorimeters (ZDC), so that the number of participants can be determined. The degree of overlap is called the centrality of the collision, with $b \sim 0$, being the most central and $b \sim 2R$, the most peripheral. The maximum time of overlap is $\tau_\circ = 2R/\gamma c$ where γ is the Lorentz factor and c is the

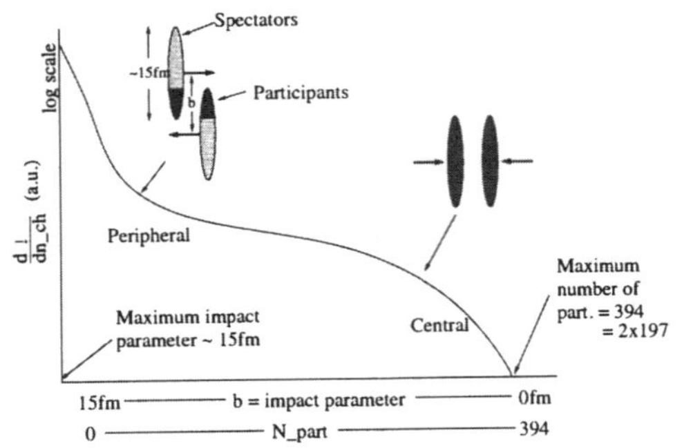

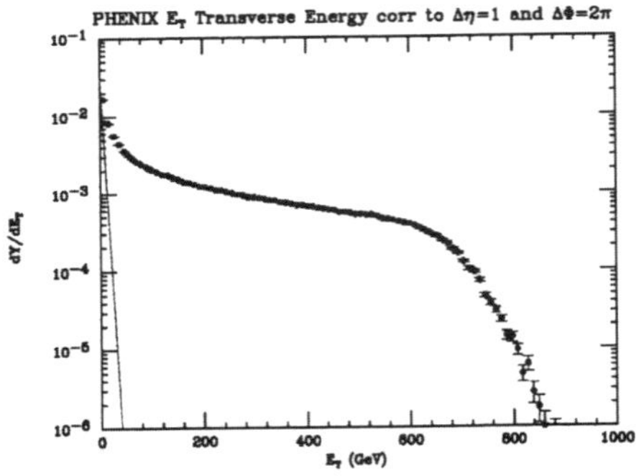

Figure 9: a) (left) Schematic of collision of two nuclei with radius R and impact parameter b. The curve with the ordinate labeled $d\sigma/dn_{\rm ch}$ represents the relative probability of charged particle multiplicity $n_{\rm ch}$ which is directly proportional to the number of participating nucleons, $N_{\rm part}$. b)(right) Transverse energy (E_T) distribution in Au+Au and p-p collisions at $\sqrt{s_{NN}} = 200$ GeV from PHENIX [25].

speed of light in vacuum.

The energy of the inelastic collision is predominantly dissipated by multiple particle production, where $N_{\rm ch}$, the number of charged particles produced, is directly proportional [9] to the number of participating nucleons ($N_{\rm part}$) as sketched on Fig. 9a. Thus, $N_{\rm ch}$ or the total transverse energy E_T in central Au+Au collisions is roughly A times larger than in a p-p collision, as shown in the measured transverse energy spectrum in the PHENIX detector for Au+Au compared to p-p (Fig. 9b) and in actual events from the STAR and PHENIX detectors at RHIC in Fig. 10. The impact parameter

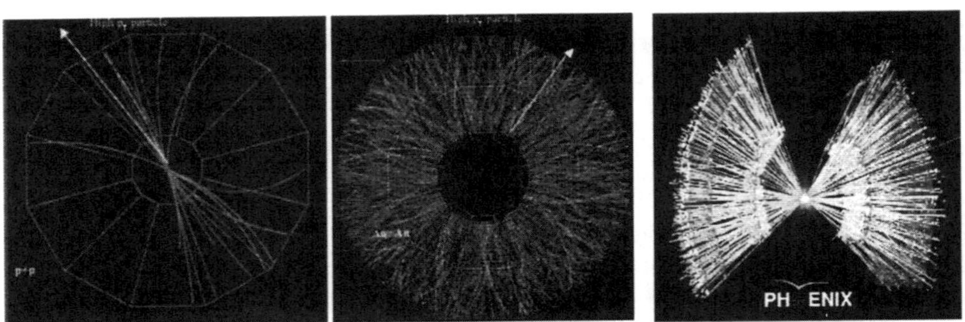

Figure 10: a) (left) A p-p collision in the STAR detector viewed along the collision axis; b) (center) Au+Au central collision at $\sqrt{s_{NN}} = 200$ GeV in STAR; c) (right) Au+Au central collision at $\sqrt{s_{NN}} = 200$ GeV in PHENIX.

b can not be measured directly, so the centrality of a collision is defined in terms of the upper percentile of n_{ch} or E_T distributions, e.g. top 10%-ile, upper 10 – 20%-ile. Unfortunately the "upper" and "-ile" are usually not mentioned which sometimes confuses the uninitiated.

In Fig. 11, measurements of the charged particle multiplicity density $dN_{ch}/d\eta$ at mid-rapidity, $|\eta| < 0.5$, relative to the number of participating nucleons, N_{part}, are shown as a function of centrality for $\sqrt{s_{NN}} = 200$ GeV Au+Au collisions at RHIC [25] together with new results this year from ALICE in $\sqrt{s_{NN}} = 2.76$ TeV Pb+Pb collisions at LHC [28]. The results are expressed as $(dN_{ch}/d\eta)/(N_{part}/2)$ for easy comparison to p-p collisions.

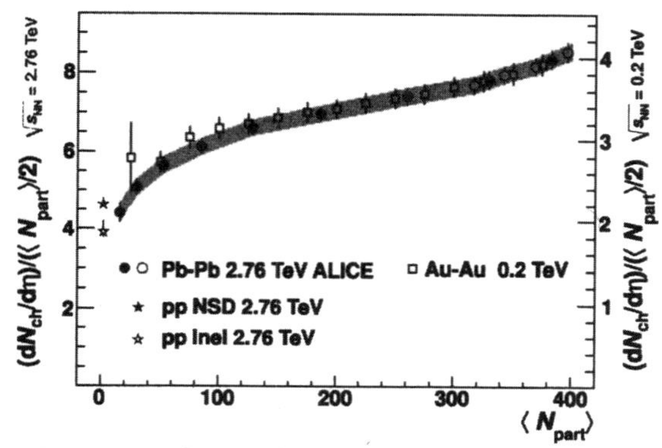

Figure 11: Dependence of $(dN_{ch}/d\eta)/(N_{part}/2)$ on the average number of participants $\langle N_{part}\rangle$ in bins of centrality, for Pb+Pb collisions at $\sqrt{s_{NN}} = 2.76$ TeV [28] and Au+Au collisions at $\sqrt{s_{NN}} = 0.200$ TeV [25]. The scale for the lower-energy data (right side) differs by a factor of 2.1 from the scale for the higher-energy data (left side).

The LHC data show the effect well known from RHIC that $dN_{ch}/d\eta$ does not depend linearly on N_{part}, since $(dN_{ch}/d\eta)/(N_{part}/2)$ is not a constant for all N_{part}. However the data also show the amazing effect that the ratio of $(dN_{ch}/d\eta)/(N_{part}/2)$ from LHC to RHIC is simply a factor of 2.1 in every centrality bin. Thus the LHC and RHIC data lie one on top of each other by simple scaling of the RHIC measurements by a factor of 2.1. This is an incredibly beautiful result which shows that in going from p-p to A+A collisions, the charged particle production is totally dominated by the nuclear geometry of the A+A collisions represented by the number of participating nucleons, N_{part}, independently of the nucleon-nucleon c.m. energy, $\sqrt{s_{NN}}$.

Since it is a huge task to reconstruct the momenta and identity of all the particles produced in these events, the initial detectors at RHIC [26] concentrated on the measurement of single-particle or multi-particle inclusive variables to analyze RHI collisions, with inspiration from the CERN ISR which emphasized those techniques before the era of jet reconstruction [27]. There are two major detectors in operation at RHIC, STAR and PHENIX, and there were also two smaller detectors, BRAHMS and PHOBOS, which have completed their program. As may be surmised from Fig. 10, STAR, which emphasizes hadron physics, is most like a conventional general purpose collider detector, a TPC to detect all charged particles over the full azimuth ($\Delta\phi = 2\pi$) and ± 1 units of pseudo-rapidity (η); while PHENIX is a very high granularity high resolution special purpose detector covering a smaller solid angle at mid-rapidity, together with a muon-detector at forward rapidity [29].

One nice feature of the STAR detector is the ability to measure the mass/charge of a particle from its momentum/charge and time of flight, and then use dE/dx measured in the TPC to determine the charge. In this way STAR has observed many anti-nuclei, notably this year the "Observation of the antimatter helium-4 nucleus" in Au+Au collisions [30] (Fig. 12a). The differential invariant yields $d^2N/(2\pi p_T dp_T dy)$ per central Au+Au col-

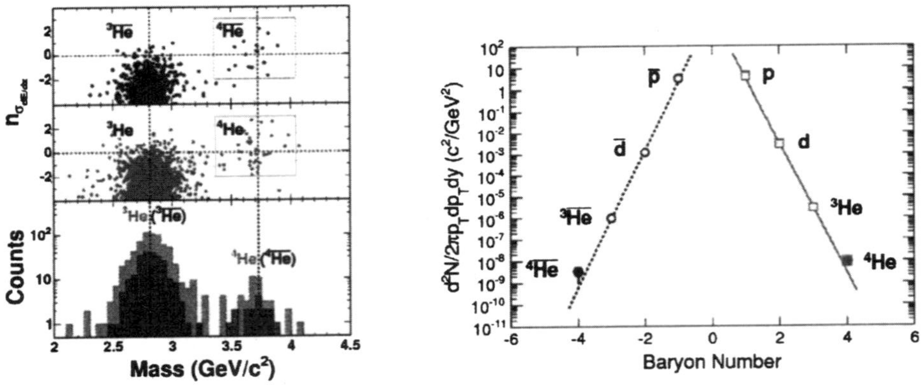

Figure 12: a)(left) Number of standard deviations, $n_{\sigma_{dE/dx}}$, of dE/dx resolution from the expected value for ^{4}He, for negative and positive particles as a function of calculated mass, and projected counts for $-2 < n_{\sigma_{dE/dx}} < 3$. b) Differential invariant yields as a function of baryon number, B.

lision at $\sqrt{s_{NN}} = 200$ GeV as a function of baryon number B, evaluated at $p_T/|B| = 0.875$ GeV/c, are shown in Fig. 12b and show a steady exponential decrease with increasing B [31]. The anti-nuclei are made by coalescence of the large number of $\bar{n}$ and $\bar{p}$ produced, an advantage of the high multiplicity.

PHENIX is designed to measure and trigger on rare processes involving leptons, photons and identified hadrons at the highest luminosities with the special features: i) a minimum of material (0.4% $X_\circ$) in the aperture to avoid photon conversions; ii) possibility of zero magnetic field on axis to prevent de-correlation of e^+e^- pairs from photon conversions; iii) Electro-Magnetic Calorimeter (EMCal) and Ring Imaging Cherenkov Counter (RICH) for $e^\pm$ identification and level-1 $e^\pm$ trigger; iv) a finely segmented EMCal ($\delta\eta$, $\delta\phi = 0.01\times\ 0.01$) to avoid overlapping showers due to the high multiplicity and for separation of single-γ and π^0 up to $p_T \sim 25$ GeV/c; v) EMCal and precision Time of Flight measurement for particle identification. Some results uniquely possible with this detector such as measurements of direct photons via internal conversion to e^+e^- pairs will be discussed below.

In addition to the large multiplicity, there are two other issues in RHI physics which are different from p-p physics: i) space-time issues, both in momentum space and coordinate space—for instance what is the spatial extent of fragmentation? is there a formation time/distance?; ii) huge azimuthal anisotropies of particle production in non-central collisions (colloquially collective flow) which are very interesting in their own right and provide much richer features than originally envisaged.

5 Collective Flow

A distinguishing feature of A+A collisions compared to either p-p or p+A collisions is the collective flow observed. This effect is seen over the full range of energies studied in heavy ion collisions, from incident kinetic energy of $100A$ MeV to c.m. energy of $\sqrt{s_{NN}} = 200$ GeV [32]. Collective flow, or simply flow, is a collective effect which can not be obtained from a superposition of independent N-N collisions. Immediately after an A+A collision, the overlap region defined by the nuclear geometry is almond shaped (see Fig 13) with the shortest axis along the impact parameter vector. Due to the reaction plane breaking the ϕ symmetry of the problem, the semi-inclusive single particle spectrum is modified by an expansion in harmonics [35] of the azimuthal angle of the particle with respect to the reaction plane, $\phi-\Phi_R$ [36], where the angle of the reaction plane Φ_R is defined to be along the impact parameter vector, the x axis in Fig. 13:

$$\frac{Ed^3N}{dp^3} = \frac{d^3N}{p_T dp_T dy d\phi} = \frac{d^3N}{2\pi\, p_T dp_T dy}\left[1 + \sum_n 2v_n \cos n(\phi - \Phi_R)\right] . \quad (1)$$

The expansion parameter v_2, called elliptical flow, is predominant at mid-rapidity. In general, the fact that flow is observed in final state hadrons shows that thermalization is rapid, so that hydrodynamics comes into play

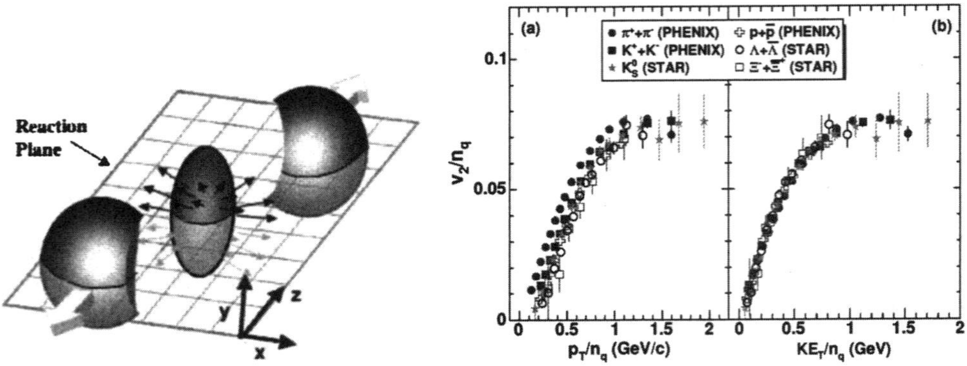

Figure 13: (left) Almond shaped overlap zone generated just after an A+A collision where the incident nuclei are moving along the $\pm z$ axis. The reaction plane by definition contains the impact parameter vector (along the x axis) [33]. (right) Measurements of elliptical-flow (v_2) for identified hadrons plotted as v_2 divided by the number of constituent quarks n_q in the hadron as a function of (a) p_T/n_q, (b) KE_T/n_q [34].

at a time, τ_0, which is before the spatial anisotropy of the overlap almond dissipates. At this early stage hadrons have not formed and it has been proposed that the constituent quarks flow [37], so that the flow should be proportional to the number of constituent quarks n_q, in which case v_2/n_q as a function of p_T/n_q would represent the constituent quark flow as a function of constituent quark transverse momentum and would be universal. However, in relativistic hydrodynamics, at mid-rapidity, the transverse kinetic energy, $m_T - m_0 = (\gamma_T - 1)m_0 \equiv KE_T$, rather than p_T, is the relevant variable; and in fact v_2/n_q as a function of KE_T/n_q seems to exhibit nearly perfect scaling [34] (Fig. 13b).

The fact that the flow persists for $p_T > 1$ GeV/c (Fig. 14a) implies that the viscosity is small [38], perhaps as small as a quantum viscosity bound from string theory [39], $\eta/s = 1/(4\pi)$ where η is the shear viscosity and s the entropy density per unit volume. This has led to the description of the "sQGP" produced at RHIC as "the perfect fluid" [10]. An estimate [40] of η/s for nuclear matter and for several common fluids, as a function of the fractional difference of the temperature from the critical temperature, at fixed pressure, is shown in Fig. 14b. This particular estimate [40] for the QGP at RHIC is quite close to the quantum bound (solid line). Also, empirically, for all common fluids η/s is a minimum at or near the critical point [41] which might suggest that the conditions at RHIC energies are near the QCD critical point.

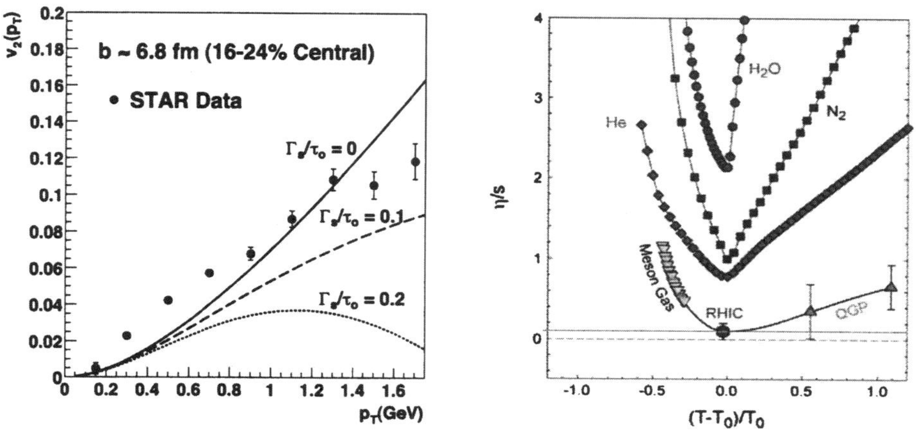

Figure 14: a) (left) Teaney's [38] predictions for $v_2(p_T)$ for ideal ($\Gamma_s/\tau_0 = 0$) and viscous hydrodynamics, where $\Gamma_s = \frac{4}{3}\frac{\eta}{sT}$ is the sound attenuation length, and τ_0 is the thermalization time. b) η/s for various fluids at fixed pressure as a function of temperature T, where T_0 is the temperature of the critical point of the liquid-gas phase transition [40, 41].

5.1 Two-Particle Correlations and Flow

In addition to measuring flow by the correlation of individual particles to the reaction plane, it is also possible to measure flow by the correlation of two-particles to each other. The advantage of this method is that one does not have to determine the reaction plane. Thus if two particles A and B are correlated to the reaction plane, but not otherwise correlated to each other,

$$\frac{dN^A}{d\phi^A} \propto 1 + \sum_n 2v_n^A \cos(n(\phi^A - \Psi_n)), \qquad \frac{dN^B}{d\phi^B} \propto 1 + \sum_n 2v_n^B \cos(n(\phi^B - \Psi_n))$$

then the correlation to the reaction plane induces a correlation of these two particles to each other which can be measured without knowledge of the reaction plane,

$$\frac{dN^{AB}}{d\phi^A d\phi^B} \propto \left[1 + 2v_2^A v_2^B \cos 2(\phi^A - \phi^B) + 2v_3^A v_3^B \cos 3(\phi^A - \phi^B) + \ldots\right]. \tag{2}$$

In p-p collisions there is no collective flow but there are strong two-particle azimuthal correlations due to di-jet production in hard-scattering (see Fig. 10a), which also exist in A+A collisions but are obscured by the large multiplicity [e.g. can you find a jet in Fig. 10b]. Before the discovery of jets, two-particle correlations were used extensively at the CERN-ISR to study hard-scattering (the production of particles with large transverse momentum) in p-p collisions which was discovered there [27]. Due to the huge multiplicities in Au+Au collisions at RHIC, where for central Au+Au collisions at $\sqrt{s_{NN}} = 200$ GeV, there is an estimated $\pi \Delta r^2 \times \frac{1}{2\pi}\frac{dE_T}{d\eta} \sim 375$ GeV of energy in one unit of the nominal jet-finding cone, $\Delta r = \sqrt{(\Delta\eta)^2 + (\Delta\phi)^2}$, two-particle correlations were used exclusively for the first 10 years to study hard scattering at RHIC.

Typical examples of the di-hadron measurements in p-p and Au+Au central (0–20%) collisions at $\sqrt{s_{NN}} = 200$ GeV are shown in Fig. 15 [42, 43] which are presented as azimuthal distributions of the conditional yields of associated particles, with p_{T_a}, with respect to trigger particles with $3 \leq p_{T_t} \leq 10$ GeV/c. The di-jet structure in p-p collisions is clearly indicated

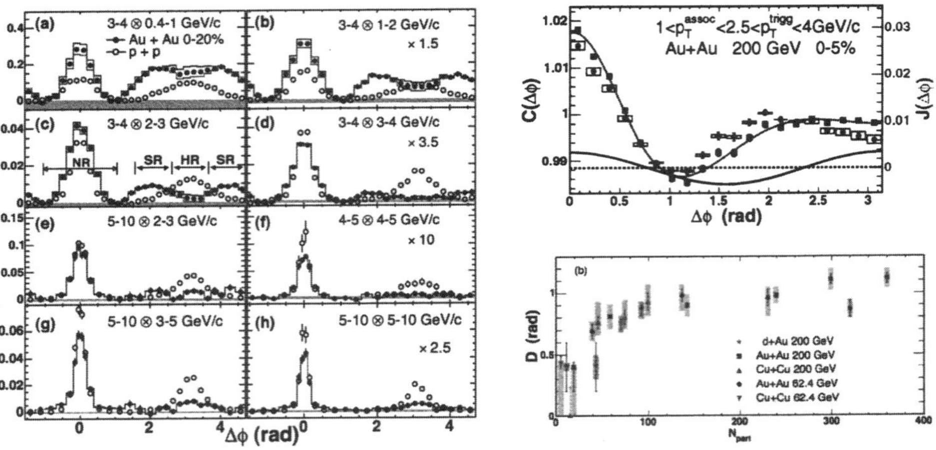

Figure 15: a-h) (left) PHENIX [42] azimuthal correlation conditional yield of associated $h^{\pm}$ particles with p_{T_a} for trigger $h^{\pm}$ with p_{T_t} for the various $p_{T_t} \otimes p_{T_a}$ combinations shown. i) (right)-(top) PHENIX [43] azimuthal correlation function $C(\Delta\phi)$ of $h^{\pm}$ with $1 \leq p_{T_a} \leq 2.5$ GeV/c with respect to a trigger $h^{\pm}$ with $2.5 \leq p_{T_t} \leq 4$ GeV/c in Au+Au central collisions, where the line with data points indicates $C(\Delta\phi)$ before correction for the azimuthally modulated (v_2) background, and the other line is the v_2 correction which is subtracted to give the jet correlation function $J(\Delta\phi)$ (data points). j) (right)-(bottom) PHENIX D parameters [43], the angular distance of the apparently displaced peak of the $J(\Delta\phi)$ distribution from the angle $\Delta\phi = \pi$ as a function of centrality, represented as the number of participants N_{part}, for the systems and c.m. energies indicated.

by the gaussian-like strong azimuthal correlation peaks on the same side ($\Delta\phi = \phi_a - \phi_t \sim 0$) and away side ($\Delta\phi \sim \pi$ rad.) relative to the trigger particle for all ranges of p_{T_t} and p_{T_a} measured. However, one of the many interesting features in Au+Au collisions is that the away side azimuthal jet-like correlations (Fig. 15c) are much wider than in p-p collisions and show a two-lobed structure ("the shoulder" (SR)) at lower p_{T_t} with a dip at 180°, reverting to the more conventional structure of a peak at 180° ("the head" (HR)) for larger p_{T_t}.

The wide away-side correlation in central Au+Au collisions is significantly obscured by the large multiparticle background which is modulated in azimuth by the v_2 collective flow of a comparable width to the jet correlation (Fig. 15i). After the v_2 correction, the double peak structure $\sim \pm 1$ radian from π, with a dip at π radians, becomes evident. The double-peak structure may indicate a reaction of the medium to a passing parton in analogy to a "sonic boom" or the wake of a boat, which was given the name "Mach Cone" [44], and has been under active study both theoretically [45] and experimentally. PHENIX characterizes this effect by the half-width D ($\sim$ 1.1 radian) of the Jet function, $J(\Delta\phi)$, the angular distance of the displaced peak of the distribution from the angle $\Delta\phi = \pi$. One of the striking features of the wide away side correlation is that the width D (Fig. 15j) does not depend on centrality, angle to the reaction plane, p_{T_a} and $\sqrt{s_{NN}}$, which always seemed problematic to me if the effect were due to a reaction to the medium. Another suspicious issue is that the same effect occurs even for auto-correlations of particles with very low p_T between 0.2 and 0.4 GeV/c where any effect of hard-scattered partons should be submerged by the predominant soft physics (Fig. 16a) [46].

In addition to the Head/Shoulder or Mach Cone effect in two-particle correlations on the away-side, same-side correlations also show a new effect in A+A collisions called "the Ridge" [47]. This is seen in two-dimensional correlations in $\Delta\eta, \Delta\phi$ (Fig. 16b) [48, 49] where the associated yield distribution can be decomposed into a narrow jet-like peak at small angular separation which has a similar shape to that found in p-p collisions, and a component that is narrow in $\Delta\phi$ but depends only weakly on $\Delta\eta$, the "ridge." However, new results this past year have dramatically changed this picture.

5.2 Triangular flow, odd harmonics

For the first 10 years of RHIC running, and dating back to the Bevalac, all the experts thought that the odd harmonics in Eq. 1 would vanish by the

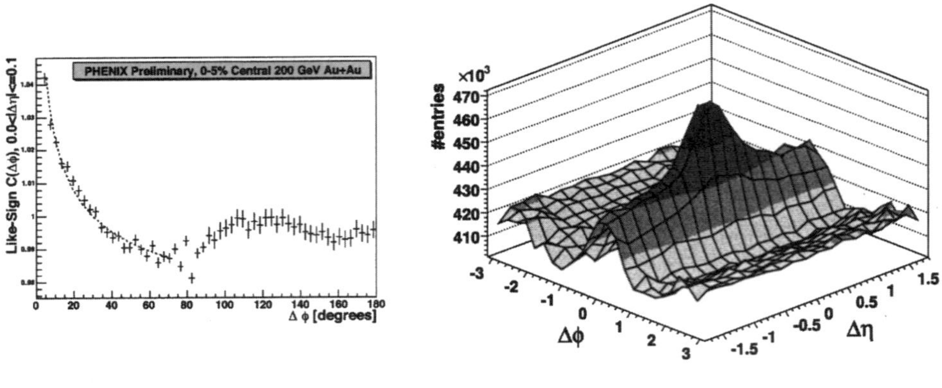

Figure 16: a) (left) Low p_T like-sign pair azimuthal correlation function for 0-5% central Au+Au collisions at $\sqrt{s_{NN}} = 200$ GeV from charged hadrons with $0.2 \leq p_{T_1}, p_{T_2} \leq 0.4$ GeV/c [46]. b) (right) "The Ridge" [48, 49].

symmetry $\phi \to \phi + \pi$ of the almond shaped overlap region [50] (Fig. 13). However, last year, 2010, an MIT graduate student an his Professor in experimental physics, seeking (at least since 2006) how to measure the fluctuations of v_2 in the PHOBOS experiment at RHIC, realized that fluctuations in the collision geometry on an event-by-event basis, i.e. the distribution of participants from event-to-event, did not respect the average symmetry. This resulted in what they called "participant triangularity" and "triangular flow", or v_3 in Eq. 1, which they measured using both PHOBOS and STAR data [51].[2]

Many experiments presented measurements of v_3 at Quark Matter 2011 this year, e.g. Fig. 17 [53], and it was one of the most exciting results of the past year. There are two striking observations from Fig. 17 which indicate that fluctuations of the initial collision geometry are driving the observed v_3: i) the centrality dependence of $v_3(p_T)$ is weak as one would expect from fluctuations, but $v_2(p_T)$ which is most sensitive to the geometry of the "almond"-shaped overlap region tracks the change in eccentricity with centrality; ii) for the most central collisions (0-10%), where the overlap region is nearly circular so that all the v_n are driven by fluctuations, $v_2(p_T)$, $v_3(p_T)$, $v_4(p_T)$ are comparable. The fact that the observed collective flow of final state particles follows the fluctuations in the initial state geometry

[2] It was pointed out by Leticia Palhares in the discussion that a Brazilian group showed in 2009 that that the ridge and the cone, i.e. v_3, does appear in an event-by-event hydrodynamics calculation without jets [52], but the MIT group [51] was the first to show it with real data.

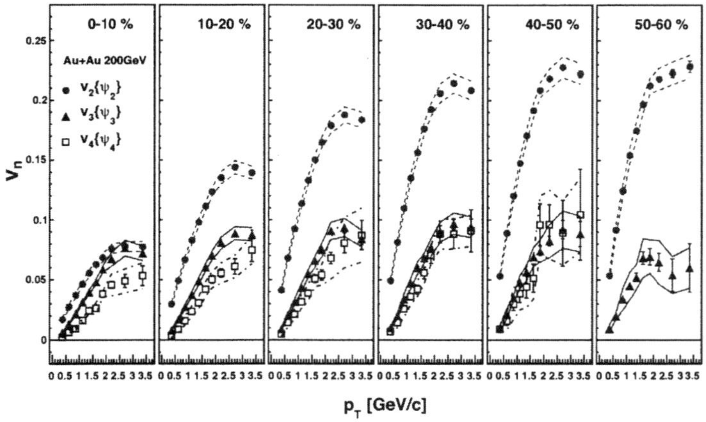

Figure 17: PHENIX [53] measurements of the v_n parameters using Eq. 1 (with the appropriate reaction plane) as a function of p_T for different centrality slices in $\sqrt{s_{NN}} = 200$ GeV Au+Au collisions.

points to real hydrodynamic flow of a nearly perfect fluid (and convinces this author of the validity of hydrodynamics in RHI collisions, of which he was quite skeptical two years ago [27]). It is evident that v_3, a $\cos 3(\Delta\phi)$ term with lobes at $\Delta\phi = 0, 2\pi/3$ and $4\pi/3 \approx 0, 2, 4$ radians, would explain the peaks at $\pi \pm D$ radian in the two-particle correlations (Fig. 15i,j) and also why $D \approx 1$ radian independent of centrality and kinematic variables; while the lobe at $\Delta\phi = 0$ explains the ridge (Fig. 16b). There is presently lots of activity to confirm in detail whether taking account of the odd harmonics in addition to v_2 and v_4 in the background of Fig. 15i will result in gaussian-like away-jet peaks in A+A collisions and the disappearance of the same-side ridge.

6 RHIC beam energy scan—In search of the critical point

The past year has seen runs at RHIC with Au+Au collisions at c.m. energies $\sqrt{s_{NN}} = 7.7, 11.5$ and 39 GeV, in addition to previous runs at 19.6, 62.4, 130 and 200 GeV, to search for the onset of large fluctuations which should occur near a critical point in the phase diagram (Fig. 1). Such fluctuations in the K/π and K/p ratio had been claimed at the CERN SPS fixed target program near $\sqrt{s_{NN}} = 8$ GeV and were presented [54] as "evidence of the onset of the deconfinement phase transition". At QM2011, STAR [55] presented many excellent results on this subject of which I show a small selection in Fig. 18.

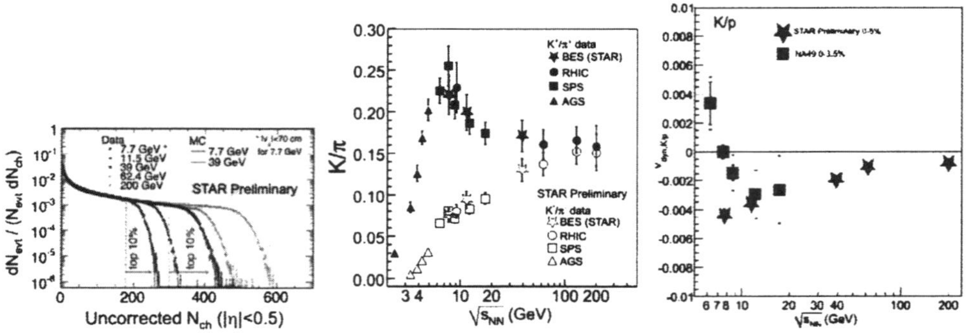

Figure 18: a) (left) $N_{\rm ch}$ distribution in the STAR detector for 5 values of $\sqrt{s_{NN}}$ [55]; b) (center) K/π ratio vs $\sqrt{s_{NN}}$ [55]; c) (right) Event-by-event fluctuation of K/p ratio [55].

Fig. 18a shows the multiplicity distribution $dN_{\rm ch}/d\eta$ which maintains the characteristic Nuclear Geometry dominated shape (as in Fig. 9), stretching to higher multiplicity, $dN_{\rm ch}/d\eta$, with increasing $\sqrt{s_{NN}}$. Fig. 18b shows the K^+/π^+ and K^-/π^- ratios over the entire range of $\sqrt{s_{NN}}$ measurements. The maximum of the K^+/π^+ ratio near $\sqrt{s_{NN}} = 8$ GeV is naturally explained [56] by the change in dominant K^+ production from $K^+\Lambda$ to K^+K^- whose smooth increase with $\sqrt{s_{NN}}$ can be seen from the K^-/π^- ratio. The famous "horn", or apparent discontinuity, at $\sqrt{s_{NN}} = 8$ GeV from the SPS data [54] is greatly smoothed when the new STAR data are added. Fig. 18c shows a smooth variation of the the STAR measurements of the fluctuations of the event-by-event K/p ratio as a function of $\sqrt{s_{NN}}$, which differs dramatically from the huge effect claimed by the SPS Fixed Target measurements below 12 GeV, notably the change from negative to positive [54]. There is no doubt in this author's mind that one must prefer the collider measurements, where the detector position at mid-rapidity in the c.m. system is constant for all values of $\sqrt{s_{NN}}$, to the fixed target measurements, where the rapidity of the c.m. system moves dramatically with respect to the the detector as $\sqrt{s_{NN}}$ varies. This is a major strength of the RHIC Beam Energy Scan program.

6.1 A press release during the school

On June 23, 2011, shortly before I was to give these lectures, a press release from LBL arrived claiming that "By comparing theory with data from STAR, Berkeley Lab scientists and their colleagues map phase changes in the QGP" [57]. Since I had criticized "physics by press-release" concerning the discovery of the QGP (above), I felt that I was obliged to review the physics behind this latest example, presumably a "Highlight from RHIC".

The subject is "Fluctuations of conserved quantities", in this case the net baryon distribution taken as $p - \bar{p}$. Since there can be no fluctuations of conserved quantities such as net charge or net baryon number in the full phase space, one has go to small intervals [59] to detect a small fraction of the protons and anti-protons which then fluctuates, i.e. varies from event to event. The argument is that, e.g. the fluctuation of one charged particle in or out of the considered interval produces a larger mean square fluctuation of the net electric charge if the system is in the hadron gas phase with integral charges than for the QGP phase with fractional charges.

However, while there are excellent statistical mechanical arguments about the utility of fluctuations of conserved quantities such as net baryon number as a probe of a critical point [58], there are, so far, no adequate treatments of the mathematical statistics of the experimental measurements. Theoretical analyses tend to be made in terms of a Taylor expansion of the free energy $F = -T \ln Z$ around the critical temperature T_c where Z is the partition function, or sum over states, which is of the form $Z \propto e^{-(E-\sum_i \mu_i Q_i)/kT}$ and μ_i are chemical potentials associated with conserved charges Q_i [58]. The terms of the Taylor expansion, which are obtained by differentiation, are called susceptibilities, denoted χ. The only connection of this method to mathematical statistics is that the cumulant generating function in mathematical statistics is also a Taylor expansion of the ln of an exponential:

$$g_x(t) = \ln \left\langle e^{tx} \right\rangle = \sum_{n=1}^{\infty} \kappa_n \frac{t^n}{n!} \qquad \kappa_m = \left. \frac{d^m g_x(t)}{dt^m} \right|_{t=0} \quad . \tag{3}$$

Thus, the susceptibilities are cumulants in mathematical statistics terms, where, in general, the cumulant κ_m represents the mth central moment with all m-fold combinations of the lower order moments subtracted.[3] For instance, $\kappa_2 = \langle (x-\mu)^2 \rangle \equiv \sigma^2$, $\kappa_3 = \langle (x-\mu)^3 \rangle$, $\kappa_4 = \langle (x-\mu)^4 \rangle - 3\kappa_2^2$, $\kappa_5 = \langle (x-\mu)^5 \rangle - 10\kappa_3\kappa_2$, where $\mu = \langle x \rangle$. Two so-called normalized or

[3] Note that factorial cumulants, also known as "Mueller moments" [60], have been used previously for multiplicity distributions.

standardized cumulants are common in this field, the skewness, $S = \kappa_3/\sigma^3$ and the kurtosis, $\kappa = \kappa_4/\sigma^4 = \langle (x-\mu)^4 \rangle / \sigma^4 - 3$.

A sample [61] of STAR measurements of the distribution of net-protons in Au+Au collisions in the small interval $0.4 \leq p_T \leq 0.8$ GeV/c, $|y| < 0.5$ for different $\sqrt{s_{NN}}$ is shown in Fig. 19a.

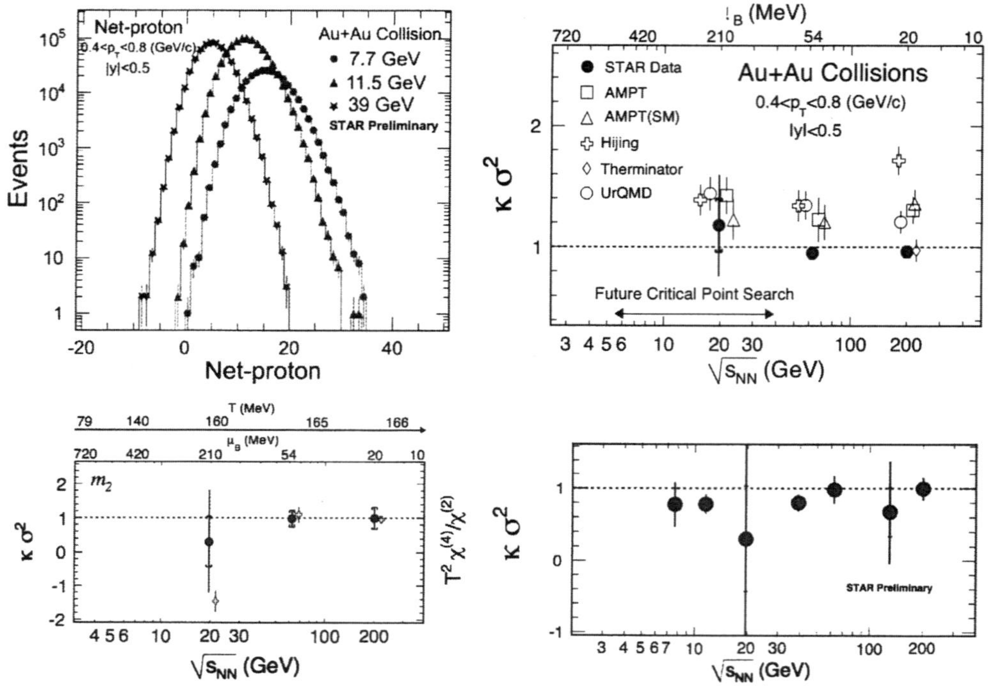

Figure 19: a) (top-left) STAR [61] distribution of event-by-event $p - \bar{p}$ at 3 values of $\sqrt{s_{NN}}$; b) (top-right) STAR published [62] measurements of $\kappa\sigma^2$; c) (bottom-left) Measurements from (b) as shown in Ref. [63] compared to the predicted ratio of susceptibilities (open crosses); d) (bottom-right) compilation [61] of STAR measurements of $\kappa\sigma^2$.

The moments in the form $\kappa\sigma^2 = \kappa_4/\kappa_2$ are shown from a previous STAR publication [62] in Fig. 19b while a plot, alleged to be of this same data, presented in the Lattice QCD theory publication that generated the press-release, is shown in Fig. 19c [63]; and a plot of the $\kappa\sigma^2$ from the data of Fig. 19c, combined with the results from Fig. 19b, is shown in Fig. 19d [61]. There are many interesting issues to be gleaned from Fig. 19.

The data point at 20 GeV in Fig. 19c is not the published one from (b), as stated in the caption [63], but the one from (d), which is different and with a much larger error. This, in my opinion, makes the data point look better compared to the predicted discontinuous value of $\kappa\sigma^2 = -1.5$ for the critical point at 20 GeV (open crosses) in contrast to the predictions of 1.0 for both 62.4 and 200 GeV. The published measurements in (b) together with the new measurements in (d) are all consistent with $\kappa\sigma^2 = 1$; but clearly indicate the need for a better measurement at $\sqrt{s_{NN}} = 20$ GeV. Apart from these issues, the main problem of comparing Lattice QCD "data" to experimental measurements is that it is like comparing peaches to a fish, since the prediction is the result of derivatives of the log of the calculated partition function of an idealized system, which may have little bearing on what is measured using finite sized nuclei in an experiment with severe kinematic cuts. Maybe this is too harsh a judgement; but since this is the first such comparison (hence the press release), perhaps the situation will improve in the future.

When I first saw the measured distributions in Fig. 19a, my immediate reaction was that STAR should fit them to Negative Binomial distributions so that they would know all the cumulants. However, I later realized that my favorite 3 distributions for integer random variables, namely, Poisson, Binomial, and Negative Binomial, are all defined only for positive integers, while the number of net-protons on an event can be negative as well as positive, especially at higher c.m. energies. This is why somebody should work out the mathematical statistics of the net proton distribution as we did in PHENIX for the distribution of the difference in foreground (opposite charge) and background (like charge) di-lepton events when both are Poisson distributed [64]. Until then, it is instructive to compare the values of $\kappa\sigma^2$ in Fig. 19 to those from the well-known distributions: Poisson, $\kappa\sigma^2 = 1$; Binomial, $\kappa\sigma^2 = 1 - 6p + 6p^2 < 1$; Negative Binomial, $\kappa\sigma^2 = 1 + 6\mu/k + 6\mu^2/k^2 > 1$; Gaussian, all cumulants=0 for $k > 2$, so $\kappa\sigma^2 = 0$. The data favor Poisson (i.e. no correlation) everywhere, with some hint of Binomial. Nevertheless, if a future measurement would show a significant huge discontinuity of $\kappa\sigma^2$ similar to the theoretical prediction at $\sqrt{s_{NN}} = 20$ GeV, then even I would admit that such a discovery would deserve a press release, maybe more!

7 Hard Scattering as a probe of the QGP

One of the best probes found at RHIC to study the QGP is the hard-scattering of quarks and gluons (the constituents of the nucleon) which can be observed via inclusive single particle production at large transverse momentum (p_T) or by two-particle correlations with a high p_T trigger. The hard-scattering takes place in the initial collision of the highly Lorentz contracted nuclei. The scattered partons which emerge near 90° to the collision axis (the sweet-spot for such observations) pass through the medium formed and then fragment to jets of particles which are detected (Fig. 20). The most likely

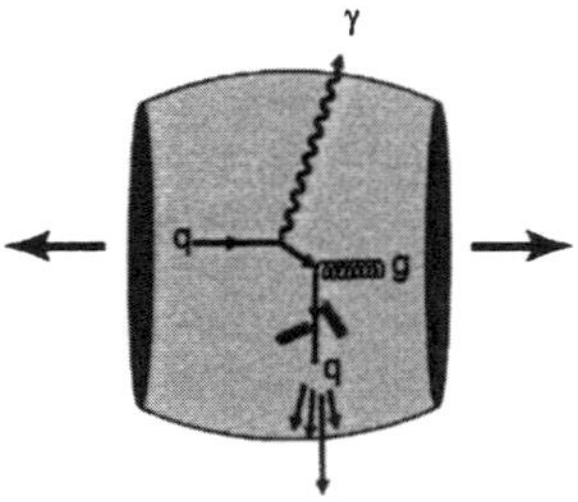

Figure 20: $g + q \to \gamma + q$ reaction with outgoing γ and q passing through the medium

constituent reactions are: $g{+}g \to g{+}g$, $g{+}q \to g{+}q$, $q{+}q \to q{+}q$, and $g{+}q \to \gamma{+}q$ (shown). This last reaction [65], called direct-γ production or the inverse QCD Compton effect, is the most beautiful because the γ-ray participates directly in the hard scattering, then emerges from the medium without interacting and can be measured with high precision. No fragmentation is involved and, in fact, these direct-γ are distinguished from e.g. γ rays from fragmenting partons because they are isolated, with no accompanying fragments. Triggering on a direct-γ of a given $\hat{p}_{T_t}$ provides a 'beam' of partons (8/1 u-quark) with exactly opposite (thus well-known) initial $p_T = -\hat{p}_{T_t}$, so that the effect of the medium can be determined by measuring, for instance, the ratio of the transverse momentum $\hat{p}_{T_a}$ of the jet from the away-parton to that of the direct-γ trigger, denoted $\hat{x}_h = \hat{p}_{T_a}/\hat{p}_{T_t}$, or equivalently, the fractional jet imbalance, $1 - \hat{x}_h$, as used by CMS at LHC [66].

Since hard-scattering at high $p_T > 2$ GeV/c is point-like, with distance scale $1/p_T < 0.1$ fm, the cross section in p+A (A+A) collisions, compared to p-p, should be larger by the relative number of possible point-like encounters, a factor of A (A^2) for p+A (A+A) minimum bias collisions. When the impact parameter or centrality of the collision is defined, the proportionality factor becomes $\langle T_{AA} \rangle$, the average overlap integral of the nuclear thickness functions.

7.1 Jet quenching—suppression of high p_T particles

The discovery, at RHIC [67], that π^0's produced at large transverse momenta are suppressed in central Au+Au collisions by roughly a factor of 5 compared to point-like scaling from p-p collisions is arguably *the* major discovery in Relativistic Heavy Ion Physics. For π^0 (Fig. 21a) [68] the hard-scattering in p-p collisions is indicated by the power law behavior p_T^{-n} for the invariant cross section, $Ed^3\sigma/dp^3$, with $n = 8.1 \pm 0.05$ for $p_T \geq 3$ GeV/c. The Au+Au data can be characterized either as shifted lower in energy relative to the point-like scaled p-p data, or down in magnitude,

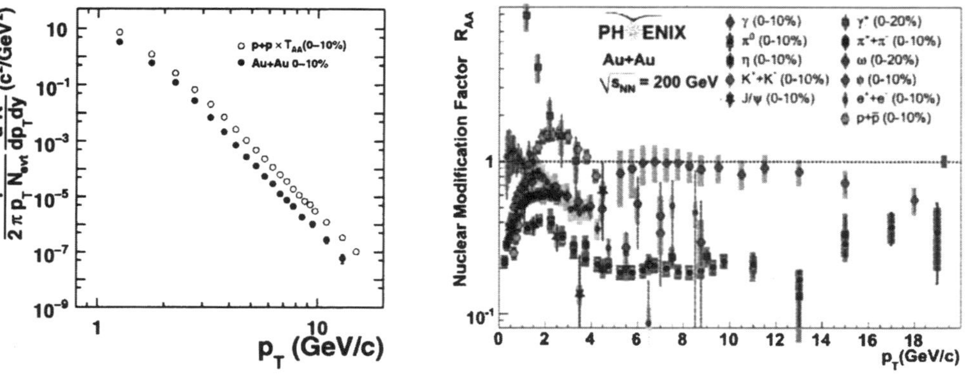

Figure 21: a) (left) Invariant yield of π^0 at $\sqrt{s_{NN}} = 200$ GeV as a function of transverse momentum p_T in p-p collisions multiplied by $\langle T_{AA} \rangle$ for Au+Au central (0–10%) collisions compared to the Au+Au measurement [68]. b) (right) $R_{AA}(p_T)$ for all identified particles so far measured by PHENIX in Au+Au central collisions at $\sqrt{s_{NN}} = 200$ GeV.

i.e. suppressed. In Fig. 21b, the suppression of the many identified particles measured by PHENIX at RHIC is presented as the Nuclear Modification Factor, $R_{AA}(p_T)$, the ratio of the yield of per central Au+Au collision (upper 10%-ile of observed multiplicity) to the point-like-scaled p-p cross section:

$$R_{AA}(p_T) = \frac{d^2 N^{\pi}_{AA}/dp_T dy N_{AA}}{\langle T_{AA} \rangle \, d^2 \sigma^{\pi}_{pp}/dp_T dy} \quad . \tag{4}$$

The striking differences of $R_{AA}(p_T)$ in central Au+Au collisions for the many particles measured by PHENIX (Fig. 21b) illustrates the importance of particle identification for understanding the physics of the medium produced at RHIC. Most notable are the equal suppression by a constant factor of 5 of π^0 and η for $4 \leq p_T \leq 15$ GeV/c, the equality of suppression of direct-single $e^{\pm}$ (from heavy quark (c, b) decay) and π^0 at $p_T \gtrsim 5$ GeV/c, the non-suppression of direct-γ for $p_T \geq 4$ GeV/c and the exponential rise of R_{AA} of direct-γ for $p_T < 2$ GeV/c [69], which is totally and dramatically different from all other particles and attributed to thermal photon production by many authors (e.g. see citations in reference [69]). For $p_T \gtrsim 4$ GeV/c, the hard-scattering region, the fact that all hadrons are suppressed but direct-γ are not suppressed, indicates that suppression is a medium effect on outgoing color-charged partons. The suppression can be explained by the energy loss of the outgoing partons in the dense color-charged medium due to coherent Landau-Pomeranchuk-Migdal radiation of gluons, predicted in QCD [70], which is sensitive to properties of the medium. Measurements of two-particle correlations (discussed below) confirm the loss of energy of the away-jet relative to the trigger jet in Au+Au central collisions compared to p-p collisions. However, as we shall see, lots of details remain to be understood.

The most interesting new results this year concern: i) direct-γ-hadron correlations at large p_T to measure the fragmentation function in p-p collisions and to search for a possible modification of the fragmentation function in Au+Au; ii) π^0-hadron correlations to compare the fractional jet imbalance at RHIC to the new LHC measurement [66]; iii) measurement of the flow (v_2) of direct-γ at low and high p_T; and iv) confirmation at the LHC of the suppression of heavy quarks comparable to that of light quarks.

7.2 Fragmentation Function and Jet Imbalance

The key to measuring the fragmentation function of the jet of particles from a hard-scattered parton is to know the energy of the original parton which fragments, as pioneered at LEP [71]. Thus, in p-p collisions, a measurement of the direct-$\gamma - h$ correlation from $g + q \to \gamma + q$, where the h represents charged hadrons opposite in azimuth to the direct-γ, is (apart from the low rate) excellent for this purpose since both the transverse momentum and identity of the jet (8/1 u-quark, maybe 8/2 if the $\bar{q} + q \to \gamma + g$ channel is included) are known to high precision. Two particle correlations are analyzed in terms of the two variables [75]: $p_{\rm out} = p_T \sin(\Delta\phi)$, the out-of-plane transverse momentum of a track; and x_E, where:

$$x_E = \frac{-\vec{p}_T \cdot \vec{p}_{Tt}}{|p_{Tt}|^2} = \frac{-p_T \cos(\Delta\phi)}{p_{Tt}} \simeq \frac{z}{z_{\rm trig}}$$

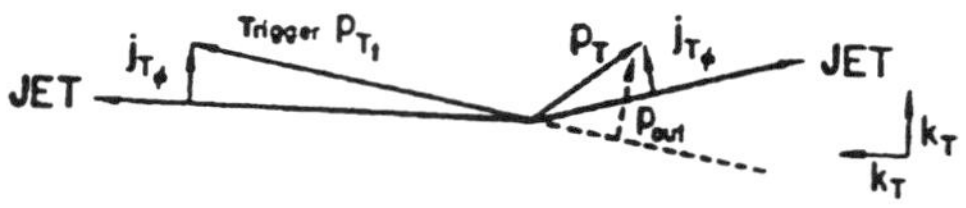

$z_{\rm trig} \simeq p_{Tt}/p_{T{\rm jet}}$ is the fragmentation variable of the trigger jet, and z is the fragmentation variable of the away jet. Note that x_E would equal the fragmenation fraction z of the away jet, for $z_{\rm trig} \to 1$, if the trigger and away jets balanced transverse momentum. The beauty of direct-γ for this purpose is that $z_{\rm trig} \equiv 1$.

Following the approach of Borghini and Wiedemann [72] who predicted the medium modification of fragmentation functions in the hump-backed or $\xi = \ln(1/z)$ representation, PHENIX measured x_E distributions in p-p collisions [73] and converted them to the $\xi = -\ln x_E$ representation (Fig. 22a) which are in quite excellent agreement with the dominant u-quark fragmen-

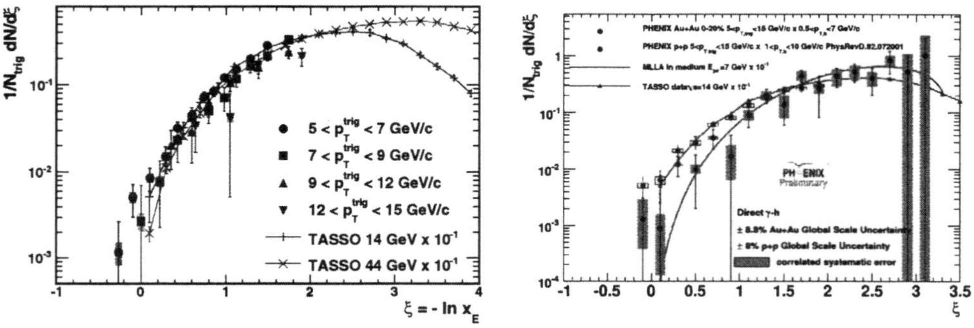

Figure 22: a) (right) $\xi = -\ln x_E$ distributions for PHENIX isolated direct-γ p-p data [73] for all p_{T_t} ranges combined, compared to e^+e^- collisions at $\sqrt{s} = 14$ and 44 GeV. b) (left) $\xi = -\ln x_E$ distribution in Au+Au (0–20%) [74]

tation functions measured in e^+e^- collisions at $\sqrt{s}/2 = 7$ and 22 GeV, which cover a comparable range in jet energy. PHENIX preliminary measurements [74] of the $\xi = -\ln x_E$ distribution in central (0–20%) Au+Au collisions, which suggest a modification consistent with Ref. [72] are shown in Fig. 22b. A firm conclusion awaits final results with improved statistics. Fragmentation functions from full jet reconstruction in A+A collisions are not yet available at RHIC.

One of the important lessons learned at RHIC [75] about fragmentation functions is that the away-side x_E distribution of particles opposite to a trigger particle (e.g. a π^0), which is itself the fragment of a jet, does not measure the fragmentation function, but, instead, measures the ratio of $\hat{p}_{T_a}$ of the away-parton to $\hat{p}_{T_t}$ of the trigger-parton and depends only on the same power n as the invariant single particle spectrum:

$$\left.\frac{dP}{dx_E}\right|_{p_{T_t}} \approx N\,(n-1)\frac{1}{\hat{x}_h}\frac{1}{(1+\frac{x_E}{\hat{x}_h})^n} \quad . \tag{5}$$

This equation gives a simple relationship between the ratio, $x_E \approx p_{T_a}/p_{T_t} \equiv z_T$, of the transverse momenta of the away-side particle to the trigger particle, and the ratio of the transverse momenta of the away-jet to the trigger-jet, $\hat{x}_h = \hat{p}_{T_a}/\hat{p}_{T_t}$. PHENIX measurements [76] of the x_E distributions of π^0-h correlations in p-p and Au+Au collisions at $\sqrt{s_{NN}} = 200$ GeV were fit to Eq. 5 (Fig. 23a) [77]. The steeper distribution in Au+Au shows that the away parton has lost energy. The results for the fitted parameters are shown on the figure. In general the values of $\hat{x}_h^{pp}$ do not equal 1 but range between $0.8 < \hat{x}_h^{pp} < 1.0$ due to k_T smearing and the range of x_E covered. In order to

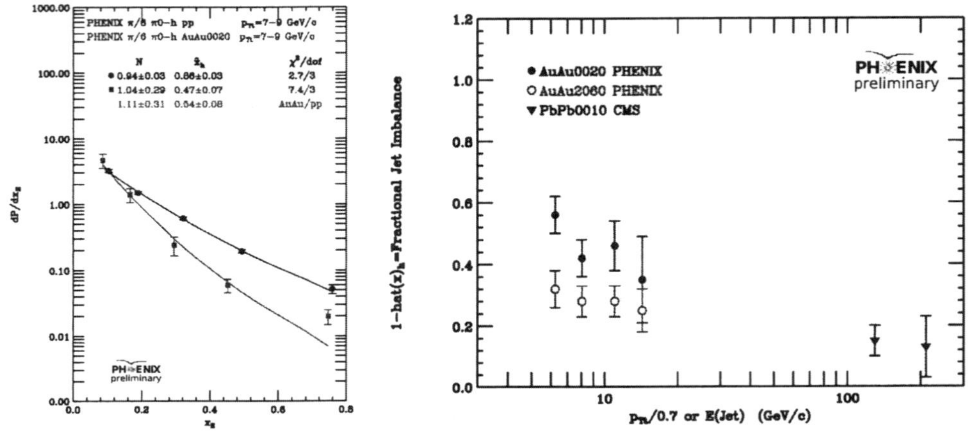

Figure 23: (left) a) x_E distributions [77] from p-p (circles) and AuAu 0-20% centrality (squares) for $p_{T_t} = 7 - 9$ GeV/c, together with fits to Eq. 5 (solid lines) with parameters indicated. The ratios of the fitted parameters for AuAu/pp are also given. b) (right) Fractional jet imbalance [77], $1 - \hat{x}_h^{AA}/\hat{x}_h^{pp}$, for RHIC and CMS data.

take account of the imbalance ($\hat{x}_h^{pp} < 1$) observed in the p-p data, the ratio $\hat{x}_h^{AA}/\hat{x}_h^{pp}$ is taken as the measure of the energy of the away jet relative to the trigger jet in A+A compared to p-p collisions. The fractional jet imbalance was also measured directly with reconstructed di-jets by the CMS collaboration at the LHC in Pb+Pb central collisions at $\sqrt{s_{\rm NN}} = 2.76$ TeV [66]. They also showed a large effect in p-p collisions. I calculated $\hat{x}_h$ from their p-p and Pb+Pb results [77] and compared these LHC values of $1 - \hat{x}_h^{AA}/\hat{x}_h^{pp}$ to those from PHENIX (Fig. 23b). The large difference in fractional jet imbalance between RHIC and LHC c.m. energies could be due to the difference in jet $\hat{p}_{T_t}$ between RHIC (~ 20 GeV/c) and LHC (~ 200 GeV/c), or different sensitivity of the direct and indirect methods, or the difference in n for the different $\sqrt{s}$, or to a difference in the properties of the medium. Future measurements will need to sort out these issues by extending both the RHIC and LHC measurements to overlapping regions of p_T.

8 Anisotropic flow (v_2) of direct-γ

Although direct-γ production [65] is the most beautiful QCD subprocess, it has a very serious problem: an overwhelming background of photons from high p_T $\pi^0 \to \gamma+\gamma$ and $\eta \to \gamma+\gamma$ decays makes it a very difficult experiment.

One must measure all the background sources: π^0, η, ..., and calculate their contributions to the inclusive γ-ray spectrum. In principle, the background can be calculated whatever the p_T distribution of the π^0 and η. However nature has been kind in that the invariant cross section for hard-scattering is a power law, $d\sigma/p_T dp_T \propto 1/p_T^n$, with $n = 8.1 \pm 0.05$ for π^0 with $p_T \geq 3$ GeV/c (Fig. 21a) [68]; also for $p_T \geq 3$ GeV/c, $\eta/\pi^0 = 0.48 \pm 0.03$ is a constant. This implies that for $\pi^0 \rightarrow \gamma + \gamma$, the spectrum of decay photons has the same power as the parent π^0 so that the ratio at any p_T is a constant:

$$\left.\frac{\gamma}{\pi^0}\right|_{\pi^0} = \frac{2}{n-1} \quad . \tag{6}$$

The resulting background inclusive γ spectrum from π^0 and η decays at $\sqrt{s_{NN}} = 200$ GeV is:

$$\gamma_{\text{background}}/\pi^0 \approx (1 + 0.48 \times 0.39) \times 2/7.1 = 1.19 \times 2/7.1 = 0.334 \tag{7}$$

where 0.39 is the branching ratio for $\eta \rightarrow \gamma + \gamma$. In PHENIX we plot what we call the double ratio:

$$R_\gamma = (\gamma_{\text{inclusive}}/\pi^0)/(\gamma_{\text{background}}/\pi^0) = \gamma_{\text{inclusive}}/\gamma_{\text{background}}$$

where it is important to see the calculated $\gamma_{\text{background}}/\pi^0$ ratio, which usually comes from some opaque Monte Carlo program, to understand whether it makes sense according to Eqs. 6 and 7.

Fig. 24a [79] shows R_γ for real photons, measured in an EM calorimeter, and virtual photons, which are e^+e^- pairs from internal conversion of the direct-γ, with $0.12 < m_{ee} < 0.30$ GeV/c^2 where there is no background from π^0 Dalitz decay. This reduces the background by a factor of $\geq 1.19/0.19 \approx 6$ [27], and allows the precision of R_γ to be greatly improved as shown. Then, using the precise virtual photon R_γ with the much higher statistics inclusive real-γ data, one can derive v_2 for direct-γ from the measured v_2 of inclusive real-γ compared to the measured v_2 of γ's from π^0 and η decay.

The result (Fig. 24b) [79] is that the v_2 of direct-γ is large in the range $1 \leq p_T \leq 3$ GeV/c but drops to zero for $p_T \geq 5$ GeV/c where the photons are produced from initial hard-scattering and do not interact with the medium so that they do not flow. Since thermal radiation is produced in the flowing medium, the observed large v_2 in the range $1 \leq p_T \leq 3$ GeV/c, where the direct-γ p_T spectrum is exponential (Fig. 21b), confirms that these γ are thermal radiation from the medium. What is very surprising is that the v_2 of the thermal photons is so large, the same or slightly greater than that of π^0's.

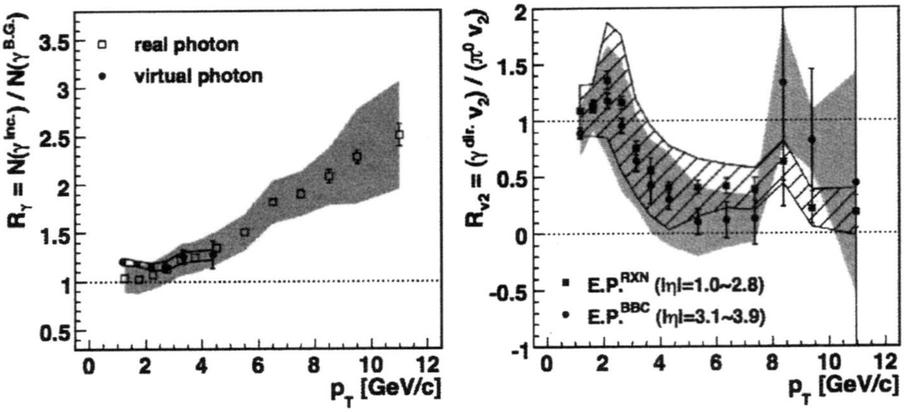

Figure 24: a) (left) R_γ vs p_T [79] for virtual photons (solid circles) [69] and real photons (open squares) [78] for minimum bias Au+Au collisions at $\sqrt{s_{NN}} = 200$ GeV. b) (right) Ratio of direct-γ v_2 to π^0 v_2 for two reaction plane detectors [79].

9 A charming surprise

This past year, the ALICE experiment at LHC [80] confirmed, with reconstructed charm mesons, the suppression of heavy quarks comparable to that of π (from light quarks) for $p_T \gtrsim 4$ GeV/c as previously observed at RHIC using direct-single-$e^\pm$ from heavy quark (c, b) decay (Fig. 25a) [81]. Also seen at RHIC is that heavy quarks exhibit collective flow (v_2) (Fig. 25b), another indication of a very strong interaction with the medium. The fact

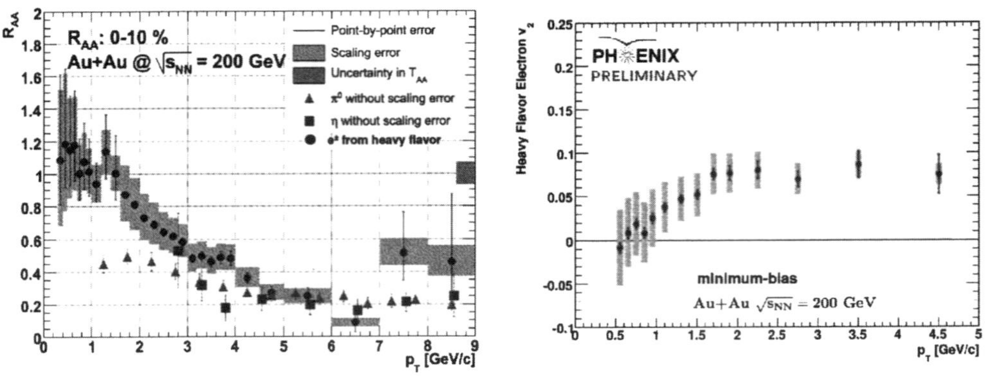

Figure 25: PHENIX [81]: a) (left) R_{AA} (central Au+Au) b) (right) v_2 (minimum bias Au+Au) as a function of p_T for direct-$e^\pm$ at $\sqrt{s_{NN}} = 200$ GeV.

that heavy quarks are suppressed the same as light quarks strongly disfavors the QCD energy-loss explanation of jet-quenching because, naively, heavy quarks should radiate much less than light quarks and gluons in the medium; but opens up a whole range of new possibilities including string theory [82].

10 Zichichi to the rescue?

In September 2007, I read an article by Nino, "Yukawa's gold mine" in the CERN Courier taken from his talk at the 2007 International Nuclear Physics meeting in Tokyo, Japan, in which he proposed: "We know that confinement produces masses of the order of a giga-electron-volt. Therefore, according to our present understanding, the QCD colourless condition cannot explain the heavy quark mass. However, since the origin of the quark masses is still not known, it cannot be excluded that in a QCD coloured world, the six quarks are all nearly massless and that the colourless condition is 'flavour' dependent."

Nino's idea really excited me even though, or perhaps because, it appeared to overturn two of the major tenets of the Standard Model since it seemed to imply that: QCD isn't flavor blind; the masses of quarks aren't given by the Higgs mechanism. Massless b and c quarks in a color-charged medium would be the simplest way to explain the apparent equality of gluon, light quark and heavy quark suppression indicated by the equality of R_{AA} for π^0 and direct single-$e^{\pm}$ in regions where both c and b quarks dominate. Furthermore RHIC and LHC-Ions are the only place in the Universe to test this idea.

Nino's idea seems much more reasonable to me than the string theory explanations of heavy-quark suppression (especially since they can't explain light-quark suppression). Nevertheless, just to be safe, I asked some distinguished theorists what they thought, among others [27], Steve Weinberg. He said that he and Lenny Susskind had a model Technicolor (or Hypercolor) that worked well in the vector boson sector but didn't give mass to the fermions.

Nino proposed to test his idea by shooting a proton beam through a QGP formed in a Pb+Pb collision at the LHC and seeing the proton 'dissolved' by the QGP. My idea is to use the new PHENIX VTX detector, installed in 2011, to map out, on an event-by-event basis, the di-hadron correlations from identified $b-\overline{b}$ di-jets and identified $c-\overline{c}$ di-jets, which do not originate from the vertex, and light quark and gluon di-jets, which originate from the vertex and can be measured with π^0-hadron correlations. A steepening of the slope of the x_E distribution of heavy-quark correlations as in Fig. 23a will confirm in detail (or falsify) whether the different flavors of quarks behave as if they have the same energy loss (hence mass) in a color-charged medium. If Nino's proposed effect is true, that the masses of fermions are not given by the Higgs particle and all quarks are nearly massless in the QGP, and we can confirm the effect at RHIC or LHC-Ions, this would be a case where we Relativistic Heavy Ion Physicists may have something unique to contribute at the most fundamental level to the Standard Model, which would constitute a "transformational discovery." Of course the LHC could falsify this idea by finding the Higgs decay to $b - \bar{b}$ at the expected rate in p-p collisions. Clearly, there are exciting years ahead of us!

References

[1] *Report of the Workshop on BeV/Nucleon Coliisions of Heavy Ions—How and Why*, Bear Mountain, NY, 29 November–1 December 1974. (BNL-50445, Upton NY, 1975).

[2] See Ref. [3] for a more extensive list of references.

[3] M. J. Tannenbaum, Rep. Prog. Phys. **69** (2006) 2005–2059.

[4] E. V. Shuryak, Phys. Repts. **61** (1980) 71–158.

[5] *The Frontiers of Nuclear Science*, NSAC Long Range Plan 2007.

[6] I. Arsene *et al.*, BRAHMS Collab. Nucl. Phys. A**757** (2005) 1–27.

[7] B. B. Back *et al.*, PHOBOS Collab. Nucl. Phys. A**757** (2005) 28–101.

[8] J. Adams *et al.*, STAR Collab. Nucl. Phys. A**757** (2005) 102–183.

[9] K. Adcox *et al.*, PHENIX Collab. Nucl. Phys. A**757** (2005) 184–283.

[10] D. Rischke and G. Levin, eds., Nucl. Phys. A**750** (2005) 1–171.

[11] Press Conference, *A New State of Matter Created at CERN* http://newstate-matter.web.cern.ch/newstate-matter/Story.html

[12] New York Times, February 10, 2000, Page 1, *Particle Physicists Getting Closer To the Bang That Started It All.*

[13] NASA/BNL Space Radiation Program, `http://www.bnl.gov/medical/NASA/LTSF.asp`

[14] E. J. Bleser, *et al.*, Nucl. Instrum. Methods A**235** (1985) 435.

[15] R. M. Bozorth, *Ferromagnetism* (VanNostrand, New York, 1951)

[16] The ALPHA Collaboration, Nature Physics **7** (2011) 558–564; G. B. Andresen, *et al.*, Nature **468** (2010) 673–676.

[17] R. Chasman, G. K. Green and E. M. Rowe, IEEE Trans Nucl. Sci. **NS-22** (1975) 1765–1767.

[18] Press Release, "Indication of Electron Neutrino Appearance at the T2K Experiment" June 15, 2011, `http://www.kek.jp/intra-e/press/2011/J-PARC_T2Kneutrino.html`

[19] NSLS-II Source Properties and Floor Layout `http://www.bnl.gov/ps/docs/pdf/SourceProperties.pdf`

[20] C. Amsler *et al.* (Particle Data Group), Phys. Lett. B**667** (2008) 1.

[21] T. Matsui and H. Satz, Phys. Lett. B**178** (1986) 416.

[22] B. Alessandro, *et al.*, NA50 Collab., Eur. Phys. J. C**39** (2005) 335–345. See also, F. Prino, Proc. XXX International Symposium on Multiparticle Dynamics, arXiv:hep-ex/0101052v1.

[23] D. M. Alde, *et al.* E772 Collab., Phys. Rev. Lett. **66** (1991) 2285–2288. See, also, M. J. Leitch, Eur. Phys. J. C**43** (2005) 157–160 and references therein.

[24] C. Albajar, *et al.* UA1 Collab., Phys. Lett. B**186** (1987) 237.

[25] S. S. Adler, *et al.* PHENIX Collab., Phys. Rev. C**71** (2005) 034908.

[26] *The Relativistic Heavy Ion Collider Project: RHIC and its Detectors*, Nucl. Instrum. Methods A**499** (2003) 235–880.

[27] See, for example, M. J. Tannenbaum, Proc. Int. School Subnuclear Physics, "The most unexpected at LHC and the status of high energy frontier", 47th Course, Erice-Sicily: 29 August - 7 September 2009, arXiv:1006.5701v1 [nucl-ex].

[28] K. Aamodt, *et al.* ALICE Collab., Phys. Rev. Lett. **106** (2011) 032301.

[29] The detector is so non-conventional that it made the cover of Physics Today, October 2003.

[30] H. Agakishiev, *et al.* STAR Collab., Nature **473** (2011) 353–356.

[31] B. I. Abelev, *et al.* STAR Collab., arXiv:0909.0566v1 [nucl-ex].

[32] R. A. Lacey, Nucl. Phys. A**774** (2006) 199–214.

[33] M. Kaneta *et al.*, PHENIX Collab., J. Phys. G**30** (2004) S1217–S1220.

[34] A. Adare, *et al.*, PHENIX Collab., Phys. Rev. Lett. **98** (2007) 162301.

[35] J.-Y. Ollitrault, Phys. Rev. D**46** (1992) 229–245; Nucl. Phys. A**638** (1998) 195c–206c.

[36] H. Heiselberg and A.-M. Levy, Phys. Rev. C**59** (1999) 2716–2727.

[37] S. A. Voloshin, Nucl. Phys. A**715** (2003) 379c–388c.

[38] D. Teaney, Phys. Rev. C**68** (2003) 034913.

[39] P. K. Kovtun, D. T. Son, and A. O. Starinets, Phys. Rev. Lett. **94** (2005) 111601.

[40] R. A. Lacey, *et al.*, Phys. Rev. Lett. **98** (2007) 092301.

[41] L. P. Csernai, J. I. Kapusta, and L. D. McLerran, Phys. Rev. Lett. **97** (2006) 152303.

[42] A. Adare, *et al.* (PHENIX Collab.), Phys. Rev. C**78** (2008) 014901; Phys. Rev. C**77** (2008) 011901(R).

[43] A. Adare, *et al.* (PHENIX Collab.), Phys. Rev. Lett. **98** (2007) 232302.

[44] J. Casalderrey-Solana, E. V. Shuryak and D. Teaney, J. Phys: Conf. Series **27** (2005) 22-31.

[45] For example, see Ref. [42] for a discussion and list of references.

[46] J. T. Mitchell, *et al.* (PHENIX Collab.), J. Phys. G**34** (2007) S911–S914.

[47] B. I. Abelev, *et al.* (STAR Collab.), Phys. Rev. C**80** (2009) 064912.

[48] J. Putschke, *et al.* (STAR Collab.), Nucl. Phys. A**783** (2007) 507c–510c.

[49] P. Jacobs, Eur. Phys. J. C**43** (2005) 467–473.

[50] B. H. Alver, C. Gombeaud, M. Luzum, and J. Y. Ollitrault, Phys. Rev. C**82** (2010) 034913.

[51] B. Alver and G. Roland, Phys. Rev. C**81** (2010) 054905.

[52] J. Takahashi, *et al.*, Phys. Rev. Lett. **103** (2009) 242301. Also, see A. P. Mishra, *et al.*, Phys. Rev. C**77** (2008) 064902.

[53] S. Esumi, *et al.* (PHENIX Collab.), arXiv:1110.3223v1 [nucl-ex].

[54] Presentation by M. Gaździcki at Quark Matter 2011. Also see M. Gaździcki, *et al.* (NA49 Collab.), J. Phys. G**30** (2004) S701–S708.

[55] Presentations by L. Kumar and B. Mohanty at Quark Matter 2011.

[56] J. Cleymans, H. Oeschler, K. Redlich and S. Wheaton, Phys. Lett. B**615** (2005) 50–54.

[57] `http://newscenter.lbl.gov/news-releases/2011/06/23/when-matter-melts/`.

[58] e.g. see V. Koch, PoS(CFRNC2006)008

[59] M. Asakawa, U. Heinz, and B. Müller, Phys. Rev. Lett. **85** (2000) 2072.

[60] A. H. Mueller, Phys. Rev. D**4** (1971) 150.

[61] Presentation by T. J. Tarnowsky at Quark Matter 2011. Also see T. J. Tarnowsky, *et al.* (STAR Collab.), arXiv:1106.6110v1 [nucl-ex].

[62] M. M. Aggarwal, *et al.* (STAR Collab.), Phys. Rev. Lett. **105** (2010) 022302.

[63] S. Gupta, X. Luo, B. Mohanty, H. G. Ritter and N. Xu, Science **332** (2011) 1525.

[64] A. Adare, *et al.* (PHENIX Collab.), arXiv:1105.1966v1 [nucl-ex].

[65] H. Fritzsch and P. Minkowski, Phys. Lett. B**69** (1977) 316.

[66] S. Chatrchyan, *et al.* (CMS Collab.), Phys. Rev. C**84** (2011) 024906.

[67] K. Adcox, *et al.* (PHENIX Collab.), Phys. Rev. Lett. **88** (2002) 022301.

[68] S. S. Adler, *et al.* (PHENIX Collab.), Phys. Rev. C**76** (2007) 034904.

[69] A. Adare, *et al.* (PHENIX Collab.), Phys. Rev. Lett. **104** (2010) 132301.

[70] See R. Baier, D. Schiff and B. G. Zakharov, Ann. Rev. Nucl. Part. Sci. **50** (2000) 37–69, and references therein.

[71] B. Adeva, *et al.* (L3 Collab.), Phys. Lett. B**259** (1991) 199–208.

[72] N. Borghini and U. A. Wiedemann, Nucl. Phys. A**774** (2006) 549–522; see also arXiv:hep-ph/0506218v1.

[73] A. Adare, *et al.* PHENIX Collab., Phys. Rev. D**82** (2010) 072001.

[74] M. Connors, *et al.* (PHENIX Collab.), Nucl. Phys. A**855** (2011) 335–338.

[75] S. S. Adler, *et al.* (PHENIX Collab.), Phys. Rev. D**74** (2006) 072002.

[76] A. Adare, *et al.* (PHENIX Collab.), Phys. Rev. Lett. **104** (2010) 252301.

[77] M. J. Tannenbaum, *et al.* (PHENIX Collab.), arXiv:1109.1760v1 [nucl-ex].

[78] S. S. Adler, *et al.* (PHENIX Collab.), Phys. Rev. Lett. **94** (2005) 232301.

[79] A. Adare, *et al.* (PHENIX Collab.), arXiv:1109.1105.4126v2 [nucl-ex].

[80] A. Dainese, *et al.* (ALICE Collab.), arXiv:1109.6.4042v2 [nucl-ex].

[81] A. Adare, *et al.*, PHENIX Collab., Phys. Rev. Lett. **98** (2007) 172301.

[82] e.g. see Ref. [81] for a list of references.

CHAIRMAN: M. J. TANNENBAUM

Scientific Secretaries: V. Begun, L. Wen

DISCUSSION I

- *Preghenella:*

If you subtract the contribution of v_3 in 2-particle correlations, you get rid of the Mach-Cone effect. What happens to the high P_T correlations? Does the away side maybe reappear?

- *Tannenbaum:*

We don't know yet. We have to reanalyze all the data. Let me show you what we did. In fact we went to higher pT where the effect goes away. On this plot, we defined previously what we call a shoulder region (SR) and a head region(HR).

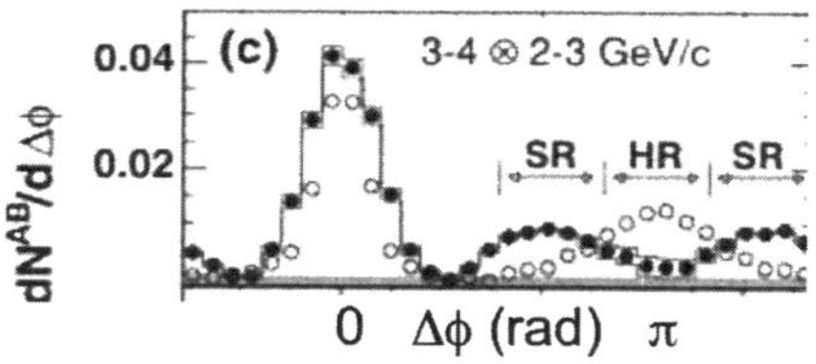

The head region contains most of the p-p away side peak (open cireles). For both AA and pp we make a cut and either analyze the head region in both cases or the full away side. We've never analyzed the shoulder region. Most of our correlation data need to be corrected. Well they probably don't need to be corrected because I don't think that the head region may change; but for precision work it's not what you think, you have to measure it again. So we definitely have to redo all our results, except for the results that I'll show tomorrow where we were high enough in pT that the background is small so the modulation doesn't matter. Basically you are right. All of our results have to be corrected for v3.

- *Preghenella:*

Last year if I remember correctly, in a RHIC highlight lecture in this school, it has been said that jet suppression at forward rapidity is due to initial state effects, like quark-gas condensate, while this morning you said the CGC does not reproduce other data. Is the CGC completely ruled out? And what could be the explanation on forward jet suppression?

- *Tannenbaum:*

It's a very good question. It's really tough to beat the Color Glass Condensate people who are all my colleagues. Basically I live in the CGC world---it's all at Brookhaven. Go to slide 130. In order to go to low x, we put two calorimeters inside a hole in the muon magnet pole to measure pizeros. They cover pseudorapidity ± 3-4. If you have a pair of pizeros, you can measure x1 and x2 from the standard kinematics formula. We measure the rate of two particles, one is at eta1, pT1, the other is at eta2, pT2. We measure the number of pairs in d+Au and in p-p and take the ratio corrected by binary-collision scaling. It is a pair-suppression ratio, analogous to RAA but for a pair of particles and we call it JdA.

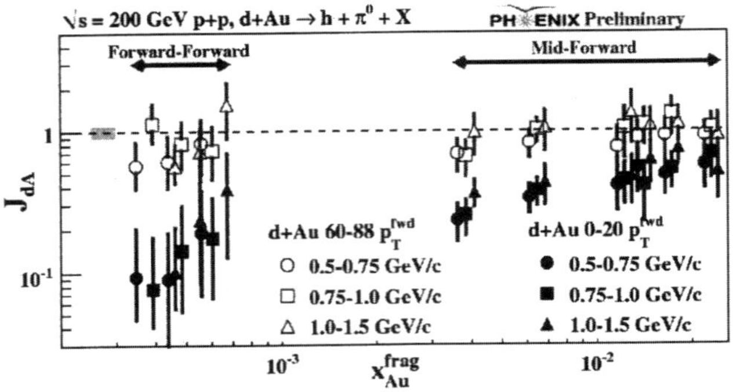

The red open points are peripheral dAu: We can measure mid-forward rapidity; or forward-forward so you can go to much lower x. For peripheral dAu, JdA is constant ~1. For 00-20% centrality d+Au, we get suppression, more suppression as we move to lower x. This is what the CGC guys predicted. HOWEVER, if we then compare this result in a new figure (triangles) to a similar measurement for the J/Psi (via a pair of muons) (filled dots) and with a calculation of the parton distribution function, EPS09, lines, we see that the J/Psi follows the cold nuclear matter (represented by EPS09) pretty well, while the two-particle JdA falls below the J/Psi data, but with large systematic errors. It's new, maybe somebody can explain it. Although the systematic error is big, for 0.001<x<0.01 the effect is way beyond the systematic error.

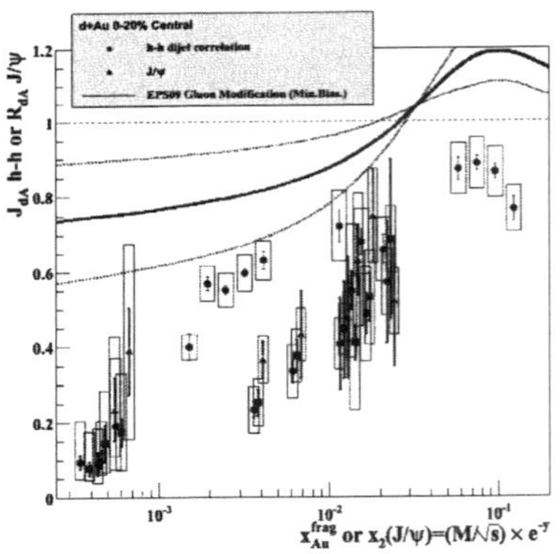

People have to look at this new result seriously. I think that the result is probably correct. Does this prove the CGC while something else disproves it? I'm sure that the CGC people would be happy to see wherever it works. But this is different from the v3 measurement, it's directly sensitive to the parton distribution function. The v3 test is indirect because they have to calculate the eccentricities and it's not clear that they know how to do it in the CGC. This is a much more serious challenge. If it can not be explained by standard DGLAP evolution, or shadowing or because the pT of the measurement it is too low (so diffraction dissociation rather than hard scattering), it could be very interesting. However it also points out the importance of measuring the gluon structure function in a nucleus for values of x in the range 10^{-3}. This can and should be done at the LHC by measurements in p-p and p+Pb of (isolated) direct-single photons at mid-rapidity with pT≥10 GeV/c, an experimental challenge.

- *Takaki:*

Can you explain the particle correlation/ridge at high P_T? The ridge disappeared at high pT. Is there a physics interpretation?

- *Tannenbaum:*

The modulation is there but is a few percent (1-2% of the background) which is comparable to the signal. That's why it has a big effect. Once you go much higher P_T, the background goes down. If you raise the trigger pT, you improve the background. It goes down like the 8th power of pT. We've not corrected any data for v_3, except for one plot. The thing that happened when we corrected for v3, was that the ridge and the cone went away, but the base level moved up—so we still have tounderstand that. If you go to very high P_T, it is Ok, but you may miss a lot of physics at low P_T. It's still at hard scattering, but the energy is low (5~7 GeV), if you go to 50GeV, you don't have these problems.

- *Takaki:*

What's the magnitude of v_3 compared to v_2? Have you done the quark number scale for these plots?

- *Tannenbaum:*

It's beautiful. We measure the magnitude of v_3 compare to v_2 for pions. For central collisions, where most are due to fluctuations, they are comparable. I'm a high energy physicist I don't know anything about Flow and Mach-Cone—they just get in my way in analyzing the correlations from jets. By the way, thousands of theorists, including String theorists, explained the Mach-Cone. It's gone, it doesn't exist, so if they explained it, what should I think of them. I have a good colleague, Roy Lacey who is a nuclear chemist. He is an expert on all this stuff, e.g. the Knudsen number, etc. In his opinion, by comparing the v_2, v_3, v_4, you are hitting the overlap region, the Quark-Gluon-Plasma and it rings, like the modes of the early universe seen in the W-map. He wants to look at the ratio of v_2, v_3, v_4, and get the speed of sound. We need some good theoretical help to exactly tell what to do. It is a good theoretical opportunity, what can you learn by measuring v_2, v_3, v_4 due to fluctuations? There's a lot to be learned. There's something new.

- *Mager:*

Are there plans to measure basic quantities like dN_{ch}/d\eta at fixed center of mass energy but different CMS rapidities?

- *Tannenbaum:*

Yes, PHOBOS did it. They are now the American group on CMS-Ions. We had two big experiments: PHENIX and STAR, two small: PHOBOS and BRAHMS. We already replaced BRAHMS by another experiment AnDY which is looking for the forward Drell Yan, but I don't think it is a very good experiment. You need at least 3 PID for muon or the electron to make sure it`s not something else. They don't have any. PHOBOS had silicon detectors that covered the full rapidity range. They have beautiful data: v2 as as function of the pseudo-rapidity; and they measured the multiplicity; and they measured a few different energies. The answer is yes, we did do that, but its old, it's eight years old.

- *Mager:*

Thank you, but it's not precisely what I meant. I would rather like to see a picture from STAR which shows you that Lorentz boost is correct. You should observe a Lorentz boost of the collision if you change the center of mass energy.

- *Tannenbaum:*

Are you checking relativity?

- *Mager:*

Precisely.

- *Tannenbaum:*

Let me tell you a better place where we did that: the ISR. That was the first hadron collider and a great beautiful machine that led a way to other great beautiful machines at CERN. All the present colliders cross head-on because they are bunched. The ISR beams crossed at an angle because they were continuous beams such that the c.m. system moved outward from the collider beam line. This meant that one side of your detector, say outside the beam had a Lorentz boost towards it, while the other side had a Lorentz boost going away from the experiment. This meant that you had to correct for this Lorentz transformation in both arms, with a check being that you get the same answer in c.m. system quantities in both arms. It worked. People have made many other measurements with muon beams. The "muon g-2" is another beautiful example. They have muons going around a ring, it precesses. It likes to decay forward because parity is violated, so as it precesses the number of electrons above a certain energy varies sinusiodally, so you see this fantastic modulations. They see hundreds oscillations. It is one of CERN's great experiments. If Lorentz invariance did not work you would not get any of that. It is really high precision measurement.

- *Palhares:*

Why do you think that the first order line in the phase diagram does not exist?

Do you have any idea of how we could rule out experimentally the existence of the transition line?

- *Tannenbaum:*

Because there is no any evidence of large fluctuations. We measured E_T, transverse energy distribution; or the multiplicity distributions. It has exactly the same shape as for low energy scan by STAR. Every ET distribution as a function of c.m. energy from $\sqrt{s_{NN}}$=5 GeV to the LHC has exactly the same shape. There is no evidence of any anomalous fluctuations, no evidence of any first order phase transition. First order phase transition is like water going to ice. We don't see any evidence for that. If you look at the phase diagram, the line of the first order phase transition could be a way out: it exists, but we do not cross it. The only way that we could missed a first order phase transition at AGS energy is that we did not excite medium to high enough temperature to cross it. That's possible, you can't tule it out. I used to say at the AGS: "is everything just outside the E802 aperture." (p.s it wasn't). But I have more other comments on fluctuations tomorrow.

- *Noferini:*

At LHC we expect to produce a higher density system. However if you compare central collisions at RHIC with semi-peripheral collisions at LHC you could obtain the same energy density? Do you think that this can be useful and what is the best observable for that?

- *Tannenbaum:*

It could be. Are you talking about suppression, about R_{AA}?

- *Noferini:*

Yes, for example.

- *Tannenbaum:*

We have measured pi0 production at every centrality at 10% intervals, it is published two years ago, it is not new [PRL 101,232301(2008)]. We measured its logarithmic slope, it is power law. At every centrality it is n=8.1±0.1 for the invariant cross section. The lines are really parallel to each other on a log-log plot. That issue has to do with how flat R_{AA} is. So, we have all those measurements. It is published. If some data are not published in detail, they are available. If you would do the same that would be quite interesting if you could match. What is really interesting, you saw the ALICE plot of R_{AA}. We have the data in the region where they strongly overlap, namely 5-20 GeV/c, and we have an issue about it, do our data those data rise or not? Clearly, ALICE R_{AA} rises, but both sets of data are pretty much one on top of another and I will show it tomorrow. Then the question is: since the P_T logarigthmic slope about 6 at ALICE, it 8 for us, it could be 5 or 6 for you. That implies that there is more fractional energy loss at ALICE than there is at RHIC, because you need more energy loss to get the same RAA in a slope that is flatter than you do in a slope that is steeper. There is a lot to be learned. We are just beginning to scratch the surface. We used to have this big problem at RHIC, our biggest problem at PHENIX was: does R_{AA} rise from 10 to 20 GeV/c. We could never see it, it is not statistically significant. Now the question is: at LHC does R_{AA} rise from 30 to 80 GeV/c? So, we have to find out. The whole point is that this science, unlike High Energy Physics, is much more complicated. It is experimentalist's game. The theorists have great ideas, but it is very difficult to get a handle on these things. So, every experimental measurement and test you can make is worthwhile. Now, I think, the next year or two will be spent on doing it: comparing different centralities at LHC and RHIC. It would be great.

CHAIRMAN: M. J. TANNENBAUM

Scientific Secretaries: M. Csanád, R. Vértesi

DISCUSSION II

- Inguglia:

You said that in 2011 a good improvement was adding a vertex detector. Can you please spend a few more words about this detector?

- Tannenbaum:

It is a silicon vertex detector naturally. It has got pixels, strip pixels. I do not know what the resolution is, but it is a high class detector. I showed the hundred micron resolution for distance from the vertex. That is what it is capable of. I do not know how many layers it has—I could look it up too. [In proof, I did look it up it has 4 layers with radii from 2.5 to 16.5 cm, the acceptance is $|\eta|<1.2$ with almost 2π in the ϕ plane] I am not the technical guy. I can tell you who built it. Yasuyuki Akiba at RIKEN was the most important, as well as_Brookhaven people. It is 9% of radiation length thick, that's what I know about it.

- Inguglia:

It is a small detector, and you have many tracks. Is there any problem with the overheating of this detector? How many events do you have per second?

- Tannenbaum:

First of all the detector has extensive cooling just for the heat from the electronics and to maintain the temperature. We worry more about ground problems than heating the electronics from collisions. We can go up to 5-10 thousand events per second, maybe more but these not all central. It's easy to know what fraction are central collisions: the upper 5% is 5%. So the answer is no for heating from collisions. The thing that you worry about is not the collisions, it's the radiation, you are worried about losing the beam, just like you heard from Bertolucci before. I should know the difference in the energy density of our beam and the LHC, but in our case it won't damage the vacuum pipe. It however would damage our detectors, if the beam hit the silicon. So that's the big worry, but we do not move it in and out. The big thing is that we are sensitive to the background around the beams. So we have to make sure the beam tune and the collimation is good. We have more trouble with our muon detectors which are forward and backward, because they are very sensitive to every background from upstream, and you get muons all over the place. Even more than you get from a central heavy ion collision.

- Burda:

I would like to ask you an inappropriate question but I hope you will excuse me because I'm a theoretician. We know that a lot of people are working on the AdS/CFT analogy. They calculate a lot of things using this gravitation analogy. Can any recent measurement from RHIC confirm or reject their calculations for real QCD?

- Tannenbaum:

Let me say, I can't know if there is confirmation or rejection, but they were big on this Mach cone; not only were they big on the Mach cone, but 90% of the top theorists in this field, were big on the mach cone, starting with Edward Shuryak, Berndt Muller, just to name a few; and it was flat out wrong. It is a tough business, that's all I want to say. You have to understand another thing about Brookhaven. You heard Sam Ting talk. He says, "I never read anybody else's paper, I never trust the background, I never trust a theoretician". I feel a bit that way, because I also spent a lot of time at Columbia University. At Brookhaven we have this great Riken-BNL spin project which I was among the people who started, along with Gerry Bunce. It was just two physicists wanting to do parity violation. We got RIKEN from Japan to support this project. They put in the money for the snakes and spin rotators and built the second muon arm. They fund it every year for maintenance and also they wanted to put in a research institute to profit from the science. So we have this RIKEN-BNL research center (RBRC) which is widely successful. They hire fellows and postdocs every few years and you get a joint appointment at university, which is wonderful for both sides, because the university only has to pay half the salary and they get a paid research project for the other half, and the new faculty are peer-reviewed. T. D. Lee was the head of the RBRC. He also is an anti-string-theorist. He thinks there is too much string theory around. First, I was also against it, especially when I asked the guys: it is great that you can explain the heavy quark losing energy, but how do you do the light quarks? They said, oh no, we don't do the light quarks, so I forgot that. Then I asked Derek Teaney, when the regular theoreticians started to do AdS/CFT calculations. I asked: why do you do that? He said they are easier than QCD. So, first I was against it. Then I thought, look: they are smart guys, you have to admit that this string theory group is smart and way out there, and if they are interested in what we are doing, it is not bad. First I thought it's bad, but it's not bad, it's great if we can stimulate them. Have they predicted anything? The eta/s, they were the ones who came out with this limit. I think it is great, it is wonderful what they are doing; have I really profited from them? No, other than thinking about eta/s.

\- *Mager:*

My first question is about particle ratios and RAA. Look at the centrality dependence, for instance I was looking at the J/Psi to Psi' ratio from NA50. One cannot explain how this is going down to the proton-proton case. Can you comment on that?

\- *Tannenbaum:*

It does not go down to the proton-proton case. I tell you what is interesting. I understood this when we were working at ISR. We were doing high pT p-p, alpha-alpha, d-d, p-alpha and pbar-p. Especially for alpha-alpha, we were looking for the Cronin effect. In fact I got into the real heavy ion field because we made an error. I did not make it – I told the guy, you are using the wrong p-p extrapolation. We did not have a p-p run at exactly the same center of mass energy, so we had to extrapolate down; and I said you are doing the extrapolation wrong. He just ignored me. I was at that time in Brookhaven doing magnets and other things, so I let it pass. When the experts in the Cronin effect said we have a factor of 64 for alpha-alpha cross section, when the right answer is 16 (it is pure A squared): so RAA was 64/16=4. Our first measurement for alpha-alpha at $\sqrt{s_{NN}}$=31 GeV was RAA=4! Everybody got excited; the chairman of my department, who was a nuclear physicist, said I should go to CERN and redo this measurement. The important thing is, that you have to get the right p-p data at the same center of mass energy, preferably the same detector, preferably at the same time, because things change from one year to the next: somebody says, why don't we put something new in there. So people who don't measure RAA but RCP, they always get the curve low compared to the correct RAA (with p-p comparison data). So that's my answer, the peripheral collisions are not really independent p-p collisions. They usually show a Cronin effect; so RCP is usually less than RAA.

\- *Mager:*

The second question would be the following. Now STAR is building the HFT. How does this compare to your plans?

- *Tannenbaum:*

It is beautiful because it is thin. I don't know if it will work any better, it may not, but it is new. The point is, that it is really thin. I had been looking for something like this. When I was your age, I used to build stuff – now I don't build stuff anymore, I fix my house and things like that, but for detectors there are all these engineers and smart guys–I was envisioning a silicon spray can. Just like they make these R.P.C's, you spread a little carbon. I was hoping you print out a beautiful design of strips and spray a silicon layer on it and you have a detector. STAR's detector is almost like that. So what is your question?

- *Mager:*

As far as I know, the HFT is not funded yet.

- *Tannenbaum:*

No, it's not, they are working on it. You know what; I'm going to give you a "Sam Ting answer": I have no idea what they are doing, and I actually don't care. More power to them. If they build something and it actually works, and suddenly they are great, then fine.

- *Preghenella:*

You have shown a plot where models cannot reproduce the direct-photon v2. I was wondering if those models are based on the same hydrodynamically inspired models that could reproduce the v2 of charged hadrons.

- *Tannenbaum:*

In the experiment, we have ~500 people, but I was on the working committee for this paper. I insisted that we plot the ratio of the direct photon v2 to the pi0 v2. It is bigger than one where the photon v2 peaks at 2.0 GeV/c. Also, this would force the theorists trying to explain the data to explain both the pizero and the thermal photons v2 at the same time. I don't know if it is the field or the age, but I grew up in an age when for the strong interactions there was no theory. The only good theory was QED and we spent a lot of time testing QED. But for all the other stuff, there was no theory. Everybody said you would not have higher mass vector mesons because they get wider with increasing mass because more channels open up. Sam Ting ignored this and found the J/Psi with a very narrow width. Back then there was no theory of strong interactions; but now, everybody is victimized by the Standard Model. The Standard Model is okay, but they are also victimized by other models for which there is no evidence at all, but people talk about them. Because of this, my rule is: never put a theoretical curve on your data---that's a theorist's job---unless you really learn something from it. If that is the case, then it is good. Also if you show a theory curve with your data, the people will ask you about the theory curve, and not about your data. I was fighting this in the paper. There were only two calculations of the thermal photon v2, and they missed it by an order of magnitude. Then, these guys, our colleagues in Finland – one of them actually works on the experiment, so he was aware of the result – tried to do it; and then Máté also. He is also on our mailing list. Maybe you should ask him, do they do the hydro, I don't know. All I know is that it is really important, because the thermal photons are made in the medium, so they have the property of the medium: if the medium flows, the photons should flow. It is also sensitive to the temperature. As the medium cools, you are going to get fewer photons or lower frequency photons. So it is obvious from a simple physics point of view. It's like when a blacksmith heats an iron poker: it glows red and then gets more white as its temperature increases. The thermal emission (black body radiation) is very sensitive to the temperature. In detail you have to do the hydro, and then Uli Heinz says "No, this has nothing to do with hydro! It's blue-shifted, it's radial flow." So everybody is arguing with each other, and I say I don't care about your arguments, just fit the data. Like I said, I did not believe in hydro. I only recently - that means, two weeks ago - believe that hydro is reasonable. Because now the v3 works, so you have this fluctuating initial collision and hydro works so that you see it in the final state. But I asked a great hydro physicist: "Tell me how come you've been doing these calculations for 10 years and you've never come up with v3!" I won't tell you who I asked. He said "Well, we didn't have event-by-event hydro calculations!" I said that wasn't the essence of my question, I meant "Why didn't you think of it?" and then tell us "Why don't you guys measure it?" Now that it's measured, they'll figure it out. The answer to your question is very interesting. We had this silly thing at Brookhaven, that we had to specify our Roles, Responsibilities, Accountabilities and Authorities. I put down that we warrant that our measurements are correct to within their stated errors in the standard meaning of experimental uncertainties; but the chairman of the department would not let me put this on my list of responsibilities; but anyway it's true. Our numbers are good to within their stated errors, and that's all I can say. This is a tough measurement, so you have to pay attention to the systematic errors quoted. The biggest problems that we had was making sure that there were no residual neutrons that we called photons and improving the measured ratio of direct to inclusive photons. But look at the paper, there is a preprint with more information (arXiv:1105.4126).

- *Preghenella:*

OK, maybe I can ask another question which is likely stupid, but anyway. Could one do something similar to that with dileptons? Like putting in, let's say, a safe mass region in the Drell-Yan spectrum and have a look at the Drell-Yan process, just trying to measure the Drell-Yan dileptons and v2.

- *Tannenbaum:*

Yes, you should! I had a plot that I did not show you, and this is very interesting. STAR has a first measurement of v2 of J/Psi. STAR is very competitive on Upsilons to us, because you only have to trigger on 4 GeV electrons, which is really easy, even though the Upsilon rate is low. It's very hard to trigger on 500 MeV electrons, there is a huge background. But for 4 GeV, it's a good trigger. So let me show this result from STAR: J/Psi v2.

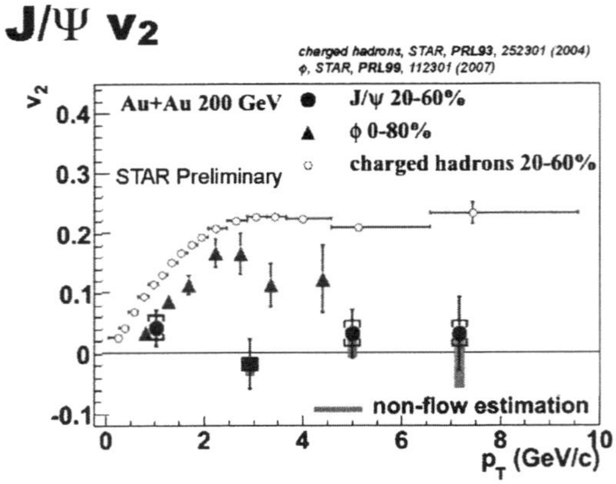

So the open circles are charged hadron v2, the squares are the phi v2, and the black dots somewhere around 0 are J/Psi v2. So they're measuring zero for the J/Psi v2. Now I find these data are incredible, because there is a huge cold nuclear effect, so you would expect the J/Psi to have a v2 from that. Because if it goes one way it is going to get more absorbed than the other way; also there is supposed to be coalescence, two charm quarks are suppose to coalesce into a J/Psi. We measured v2 for charm and it's pretty big! And yet there's absolutely no v2 there! So, the answer is yes. But I have another thing. You didn't really ask this question, but there's this other thing that was a big deal last year if you were here, called local parity violation from Chern-Simons magnetic fields. This data was a big deal, it's a very hard measurement, but now I think nobody believes in it anymore. But I didn't discuss that: LHC has a null result, and other people can explain it just by ordinary things. It's like a one part in a thousand effect. But if there is this strong magnetic field of this parity violation, then when you have an internal conversion where does it get born? If it's literally in the medium, then if there were a strong magnetic field, you would expect the pairs to always open up perpendicular to the magnetic field, right? So you might see, not only the v2 of the photons, you might see a correlation between the plane of the pair and the magnetic field axis, right? There's your magnetic field e.g. upwards, then you might see a correlation horizontally. That would be wild! So we'll be looking for that! The whole problem with this is, when I make an internal conversion, a virtual photon with, let's say 2 GeV, and it internally converts: how many fm distance is needed for that? Is it in the medium when it internally converts? (Nobody answers). But these are the problems we have been working on. There are a lot of wonderful measurements you can make and you have to try everything. So, yes, we are going to do Drell-Yan as well. It's not the hardest experiment; it's actually fairly easy for us. You know, 10 GeV lepton pairs. But it's the lowest cross section because it is basically all electromagnetic. There is no alpha strong.

- *Begun:*

It seems that RHIC measurements confirm NA49 data about suppression. Do you agree that at 20 or at 30 GeV lab energy there is place where Quark Gluon Plasma starts to appear? Or if you do not agree then what is your opinion, where does QGP appear? Because CERN discovered QGP and RHIC also discovered QGP.

- *Tannenbaum:*

Oh, the CERN results had nothing to do with what we observed, nothing at all. They just said that some hadron model didn't work and therefore there must be something else and then they said, we found J/Psi suppression which was predicted for the QGP so it must be the QGP. It's like, I'm not in NY, I'm not in Chicago, I'm not in Los Angeles, so I must be somewhere else. So if I'm somewhere else I can't be here, right? It's the same logic. But the point is that we are looking for it. We know that at CERN, at 17 GeV there is a Cronin effect. You saw that one actually, it was on the cover of Phys Rev Letters that I showed in my talk. We're also going down in energy, we still see suppression at 39 GeV. We don't see suppression at 24 GeV in Cu+Cu. But yes, we are interested to see where the particle suppression starts. Now one of the problems is, as you reduce the c.m. energy to below 20 GeV, there is no hard scattering any more. So you don't know whether the medium is gone, or the hard scattering is gone? In PHENIX we have more of an interest in seeing where the pi0 suppression starts. And we have it published down to 24 GeV and there is no suppression at 24 GeV in Cu+Cu central collisions [PHENIX, PRL 101, 162301 (2008)]

- *Csanad:*

Since the time is running out I would just like to make a comment about the v2 and the hydro models. I think it's very important that the community doesn't make the mistake again as with the RHIC HBT puzzle. A puzzle was created from some of the models not describing a measurement. And some people saw that hydro fails and does not describe the data, but there were models that described it. Same here with the v2.

- *Tannenbaum:*

All I can say is that we had this result which we knew is very interesting because there are only two calculations in the literature and they are an order of magnitude smaller than what we measure. But in fact Charles Gale was one of the guys who did that and he's one of the guys I really respect. So right before the talk at Quark Matter, I happened to go into the meeting, and he was sitting there, so I said: "Charles, you know we're going to announce we found v2 flow of direct photons at the same level as pi0s, at pt of 1 or 2 GeV". And he said "It's impossible." So that was great, but Kari Eskola, got very excited about the result and quickly did a calculation. So this field is hard! You can't really test it, like you can with these beautiful fermion-pair bose condensation states, the cold things where you can play around and change everything. No, we can't do that, so it's very hard. You have to do the experiment and the theory says: Well, if that's right, you should try this!. And that's the way we do it. It's very hard.

- *Palhares:*

I also would like to make a comment. There are also event-by-event fluctuations in hydro. There's a group that has been putting this forward in the last ten years, and they really have shown that it's important for full description and it was like ten years ago. I can show you the reference.

- *Tannenbaum:*

So why did they not find v3? Anyway, I have been studying event-by-event fluctuations since the 70's. Every ET-distribution is an event-by-event fluctuation. But the point is, as my wife always tells me: "If you're so smart, why aren't you rich?" And it's the same thing. If those hydro event-by-event guys were so smart, why didn't they calculate the harmonics that aren't there except if the initial state symmetry fluctuates? It would have saved me personal aggravation. [added in proof. The Brazalian group did show in 2009 that the ridge and the cone, i.e. v3, does appear in an event-by-event hydro calculation without jets PRL103(2009)242301, but the MIT group was the first to show it with real data, PRC 81, 054905(2010)]. Those MIT graduate students, that's what the other lesson is. That's why we need young blood in the field. Because they don't know of things that were said to be impossible, they are not weighed down by all those fancy theories in their heads or even fancy experiments, they think independently. And with a good thesis advisor they make the progress. And this is a perfect example. So you will be next.

- *G. t'Hooft:*

My answer to that question is that if you are smart you don't want to be rich, you want to be happy.

- *Tannenbaum:*

Well, she knows that I'm happy.

- *t'Hooft:*

Anyway, this brings today's discussions to an end, and I would like to thank Professor Tannenbaum for these very lively discussions.

Highlights from Gran Sasso

Lucia Votano
Laboratori Nazionali del Gran Sasso dell'Istituto Nazionale di Fisica Nucleare, Italy
E-mail: lucia.votano@lngs.infn.it

1. Introduction

The last decades have seen enormous successes in terms of knowledge and understanding of the natural phenomena in physics which stand at the crossroads among elementary particle physics, astrophysics and cosmology. An ever increasing number of physicists has started working in the astroparticle sector charmed by the prospect of getting answers to fundamental questions in particle physics and astrophysics, by studying interactions at very high energies which are still inaccessible even to the most powerful particle accelerators. We are also experiencing the growth of complexity and performances of new generation of experiments in the underground laboratories.

Technologies developed in accelerator apparatuses were first adopted; only later the search for very rare events, the need of increasing sensitivity and efficiency and the complexity of the analysis have called for the development of ever more cutting edge technologies. At present experimental apparatuses dedicated to astroparticle physics have mass and dimensions and technological complexity comparable to that used in the LHC. An overall view of the last 30-year scientific production clearly shows how the presence of astroparticle physics has increased in the worldwide most quoted scientific journals.

Astroparticle physics would not have been able to make such a massive and rapid growth without the outstanding infrastructures that only underground laboratories can offer and that turn to be essential for this kind of studies. Having planned and built such a large and well equipped laboratory as early as the late 1970s, Italy has preserved, since then, a leading role in this process.

At present the scientific motivations for such big and complex underground infrastructures are even more valid and new halls are being excavated in existing laboratories and new projects are under considerations.

Underground laboratories are the main infrastructures for astroparticle and neutrino physics as they allow to explore the highest energy scales which are inaccessible to accelerators, by searching for extremely rare phenomena. The searches for rare events like $2\beta0\nu$ or proton decay, the study of weak interactions from cosmic or neutrinos from accelerators, the detection of dark matter candidates and the nuclear astrophysics studies require low-background environments that can only be achieved in a full shielded underground environment. Thanks to the rock coverage and to the corresponding reduction in the cosmic ray flux and c.r.-spallation induced neutrons, underground laboratories provide the necessary low-background environment to investigate these processes. The main characteristics of the Gran Sasso Laboratory will be described together with an overview of its broad

Figure 1: Sketch of the underground Laboratory and the Gran Sasso massif.

scientific activities.

2. Gran Sasso National Laboratory

INFN Gran Sasso National Laboratory (LNGS) is the largest underground laboratory in the world devoted to astroparticle physics. It is one of the four INFN National Laboratories and it is a worldwide facility for scientists working in one of the twenty experiments currently set there. Located between L'Aquila and Teramo - at about 120 km far from Rome - the underground structures are on one side of the highway tunnel (10 km long) which crosses the Gran Sasso massif (A24 Teramo-Rome Highway) and consist of three huge experimental halls (each one 100 m long, 20 m large and 18 m high) linked by service tunnels, for a total volume of $\sim$180.000 m^3 and a surface of $\sim$18.000 m^2. Access to experimental halls is horizontal and it is made easier by the highway tunnel. Halls are equipped with all technical and safety plants required for the experimental activities and to ensure proper working conditions. For instance ventilation plants allow, in normal conditions, a change of air in less than 3 hours, and air radon content is kept in the 20-80 Bq/m^3 range. The 1400 metre-rock thickness above provides a cosmic ray flux reduction by one million times; moreover, the flux of neutrons in the underground halls is about thousand times less than on the surface, due to the very small amount of uranium and thorium of the Dolomite calcareous rock of the mountain. Outside, next to the highway tollgate of Assergi, an area of more than 23 acres hosts the external laboratories, the Computing Centre, the Directorate and various Offices. Presently LNGS staff consists of 90 people; besides, more than 950 scientists from 29 different Countries take part in its experimental activities.

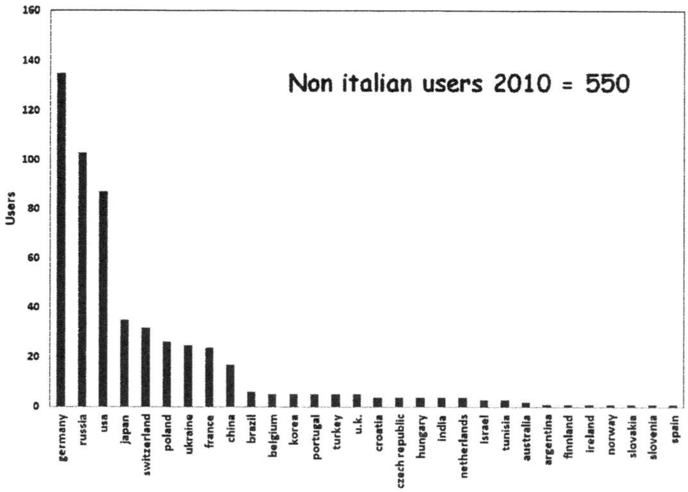

Figure 2: Country of the researchers involved in the experiments at Gran Sasso.

3. Physics at LNGS

Many of the experiments currently under way or soon to be started at LNGS are intended to search for fundamental and fascinating questions, still unanswered. What is the Dark Matter in the Universe? Do we really know the nature and the characteristics of neutrinos and what's the role they have played in the evolution of the Universe? What do we know about the interior of the Sun and stars, of the mechanisms that produce energy and how they evolve and die? What do we know about the interior of our own Earth, can we find out more through the study of neutrinos?

4. Neutrino project at Gran Sasso

The study of the intrinsic properties of neutrino is of prime interest in particle physics and one of the main research topics of the present scientific program of the Laboratory where various neutrino sources, both natural (the Sun, the stars and the Earth) and artificial (particle accelerators) are used.

4.1. Borexino

Borexino is a large unsegmented liquid scintillator detector built for real time measurement of low-energy (below 1 MeV) solar neutrinos. The inner part of the detector is a

Stainless Steel Sphere (SSS) 13.7 m in diameter that represents both the container of the scintillator and the mechanical support of the 2200 PMTs for the light collection. Within the sphere, two nylon vessels separate the scintillator volume in three shells of increasing radiopurity crucial for the sensitivity of the experiment. The SSS is enclosed within a large tank filled with ultrapure water as shielding against external background and as a Cherenkov muon tracker. The experiment can then rely on an optimal energy and spatial resolution together with a very low energy threshold (40 KeV).

The main goal of Borexino is the direct detection of the monochromatic ^7Be solar neutrinos: in 2011 Borexino has published the measurement of the ^7Be rate with an accuracy better than 5% rejecting the hypothesis of no oscillation for ^7Be solar neutrinos at 4.9σ C.L. [1]. Borexino is also able of probing a quite interesting features of solar neutrino oscillations, the transition from vacuum-dominated to matter-enhanced oscillations, proper of the MSW-LMA solution to the solar neutrino problem. The experiment has also searched for a day-night asymmetry of the ^7Be rate, showing the absence of a significant asymmetry. This result, combined with the other solar neutrino data, allows a rejection of the MSW-LOW solution without relying on the assumption of CPT symmetry in the neutrino sector.

Thanks to the extreme high radiopurity achieved, the high photon yield, and the large number of free target protons, Borexino is now aiming at the spectral study of other solar neutrino low energy components such as pep, CNO, and pp. The apparatus is a very sensitive detector for geo-neutrinos too. Geo-neutrinos are electron antineutrinos produced in β decays of ^{40}K and of several nuclides in the chains of long-lived ^{238}U and ^{232}Th present in the Earth crust. They are direct messengers of the abundances and distribution of radioactive elements within our planet.

4.2. CNGS project: OPERA and ICARUS

The CNGS project consists of an artificial neutrino beam, produced by the protons accelerator SPS of CERN and directed towards Gran Sasso. The main experiment of Gran Sasso National Laboratory devoted to CNGS neutrino detection is OPERA. The goal of the experiment is the detection of neutrino oscillations in direct appearance mode through the study of $\nu_\mu \leftrightarrow \nu_\tau$ channel.

Neutrino oscillations at the 'atmospheric scale' have been studied up to now through atmospheric and accelerator neutrinos in disappearance mode, while the direct observation of flavour transition through the detection of the corresponding lepton has never been observed before OPERA first evidence. The challenge of the OPERA experiment is the detection of tau neutrinos in the beam CNGS, originally constituted by muon neutrinos only, providing the first direct evidence of the so called 'oscillation' mechanism of these particles [2]. The ν_μ beam from CERN is optimized for the observation of ν_τ-CC interactions, the average energy being 17 GeV. With a total CNGS beam intensity of 22.5×10^{19} protons on target, about 24.300 neutrino events would be collected. The experiment should observe ~ 10 ν_τ-CC events for the present Δm^2_{23} allowed region with a background of less than one event.

The detector consists of 150.000 'bricks' made up of lead layers interleaved with nu-

Event type	detected	expected
ν_μ CC	94	98
ν NC	32	31
Total CC	94	98

Table 1: Number of collected interactions compared with number of interactions predicted ((2.6νCC + 0.86νNC) 10^{-17}/pot), in the whole energy range up to 100 GeV, corrected by fiducial volume and DAQ dead-time.

clear emulsions, historically called Emulsion Cloud Chamber (ECC). OPERA is an hybrid detector made of a veto plane followed by two identical super modules (SM) each one consisting of a target section of about 625 tons made of 75.000 bricks, and of a scintillator Target Tracker Detector (TT) to localize neutrino interactions within the target, followed by a muon spectrometer. The observation of a first ν_τ candidate event in the experiment has been reported in June 2010 based on the analysis of part of the data taken during 2008 and 2009 runs. OPERA data taking will continue at least up to 2012.

Another experiment able to detect CNGS beam is ICARUS an innovative apparatus consisting of a big mass (about 600 tons) of liquid Argon, at a temperature of -186 ^oC. In particular conditions and by means of proper devices, this liquefied gas is able to act as an extraordinary particle detector, allowing a 3D reproduction of any interactions of charged particles inside its volume. The commissioning of ICARUS was successfully completed in 2010 and in May the first CNGS neutrino events were recorded. ICARUS is now continously collecting data from CNGS. The 2010 data have been analyzed: the sample consists of 1494 CNGS triggers (4.54×10^{18} pot equal to 78 % out of whole sample) classified by visual scanning into fiducial volume of 434 tons. In table 1 the comparison between the collected and predicted interactions are reported.

Such a massive liquid argon experiment running in an underground laboratory is, so far, the most important milestone for the LAr-TPC technology towards the design of a much more massive multikiloton LAr detector with unique imaging capability and spatial/calorimetric resolutions.

4.3. Very rare decays and neutrino properties. GERDA and CUORE

Neutrino oscillation experiments have clearly shown that neutrinos do oscillate and that they are massive, though this is not enough to determine the nature of such mass (Dirac or Majorana) and to determine the absolute mass scale. In fact neutrino oscillation experiments can only measure the absolute value of the difference of the square of the neutrino masses while two different hierarchical mass arrangements of neutrino masses (Direct and Inverted), besides the obvious quasi-degenerate option, are possible. Many theories beyond the standard Model suggest a mass generation mechanism that implies a Majorana character of neutrinos. Neutrinoless double-beta decay is a process by which two neutrons in a nucleus undergo beta decay by exchanging a virtual Majorana neutrino, emitting an electron each. This would violate lepton number conservation ($\Delta L = 2$) and would necessarily require neutrinos to be Majorana particles; therefore this represents a unique tool to test this hypothesis and nowadays, thanks to the discovery of neutrino oscillations, this makes it the object of a renewed interest. The LNGS program exploits complementary approaches concerning isotopes and technique.

GERDA is designed to search for $2\beta0\nu$-decay of ^{76}Ge using high purity germanium detectors (HPGe), enriched ($\sim 85\%$) in ^{76}Ge, directly immersed in LAr which acts both as shield and as cooling medium. The cryostat is located in a stainless steel water tank providing an additional shield against external background. The installation of the experiment in Gran Sasso Laboratory has been accomplished and now the experiment is in the commissioning phase. The first non-enriched Ge detectors have been deployed for a technical run aimed to perform background evaluation.

The CUORE experiment (Cryogenic Underground Detector for Rare Events) aims to detect $2\beta0\nu$ decay of ^{130}Te by using the cryogenic bolometer detectors made of TeO_2 crystals: the energy from particle interactions is converted into heat and measured via the resulting rise in temperature. The prototype CUORICINO, already ran at LNGS, demonstrated the feasibility of the large scale experiment CUORE which will consist of a closely packed array of 988 detectors arranged in 19 towers.

As well as being protected by the Gran Sasso rock, the apparatus is protected by a series of shields, including ancient Roman lead for the utmost reduction of residual radiation in the Laboratory and to allow the detection of the very rare decay. Very recently the Laboratory has received additional 120 ingots (4 tons) from an ancient Roman ship that sunk off of the coast of Sardinia 2.000 years ago. CUORE is presently in the construction phase in the Gran Sasso National Laboratory and it is expected to start data taking in 2014.

5. Study of dark matter

Convincing experimental results (the first and most famous is the observation of the spiral galaxies rotation speed) indicate the existence in the Universe of a non-baryonic and non-relativistic massive component larger than the one observable through telescopes. It is called Dark Matter because it neither emits nor absorbs radiation and thus it is invisible to instruments sensitive to electromagnetic radiation. It is supposed to be five times more abundant than ordinary matter, which constitutes only 5% of our Universe.

Astroparticle physicists assume the hypotheses of the existence of very massive particles (with respect to protons) that constitute a widespread halo permeating our galaxy as well as others. Such particles, whether they exist or not, interact very weakly with the ordinary matter, therefore their detection could exclusively occur in apparatus shielded against cosmic rays and natural radioactivity, as in LNGS underground halls, with experiments very sensitive to the weak interactions among these particles and the ordinary matter atoms.

In the next years LNGS will be characterized by a rich experimental program for dark matter search. In fact, four experiments are aimed - in various phases and making use of different experimental techniques - to the direct detection of dark matter candidates. These experiments put Gran Sasso Laboratory at the forefront in this kind of study.

5.1. DAMA/LIBRA

DAMA/LIBRA experiment is based on the development and use of extremely radiopure NaI(Tl) crystals. The former DAMA and the present DAMA/LIBRA experiment aim to the Dark matter detection exploiting the model independent DM annual modulation signal. Since 2003 the detector makes use of 250 kg NaI(Tl) crystals. The most recent results confirm the annual modulation of the very low energy signals induced in the detector, already observed in the previous experiment DAMA of lower mass. Such modulation is identical to the one that the relative motion of the Earth through the huge amount of dark matter halo of our galaxy is supposed to cause [3].

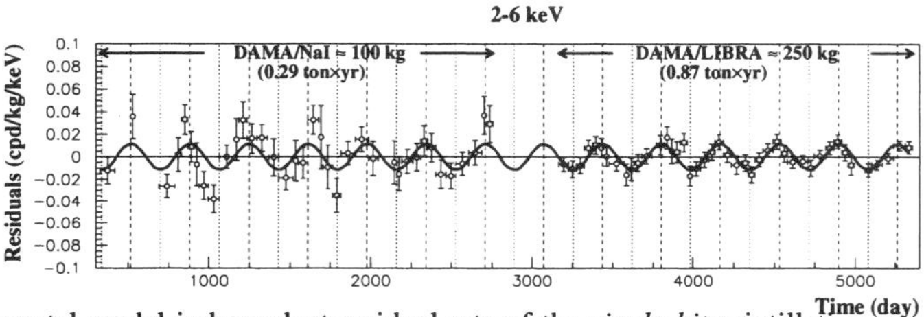

Figure 3: Experimental model-independent residual rate of the *single-hit* scintillation events, measured by DAMA/NaI over seven and by DAMA/LIBRA over six annual cycles in the (2 – 6) keV energy interval as a function of the time. The zero of the time scale is January 1st of the first year of data taking of the former DAMA/NaI experiment. The experimental points present the errors as vertical bars and the associated time bin width as horizontal bars. The superimposed curve is the cosinusoidal function behavior $A \cos \omega(t - t_0)$ with a period $T = \frac{2\pi}{\omega} = 1$ yr, with a phase $t_0 = 152.5$ day (June 2nd) and with modulation amplitude, A, equal to the central value obtained by best fit over the whole data: cumulative exposure is 1.17 ton × yr. The dashed vertical lines correspond to the maximum expected for the DM signal (June 2nd), while the dotted vertical lines correspond to the minimum. When all the three parameters are kept free, one gets from the data: $A = (0.0116 \pm 0.0013)$ cpd/kg/keV, $T = (0.999 \pm 0.002)$ yr and $t_0 = (146 \pm 7)$ day, values well in agreement with expectations for DM signal.

No systematic or side process able to simultaneously satisfy all the many peculiarities of the signature and to account for the whole measured modulation amplitude is available. In particular, Fig. 3 shows the time behaviour of the experimental residual rates of the *single-hit* events collected by DAMA/NaI and by DAMA/LIBRA in the (2–6) keV energy interval. The cumulative exposure with the former DAMA set-up is 1.17 ton × yr – corresponding to 13 annual cycles in total – which is orders of magnitude larger than the exposures typically considered in the field. Very recently great excitement in DM field produced by the CoGeNT(SUL) results of 15 months of data showing indication at 2.8σ of annual modulation in a P-type Ge-detector. DAMA and CoGeNT results are consistent with a DM particle with mass m_X in the range $\sim$10 GeV and elastic $\sigma \sim 10^{-40}$cm^2. Phase, period and amplitude look also consistent in the two experiments.

5.2. XENON100

The experiment Xenon100 is a two-phase detector (liquid and gas) time projection chamber of large mass and low background able to measure concurrently the scintillation and ionization produced by the interaction of the particles with the xenon contained in the apparatus. The apparatus contains 170 kg of xenon, 65 kg of which constitute the active part while the remaining ones act as a shield. A key feature of the detector is its ability to localize events with millimeter resolution in 3 dimensions which further allows to select only the innermost 48 Kg as ultra-low background.

In 100.9 live days of 2010 no evidence of dark matter signals were found. Three candidate events were identified in a previously well defined signal region for the expected signals, consistent with an expected background of (1.8 +/- 0.6) events [4]. This new result puts an upper limit on spin- elastic WIMP-nucleon scattering cross section i of $7.0 \times 10^{-45} cm^2$ (90 % C.L.) at a WIMP mass of 50 GeV/c^2 Xenon 100 is the current operating phase of the Xenon dark matter program at LNGS started with Xenon10.

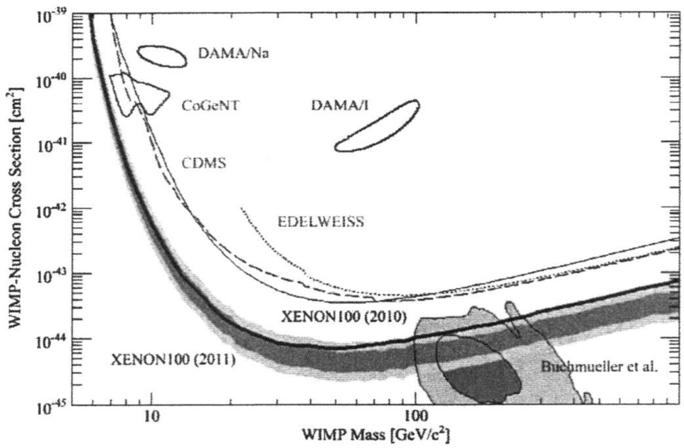

Figure 4: 2011 Results: XENON100 Dark Matter Limit (90% CL)

The experiment Xenon100 will run for at least one year and will search for the annual modulation signal observed by the DAMA/LIBRA experiment. In order to reach a lower sensitivity of about $5 \times 10^{-47} cm^2$ and/or to conform a possible detection in Xenon100, the collaboration proposes to build a detector with a total mass of 2.4 tons of LXe at LNGS.

5.3. WARP

The experiment Warp 100 is a cryogenic detector that uses two-phase argon (liquid and gas) as active volume and is based on a double detection technique, the scintillation in liquid argon that produces the primary signal and the ionization in the gaseous state. The ratio between the two different signals produced and the analysis of the primary signal shapes are means to discriminate between the signal produced by a dark matter particle and other background signals. The construction and assembly of the WArP-100 has been completed and the commissioning of the detector has been started.

5.4. CRESST

The experiment CRESST aims at the detection of WIMPs via elastic scattering off nuclei in $CaWO_4$ crystals is based on the bolometer technique with $CaWO_4$ crystals operated at a temperature of 10 mK and exploits the simultaneous detection of scintillation light and the heat resulting by the interaction of a particle with the crystals. The scintillating crystals equipped with a superconducting tungsten phase-transition thermometer for the detection of the phonons created by a particle interaction in the scintillating crystal. A small fraction of $\sim 1\%$ of the deposited energy is emitted as scintillation light, which is measured with a separate cryogenic detector, optimized for light detection.

The experiment started a new run in 2009 with 10 complete detector modules and is running stably. The analysis of the first 330 kg per day of data has started and preliminary results look promising. Furthermore research and development activity goes on in an ad hoc cryostat, meant to finalize an experiment proposal of larger mass.

6. Conclusion

The Gran Sasso National Laboratory of INFN, the largest underground laboratory in the world, holds the leadership in massive experiments with record performance and low-level background. The present scientific program of LNGS includes a very broad spectrum of competitive experiments (astroparticle, particle and nuclear physics), including world-leading ones in the fields of solar neutrinos, accelerator neutrinos, double beta decay, dark matter and nuclear astrophysics. The Gran Sasso National Laboratory aims at preserving its scientific excellence in the next years by an extensive physics program, by new experiments and by upgrades of the present ones. After the end of the CNGS program (2013-2015), underground space (OPERA and ICARUS) could be made available for new and innovative experiments.

7. References

1) G. Bellini et al., Phys. Rev. Lett. **108**(2012), 051302
2) N. Agafonova et al., Physics Letters B **691**(2010), 138-145
3) R. Barnabei et al., Nucl. Phys. B **307**(2011), 212-213
4) E. Aprile et al., Phys. Rev. Lett. **107**(2011), 131302

CHAIRMAN: L. VOTANO

Scientific Secretaries: E. Krylov, A. Ter-Minasyan

DISCUSSION

- *Alba:*

How is it possible to distinguish different types of neutrinos: solar, atmospheric?

- *Votano:*

Solar neutrinos have low energies and atmospheric or CNGS neutrinos have higher levels of energy and different energy spectra. Each experiment is optimized for different ranges of energies.

- *Alba:*

What is the plan to find out which neutrinos are Dirac or Majorana?

- *Votano:*

We are looking for neutrino-less double beta decays. This decay is only possible if neutrinos are Majorana.

- *Dunin-Barkovski:*

What is the lifetime of the OPERA experiment and how many tau neutrino events do you expect ultimately?

- *Votano:*

The OPERA experiment has got a significant number of neutrinos from CERN starting only from 2008. In 5 running years at the CNGS nominal intensity, the experiment should see approximately 10 tau neutrinos with less than 1 event of background. The present CERN plan foresees a long shutdown starting in 2013. If OPERA stops definitively the data taking, it will require approximately two additional years to accomplish the data analysis.

- *Burda:*

What is the key idea of the CUORE experiment? How will it be able to probe the neutrino mass hierarchy? How exactly do we use bolometers in such measurements?

- *Votano:*

CUORE is the latest development of the "TeO_2bolometer" technique: the energy from particle interactions is converted into heat and measured via the resulting rise in temperature. 0vDBD is able to test neutrino mass, the rate of this process is related to the absolute neutrino mass.Various experiments use different decaying isotopes.

- *Inguglia:*

What are the hits about a new family of neutrinos? Is that sterile or 4th generation?

- *Votano:*

Hints of sterile neutrinos come from MiniBoone, Minos and a re-analysis of reactors experiments.

- *Inguglia:*

In 2010 you expected 0.5 events of neutrino oscillations but you found 1…

- *Votano:*

The OPERA paper has been published in 2010 using the 2008 and 2009 data.The data analysis in OPERA is quite complicated because of the use of the nuclear emulsions. Now we are analyzing the 2010 data.There is approximately one year of delay from the data taking.

- *Wen:*

What is the new idea for Gran Sasso to solve sterile neutrino puzzle? Gran Sasso has a lot of low background or Ultra Low Background detectors.

- *Votano:*

Two ideas related to Gran Sasso.The first idea foresees the ICARUS detector at CERN at appropriate distance on the CERN PS neutrino beam, the second one the use of mega Curie neutrino and antineutrino sources inside the Borexino detector.

- *Wen:*

There are detectors in Grand Sasso with extremely low background. Could you briefly summarize their unique advantages?

- *Votano:*

The background in the Gran Sasso Laboratory is quite low due to its depth and the very low level of radioactivity of the rocks. Furthermore LNGS has its own low background test-facility for material selection of the detectors. Each experiment uses also passive and active shields to suppress background, for example CUORE usesalso ancient roman lead as shield.

- *Dazasti:*

Do you know the cause of the annual modulation seen in the DAMA experiments?Can they be just fake signals according to atmospheric muons?

- *Votano:*

First of all, we know a lot about annual modulations of cosmic rays to distinguish it. Secondly, in DAMA they find modulation only on single hit and low energies. The phasealso should be different and if this modulation is due to the annual modulation of cosmic rays, it must be on all energies and in multihit signals. All possible and known sources of this effect have beencarefully studied inside the collaboration: there is no obvious explanation.

- *Dazasti:*

Can we install experiments in the south hemispheric to reject the seasonal effects?

- *Votano:*

It is a good idea. But it should be another group with another technology.

- *Manoni:*

It is due to that fact that DAMA/Nal and DAMA/LIBRA were using different crystals as targets in the detector?

- *Votano:*

DAMA is performing research with different targets.

- *Biassoni:*

Why is the limit of Cogent in DAMA/LIBRA not overlapping? Does that mean that parameters feat of experiments are different or do you just try to interpret data from different experimental setup?

- *Votano:*

The result of Cogent has the same phase and period of DAMA as it is observed also at low energies. The CDMS detector has a technology similar to Cogent, but the threshold is different.

- *de With*:

If a fourth neutrino would exist? Which one of the detectors at Gran Sasso could see it?

- *Votano:*

No experiment is optimized for this. ICARUS could try to say something.

- *Palhares:*

Recently, an anti-neutrino burst was proposed as a signal of quark matter formation in supernovae explosions: would Gran Sasso be able to measure it?

- *Votano:*

In Gran Sasso LVD is the experiment devoted to the detection of bursts in supernovae explosions, but also Borexino and ICARUS can detect them. These experiments cannot only detect bursts but also measure the timing and dynamics of the explosion.

- *Martinez:*

What is the isospin violating in elastic dark matter scattering? And how can we reconcile the different experiments?

- *Votano:*

The hypothesis is an isospin-violating inelastic scattering; it means that the scattering depends on the different proton-neutron combination of the targets.

On combinatorial expansion of the conformal blocks arising from AGT conjecture

Vasyl A. Alba[1,2,3,4,5]

[1] *Landau Institute for Theoretical Physics, 142432 Chernogolovka, Russia*

[2] *Bogolyubov Institute for Theoretical Physics NASU, 03680 Kyiv, Ukraine*

[3] *Department of General and Applied Physics, Moscow Institute of Physics and Technology, 141700 Dolgoprudny, Russia*

[4] *Institute for Theoretical and Experimental Physics, 117218 Moscow, Russia*

[5] *Departement of Physics, Princeton University, Jadwin Hall, 08544 Princeton, NJ, USA*

Abstract

In their recent paper [1] Alday, Gaiotto and Tachikawa proposed a relation between $\mathcal{N} = 2$ four-dimensional supersymmetric gauge theories and two-dimensional conformal field theories. As part of their conjecture they gave an explicit combinatorial formula for the expansion of the conformal blocks inspired from the exact form of instanton part of the Nekrasov partition function. In this paper we study the origin of such an expansion from a CFT point of view. We consider the algebra $\mathcal{A} = \mathsf{Vir}\otimes\mathcal{H}$ which is the tensor product of mutually commuting Virasoro and Heisenberg algebras and discover the special orthogonal basis of states in the highest weight representations of $\mathcal{A}$. The matrix elements of primary fields in this basis have a very simple factorized form and coincide with the function called Z_{bif} appearing in the instanton counting literature. Having such a simple basis, the problem of computation of the conformal blocks simplifies drastically and can be shown to lead to the expansion proposed in [1]. This paper is review of the work [28].

1 Introduction

The bootstrap approach to two-dimensional conformal field theory was suggested in the seminal paper [2] by Belavin, Polyakov and Zamolodchikov. Their main idea was to use simultaneously the conformal symmetry of the theory and the hypothesis about the operator algebra of local fields [3]. Namely, if one supposes the existence of the complete set of local fields $\{\mathcal{O}_k(\xi)\}$ in the theory then the completeness of this set is equivalent to the operator algebra (OPE)

$$\mathcal{O}_i(\xi)\mathcal{O}_j(0) = \sum_k C_{ij}^k(\xi)\mathcal{O}_k(0). \tag{1.1}$$

The structure constants $C_{ij}^{k}(\xi)$ are some single-valued functions which are subject to the infinite system of equations following from the condition of the associativity of the operator algebra (1.1). In general, this system of equations is rather complicated to be solved exactly and the OPE would be of limited usefulness. However, in two-dimensional CFT one can proceed further since the conformal group is infinite-dimensional in this case and it implies the strong restriction on the possible form of the structure constants $C_{ij}^{k}(\xi)$. One can show that the complete set of fields $\{\mathcal{O}_k(\xi)\}$ decomposes in this case into direct sum of conformal families

$$\{\mathcal{O}_k(\xi)\} = \sum_n [\Phi_n]. \tag{1.2}$$

The ancestor of each family Φ_n is called primary field. It transforms in the simple manner

$$\Phi_n(z,\bar{z}) \longrightarrow \left(\frac{dw}{dz}\right)^{\Delta_n} \left(\frac{d\bar{w}}{d\bar{z}}\right)^{\bar{\Delta}_n} \Phi_n(w,\bar{w}), \tag{1.3}$$

under the conformal transformations

$$z \to w(z), \quad \bar{z} \to \bar{w}(\bar{z}).$$

The quantum numbers Δ_n and $\bar{\Delta}_n$ are called conformal dimensions. Other representatives of the conformal family $[\Phi_n]$ are usually referred as descendant fields. Their conformal dimensions constitute an infinite integer sequence

$$\Delta_n^{(k)} = \Delta_n + k, \qquad \bar{\Delta}_n^{(\bar{k})} = \bar{\Delta}_n + \bar{k},$$

and each conformal family corresponds to some highest weight representation of the conformal group. In two dimensions the conformal group is the tensor product of holomorphic and antiholomorphic Virasoro algebras

$$\begin{aligned} [L_n, L_m] &= (n-m)L_{n+m} + \frac{c}{12}(n^3-n)\,\delta_{n+m,0}, \\ [\bar{L}_n, \bar{L}_m] &= (n-m)\bar{L}_{n+m} + \frac{c}{12}(n^3-n)\,\delta_{n+m,0}, \end{aligned} \tag{1.4}$$

and hence the conformal family is a tensor product $[\Phi_n] = \pi_n \otimes \bar{\pi}_n$ of two Verma modules over the Virasoro algebra. The parameter c in (1.4) is an important characteristic of CFT called central charge. Moreover, it can be shown that all the structure constants $C_{ij}^{k}(\xi)$ can be computed in terms of structure constants $\mathbb{C}_{ij}^{k}$ of primary fields [2].

This simple structure of the space of fields in two-dimensional CFT leads to the introduction of the notion of the *conformal blocks*. They represent holomorphic contributions to the multi-point correlation function of primary fields

$$\langle \Phi_1(z_1,\bar{z}_1)\ldots\Phi_n(z_n,\bar{z}_n)\rangle \tag{1.5}$$

picking given conformal families $\tilde{\Delta}_j$ in the intermediate channels. It is convenient to represent the $n-$point conformal block by the following picture [2][1]

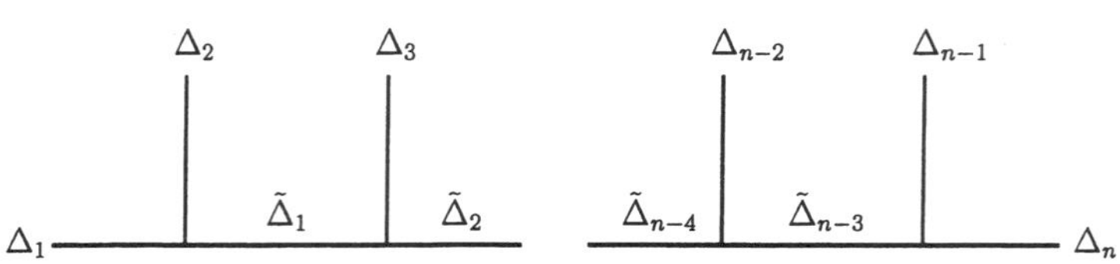

(1.6)

[1]Throughout this paper we will consider only conformal blocks on a sphere i.e. on a surface of genus 0.

which encodes the way how the OPE in (1.5) is performed. The $n-$point conformal block (1.6) is a function of holomorphic coordinates $z_1, \ldots, z_n$, of external dimensions $\Delta_1, \ldots, \Delta_n$, intermediate dimensions $\tilde{\Delta}_1, \ldots, \tilde{\Delta}_{n-3}$ and the central charge c. To be more specific it is convenient to use the projective invariance and fix $z_1 = 0$, $z_{n-1} = 1$ and $z_n = \infty$. It is also convenient to choose

$$z_{i+1} = q_i q_{i+1} \cdots q_{n-3} \quad \text{for} \quad 1 \leq i \leq n-3,$$

then the conformal block corresponding to the picture (1.6) is a power series expansion [2]

$$\mathcal{F}(q|\Delta_i, \tilde{\Delta}_j, c) = 1 + \sum_{\vec{k}} q_1^{k_1} q_2^{k_2} \ldots q_{n-3}^{k_{n-3}} \mathcal{F}_{\vec{k}}(\Delta_i, \tilde{\Delta}_j, c), \tag{1.7}$$

where sum goes over all set of positive integers $\vec{k} = (k_1, \ldots, k_{n-3})$ and the coefficients $\mathcal{F}_{\vec{k}}(\Delta_i, \tilde{\Delta}_j, c)$ are some rational functions of Δ_i, $\tilde{\Delta}_j$ and the central charge c which are completely determined (in principle) by the conformal symmetry [2]. The idea is to "cut" all the intermediate necks of the conformal block (1.6) and to "insert" the complete set of states [4]. Then the problem of computing the coefficients $\mathcal{F}_{\vec{k}}(\Delta_i, \tilde{\Delta}_j, c)$ in (1.7) is equivalent to the problem of computing the normalized matrix elements of arbitrary primary field between CFT states

$$\frac{\langle i| L_{k_1'} \ldots L_{k_m'} \Phi_k(1)\, L_{-k_n} \ldots L_{-k_1} |j\rangle}{\langle i| \Phi_k(1) |j\rangle}. \tag{1.8}$$

For any two given sets of integers $(k_1, \ldots, k_n)$ and $(k_1', \ldots, k_m')$ it is a matter of algebraic manipulations to compute the matrix element (1.8) which is in fact some polynomial of conformal dimensions Δ_i, Δ_j and Δ_k. However, this computation becomes tedious for higher levels. It can be facilitated if one distinguishes between "conformal fields" and "derivatives" [5], but still it is rather difficult to find a closed form expression for all the matrix elements (1.8). Explicit expressions obtained by symbolic algebra program for lower levels show no signs of general pattern. So, it would be desirable to have more efficient algorithm for computation of the expansion (1.7). One possibility known as the recursive procedure was suggested by Alyosha Zamolodchikov in [6]. The method developed in [6] was originally applied to four-point conformal block on a sphere and although the generalization of this method for more general case seems to be possible there is no much available results in the existing literature[2].

A renewed interest in conformal field theory appeared after the paper [1] of Alday, Gaiotto and Tachikawa where they proposed a relation between two-dimensional conformal field theories and $\mathcal{N} = 2$ four-dimensional supersymmetric gauge theories (it is usually referred as the AGT conjecture). In particular, they related the $n-$point conformal block on a sphere (1.7) with the instanton part of the Nekrasov partition function [8–10] for the gauge theory with gauge group $U(2)_1 \otimes \cdots \otimes U(2)_{n-3}$ and with special matter content which is either in (anti-)fundamental representation of $U(2)_1$ or $U(2)_{n-3}$ or in bifundamental representation of $U(2)_i \otimes U(2)_{i+1}$ for $i = 1, \ldots, n-2$. Theories of such a type are usually called linear quiver gauge theories [11–14]. In order to formulate the result of [1] we define the new function

$$Z(q|\Delta_i, \tilde{\Delta}_j, c) \stackrel{\text{def}}{=} \prod_{k=1}^{n-3} \prod_{m=k}^{n-3} (1 - q_k \cdots q_m)^{2\alpha_{k+1}(Q-\alpha_{m+2})} \mathcal{F}(q|\Delta_i, \tilde{\Delta}_j, c), \tag{1.9}$$

where the parameters α_k and Q were introduced to parametrize the external conformal dimensions Δ_k and the central charge c as

$$\Delta_k = \alpha_k(Q - \alpha_k), \quad c = 1 + 6Q^2, \qquad Q = b + \frac{1}{b}. \tag{1.10}$$

[2]Recently, the recursive procedure of [6] was generalized to the case of one-point conformal block on a torus in [7].

It was conjectured in [1] that the function $Z(q|\Delta_i, \tilde{\Delta}_j, c)$ defined by (1.9) possesses the nice expansion

$$Z(q|\Delta_i, \tilde{\Delta}_j, c) = 1 + \sum_{\vec{k}} q_1^{k_1} q_2^{k_2} \dots q_{n-3}^{k_{n-3}} Z_{\vec{k}}(\Delta_i, \tilde{\Delta}_j, c), \tag{1.11}$$

with the coefficients $Z_{\vec{k}}(\Delta_i, \tilde{\Delta}_j, c)$ having explicit combinatorial expression

$$\begin{aligned} Z_{\vec{k}}(\Delta_i, \tilde{\Delta}_j, c) = \sum_{\vec{\lambda}_1, \dots, \vec{\lambda}_{n-3}} & Z_{\mathsf{vec}}(P_1, \vec{\lambda}_1) \dots Z_{\mathsf{vec}}(P_{n-3}, \vec{\lambda}_{n-3}) \times \\ & \times Z_{\mathsf{bif}}(\alpha_2|P, \varnothing; P_1, \vec{\lambda}_1) Z_{\mathsf{bif}}(\alpha_3|P_1, \vec{\lambda}_1; P_2, \vec{\lambda}_2) Z_{\mathsf{bif}}(\alpha_4|P_2, \vec{\lambda}_2; P_3, \vec{\lambda}_3) \times \dots \\ & \dots \times Z_{\mathsf{bif}}(\alpha_{n-3}|P_{n-5}, \vec{\lambda}_{n-5}; P_{n-4}, \vec{\lambda}_{n-4}) \times \\ & \times Z_{\mathsf{bif}}(\alpha_{n-2}|P_{n-4}, \vec{\lambda}_{n-4}; P_{n-3}, \vec{\lambda}_{n-3}) Z_{\mathsf{bif}}(\alpha_{n-1}|P_{n-3}, \vec{\lambda}_{n-3}; \hat{P}, \varnothing). \end{aligned} \tag{1.12}$$

The sum in (1.12) goes over the pairs $\vec{\lambda} = (\lambda_1, \lambda_2)$ of Young diagrams (bipartitions) such that $|\vec{\lambda}_j| = k_j$ where $|\vec{\lambda}_j|$ is the total number of boxes in the pair $\vec{\lambda}_j$. The parameters P, $\hat{P}$ and P_j in (1.12) are related with the external dimensions Δ_1, Δ_n and intermediate dimensions $\tilde{\Delta}_j$ by

$$\Delta_1 = \frac{Q^2}{4} - P^2, \quad \Delta_n = \frac{Q^2}{4} - \hat{P}^2 \quad \text{and} \quad \tilde{\Delta}_j = \frac{Q^2}{4} - P_j^2.$$

The explicit form of the functions Z_{bif} and Z_{vec} were derived in [15–17]. The function Z_{bif} is given by

$$\begin{aligned} Z_{\mathsf{bif}}(\alpha|P', \vec{\mu}; P, \vec{\lambda}) = \prod_{i,j=1}^{2} \prod_{s \in \lambda_i} \left(Q - E_{\lambda_i, \mu_j}(P_i - P'_j|s) - \alpha\right) \times \\ \times \prod_{t \in \mu_j} \left(E_{\mu_j, \lambda_i}(P'_j - P_i|t) - \alpha\right), \end{aligned} \tag{1.13}$$

where $\vec{P} = (P, -P)$, $\vec{P}' = (P', -P')$ and

$$E_{\lambda,\mu}(P|s) = P - b\, l_\mu(s) + b^{-1}(a_\lambda(s) + 1). \tag{1.13a}$$

In (1.13a) $a_\lambda(s)$ and $l_\mu(s)$ are correspondingly the arm length of the square s in the partition λ and the leg length of the square s in the partition μ. We choose the English convention to draw partitions $\lambda = (\lambda_1 \geq \lambda_2 \geq \dots)$. For example the partition $\lambda = (4, 3, 2, 1, 1)$ is drawn as follows

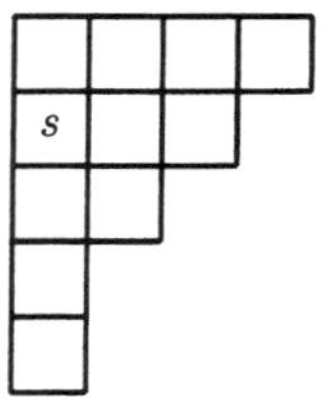

where the number of filled circles is equal to the arm length $a_\lambda(s)$ while the number of unfilled ones to the leg length $l_\lambda(s)$. We note that in (1.13a) the arm and leg lengths are computed with respect to different partitions λ and μ and the square s always belongs to the partition λ which is the first subscript of $E_{\lambda,\mu}(P|s)$. The function Z_{vec} is defined by

$$Z_{\mathsf{vec}}(P,\vec{\lambda}) = \frac{1}{Z_{\mathsf{bif}}(0|P,\vec{\lambda};P,\vec{\lambda})}. \tag{1.14}$$

The AGT conjecture [1] attracts an attention of both specialists in gauge theories and conformal field theory. In particular, we note that the combinatorial expansion (1.11)-(1.12) was rather unexpected from CFT point of view. In a series of papers [18–23] this expansion was checked for some particular cases against CFT predictions. However, general mathematically rigorous proof of (1.11)-(1.12) was lacking.

The factor in the r.h.s. in (1.9) was called in [1] the "$U(1)$ factor" which presumably corresponds to the "stripping off" the $U(1)$ part from the Nekrasov partition function which is computed for $U(2)$ groups, rather than $SU(2)$. It looks natural to express the "$U(1)$ factor" in terms of correlation function of chiral vertex operators of some free bosonic field. We found that the introduction of the auxiliary bosonic field is not only the convenient way to represent the "$U(1)$ factor" but it plays a crucial rôle in the whole construction. We consider the algebra $\mathcal{A} = \mathsf{Vir}\otimes\mathcal{H}$ which is the tensor product of Virasoro and Heisenberg algebras and construct the special orthogonal basis in the highest weight representations of this algebra. The matrix elements of primary fields between any two states from this basis have particularly simple form which coincides with Z_{bif} defined above. The norm of these states is equal to $1/Z_{\mathsf{vec}}$. It is clear that such a basis essentially leads to the expansion (1.11)–(1.12). Similar idea was proposed by Alday and Tachikawa in [24]. In this paper we prove the existence and uniqueness of such a basis.

2 Special basis of states in Vir$\otimes\mathcal{H}$

We consider the algebra $\mathcal{A} = \mathsf{Vir}\otimes\mathcal{H}$ which is the tensor product of Virasoro and Heisenberg algebras with commutation relations

$$\begin{aligned} &[L_n, L_m] = (n-m)L_{n+m} + \frac{c}{12}(n^3-n)\,\delta_{n+m,0}, \\ &[a_n, a_m] = \frac{n}{2}\,\delta_{n+m,0}, \qquad [L_n, a_m] = 0. \end{aligned} \tag{2.1}$$

We will parametrize the central charge c of Virasoro algebra in a Liouville manner as

$$c = 1 + 6Q^2, \qquad \text{where} \quad Q = b + \frac{1}{b}, \tag{2.2}$$

and define the primary field V_α as

$$V_\alpha \stackrel{\text{def}}{=} \mathcal{V}_\alpha \cdot V_\alpha^{\mathrm{L}}, \tag{2.3}$$

where V_α^{L} is the primary field of Virasoro algebra with conformal dimension $\Delta(\alpha) = \alpha(Q-\alpha)$ and $\mathcal{V}_\alpha$ is a free exponential[3]

$$\mathcal{V}_\alpha = e^{2(\alpha-Q)\varphi_-}e^{2\alpha\varphi_+}, \tag{2.4}$$

[3]We note that this "strange" form of the vertex operator (2.4) was suggested in somewhat different but related context by Carlsson and Okounkov in [25].

with $\varphi_+(z) = i\sum_{n>0}\frac{a_n}{n}z^{-n}$ and $\varphi_-(z) = i\sum_{n<0}\frac{a_n}{n}z^{-n}$. The commutation relations of the primary field $V_\alpha(z)$ with generators L_m and a_n can be summarized as

$$\begin{aligned}
&[L_m, V_\alpha^{\mathrm{L}}(z)] = \left(z^{m+1}\partial_z + (m+1)\Delta(\alpha)z^m\right)V_\alpha^{\mathrm{L}}(z),\\
&[a_n, \mathcal{V}_\alpha(z)] = -i\alpha z^n \mathcal{V}_\alpha(z), \qquad \text{for} \quad n<0,\\
&[a_n, \mathcal{V}_\alpha(z)] = i(Q-\alpha)z^n \mathcal{V}_\alpha(z), \qquad \text{for} \quad n>0,\\
&[L_m, \mathcal{V}_\alpha(z)] = [a_n, V_\alpha^{\mathrm{L}}(z)] = 0.
\end{aligned} \tag{2.5}$$

There is a natural basis in the space of states

$$a_{-l_m}\dots a_{-l_1}L_{-k_n}\dots L_{-k_1}|P\rangle, \quad k_1 \geq k_2 \geq \dots \geq k_n, \quad l_1 \geq l_2 \geq \dots \geq l_m, \tag{2.6}$$

where P parametrizes the Virasoro conformal dimension as $\Delta(P) = \frac{Q^2}{4} - P^2$ and $|P\rangle$ is the vacuum state which is defined by

$$L_n|P\rangle = a_n|P\rangle = 0, \quad \text{for} \quad n>0, \qquad L_0|P\rangle = \Delta(P)|P\rangle, \qquad \langle P|P\rangle = 1.$$

The matrix elements[4]

$$\frac{\langle P'|L_{k_1'}\dots L_{k_{n'}'}a_{l_1'}\dots a_{l_{m'}'}\,V_\alpha(1)\,a_{-l_m}\dots a_{-l_1}L_{-k_n}\dots L_{-k_1}|P\rangle}{\langle P'|V_\alpha(1)|P\rangle}, \tag{2.7}$$

which are some polynomials in α, P and P' can be computed using commutation rules (2.5) and the explicit form of the coordinate dependence of the matrix element[5]

$$\langle P'|V_\alpha^L(z)|P\rangle \sim z^{P^2-P'^2-\Delta(\alpha)}. \tag{2.8}$$

We note that the states (2.6) are the eigenstates of the operator $L_0 + 2\sum_{k>0}a_{-k}a_k$ with the eigenvalues given by

$$\Delta^{(k+l)}(P) \overset{\text{def}}{=} \Delta(P) + k + l, \quad k = \sum_{i=1}^{n} k_i, \quad l = \sum_{j=1}^{m} l_j. \tag{2.9}$$

For general values of the momenta P the module $|P\rangle$ is irreducible and the number of states with given value of $\Delta^{(N)}(P)$ is equal to the number of pairs of Young diagrams $\vec{\lambda} = (\lambda_1, \lambda_2)$ with $|\vec{\lambda}| = N$. So, it is natural to define

$$\hat{a}_{-\lambda_1}\hat{L}_{-\lambda_2}|P\rangle \overset{\text{def}}{=} a_{-l_m}\dots a_{-l_1}L_{-k_n}\dots L_{-k_1}|P\rangle, \tag{2.10}$$

where $\lambda_1 = (l_1, \dots, l_m)$, $\lambda_2 = (k_1, \dots, k_n)$. We find it convenient to use different basis $|P\rangle_{\vec{\lambda}}$ instead of using naive one (2.10)

$$|P\rangle_{\vec{\lambda}} = \sum_{|\vec{\mu}|=|\vec{\lambda}|} C_{\vec{\lambda}}^{\mu_1,\mu_2}(P)\,\hat{a}_{-\mu_1}\hat{L}_{-\mu_2}|P\rangle, \tag{2.11}$$

where the sum goes over all pairs $\vec{\mu} = (\mu_1, \mu_2)$ of Young diagrams such that $|\vec{\mu}| = |\vec{\lambda}|$ and $C_{\vec{\lambda}}^{\mu_1,\mu_2}(P)$ are some unknown coefficients. The meaning of pairs $\vec{\lambda}$ in (2.11) is different from one in (2.10) and will become clear below. We stress that the conjugation in the algebra $\mathcal{A}$ is defined as

$$(L_{-k_n}\dots L_{-k_1})^+ = L_{k_1}\dots L_{k_n}, \qquad (a_{-n})^+ = a_n, \tag{2.12}$$

[4]Below we will hide the dependence on the insertion point when possible. So, in any expression below V_α means $V_\alpha(1)$.
[5]The matrix element (2.7) is recovered in the limit $z \to 1$.

and the conjugation of the state $|P\rangle_{\vec{\lambda}}$ does not involve complex conjugation of its coefficients, i.e. for $|P\rangle_{\vec{\lambda}}$ given by (2.11) we define conjugated state ${}_{\vec{\lambda}}\langle P|$ by

$$ {}_{\vec{\lambda}}\langle P| = \sum_{|\vec{\mu}|=|\vec{\lambda}|} C_{\vec{\lambda}}^{\mu_1,\mu_2}(P)\,\langle P|(\hat{a}_{-\mu_1})^{+}(\hat{L}_{-\mu_2})^{+}. \tag{2.13}$$

As it will be shown below, all the coefficients $C_{\vec{\lambda}}^{\mu_1,\mu_2}(P)$ in (2.11) are uniquely determined from the requirement that the matrix elements of primary field (2.3) between any two states (2.11) coincide with Z_{bif} defined above (equation (1.13)). Here we arrive to the proposition which is the main result of our paper:

Proposition 2.1 There exists unique orthogonal basis $|P\rangle_{\vec{\lambda}}$ such that

$$\frac{{}_{\vec{\mu}}\langle P'|V_\alpha|P\rangle_{\vec{\lambda}}}{\langle P'|V_\alpha|P\rangle} = Z_{\text{bif}}(\alpha|P',\vec{\mu};P,\vec{\lambda}). \tag{2.14}$$

The proof one can find in [28].

3 Sketch of the proof

Proposition 3.1 The matrix elements between the states $|P\rangle_{\lambda,\varnothing}$ and ${}_{\mu,\varnothing}\langle P'|$ defined by

$$|P\rangle_{\lambda,\varnothing} = \Omega_\lambda(P)\,\mathbf{J}_\lambda^{(1/g)}(x)|P\rangle, \qquad {}_{\mu,\varnothing}\langle P'| = \Omega_\mu(P')\,\langle P'|\,\mathbf{J}_\mu^{(1/g)}(y), \tag{3.1}$$

are given by $Z_{\text{bif}}(\alpha|P',(\mu,\varnothing);P,(\lambda,\varnothing))$. Here $g=-b^2$,

$$a_{-k}-c_{-k} = -ib\,p_k(x), \qquad a_k+c_k = -ib\,p_k(y),$$

with $p_k(x)$ being k-th power sum symmetric polynomial $p_k(x)=\sum_j x_j^k$ and $\mathbf{J}_\lambda^{(1/g)}(x)$ is the Jack polynomial associated with the Young diagram λ normalized as ("integral form" normalization [26])

$$\mathbf{J}_\lambda^{(1/g)}(x) = |\lambda|!\,m_{[1,\dots,1]}(x)+\dots,$$

where $m_{[\nu_1,\dots,\nu_n]}(x)$ is the monomial symmetric polynomial.

In (3.1) the factor $\Omega_\lambda(P)$ is defined by

$$\Omega_\lambda(P) = (-b)^{|\lambda|}\prod_{(i,j)\in\lambda}(2P+ib+jb^{-1}), \tag{3.2}$$

index i runs vertically and j runs horizontally over the diagram λ.

Proposition 3.2 Let $\lambda=(\lambda_1,\lambda_2,\dots)$ a partition and $|P\rangle_{\lambda,\varnothing}$ the state defined by (3.1) then the state $|P=P_{m,n}\rangle_{\lambda,\varnothing}$ for $(m,n)\in\lambda$ has a factorized form

$$|P_{m,n}\rangle_{\lambda,\varnothing} = (-1)^{mn}X_\lambda^{(m,n)}\,D_{m,n}|P_{m,n}\rangle \quad \text{for} \quad (m,n)\in\lambda, \tag{3.3}$$

where

$$X_\lambda^{(m,n)} = \sum_{|\vec{\sigma}|=|\lambda|-mn} C_{\lambda;\vec{\sigma}}^{(m,n)}\,\hat{a}_{-\sigma_1}\hat{L}_{-\sigma_2},$$

is the operator which satisfies

$$\frac{{}_{\mu,\varnothing}\langle P'|V_\alpha X_\lambda^{(m,n)}|P_{m,-n}\rangle}{\langle P'|V_\alpha|P_{m,-n}\rangle} = Z_{\mathsf{bif}}(\alpha|P',(\mu,\varnothing);P_{m,-n},(\rho,\nu)), \tag{3.4}$$

and the pair of partitions (ρ,ν) is defined as $\rho=(\lambda_1-n,\ldots,\lambda_m-n)$ and $\nu=(\lambda_{m+1},\lambda_{m+2},\ldots)$.

Proposition 3.3 For any pair of Young diagrams (ρ,ν) there exists unique operator

$$\mathbf{X}_{\rho,\nu}(P) = \sum_{|\vec{\sigma}|=|\rho|+|\nu|} C^{\vec{\sigma}}_{\rho,\nu}(P)\,\hat{a}_{-\sigma_1}\hat{L}_{-\sigma_2}, \tag{3.5}$$

such that

$$\frac{{}_{\mu,\varnothing}\langle P'|V_\alpha \mathbf{X}_{\rho,\nu}(P)|P\rangle}{\langle P'|V_\alpha|P\rangle} = Z_{\mathsf{bif}}(\alpha|P',(\mu,\varnothing);P,(\rho,\nu)), \quad \forall\,\mu \tag{3.6a}$$

where ${}_{\mu,\varnothing}\langle P'|$ is defined by (3.1). Moreover, for $m\geq l(\rho)$, $n\geq \nu_1-\rho_m$

$$\mathbf{X}_{\rho,\nu}(P_{m,-n}) = X_\lambda^{(m,n)}, \tag{3.6b}$$

where $\lambda=(\rho_1+n,\ldots,\rho_m+n,\nu_1,\nu_2,\ldots)$ and $X_\lambda^{(m,n)}$ is the operator defined in Proposition 3.2.

Proposition 3.4 For any two pairs of Young diagrams $\vec{\lambda}=(\lambda_1,\lambda_2)$ and $\vec{\mu}=(\mu_1,\mu_2)$

$$\frac{\langle P'|\mathbf{X}^{+}_{\vec{\mu}}(P')\,V_\alpha\,\mathbf{X}_{\vec{\lambda}}(P)|P\rangle}{\langle P'|V_\alpha|P\rangle} = Z_{\mathsf{bif}}(\alpha|P',\vec{\mu};P,\vec{\lambda}). \tag{3.7}$$

Proposition 3.5 The basis of states $\mathbf{X}_{\vec{\lambda}}(P)|P\rangle$ defined above is the unique basis satisfying

$$\frac{\langle P'|\mathbf{X}^{+}_{\vec{\mu}}(P')\,V_\alpha\,\mathbf{X}_{\vec{\lambda}}(P)|P\rangle}{\langle P'|V_\alpha|P\rangle} = Z_{\mathsf{bif}}(\alpha|P',\vec{\mu};P,\vec{\lambda}). \tag{3.8}$$

Acknowledgments

The idea about the existence of an orthogonal basis diagonalizing an infinite system of Integrals of Motion in algebra $\mathcal{A}$ which can explain the AGT conjecture was proposed by Boris Feigin in a more general context. We thank him for sharing his insights with us.

The authors are grateful to Alexander Belavin, Mikhail Bershtein, Sergei Parkhomenko, Yaroslav Pugai and Jun'ichi Shiraishi for numerous discussions. The greatest acknowledgment is to Lesha Litvinov whose great intelligent and genius led us to final result. The research of V. A. is also supported by RFBR grant 10-02-00499 and Goskontrakt No. 14.740.11.0081.

References

[1] L. F. Alday, D. Gaiotto, and Y. Tachikawa, *Liouville Correlation Functions from Four-dimensional Gauge Theories*, *Lett. Math. Phys.* **91** (2010) 167–197, [arXiv:0906.3219].

[2] A. A. Belavin, A. M. Polyakov, and A. B. Zamolodchikov, *Infinite conformal symmetry in two-dimensional quantum field theory*, *Nucl. Phys.* **B241** (1984) 333–380.

[3] A. M. Polyakov, *Nonhamiltonian approach to conformal quantum field theory*, Zh. Eksp. Teor. Fiz. **66** (1974) 23–42.

[4] G. W. Moore and N. Seiberg, *Classical and Quantum Conformal Field Theory*, *Commun. Math. Phys.* **123** (1989) 177.

[5] V. P. Yurov and A. B. Zamolodchikov, *Truncated conformal space approach to scaling Lee-Yang model*, *Int. J. Mod. Phys.* **A5** (1990) 3221–3246.

[6] Al. B. Zamolodchikov, *Conformal symmetry in two-dimensions: an explicit reccurence formula for the conformal partial wave amplitude*, *Commun. Math. Phys.* **96** (1984) 419–422.

[7] L. Hadasz, Z. Jaskolski, and P. Suchanek, *Recursive representation of the torus 1-point conformal block*, *JHEP* **01** (2010) 063, [arXiv:0911.2353].

[8] G. W. Moore, N. Nekrasov, and S. Shatashvili, *Integrating over Higgs branches*, *Commun. Math. Phys.* **209** (2000) 97–121, [hep-th/9712241].

[9] N. A. Nekrasov, *Seiberg-Witten Prepotential From Instanton Counting*, *Adv. Theor. Math. Phys.* **7** (2004) 831–864, [hep-th/0206161].

[10] N. Nekrasov and A. Okounkov, *Seiberg-Witten theory and random partitions*, hep-th/0306238.

[11] M. R. Douglas and G. W. Moore, *D-branes, Quivers, and ALE Instantons*, hep-th/9603167.

[12] D. Gaiotto and J. Maldacena, *The gravity duals of N=2 superconformal field theories*, arXiv:0904.4466.

[13] D. Gaiotto, *N=2 dualities*, arXiv:0904.2715.

[14] F. Benini, S. Benvenuti, and Y. Tachikawa, *Webs of five-branes and N=2 superconformal field theories*, *JHEP* **09** (2009) 052, [arXiv:0906.0359].

[15] F. Fucito, J. F. Morales, and R. Poghossian, *Instantons on quivers and orientifolds*, *JHEP* **10** (2004) 037, [hep-th/0408090].

[16] R. Flume and R. Poghossian, *An algorithm for the microscopic evaluation of the coefficients of the Seiberg-Witten prepotential*, *Int. J. Mod. Phys.* **A18** (2003) 2541, [hep-th/0208176].

[17] S. Shadchin, *Cubic curves from instanton counting*, *JHEP* **03** (2006) 046, [hep-th/0511132].

[18] A. Mironov and A. Morozov, *The Power of Nekrasov Functions*, *Phys. Lett.* **B680** (2009) 188–194, [arXiv:0908.2190].

[19] A. Marshakov, A. Mironov, and A. Morozov, *Zamolodchikov asymptotic formula and instanton expansion in $N = 2$ SUSY $N_f = 2N_c$ QCD*, *JHEP* **11** (2009) 048, [arXiv:0909.3338].

[20] A. Mironov and A. Morozov, *Proving AGT relations in the large-c limit*, *Phys. Lett.* **B682** (2009) 118–124, [arXiv:0909.3531].

[21] V. Alba and A. Morozov, *Check of AGT Relation for Conformal Blocks on Sphere*, *Nucl. Phys.* **B840** (2010) 441–468, [arXiv:0912.2535].

[22] V. A. Fateev and A. V. Litvinov, *On AGT conjecture*, *JHEP* **02** (2010) 014, [arXiv:0912.0504].

[23] L. Hadasz, Z. Jaskolski, and P. Suchanek, *Proving the AGT relation for* $N_f = 0, 1, 2$ *antifundamentals*, *JHEP* **06** (2010) 046, [arXiv:1004.1841].

[24] L. F. Alday and Y. Tachikawa, *Affine SL(2) conformal blocks from 4d gauge theories*, *Lett. Math. Phys.* **94** (2010) 87–114, [arXiv:1005.4469].

[25] E. Carlsson and A. Okounkov, *Exts and vertex operators*, arXiv:0801.2565.

[26] I. G. Macdonald, *Symmetric functions and Hall polynomials.* Oxford University Press, 1995.

[27] A. B. Zamolodchikov and Al. B. Zamolodchikov, *Structure constants and conformal bootstrap in Liouville field theory*, *Nucl. Phys.* **B477** (1996) 577–605, [hep-th/9506136].

[28] V. A. Alba, V. A. Fateev, A. V. Litvinov and G. M. Tarnopolskiy, "On combinatorial expansion of the conformal blocks arising from AGT conjecture," Lett. Math. Phys. **98** (2011) 33 [arXiv:1012.1312 [hep-th]].

Decaying Gravitino Dark Matter

Presentation at the 49th International School of Subnuclear Physics,
Erice, June 30 2011

N.-E. Bomark

Department of Physics and Technology,
University of Bergen, N-5020 Bergen, Norway

Abstract

Supersymmetry is a great idea for new physics, but why should this strange symmetry called R-Parity be exactly conserved?

In fact, even if it is not, we have a very good dark matter candidate in the gravitino and if we are really lucky, the decay products of the gravitino might be detectable by cosmic ray experiments.

In this presentation it is shown that the recent cosmic ray data on positron fraction from PAMELA as well as the electron plus positron data from Fermi LAT, can be well described with decay products from decaying gravitino dark matter. However, if this was really the reason the data looks like it does, one would expect to see something in gamma rays as well. According to Fermi LAT, that something is not seen and although this is not enough to rule out decaying gravitinos as the source of the extra electron-positron pairs seen by PAMELA and Fermi LAT, this explanation is significantly weakened.

1 Introduction

Supersymmetry has for a long time been the leading candidate for physics beyond the Standard Model of particle physics (SM). When reading the literature, there are a large number of motivations why supersymmetry should exist in nature. However, from the beginning, the motivation was simply to find the most general extension of the Lorentz group of special relativity [1].

Furthermore, string theory, one of the leading candidates for quantum gravity, requires supersymmetry to work and therefore the study of supersymmetry is of great importance also to this field.

Also for phenomenology, supersymmetry offers plenty of opportunities; the large parameter spaces of the models present an ever lasting source of new types of scenarios for LHC and cosmology!

Some of these rich possibilities are contained in the B and L violating operators [2],

$$\lambda_{ijk} L_i L_j \bar{E}_k + \lambda'_{ijk} L_i Q_j \bar{D}_k + \lambda''_{ijk} \bar{U}_i \bar{D}_j \bar{D}_k + \mu_i H L_i, \tag{1.1}$$

where L_i, Q_i, H are lefthanded lepton, quark and Higgs superfields and E_i, D_i, U_i are righthanded lepton, down quark and up quark superfields. The bilinear couplings, $\mu_i H L_i$, will not be discussed here.

The operators of Eq. (1.1) are problematic due to their violation of baryon and lepton number. If all such operators are allowed, it leads to a much too short lifetime for the proton [3]. One can also get problems with flavour changing neutral currents [4], double nucleon decay and neutron-antineutron oscillations as well as other processes [5]. The most common solution is to introduce R-parity; a $\mathbb{Z}_2$ symmetry that is designed to forbid all the couplings of Eq. (1.1).

Technically we assign R-parity 1 to all SM particles and -1 to all their new superpartners. This has the bonus that if R-parity is conserved, the lightest partner will be absolutely stable and hence a good dark matter candidate. Do we have any reason to depart from this picture?

We could ask; why should the particles we have discovered (plus the Higgs) be different from the ones we have not yet discovered, in that they have different R-parity? It is hard to find a good motivation for that.

Furthermore, parities introduced in the past (almost) always turn out to be violated at some level. Good examples are C, P and CP but also lepton flavour numbers are violated by neutrino oscillations. Even B and L are violated in the SM due to non-trivial vacuum transformations [6] although this has not yet been observed by experiments. Why should R-parity be different?

Of course, the symmetries mentioned above, do still serve as useful approximations even if they are not entirely conserved, maybe that is the role of R-parity too? This however, does not work if we want the lightest superpartner to be dark matter; for the neutralino (the usual dark matter candidate) to be long-lived enough to be dark matter and avoid conflict with cosmic ray measurements due to its decay products, the trilinear couplings of Eq. (1.1) have to be $\lesssim 10^{-23}$, which is perhaps more difficult to accept than a simple zero.

What then about dark matter? If R-parity is violated, all new particles decay, right?

They do indeed; but there is one, the gravitino, that due to the gravitational nature of its couplings to matter, has a naturally long lifetime. It is not very difficult to have a gravitino with long enough lifetime to be dark matter even if R-parity is violated [7, 8].

In contrast to the R-parity conserving scenario with gravitino dark matter, where the gravitinos are practically undetectable due to their small interactions, the violation of R-parity and the slow decay of the gravitinos that follow, actually provides a possibility of detecting cosmic rays from decaying gravitinos.

2 Cosmic ray anomalies

Having identified cosmic rays as the way of detecting gravitino dark matter, it is appropriate to discuss the status of cosmic ray measurements with respect to dark matter.

The most discussed set of data in this respect was published in 2008 by the PAMELA collaboration [9]. In figure 1 we see their data on the positron fraction (i.e. positron divided by electron + positron flux), that caused all the excitement. The "amazing" feature of this plot is of course the complete lack of correlation between the data and the plotted theoretical expectation.

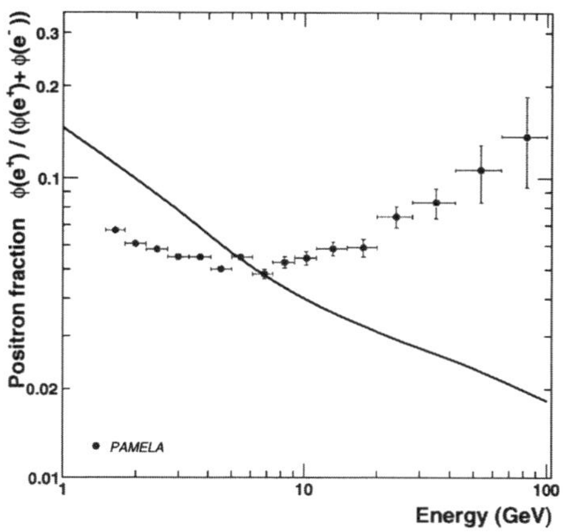

Figure 1: Measurement of positron fraction from PAMELA, compared with a theoretical prediction by Strong and Moskalenko [10]. Courtesy of the PAMELA collaboration, figure from [9].

The discrepancies between theory and experiment can here be divided into two distinct ones; one below 10 GeV and another one above 10 GeV. The low energy discrepancy can then be understood as an effect of solar modulations, i.e. changes in the direction and magnitude of the solar magnetic field. This makes sense, since older data, collected in a different solar magnetic field, are also in conflict with the PAMELA data and do in fact agree with the theoretical model [9]. For the rest of this discussion we shall therefore regard the low energy discrepancy as understood and focus on the high energy one.

It is indeed the discrepancy above 10 GeV that has caught most of the attention. The reason is that there is no clear explanation for it; the theoretical model assumes that all positrons come from collisions between cosmic ray protons and interstellar gas molecules and although there is a lot of uncertainty about exactly what spectrum that assumption gives, it is generally accepted that the spectrum should decrease with energy [11]. The data from PAMELA on the other hand shows an increase with energy.

The conclusion is that we need a new source of positrons (most likely electron-positron pairs) to explain the data.

One possible source of these new electron-positron pairs could be dark matter, either annihilating WIMPs [12] or maybe our decaying gravitinos. But before being carried away by this, we should have a look at other possibly correlated cosmic ray measurements.

The first thing to notice is that PAMELA sees no excess in anti-protons [13] and therefore we have to conclude that whatever is producing our positrons, is probably not producing any anti-protons. This is a bit problematic for most dark matter models.

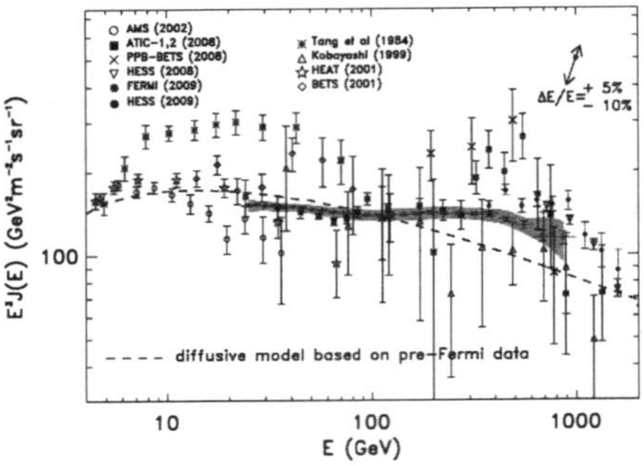

Figure 2: The electron plus positron spectrum as measured by Fermi LAT is shown in red and compared to some older measurements. The blue dashed curve is a theoretical prediction for the spectrum. Courtesy of the Fermi LAT collaboration, figure from [14].

Next we should look for other measurements of cosmic ray leptons. The appropriate data here is the electron plus positron spectrum from Fermi LAT [14] which is shown in figure 2. Here we can see that the situation is much less clear than for the PAMELA positron fraction data. It is in fact possible to fit the data with no so far unknown component. However, if the fit should also be consistent with the PAMELA positron data, it is clear that a new ingredient is needed [11].

This new component needs to start at around 10 GeV and probably has a cutoff around 1 TeV; at least there are some hints of that in the Fermi LAT data and the HESS data [15] strongly suggest such a cutoff.

3 Decaying gravitinos

The obvious question to ask now is whether decaying gravitinos could be this new source of positrons that we need to understand the data.

The simple answer is yes [16, 17]. But let us be a bit more detailed.

Calculating the charged particle spectrum expected on earth due to decaying gravitinos is not a trivial matter. First we need to know the expected spectrum of the various kinds of cosmic rays from our decaying gravitinos. This is not entirely trivial, especially if we want to include operators with quarks in the analysis. To deal with this we simulate the decay of the gravitinos using PYTHIA 6.4 [18].

The next difficulty is the propagation of the cosmic rays through the galaxy, this is affected by a large number of effects including diffusion by galactic magnetic fields, collisions with photons from the cosmic microwave background and various plasma phenomena [19]. We also need to know the background from astrophysical sources. Fortunately, all this is well handled by the GALPROP [20] code.

We also need to have a guess for the distribution of dark matter within our galaxy and

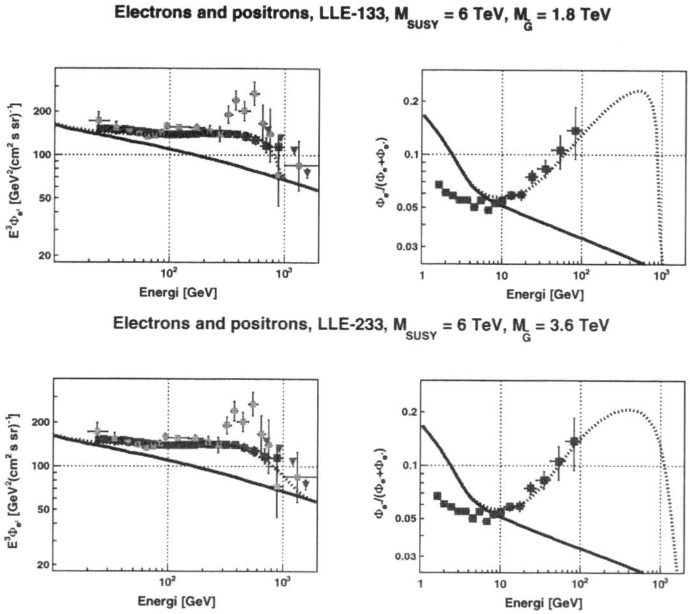

Figure 3: Electron and positron data compared to expected fluxes from the two best models of decaying gravitinos, couplings and masses as given in the figure ($M_{\rm SUSY}$ is a common mass for all sparticles). The solid blue lines are background from GALPROP and the dotted blue lines also include the signal. The left panels show electron plus positron flux compared with data from Fermi LAT [14] (red circles), ATIC [22] (green squares) and HESS [15] (violet triangles). The right panels show positron fraction compared with the PAMELA [9] data in red. Figure from [16]

the best estimate we have for that is the NFW profile [21] given by:

$$\rho_{\rm Halo}(r) = \frac{\rho_0}{(r/r_c)(1+r/r_c)^2}, \tag{3.2}$$

where r denotes the distance to the center of the galaxy, $r_c = 20$ kpc and $\rho_0 = 0.33$ GeV cm^{-3}.

The conclusion from this study is that the anomalies discussed above can be well explained by gravitinos decaying through some suitably chosen $LL\bar{E}$ type operator. The two best fits are shown in figure 3. For similar studies of gravitinos decaying through bilinear R-parity violating couplings, see [23–29].

We see that the fit to the Fermi LAT as well as the PAMELA data, is in fact very good for both our models. We ignore here the discrepancy in positron fraction below 10 GeV since this is supposed to be a question of solar modulation rather than new physics and

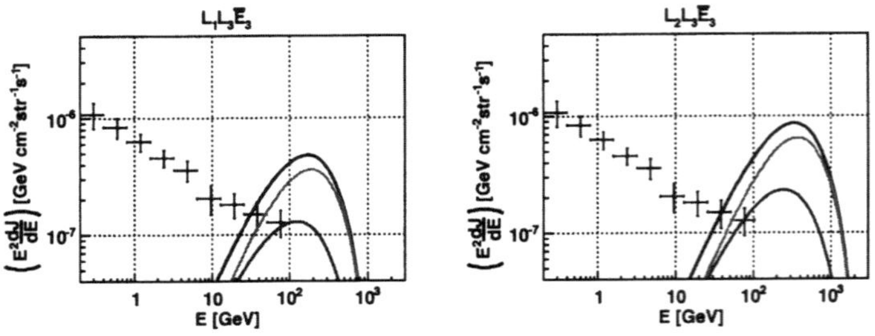

Figure 4: Gamma ray fluxes for the best fit models from figure 3, compared to the Fermi LAT isotropic diffuse emission [32]. The red curves show the red-shifted extragalactic flux while the green curves show the flux from the galactic halo averaged over all directions; the blue curves show the sum of the other two.

therefore not of primary importance to us.

The prize to pay for this good agreement is most importantly a disturbingly high gravitino mass, especially for $L_2L_3\bar{E}_3$ the lack of directly produced electrons and positrons forces us to use $m_{\tilde{G}} = 3.6$ TeV in order for the spectrum to reach the high end of the Fermi LAT data. This would be very bad news for the LHC since all other sparticles have to be heavier than the gravitino, otherwise there is good hope to detect R-parity violating supersymmetry at the LHC [30].

We should also take a look at other decay products than leptons. No matter how a particle is decaying, there will be photons in the final state, so also for our gravitinos [31]. Since both of our best couplings contain taus, most of the gamma rays will come from π^0s that stem from the taus.

The spectrum from the two models used in figure 3 are shown in figure 4 together with the isotropic diffuse emission data from Fermi LAT [32]. The interpretation of this plot is not entirely straightforward. The data is only the isotropic diffuse flux and the only part of the flux from decaying gravitinos that is truly isotropic is the extragalactic part (red curves in figure 4), however, we expect some fraction of the flux from the galactic halo (green curves in figure 4) to also be part of the isotropic flux. How much is not possible to say without a detailed study of the fitting procedure used by Fermi LAT to calculate the data shown in figure 4. We should also remember that we expect an exponentially falling background in addition to a possible contribution from decaying gravitinos.

All in all we cannot exclude either of these two models based on the isotropic diffuse data from Fermi LAT, but it is enough to impose significant doubt on decaying gravitinos as the explanation for the electron/positron anomalies seen by PAMELA and Fermi LAT. It is also worth pointing out that other dark matter explanations of these anomalies, especially annihilating dark matter, are in at least as much trouble as our models [33]. It seems likely that the solution to the puzzle will come from astrophysics rather than particle physics.

4 Summary

Supersymmetry is for good reasons a leading candidate for physics beyond the standard model. It is important to remember that this is true also if R-Parity is not conserved.

If R-Parity is indeed violated, we still have a very good dark matter candidate in the gravitino and the decay products of such dark matter gravitinos may be possible to detect in cosmic rays.

The recent cosmic ray features from PAMELA and Fermi LAT can be well described by decaying gravitinos but the complete absence of any corresponding signature in gamma rays clearly disfavours such an explanation.

References

[1] R. Haag, J. T. Lopuszanski, M. Sohnius, Nucl. Phys. **B88** (1975) 257.

[2] R. Barbier *et al.*, Phys. Rept. **420** (2005) 1 [arXiv:hep-ph/0406039].

[3] A. Y. Smirnov and F. Vissani, Phys. Lett. B **380** (1996) 317 [arXiv:hep-ph/9601387],

[4] D. Choudhury and P. Roy, Phys. Lett. B **378**, 153 (1996) [arXiv:hep-ph/9603363]. D. Chakraverty and D. Choudhury, Phys. Rev. D **63**, 112002 (2001) [arXiv:hep-ph/0012309];

[5] G. Bhattacharyya, arXiv:hep-ph/9709395;

[6] S. L. Adler, Phys. Rev. **177** (1969) 2426-2438.

[7] F. Takayama and M. Yamaguchi, Phys. Lett. B **485** (2000) 388 [arXiv:hep-ph/0005214].

[8] G. Moreau and M. Chemtob, Phys. Rev. D **65** (2002) 024033 [arXiv:hep-ph/0107286].

[9] O. Adriani *et al.* [PAMELA Collaboration], Nature **458** (2009) 607 [arXiv:0810.4995 [astro-ph]].

[10] I. V. Moskalenko and A. W. Strong, Astrophys. J. **493** (1998) 694 [arXiv:astro-ph/9710124].

[11] D. Grasso *et al.* [FERMI-LAT Collaboration], Astropart. Phys. **32**, 140 (2009) [arXiv:0905.0636 [astro-ph.HE]].

[12] P. Meade, M. Papucci, A. Strumia and T. Volansky, arXiv:0905.0480 [hep-ph].

[13] O. Adriani *et al.*, Phys. Rev. Lett. **102** (2009) 051101 [arXiv:0810.4994 [astro-ph]].

[14] A. A. Abdo *et al.* [The Fermi LAT Collaboration], Phys. Rev. Lett. **102**, 181101 (2009) [arXiv:0905.0025 [astro-ph.HE]].

[15] F. Aharonian *et al.* [H.E.S.S. Collaboration], Phys. Rev. Lett. **101** (2008) 261104 [arXiv:0811.3894 [astro-ph]].

[16] N.-E. Bomark, S. Lola, P. Osland and A. R. Raklev, Phys. Lett. **B686** (2010) 152-161. [arXiv:0911.3376 [hep-ph]].

[17] N.-E. Bomark, S. Lola, P. Osland, A. R. Raklev, PoS **EPS-HEP2009** (2009) 098. [arXiv:0911.3571 [hep-ph]].

[18] T. Sjostrand, S. Mrenna and P. Skands, JHEP **0605** (2006) 026 [arXiv:hep-ph/0603175].

[19] A. W. Strong, I. V. Moskalenko and V. S. Ptuskin, Ann. Rev. Nucl. Part. Sci. **57** (2007) 285 [arXiv:astro-ph/0701517].

[20] A. W. Strong and I. V. Moskalenko, Adv. Space Res. **27** (2001) 717 [arXiv:astro-ph/0101068].

[21] J. F. Navarro, C. S. Frenk and S. D. M. White, Astrophys. J. **490** (1997) 493 [arXiv:astro-ph/9611107].

[22] J. Chang *et al.*, Nature **456** (2008) 362.

[23] W. Buchmuller, L. Covi, K. Hamaguchi, A. Ibarra and T. Yanagida, JHEP **0703** (2007) 037 [arXiv:hep-ph/0702184].

[24] A. Ibarra and D. Tran, Phys. Rev. Lett. **100** (2008) 061301 [arXiv:0709.4593 [astro-ph]].

[25] A. Ibarra and D. Tran, JCAP **0807**, 002 (2008) [arXiv:0804.4596 [astro-ph]]; JCAP **0902**, 021 (2009) [arXiv:0811.1555 [hep-ph]].

[26] W. Buchmuller, A. Ibarra, T. Shindou, F. Takayama and D. Tran, JCAP **0909**, 021 (2009) [arXiv:0906.1187 [hep-ph]].

[27] K. Ishiwata, S. Matsumoto and T. Moroi, Phys. Rev. D **78**, 063505 (2008) [arXiv:0805.1133 [hep-ph]].

[28] K. Ishiwata, S. Matsumoto and T. Moroi, Phys. Lett. B **675**, 446 (2009) [arXiv:0811.0250 [hep-ph]].

[29] S. L. Chen, R. N. Mohapatra, S. Nussinov and Y. Zhang, Phys. Lett. B **677**, 311 (2009) [arXiv:0903.2562 [hep-ph]].

[30] N.-E. Bomark, D. Choudhury, S. Lola, P. Osland, JHEP **1107** (2011) 070. [arXiv:1105.4022 [hep-ph]].

[31] N.-E. Bomark, S. Lola, P. Osland and A. R. Raklev, Phys. Lett. B **677** (2009) 62 arXiv:0811.2969 [hep-ph].

[32] A. A. Abdo *et al.* [The Fermi-LAT collaboration], Phys. Rev. Lett. **104** (2010) 101101 [arXiv:1002.3603 [astro-ph.HE]].

[33] A. A. Abdo *et al.* [Fermi-LAT Collaboration], JCAP **1004** (2010) 014 [arXiv:1002.4415 [astro-ph.CO]].

Cosmological Constant and Casimir Effect

Philipp Burda[1]

ITEP, B. Cheremushkinskaya, 25, Moscow, Russia 117218

Abstract

Relation between Cosmological Constant and Casimir effect is studied. It is shown that naive calculation for the energy between two plates in flat spacetime has a similar structure to the Casimir energy between two plates in de Sitter spacetime. Last observaton provides us with one more argument in a favor of the idea, that cosmological constant is made of zero modes.

1 Introduction

1.1 Cosmological Constant Problem

Contemprorary observations of the expansion rate of the Universe could be explained by introducing vacuum energy, i.e. cosmological term into the Einstein equations. Experimantal data is in agreement with the following value of vacuum energy

$$E_{observ} = \mu^2 \tag{1}$$

where $\sqrt{\mu} \simeq 10^{-3} eV$ [1].

The first candidate for this quantity is the energy of zero modes of quantum fields. It is precisely this that explains the term "vacuum energy" is used. But, if one naively compute energy of zero modes of some simple scalar field by treating it as an infinite set of quantum oscillators, then one will find that this energy diverge:

$$E = \frac{1}{2} \sum_{\vec{k}} \omega_k \tag{2}$$

After introducing a cut-off at the energy scale, where our theory breaks down (for quantum field theory it is usually Planck scale) we arrive at the following expression for the zero-modes energy

$$E = \frac{1}{2} \sum_{\vec{k}} \omega_k \to \int_{k<\sqrt{\Lambda}} \frac{d^3k}{(2\pi)^3} \omega_k \sim \Lambda^2 \tag{3}$$

And by imposing $\sqrt{\Lambda} \sim M_{pl} = 10^{19} GeV$ we obtain the following ratio for the theoretical and observed cosmological constants

$$\frac{E_{theory}}{E_{observ}} \simeq= \left(\frac{\Lambda}{\mu}\right)^2 \simeq 10^{124} \tag{4}$$

This is a famouse Cosmological Constant Problem, which is sometimes also called a worse theoretical prediction ever made. But as we would see, maybe it wasn't such a bad idea to assume that cosmological constant is made from zero modes of quantum fields.

1.2 Casimir Effect

Now let's recall basics of the famous Casimir Effect. This effect was firstly discovered theoretically by Henrick Casimir [2] and then confirmed experimentally [3]. The phenomenon is that two uncherged and conducting plates, separated by distance L, are atracted to eash other by the force proportional to the L^{-4}. This is direct manifestation of the fact, that zero modes do affect physics.

There are a lot of different ways to obtain finite and expereimantally verified expression for Casimir energy, one of them will be presented here later. But for the beginning I would like to present simplest calculation, based again on simple counting the energy of zero modes between two plates. In this case not all wave vectors

[1] burda@itep.ru

are permited, because the field itself should satisfy the Dirichlet boundary conditions on the top of the plates ($\phi_{z=0} = \phi_{z=L} = 0$). Therefore, only the following wave vectors is allowed

$$\vec{k}^2 = k_x^2 + k_y^2 + \left(\frac{\pi n}{L}\right), \qquad n = 1, 2, 3, ... \tag{5}$$

this is correspond to the case, when plates situated orthogonaly to the z-axies.

Then the vacuum energy between the plates is

$$E_{naive} \sim \sum_n \int_{|\vec{k}|<\sqrt{\Lambda}} dk_x dk_y \sqrt{kx^2 + k_y^2 + \left(\frac{\pi n}{L}\right)^2} \sim \frac{1}{288}\frac{\Lambda}{L} - \frac{1}{1440}\frac{\pi^2}{L^3} + \alpha\Lambda^2 L + \beta\Lambda^{3/2} \tag{6}$$

here I want to concentrate an attention on the first two terms which will reappear in the following calculation of the Casimir energy in de Sitter spacetime, when Λ will be standing for the cosmological constant. In other words, naive summation of zero modes provide us the answer with the same structure as the proper calculation of Casimir energy in a spacetime with cosmological constant.

2 Casimir Effect in de Sitter Spacetime

Let's now assume that we live in a spacetime with positive cosmological constant of unknown nature. Actually, last statement is in a exact agreement with the nowadays observations. We want to study the renormalized v.e.v. of the stress-energy tensor for massless minimally coupled scalar field between two perfectly conducting parallel planes, situated orthogonally to the z axis at $z = 0$ and $z = L$. The action

$$S = \int d^4x \left\{ \sqrt{-g(x)} g^{\mu\nu}(x) \partial_\mu \phi \partial_\nu \phi \right\}, \tag{7}$$

gives rise to the following expression for the stress-energy tensor

$$T_{\mu\nu} = \partial_\mu \phi \partial_\nu \phi - \frac{g_{\mu\nu}}{2} \partial_\rho \phi \partial^\rho \phi = \left[\left\{ \partial_{x^\mu} \partial_{y^\nu} - \frac{g_{\mu\nu}}{2} \partial_{x^\rho} \partial^{y_\rho} \right\} \phi(x)\phi(y) \right]_{y=x} \tag{8}$$

Therefore, in order to calculate its v.e.v. one could just act by the specified differential operator on the appropriate Green function (which in our case should satisfy Direchlet boundary conditions at $z = 0$ and $z = L$ — $\phi_{z=0} = \phi_{z=L} = 0$) and then take the limit of coincident points.

$$\left\langle T_{\mu\nu} \right\rangle = \left[\left\{ \partial_{x^\mu} \partial_{y^\nu} - \frac{g_{\mu\nu}}{2} \partial_{x^\rho} \partial^{y_\rho} \right\} \left\langle \phi(x)\phi(y) \right\rangle_{D(0,L)} \right]_{y=x} = \left[\mathcal{T}_{\mu\nu} \cdot \left\langle \phi(x)\phi(y) \right\rangle_{D(0,L)} \right]_{y=x} \tag{9}$$

To get rid of additional and not interesting divergencies it is convenient to use, so-called, symmetric two-point function $G^{(1)}(x, y)$, which by definition is equal to the v.e.v. of anti-commutator of the fields

$$G^{(1)}(x, y) = \left\langle \{\phi(x), \phi(y)\} \right\rangle = \left\langle \phi(x)\phi(y) + \phi(y)\phi(x) \right\rangle \tag{10}$$

In the limit of coincident points it gives only the additional factor of 2

$$\left\langle \phi(x)\phi(y) \right\rangle\Big|_{y=x} = \frac{1}{2}\, G^{(1)}(x, y)\Big|_{y=x} \tag{11}$$

but the pleasant thing is that expression for $G^{(1)}(x, y)$ is simpler than for $\langle \phi(x)\phi(y) \rangle$.

Thus, we have the following expression for the v.e.v. of the stress-energy tensor:

$$\left\langle T_{\mu\nu} \right\rangle = \frac{1}{2} \left[\mathcal{T}_{\mu\nu} \cdot G^{(1)}(x, y)_{D(0,L)} \right]_{y=x} \tag{12}$$

The problem is reduced now to the construction of the Green function $G^{(1)}(x, y)_{D(0,L)}$ for the theory with two perfectly conducting plates at $z = 0$ and $z = L$. Such a Green function could be obtained from the empty space Green function $G^{(1)}(x, y)$, using the well known high–school trick, called *method of images*. One only needs

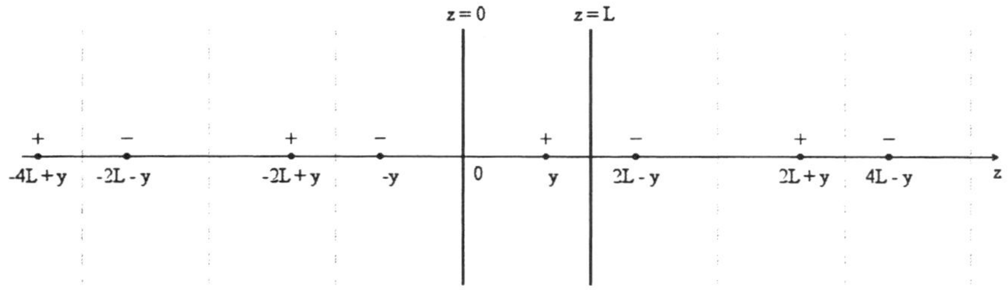

Figure 1: Positions of mirror sources. Building method: starting from real source with positive "charge" at the point y, one needs to built its images in the "mirrors" at $z = 0$ and $z = L$, but already with negative charge, and then, because each charge (real or image) should be reflected in each mirror, one should reflect left image $(-y)$ in the right mirror, and right image $(2L - y)$ – in the left one. In this way one obtains second generation of images (they again have positive "charge"), which in turn produce third generation of images and so on.

to add an infinite number of mirror sources, as shown and explained on the Figure 1, and then the boundary conditions will be automatically satisfied.

$$G^{(1)}(x,y)_{D(0,L)} = \sum_{n=-\infty}^{\infty} \left[G^{(1)}(x,y_n^+) - G^{(1)}(x,y_n^-) \right], \qquad (y_n^\pm)_z = \pm 2nL \pm y_z \tag{13}$$

Substituting this expansion into (12) and changing the order of the action of the operator $\mathcal{T}_{\mu\nu}$ and the summation

$$\left\langle T_{\mu\nu} \right\rangle = \frac{1}{2} \sum_{n=-\infty}^{\infty} \left[\mathcal{T}_{\mu\nu} \cdot G^{(1)}(x,y_n^+) - \mathcal{T}_{\mu\nu} \cdot G^{(1)}(x,y_n^-) \right]_{y=x} = \frac{1}{2} \sum_{n=-\infty}^{\infty} \left[t_{\mu\nu}^+(x,n) - t_{\mu\nu}^-(x,n) \right] \tag{14}$$

we arrive at the almost final desirable expression for $\langle T_{\mu\nu} \rangle$, the only its disadvantage is that this expression is divergent. The most straightforward and simple method of curing out such a divergence, based on subtraction of an empty space part of the v.e.v, works perfectly well also in the de Sitter spacetime (because of its high symmetry). Hence, we apply this, standard for Minkowski spacetime, subtraction scheme

$$\left\langle \ldots \right\rangle_{ren} = \left\langle \ldots \right\rangle_{with\, b.c.} - \left\langle \ldots \right\rangle_{empty\, space}. \tag{15}$$

In our notations this regularization procedure is equivalent to the subtraction of the term $t_{\mu\nu}^+(x,0)$ with the source at the point y_0^+ from (14). Hence, finally we obtain

$$\left\langle T_{\mu\nu} \right\rangle_{ren} = \frac{1}{2} \left[{\sum_{n=-\infty}^{\infty}}' \, t_{\mu\nu}^+(x,n) - \sum_{n=-\infty}^{\infty} t_{\mu\nu}^-(x,n) \right] \tag{16}$$

Every component of this expression can be easily calculated, but for Minkowski and de Sitter spacetimes one can also sum them up and obtain an explicit answer for $\langle T_{\mu\nu} \rangle$.

For the sake of brevity in all the examples we will consider not the stress tensor itself, but the energy per unit area of the field between the plates. Expression for the energy can be easily obtained from T_{00} component of the stress tensor by the simple integration over z

$$\begin{aligned} E &= \int_0^L dz \left\langle T_{00} \right\rangle_{ren} = \frac{1}{2} \int_0^L dz \left[{\sum_{n=-\infty}^{\infty}}' \, t_{00}^+(x,n) - \sum_{n=-\infty}^{\infty} t_{00}^-(x,n) \right] = \\ &= \frac{1}{2} \left[{\sum_{n=-\infty}^{\infty}}' \int_0^L t_{00}^+(x,n) dz - \sum_{n=-\infty}^{\infty} \int_0^L t_{00}^-(x,n) dz \right] = \frac{1}{2} \left[{\sum_{n=-\infty}^{\infty}}' \, \varepsilon_n^+ - \sum_{n=-\infty}^{\infty} \varepsilon_n^- \right] \end{aligned} \tag{17}$$

Such an interchange of the summation and integration has also some physical meaning, because as we will show properly treated contribution of all negative charges (y_n^-) to the energy of the field vanishes. Whereas, their contribution to the energy density is some non-trivial function of z, which is divergent at the boundaries. Of course, there are also divergencies in some of ε_n^-, but there is an obvious way to handle them without any complication of the problem.

2.1 Example of Minkowski spacetime

In this section we apply our general strategy, developed in the previous section, to the Minkowski spacetime to check its consistency and discuss everything in details on the concrete example.

Using expression for the symmetric function in empty and flat spacetime

$$G^{(1)}(x,y) = -\frac{1}{2\pi^2}\frac{1}{(t_x - t_y)^2 - (\vec{x}-\vec{y})^2} \tag{18}$$

after some trivial calculations we obtain

$$\varepsilon_n^+ = -\frac{1}{16\pi^2 L^3}\frac{1}{n^4} \tag{19}$$

As we expect, expression for ε_0^+ diverges. Nature of this divergence is clear, because, by definition, ε_0^+ corresponds to the energy in empty space without any boundaries. Character of the divergence ($\varepsilon_0^+ \sim \alpha^{-4}$, $\alpha \to 0$) also supports this observation.

For ε_n^- we get

$$\varepsilon_n^- = \frac{1}{24\pi^2 L^3}\frac{3n^2+3n+1}{n^3(n+1)^3} \tag{20}$$

This expression is also divergent for $n = 0$ and $n = -1$. But it is worth stressing here that the sum $\varepsilon_{-1}^- + \varepsilon_0^-$ could be made finite, if one approaches divergencies simultaneously. From the fact, that Casimir effect is experimentally confirmed, we conclude that we act properly and these divergencies should cancel each other. Nevertheless, it is interesting to understand their nature. Notice, that ε_{-1}^- and ε_0^- correspond to the sources at points $2L - y$ and $-y$ respectively. In our notation, this is the first generation of images, i.e. they were obtained by reflecting "real" source in the mirrors. It is quite interesting, that divergent expressions arise only when one builds an image of original source in the mirrors of the systems. The same holds for de Sitter spacetime and again divergencies cancel each other. It seems, that such divergencies always arise in the presence of any boundaries and one, probably, should search their roots in *Heisenberg's uncertainty principle* — in the problem in question one tries to specify values of the filed on the boundary, hence momentum of the field (stress-energy tensor) diverges. As far as we know, such picture was proposed in [4]. According to our results it is also clear, that some smearing procedure should be applied to each boundary, when they are considered separately. But as we have seen, these divergencies eliminate each other, if we consider both boundaries simultaneously. Therefore, if we can think that we understand the nature of these divergencies, at least qualitatively, the question about their cancelation still remains. Why does such cancelation occur and does it happen in any spacetime, for any geometry and for any external conditions (temperature, presence of matter)?

Let's now return to the Casimir effect for Minkowski spacetime, and, using expressions for ε_n^+ and ε_n^-, we get well-known result for the Casimir energy between two plates at distance L

$$E_M = -\frac{1}{1440}\frac{\pi^2}{L^3} \tag{21}$$

As we have mentioned at the end of the previous section, it appears that properly treated sum of all negative mirror sources gives zero contribution to the final answer for the energy, i.e.

$$\sum_{n=-\infty}^{\infty} \varepsilon_n^- = 0 \tag{22}$$

The same is true for de Sitter as well. This fact should have some trivial explanation, but at the moment we cannot provide it. There is quite simple mathematical hint and some speculations about electromagnetic field in [5], but we actually seek for some physical reason for such a vanishing in the case of massless scalar field.

3 de Sitter spacetime

In this section we proceed with the consideration of the Casimir effect for two parallel plates in de Sitter spacetime. But, at first, we make some comments about the geometry of de Sitter spacetime and say a few words about QFT on such a background [6].

de Sitter spacetime is a vacuum solution of Einstein equations with the positive cosmological constant. It is convenient to consider it as a hypersurface in a flat five-dimensional space $\eta_{MN} = \mathrm{diag}(1,-1,-1,-1,-1)$, satisfying the following equation

$$X^M X^N \eta_{MN} = -\frac{1}{H^2}. \tag{23}$$

Introducing different parametrizations of this hypersurface one could obtain different metrics for de Sitter. Convenient for our purposes coordinate system (usually referred to as planar) is specified by the following parametrization

$$\begin{aligned} X^0 &= \frac{\sinh(Ht)}{H} + \frac{H}{2}e^{Ht}|\vec{x}|^2 \\ X^i &= x^i e^{Ht}, \quad i = 1,2,3 \\ X^4 &= \frac{\cosh(Ht)}{H} - \frac{H}{2}e^{Ht}|\vec{x}|^2, \quad -\infty < t, x_i < \infty \end{aligned} \tag{24}$$

It covers only half of the hyperboloid, but this is not a problem, because it can be resolved by further transformations. The line element in the above coordinates has a quite simple form $ds^2 = dt^2 - e^{2Ht}\left(dx_1^2 + dx_2^2 + dx_3^2\right)$, from which it is easy to observe the expanding nature of the de Sitter universe. Making a further transformation and proceeding from, so called, cosmic time t to the conformal time $\tau = -\frac{1}{H}e^{-Ht}$ $(-\infty < \tau < 0)$, one obtains conformal form of the de Sitter metric

$$ds^2 = \frac{1}{H^2\tau^2}\left(d\tau^2 - dx_1^2 - dx_2^2 - dx_3^2\right) \tag{25}$$

which we will use for calculations. Notice, that one could cover whole de Sitter space by expanding the range of τ to the whole real line.

The only thing we want to stress about QFT on de Sitter background, is that there is no de Sitter invariant vacuum state for the case of massless minimally coupled scalar field [7]. The common strategy is to look for states, that break de Sitter invariance as little as possible [8]. We choose the $E(3)$ invariant vacuum, because it corresponds to the Bunch-Davies vacuum for the massive case, and also in such a vacuum $\langle\phi^2\rangle = const$ (i.e. does not depend on time) for flat spacelike surfaces, which matches with our geometry. For such a vacuum in the conformal coordinates function $G^{(1)}(x,y)$ reads [7]

$$G^{(1)}(x,y) = \frac{H^2}{4\pi^2}\left(\frac{2\tau_x\tau_y}{(\vec{x}-\vec{y})^2 - (\tau_x-\tau_y)^2} - \log\left[(\vec{x}-\vec{y})^2 - (\tau_x-\tau_y)^2\right]\right) \tag{26}$$

Computations for positive sources give

$$\varepsilon_n^+ = \frac{H^2}{16}\frac{n^2L^2 - \tau^2}{\pi^2 L^3 n^4} = \frac{H^2}{16\pi^2 L}\frac{1}{n^2} - \frac{H^2\tau^2}{16\pi^2 L^3}\frac{1}{n^4} \tag{27}$$

One can see that two type of divergencies at $n = 0$ are present here. The first term appears because we consider non-conformal coupling of scalar field. It has special for de Sitter spacetime divergency, which is related to the fact, that chosen vacuum breaks the de Sitter invariance. This term also has unusual for Casimir energy dependence from the distance between the plates $\sim L^{-1}$. The second term has standard Minkowskian divergency and, in fact, it is equal to the ε_n^+ for Minkowski spacetime multiplied by the expansion factor $H^2\tau^2$. Notice, that the first term has an opposite sign, hence it compensates contribution of the second term. Its presence is the manifestation of antigravitational nature of Λ in de Sitter universe.

Contribution of the negative charges

$$\varepsilon_n^+ = \frac{H^2}{48\pi^2 L^3}\frac{9n^2L^2(n+1)^2 + 2\tau^2(3n^2+3n+1)}{n^3(n+1)^3} = \frac{3H^2}{16\pi^2 L}\frac{1}{n(n+1)} + \frac{H^2\tau^2}{24\pi^2 L^3}\frac{3n^2+3n+1}{n^3(n+1)^3} \tag{28}$$

also has two analogous terms. They both diverge at $n = -1$ and $n = 0$ and for both of them these divergencies cancel each other.

Finally, using (17) for the Casimir energy between two plates at distance L in de Siter spacetime, we obtain

$$E_{dS} = \frac{H^2}{1440}\frac{15L^2 - \pi^2\tau^2}{L^3} = \frac{H^2}{96L} - \frac{H^2\tau^2\pi^2}{1440L^3} \tag{29}$$

Returning to the cosmic time t and expressing everything through the physically measurable distance in the expanding unverse

$$L_{phys} = \int \sqrt{\gamma_{\alpha\beta}dx^\alpha dx^\beta} = \int \sqrt{-g_{\alpha\beta}dx^\alpha dx^\beta} = \int_0^L \sqrt{-g_{zz}}\, dz = -\frac{L}{H\tau} = Le^{Ht} \tag{30}$$

one could rewrite the expression for the Casimir energy in physically meaningful terms

$$E_{dS} = \left(\frac{H^2}{96L_{phys}} - \frac{\pi^2}{1440L^3_{phys}}\right)e^{Ht} \tag{31}$$

And after proceeding from Hubble constant to the Cosmological Constant ($H = \sqrt{\Lambda/3}$) in first term we arrive to the final answer

$$E_{dS} = \left(\frac{\Lambda}{288L_{phys}} - \frac{\pi^2}{1440L^3_{phys}}\right)e^{Ht} \tag{32}$$

If now one recall naive answer for the empty space (6) and compare it with the result for de Sitter spacetime (32) one would see that they have the same structure. Of course it is not very strong bet still yet another indication that cosmological constant is made of zero modes of quantum fields.

4 Conclusions

In this small note we have recognized, that summation of energy of zero modes between two plates in empty spacetime with cut-off regularization at some scale have the same structure as casimir energy between two plates in de Sitter spacetime. One just need to identify cosmological constsant to cut-off scale in the first calculation.

This observation is a small evidance of the very natural assumption, that cosmological constant is made of zero modes. Such huge distinction between theoretical prediction and current observed value (4) is another problem. There is a obvious and natural scenario for the solution of this problem.

First, enormous theoretical number is a prediction for cosmological constant for the very early stage of the evolution of the Universe. In particular, such huge value of prehistoric cosmological constant could be used as a fuel for the inflation, instead of using artificial inflaton field. It is always better to explain puzzles without inventing a new quantity for each puzzle.

Second, it is a well know fact that there is a particle production in de Sitter spacetime [10]. In other words, positive cosmological constant is a source of the particle production, the bigger it is the bigger is rate of the production of particles. And there are a lot of indications that initial, or bare, cosmological constant is screened off by these produced particles [11, 12, 13]. So, too big cosmological is not a problem, one just need a mechanism of its relaxation. I deeply belive that it is a worth problem to address.

5 Acknowledgments

I would like to thank E.Akhmedov for initiation of this work and very valuable discussions. I also would like to acknowledge the organisers of the Erice School for the opportunity to present this talk and for hospitality. This work is partly supported by the Dynasty Foundation.

References

[1] B. Ravndal, arXiv:hep-ph/0009208v1

[2] H.B.G. Casimir, Proc. K. Ned. Akad. Wet. 51, 793 (1948)

[3] S.K. Lamoreaux, Phys. Rev. Lett. 78, 5 (1997)

[4] L. H. Ford, N. F. Svaiter, Phys. Rev. D **58** 065007 (1998)

[5] L. C. de Albuquerque, C. Farina, L. G. A. Theodoro, Braz. J. Phys. **27** 488 (1997)

[6] N. D. Birrell, P. C. V. Davies, *Quantum Fields in Curved Space* (Cambridge University Press, Cambridge, England, 1982)

[7] B. Allen, Phys. Rev. D **32** 3136 (1985)

[8] B. Allen, A. Folacci, Phys. Rev. D **35** 3771 (1987)

[9] R. Bousso, arXiv:hep-th/0205177

[10] E. Mottola, Phys. Rev. D **31** 4 (1985)

[11] A. M. Polyakov, Nucl. Phys. B **797**, 199 (2008) [arXiv:hep-th/0709.2899]

[12] D. Krotov, A. M. Polyakov, arXiv:hep-th/1012.2107

[13] E. T. Akhmedov, P. V. Buividovich, Phys. Rev. D **78** 104005 (2008) [arXiv:hep-th/0808.4106];
E. T. Akhmedov, P. V. Buividovich and D. A. Singleton, arXiv:gr-qc/0905.2742

AEI-2011-074

Higher Spins in $D = 2 + 1$

ANDREA CAMPOLEONI

Max-Planck-Institut für Gravitationsphysik
Albert-Einstein-Institut
Am Mühlenberg 1
14476 Golm, GERMANY

andrea.campoleoni@aei.mpg.de

Abstract

We give a brief overview of some three-dimensional toy models for higher-spin interactions. We first review the construction of pure higher-spin gauge theories in terms of Chern-Simons theories. We then discuss how this setup could be modified along the lines of the known topologically massive theories.

Based on the talk presented at the 49th Course of the International School of Subnuclear Physics, Erice, Italy, 24 June - 3 July 2011

1 Introduction

The study of field theories for particles of arbitrary spin has a long and variegated history,[1] but in the last years most of the efforts focussed on gauge theories, to wit on massless particles. The motivations come from String Theory – whose higher-spin excitations could acquire their masses via the breaking of a "hidden" gauge symmetry (see e.g. [2] and refs therein) – and, more recently, from the study of holographic dualities [3]. We are still far from having a full control on these proposals but – in spite of various renowned no-go theorems (see e.g. [4] for a review) – we now have examples of consistent interactions to analyse. The old negative results thus only signal that higher-spin interactions are "unconventional", mainly on account of the higher-derivative nature of the couplings [5, 2]. In flat space the

[1]For a more complete overview on higher-spin field theories see, for instance, [1].

construction of a complete higher-spin gauge theory is still ongoing [6], but in the presence of a cosmological constant we already have at our disposal the Vasiliev equations [7], which are a set of non-linear field equations for an infinite tower of gauge fields of increasing spin. The need for a cosmological constant, however, makes it difficult to compare Vasiliev's models with String Theory. On the other hand, it suggested to consider them in the AdS/CFT context [3]. In this case the comparison with the candidate boundary duals has to face the "unconventional" formulation of the Vasiliev equations, and thus often requires the development of ad hoc techniques (see e.g. [8] and refs therein).

These open questions call for models where one can address at least some of them in a simplified setup. Various low-dimensional field theories fit in this scheme even if, at a first glance, they seem to have no chance to account for higher spins. For instance, in a space-time of dimension $D = 2+1$ the Poincaré group does not admit massless representations of arbitrary spin [9]. Nevertheless, one can still consider the Fronsdal equation[2]

$$\Box\, \varphi_{\mu_1 \dots \mu_s} - \partial_{(\mu_1} \partial^{\lambda} \varphi_{\mu_2 \dots \mu_s)\lambda} + \partial_{(\mu_1} \partial_{\mu_2} \varphi_{\mu_3 \dots \mu_s)\lambda}{}^{\lambda} = 0\,, \tag{1.1}$$

whose solutions describe, for $D > 3$, the free propagation of a massless spin-s particle [10]. Here $\varphi_{\mu_1 \dots \mu_s}$ is a symmetric tensor of rank s, and eq. (1.1) is left invariant by the gauge transformations

$$\delta \varphi_{\mu_1 \dots \mu_s} = \partial_{(\mu_1} \xi_{\mu_2 \dots \mu_s)} \qquad \text{with} \quad \xi_{\mu_1 \dots \mu_{s-3}\lambda}{}^{\lambda} = 0\,, \tag{1.2}$$

that are crucial to ensure the decoupling of unphysical polarisations. If one wants to derive (1.1) from a Lagrangian, one also has to force the double trace of $\varphi_{\mu_1 \dots \mu_s}$ to vanish, either directly or by adding a Lagrange multiplier (see e.g. [1]). One of the main goals of the current research on higher spins is the classification of the consistent non-linear deformations of (1.1) and (1.2), and their comparison with Vasiliev's models.[3] One can pursue both tasks even in $D = 2+1$: the price to pay is that (1.1) does not propagate local degrees of freedom when $s > 1$, while the advantage is the opportunity to deal with questions that are beyond reach in higher space-time dimensions.

---

[2]Here and in the following a couple of parentheses (brackets) denotes the (anti)symmetrisation of the indices it encloses, with the minimum number of terms and without any normalisation factor. Repeated indices denote a summation, and we adopt the mostly plus convention for the space-time metric.

[3]We focussed on (1.1) for simplicity, but for $D > 5$ the description of all irreps of the Poincaré group requires fields with multiple groups of symmetrised indices (see e.g. [11]).

The possible three-dimensional deformations include a class of models that is particularly appealing for its simplicity: the full non-linear theory is described by a Chern-Simons action [12] (see also [13]). In sect. 2 we review how to select the relevant Chern-Simons actions and in sect. 3 we discuss how one could extend this setup by adding topological mass terms. Along the way we briefly recall some recent applications of these models.

2 Chern-Simons theories

For $s = 2$ the Fronsdal equation (1.1) imposes the vanishing of the linearised Ricci tensor. The study of its deformations thus aims at understanding higher-spin interactions along the lines of the metric formulation of gravity. However, alternative descriptions of the free dynamics are available [1]. For instance, one can mimic the frame formulation of gravity and associate to each particle a couple of differential forms [14]. The second-order Fronsdal action for the symmetric field $\varphi_{\mu_1\dots\mu_s}$ is replaced with a first-order action for a vielbein-like field

$$e^{a_1\dots a_{s-1}} = e_\mu{}^{a_1\dots a_{s-1}}\, dx^\mu\,, \tag{2.1}$$

and an auxiliary field

$$\omega^{a_1\dots a_{s-1}\,,\,b} = \omega_\mu{}^{a_1\dots a_{s-1}\,,\,b}\, dx^\mu\,, \tag{2.2}$$

which generalises the spin connection. The vielbein-like field is traceless and fully symmetric in the fiber indices, while the auxiliary field is traceless and satisfies the condition $\omega^{(a_1\dots a_{s-1}\,,\,b)} = 0$. For more details we refer to [13]. For the purpose of this note it is sufficient to recall that the free action is invariant under the transformations

$$\delta\, e_\mu{}^{a_1\dots a_{s-1}} = \partial_\mu\, \xi^{a_1\dots a_{s-1}} + \bar{e}_{\mu\,,\,b}\, \Lambda^{a_1\dots a_{s-1}\,,\,b}\,, \tag{2.3a}$$

$$\delta\, \omega_\mu{}^{a_1\dots a_{s-1}\,,\,b} = \partial_\mu\, \Lambda^{a_1\dots a_{s-1}\,,\,b}\,, \tag{2.3b}$$

where $\bar{e}_\mu{}^a$ denotes the background vielbein. The gauge transformations generated by Λ allow one to recover the Fronsdal field by gauging away the non-symmetric component of the vielbein. Those generated by ξ then reproduce the gauge transformations (1.2) (see e.g. [13]). Also in this context one can classify the non-linear deformations of the free action that admit a gauge symmetry reducing to (2.3) at the linearised level.

In $D = 2 + 1$ there is a simple class of deformations that is not available in higher space-time dimensions [12], and all ingredients to build it are already visible in the spin-two case. To spot them, it is convenient to combine dreibein and spin connection in a one-form taking values in the algebra of isometries of the maximally symmetric solution of the field equations,

$$A = e + \omega := \left(e_\mu{}^a P_a + \frac{1}{2}\, \omega_\mu{}^{a,b}\, M_{ab} \right) dx^\mu \,. \tag{2.4}$$

Here the M_{ab} denote the generators of the Lorentz group, while the P_a denote the translation generators:

$$\begin{aligned} [\, P_a \,,\, P_b \,] &= \frac{1}{l^2}\, M_{ab} \,, \qquad [\, P_a \,,\, M_{bc} \,] = \eta_{a[b} P_{c]} \,, \\ [\, M_{ab} \,,\, M_{cd} \,] &= \eta_{a[c} M_{d]b} \;-\; \eta_{b[c} M_{d]a} \,. \end{aligned} \tag{2.5}$$

The factor l is related to the cosmological constant by $l = 1/\sqrt{-\Lambda}$, so that (2.5) gives the Poincaré algebra for $l \to \infty$ and either the $so(2, D-1)$ or $so(1, D)$ algebras for positive or negative values of l^2. A Poincaré – or (Anti) de Sitter – gauge transformation $\delta A = d\lambda + [A, \lambda]$ then induces

$$\delta e \;=\; d\xi \,+\, [\,\omega \,,\, \xi\,] \,+\, [\,e \,,\, \Lambda\,] \;=\; \mathcal{D}\xi \,+\, [\,e \,,\, \Lambda\,] \,, \tag{2.6a}$$

$$\delta \omega \;=\; d\Lambda \,+\, [\,\omega \,,\, \Lambda\,] \,+\, [\,e \,,\, \xi\,] \;=\; \mathcal{D}\Lambda \,+\, [\,e \,,\, \xi\,] \,, \tag{2.6b}$$

where we split the gauge parameter as

$$\lambda \;=\; \xi \,+\, \Lambda := \xi^a P_a \,+\, \frac{1}{2}\, \Lambda^{a,b} M_{ab} \,, \tag{2.7}$$

and we denoted by $\mathcal{D}$ the covariant exterior derivative. At the linearised level (2.6) reduces to (2.3) or to its (A)dS counterpart. However, in a space-time of generic dimension the Einstein-Hilbert action is *not* invariant under the full set of transformations (2.6), but only under local Lorentz transformations. The non-linear deformation of (2.3a) leads to diffeomorphisms rather than local translations.

On the other hand, in $D = 2 + 1$ the Einstein-Hilbert action reads

$$S_{EH} \;=\; \frac{1}{16\pi G} \int \epsilon_{abc} \left(e^a \wedge R^{bc} \,+\, \frac{1}{3\, l^2}\, e^a \wedge e^b \wedge e^c \right) , \tag{2.8}$$

where we introduced the Riemann curvature

$$R := d\omega + \omega \wedge \omega \,. \tag{2.9}$$

Up to boundary terms, the variation of (2.8) under (2.6) is

$$\delta S_{EH} \;=\; -\frac{1}{16\pi G}\int \epsilon_{abc}\left(\xi^a \mathcal{D}R^{bc}\right) \;=\; 0\,. \tag{2.10}$$

The remainder vanishes on account of the Bianchi identity, but setting the dimension of space-time to $D = 2+1$ is instrumental in obtaining (2.10). For instance, the variation of the Einstein-Hilbert action in $D = 3+1$ is

$$\delta S_{EH} \sim \int \epsilon_{abcd}\,\mathcal{D}\xi^a \wedge e^b \wedge R^{cd} \;=\; -\int \epsilon_{abcd}\,\xi^a\,T^b \wedge R^{cd} \;\neq\; 0\,, \tag{2.11}$$

and does not vanish off shell due to the presence of the torsion two-form,

$$T := \mathcal{D}e \;=\; de + (\omega \wedge e + e \wedge \omega)\,. \tag{2.12}$$

As a result, *only* in three dimensions Einstein gravity is a gauge theory for the full group of isometries of the vacuum. Moreover, the field equations following from (2.8) are

$$R + \frac{1}{l^2}\,e \wedge e \;=\; 0\,, \qquad\qquad T \;=\; 0\,, \tag{2.13}$$

and imply that the curvature of A vanishes. One can thus conclude that, up to boundary terms, (2.8) should be equivalent to a Chern-Simons action with gauge algebra (2.5). This was indeed realised more than twenty years ago in [15], where (2.8) was recovered from

$$S_{CS} = \frac{k}{4\pi}\int \mathrm{tr}\left(A \wedge dA + \frac{2}{3}A \wedge A \wedge A\right) \quad \text{with } k \;=\; \frac{1}{4G}\,, \tag{2.14}$$

using the following bilinear invariant form of (2.5):[4]

$$\mathrm{tr}\,(P_aP_b) \;=\; 0\,, \quad \mathrm{tr}\,(P_aM_{bc}) \;=\; \epsilon_{abc}\,, \quad \mathrm{tr}\,(M_{ab}M_{cd}) \;=\; 0\,. \tag{2.15}$$

The invariance of (2.8) under local translations (i.e. the transformations generated by ξ in (2.6)) does not imply that its gauge group is bigger than expected. In fact, *on shell* a local translation generated by $\xi^a = v^\mu e_\mu{}^a$ is equivalent to the Lie derivative along v^μ [15]. This is the key to three-dimensional

[4] This invariant form exists only in $D = 2+1$, where (2.5) is not simple even for finite values of l [15].

simplicity: one can trade diffeomorphisms for gauge translations! In the higher-spin context it is not yet clear what the precise analogue of diffeomorphism invariance is, but it is relatively straightforward to extend the Chern-Simons formulation of the dynamics [12].

To this end, one should notice that in $D = 2+1$ the higher-spin vielbein and spin connection (2.1) and (2.2) have the same number of independent components. For instance, in the $s = 2$ case this can be made manifest by

$$\omega^a = \frac{1}{2}\,\epsilon^{abc}\,\omega_{bc} \quad \leftrightarrow \quad M_a = -\frac{1}{2}\,\epsilon_{abc}\,M^{bc}\,, \tag{2.16}$$

that induces a rewriting of (2.5) as

$$[\,P_a\,,\,P_b\,] = \frac{1}{l^2}\,\epsilon_{abc}\,M^c, \quad [\,M_a\,,\,P_b\,] = \epsilon_{abc}\,P^c, \quad [\,M_a\,,\,M_b\,] = \epsilon_{abc}\,M^c. \tag{2.17}$$

The possibility to deal with vielbeins and spin connections with the same structure suggests to build higher-spin gauge theories out of Chern-Simons theories with a non-compact gauge algebra of the form

$$[\,P_A\,,\,P_B\,] = \frac{1}{l^2}\,f_{AB}{}^C\,M_C\,, \tag{2.18a}$$

$$[\,M_A\,,\,P_B\,] = f_{AB}{}^C\,P_C\,, \tag{2.18b}$$

$$[\,M_A\,,\,M_B\,] = f_{AB}{}^C\,M_C\,. \tag{2.18c}$$

The structure constants that appear in the three sets of commutators are identical, and thus there is an equal number of P_A and M_A to be contracted with (2.1) and (2.2). Any $sl(2,\mathbb{R})$ subalgebra of (2.18c) can then play the role of the Lorentz algebra since $so(1,2) \sim sl(2,\mathbb{R})$. Once one chooses it, the adjoint action of the Lorentz generators $M_a \subset M_A$ branches the P_A and M_A into identical $sl(2,\mathbb{R})$-irreducible subsets.[5] Each of them has dimension $2s+1$ with $2s \in \mathbb{N}$ and, for integer s, this number corresponds to the number of independent components of a traceless symmetric tensor of rank s. This suffices to identify the corresponding vielbeins and spin connections with (2.1) and (2.2). In the linearised regime (2.6) then induces the transformations (2.3) on all fields emerging from this branching [13] (see also [16]

[5]This is the crucial condition that one has to impose on the gauge algebra. One can generalise (2.18) provided the two sets of P_A and M_A are still branched into identical $sl(2,\mathbb{R})$-irreducible subsets. An example can be found in sect. 2.3 of [13].

for the treatment of half-integer values of s). This completes the argument that any Chern-Simons action based on an algebra of the form (2.18) gives a non-linear deformation of the free actions of a proper set of higher-spin gauge fields.

A key feature of (2.18) is the enlargement of the Lorentz algebra to (2.18c). A field $\varphi_{\mu_1 \dots \mu_s}$ must be invariant under Lorentz-like transformations since they do not have any counterpart in the Fronsdal approach. In the case of the metric, the only combination of the vielbeins that is invariant under generalised Lorentz transformations is

$$g_{\mu\nu} \sim l^2 \operatorname{tr}(e_\mu e_\nu) = l^2 \, e_\mu{}^A e_\nu{}^B \operatorname{tr}(P_A P_B) \,. \tag{2.19}$$

Here e is the full higher-spin vielbein $e^A P_A$ [13], so that the metric depends on the higher-spin vielbeins! Unfortunately, requiring Lorentz-like invariance does not suffice to fix the form of higher-rank fields, but additional results can be found in [16]. In spite of this limitation, (2.19) already played a role in the study of BTZ-like solutions with higher-spin hairs and of the effects of higher-spin gauge transformations on the causal structure of space-time [17].

To conclude, let us recall that for $l^2 > 0$ there is a simple way to satisfy (2.18). It is sufficient to consider algebras of the form $\mathfrak{g} \oplus \mathfrak{g}$ and to define the P_A and M_A as

$$P_A = \frac{1}{l}\left(T_A - \widetilde{T}_A\right), \qquad M_A = T_A + \widetilde{T}_A \,, \tag{2.20}$$

where T_A and $\widetilde{T}_A$ denote the generators of the two copies of $\mathfrak{g}$. This setup was used in [18, 13] to show that the asymptotic symmetries of three-dimensional higher-spin gauge theories on asymptotically (A)dS backgrounds are given by non-linear $\mathcal{W}$-algebras (see also [19, 16]). In the previous discussion nothing prevents the algebras $\mathfrak{g}$ from having a finite dimension, so that in $D = 2+1$ one can "truncate" the spectra of higher-spin gauge theories. An example of this type is given by $\mathfrak{g} \equiv sl(N, \mathbb{R})$, where the Lorentz algebra is identified with the principal embedding of $sl(2, \mathbb{R})$ in the diagonal $sl(N, \mathbb{R})$ generated by the M_A. In this case, the Chern-Simons theory describes the interactions of a tower of symmetric tensors of ranks $2, 3, \dots, N$ (see also [20] for a discussion of the quantum spectrum). On the other hand, it is possible to consider infinite-dimensional algebras $\mathfrak{g}$ so as to mimic the gauge algebras underlying Vasiliev's construction. In this context, Gaberdiel and Gopakumar proposed a holographic duality between Vasiliev's models in $D = 2+1$ [21] and suitable limits of minimal model CFT's with $\mathcal{W}_N$ symmetry [22] (see also [23]).

The construction (2.20) also allows an interesting rewriting of the action (2.14). A generic bilinear invariant form on $\mathfrak{g} \oplus \mathfrak{g}$ can be cast in the form

$$\operatorname{tr}(T_A T_B) = c_1 \, \eta_{AB} \,, \quad \operatorname{tr}(T_A \widetilde{T}_B) = 0 \,, \quad \operatorname{tr}(\widetilde{T}_A \widetilde{T}_B) = c_2 \, \eta_{AB} \,, \tag{2.21}$$

so that (2.14) can be rewritten as

$$S_{\mathfrak{g}\oplus\mathfrak{g}} = c_1 \, S_{CS}[A] + c_2 \, S_{CS}[\widetilde{A}] \,, \tag{2.22}$$

where now both A and $\widetilde{A}$ take values in a single copy of $\mathfrak{g}$: $A = A^B T_B$ and $\widetilde{A} = \widetilde{A}^B T_B$. If one defines

$$e = \frac{l}{2}\left(A - \widetilde{A}\right) , \qquad \omega = \frac{1}{2}\left(A + \widetilde{A}\right) , \tag{2.23}$$

and chooses $c_1 = l$ and $c_2 = -\,l$, one then obtains

$$S_{\mathfrak{g}\oplus\mathfrak{g}} = \frac{1}{8\pi G}\int \operatorname{tr}\left(e \wedge R + \frac{1}{3l^2}\, e \wedge e \wedge e \right) , \tag{2.24}$$

where R is defined as in (2.9), all fields take values in a single copy of $\mathfrak{g}$ and the trace is defined by $\operatorname{tr}(T_A T_B) = \eta_{AB}$. This is the direct generalisation of (2.8) for a generic higher-spin gauge theory. In this respect, the Chern-Simons form of the action is just a convenient tool to handle these models, but it is not crucial. One could also deal directly with (2.24), and this opens the way to the generalisations of (2.14) that we shall briefly consider in the following section.

3 Topologically massive theories

The actions (2.24) do not propagate local degrees of freedom, but in the $s = 2$ case it is known how to make the graviton massive and propagating without breaking diffeomorphism invariance [24]. In a first-order formalism this goal is achieved with the action [25]

$$S = S_{EH} + \frac{\mu^{-1}}{16\pi G}\int \operatorname{tr}\left(\omega \wedge d\omega + \frac{2}{3}\,\omega \wedge \omega \wedge \omega\right) + \int \operatorname{tr}\left(\beta \wedge T\right) , \tag{3.1}$$

where we continue to follow the notation introduced in (2.24). Therefore, all fields are one-forms taking values in $sl(2,\mathbb{R}) \sim so(1,2)$. The one-form β is a

Lagrange multiplier enforcing the torsion constraint, so that the elimination of all auxiliary fields would lead to the three-derivative action of [24].[6] Moreover, (3.1) is manifestly invariant under diffeomorphisms and local Lorentz transformations.

It is thus natural to preserve the structure (3.1) and try and enlarge the tangent algebra $sl(2,\mathbb{R})$ to a generic semisimple $\mathfrak{g}$. However, the resulting action is still manifestly invariant under diffeomorphisms and extended Lorentz transformations of the form

$$\delta e \,=\, [\, e \,,\, \Lambda\,]\,, \qquad \delta\omega \,=\, \mathcal{D}\Lambda\,, \qquad \delta\beta \,=\, [\,\beta\,,\, \Lambda\,]\,, \tag{3.2}$$

but it is not possible to tune the gauge variation of β to cancel the variation of (3.1) under extended local translations of the form

$$\delta e \,=\, \mathcal{D}\xi\,, \qquad \delta\omega \,=\, \frac{1}{l^2}\,[\, e\,,\, \xi\,]\,. \tag{3.3}$$

In both (3.2) and (3.3) fields and gauge parameters take values in $\mathfrak{g}$, so that we had to exhibit the l-dependence in (3.3). The gauge variation of (3.1) is

$$\delta S \,=\, \int \mathrm{tr}\left(\,\delta\beta\wedge T \,+\, \beta\wedge[\,R \,+\, \frac{1}{l^2}\,e\wedge e\,,\,\xi\,]\right)\,. \tag{3.4}$$

At the linearised level $\delta\beta = \mathcal{D}\xi$ implies the vanishing of (3.4): the remainder is proportional to the equations that are satisfied by the background dreibein and spin connection. At the non-linear level, however, one has at least to deform (3.3) and $\delta\beta = \mathcal{D}\xi$ in order to preserve the linearised gauge symmetry. In the $s = 2$ case we know how: the non-linear deformation leads to diffeomorphisms. In the higher-spin setup it is not clear whether (3.1) admits a larger symmetry group with respect to $\mathit{diff} \times \mathfrak{g}$ and, possibly, what its precise structure is.

The enhancement of $sl(2,\mathbb{R})$ to $sl(N,\mathbb{R})$ was already discussed in [28], where the authors studied the linearised spectrum of the resulting theory. A similar analysis was performed in [29] starting from the (A)dS generalisation of a quadratic action first proposed in [30]. Since this is the most general

[6] Without the term $\beta\wedge T$ the action (3.1) is instead equivalent to (2.24) [15]. On the other hand, adding a term $e\wedge T$ leads to a higher-spin generalisation of the Mielke-Baekler model [26]. The resulting action is still expressible as the difference of two Chern-Simons actions (by extending [27]), but the higher-spin torsion T no longer vanishes on shell.

quadratic action with three derivatives displaying the Fronsdal gauge symmetry, the authors argued that it should be related to the linearisation of an action of the type (3.1). A comparison of these results with a direct analysis of the constraint algebra of (3.1) would further clarify the nature of the symmetries of (3.1), or open the way to possible completions preserving all gauge symmetries of (2.24).[7] The same result could be achieved by deforming the symmetries (3.3) within the Noether procedure.

Higher-order gauge invariant massive theories for higher spins were also considered in [32] in a metric-like formalism. In this case the linearised gauge transformations still have the form (1.2), but the gauge parameter is not traceless. As a result, this proposal is not directly related to (2.24), that at the linearised level reproduces the Fronsdal action if one eliminates the spin connections via the generalised torsion constraint [13].

Acknowledgments: I am grateful to S. Fredenhagen, S. Pfenninger and S. Theisen for extensive discussions and collaboration on the material discussed in this note. I also would like to thank A. Anabalón, X. Bekaert, D. Francia, E. Joung, I. V. Melnikov, A. Sagnotti, L. Sindoni, M. Sivakumar and M. Taronna for stimulating discussions.

References

[1] "Higher-Spin Gauge Theories", Proceedings of the First Solvay Workshop, held in Brussels on May 12-14, 2004, eds. R. Argurio, G. Barnich, G. Bonelli and M. Grigoriev (Int. Solvay Institutes, 2006).
This collection contains some contributions related, in spirit, to the present work, including: N. Bouatta, G. Compère, A. Sagnotti, hep-th/0409068; A. Sagnotti, E. Sezgin, P. Sundell, hep-th/0501156; X. Bekaert, S. Cnockaert, C. Iazeolla, M. A. Vasiliev, hep-th/0503128.
Other recent reviews include: D. Sorokin, AIP Conf. Proc. **767** (2005) 172 [hep-th/0405069]; D. Francia, A. Sagnotti, J. Phys. Conf. Ser. **33** (2006) 57 [hep-th/0601199]; A. Fotopoulos, M. Tsulaia, Int. J. Mod. Phys. **A24** (2009) 1 [arXiv:0805.1346]; C. Iazeolla, arXiv:0807.0406; A. Campoleoni, Riv. Nuovo Cim. **033** (2010) 123 [arXiv:0910.3155]; A. Sagnotti, arXiv:1002.3388; D. Francia, Prog. Theor. Phys. Suppl. **188** (2011) 94 [arXiv:1103.0683].

[7] An analysis of fermionic higher-spin topologically massive theories was performed in [31], but the authors spotted some problems in coupling these theories to gravity.

[2] M. Taronna, arXiv:1005.3061; A. Sagnotti, M. Taronna, Nucl. Phys. **B842** (2011) 299 [arXiv:1006.5242]; A. Fotopoulos, M. Tsulaia, JHEP **1011** (2010) 086 [arXiv:1009.0727].

[3] E. Sezgin, P. Sundell, Nucl. Phys. **B644** (2002) 303 [hep-th/0205131]; I. R. Klebanov, A. M. Polyakov, Phys. Lett. **B550** (2002) 213 [hep-th/0210114].

[4] X. Bekaert, N. Boulanger, P. Sundell, arXiv:1007.0435.

[5] R. R. Metsaev, Nucl. Phys. **B759** (2006) 147 [hep-th/0512342]; R. Manvelyan, K. Mkrtchyan, W. Ruhl, Nucl. Phys. **B836** (2010) 204 [arXiv:1003.2877].

[6] M. Taronna, arXiv:1107.5843.

[7] M. A. Vasiliev, Phys. Lett. **B243** (1990) 378, Phys. Lett. **B567** (2003) 139 [hep-th/0304049].

[8] S. Giombi, X. Yin, arXiv:1105.4011.

[9] B. Binegar, J. Math. Phys. **23** (1982) 1511.

[10] C. Fronsdal, Phys. Rev. **D18** (1978) 3624.

[11] A. Campoleoni, D. Francia, J. Mourad, A. Sagnotti, Nucl. Phys. **B815** (2009) 289 [arXiv:0810.4350], Nucl. Phys. **B828** (2010) 405 [arXiv:0904.4447].

[12] M. P. Blencowe, Class. Quant. Grav. **6** (1989) 443; E. Bergshoeff, M. P. Blencowe, K. S. Stelle, Commun. Math. Phys. **128** (1990) 213.

[13] A. Campoleoni, S. Fredenhagen, S. Pfenninger, S. Theisen, JHEP **1011** (2010) 007 [arXiv:1008.4744].

[14] M. A. Vasiliev, Yad. Fiz. **32** (1980) 855 [Sov. J. Nucl. Phys. **32** (1980) 439]; V. E. Lopatin, M. A. Vasiliev, Mod. Phys. Lett. **A3** (1988) 257.

[15] A. Achúcarro, P. K. Townsend, Phys. Lett. **B180** (1986) 89; E. Witten, Nucl. Phys. **B311** (1988) 46.

[16] A. Campoleoni, S. Fredenhagen, S. Pfenninger, JHEP **1109** (2011) 113 [arXiv:1107.0290].

[17] M. Gutperle, P. Kraus, JHEP **1105** (2011) 022 [arXiv:1103.4304]; M. Ammon, M. Gutperle, P. Kraus, E. Perlmutter, arXiv:1106.4788; A. Castro, E. Hijano, A. Lepage-Jutier, A. Maloney, arXiv:1110.4117.

[18] M. Henneaux, S.-J. Rey, JHEP **1012** (2010) 007 [arXiv:1008.4579].

[19] M. R. Gaberdiel, R. Gopakumar, A. Saha, JHEP **1102** (2011) 004 [arXiv:1009.6087].

[20] A. Castro, A. Lepage-Jutier, A. Maloney, JHEP **1101** (2011) 142 [arXiv:1012.0598]; A. Castro, T. Hartman, A. Maloney, Class. Quant. Grav. **28** (2011) 195012 [arXiv:1107.5098].

[21] S. F. Prokushkin, M. A. Vasiliev, Nucl. Phys. **B545** (1999) 385 [hep-th/9806236].

[22] M. R. Gaberdiel, R. Gopakumar, Phys. Rev. **D83** (2011) 066007 [arXiv:1011.2986].

[23] M. R. Gaberdiel, T. Hartman, JHEP **1105** (2011) 031 [arXiv:1101.2910]; M. R. Gaberdiel, R. Gopakumar, T. Hartman, S. Raju, JHEP **1108** (2011) 077 [arXiv:1106.1897]; C.-M. Chang, X. Yin, arXiv:1106.2580; P. Kraus, E. Perlmutter, arXiv:1108.2567; K. Papadodimas, S. Raju, arXiv:1108.3077.

[24] S. Deser, R. Jackiw, S. Templeton, Phys. Rev. Lett. **48** (1982) 975, Annals Phys. **140** (1982) 372.

[25] S. Deser, X. Xiang, Phys. Lett. **B263** (1991) 39; S. Carlip, Nucl. Phys. **B362** (1991) 111.

[26] E. W. Mielke, P. Baekler, Phys. Lett. **A156** (1991) 399.

[27] M. Blagojević, M. Vasilić, Phys. Rev. **D68** (2003) 104023 [gr-qc/0307078].

[28] B. Chen, J. Long, J.-b. Wu, arXiv:1106.5141; B. Chen, J. Long, arXiv:1110.5113.

[29] A. Bagchi, S. Lal, A. Saha, B. Sahoo, arXiv:1107.0915, arXiv:1107.2063.

[30] T. Damour, S. Deser, Annales Poincaré Phys. Theor. **47** (1987) 277.

[31] C. Aragone, S. Deser, Class. Quant. Grav. **1** (1984) 331, Class. Quant. Grav. **2** (1985) L29.

[32] E. A. Bergshoeff, O. Hohm, P. K. Townsend, Annals Phys. **325** (2010) 1118. [arXiv:0911.3061]; E. A. Bergshoeff, M. Kovacevic, J. Rosseel, P. K. Townsend, Y. Yin, arXiv:1109.0382.

Gluons for (almost) nothing, gravitons for free.

John Joseph M. Carrasco
Stanford Institute for Theoretical Physics and Department of Physics, Stanford University, Stanford, CA 94305-4060, USA

In this talk I describe a new method for organizing Yang-Mills scattering amplitudes that allow the definition of an entire multi-loop scattering amplitude in terms of a small number of "master" graphs. A small amount of information is required from the theory, and constraints propagate this information to the full amplitude. When organized in such away corresponding gravitational amplitudes are trivially found. This talk is based on work[1–4] done in collaboration with Zvi Bern, Lance Dixon, Henrik Johansson, and Radu Roiban, and follows closely the presentation given in ref. [5].

I. INTRODUCTION

There is a relationship that is absolutely fascinating. It has been known since the 1980s that classical (tree-level) gluon scattering amplitudes in Yang-Mills theories encode all the requisite information to write down tree-level graviton scattering amplitudes in related gravity theories. This was first demonstrated in string theory by a set of relations by Kawai, Lewellen, and Tye [6]. These relations are intricate and, until fairly recently, were somewhat mysterious—involving increasingly complicated products between momentum invariants and permutations of gauge-invariant "color-stripped" Yang-Mills amplitudes. Recently we discovered it was possible to extract gravity information from gauge theory amplitudes in a very direct way when organized around certain cubic (trivalent) graphs in a particularly constrained form [1–4], see also refs. [5, 7] for reviews. The constrained gauge theory representation (the so-called BCJ representation) is one where the kinematic weights of contributing graphs manifest the same algebra as the color factors of the graphs. We refer to this as manifesting a duality or correspondence between color and kinematics. When the gauge theory amplitude is so organized, gravity amplitudes are generated trivially by taking a double copy of the Yang-Mills kinematic factor. Schematically,

$$\text{YM} \propto \sum_{g\in\text{graphs}} \frac{n(g)c(g)}{p(g)} \Rightarrow \text{GR} \propto \sum_{g\in\text{graphs}} \frac{n(g)\tilde{n}(g)}{p(g)},$$

where $n(g)$ are the kinematic numerator factors, $c(g)$ are the color factors, $p(g)$ are the propagators of the graphs given as for a scalar ϕ^3 theory, and $\tilde{n}(g)$ is simply another copy of the Yang-Mills factor $n(g)$. There is something very suggestive here, that gravitons are gluons but with a dynamic gauge group.

This relationship between kinematic numerators of gauge and gravity theories seems to generalize seamlessly to quantum (loop-level) corrections at the integrand level. Quantum corrections involve integrating over momenta not constrained by the in and out states. For the theories under consideration one can define the relevant integrands by the set of expressions which satisfy all D-dimensional unitarity cuts. Integrands are testable independently of the regularization scheme and we can use them to probe emergent quantum gravitational structure without needing to understand how ultraviolet divergences are ultimately regularized. We have verified this double-copy construction explicitly through four-loops at four-point and through three-loops at five-point in the maximally supersymmetric theories. It should be pointed out that these are the only full multi-loop amplitudes known in the maximally supersymmetric theories. Furthermore, these relationships have been confirmed through a variety of lower-loop calculations in less supersymmetric theories – even the all plus amplitude of two-loops pure-Yang-Mills, demonstrating its generality beyond ideas of supersymmetry.

What has been astonishing is that these relationships between kinematic factors have revolutionized the manner in which we now can calculate multi-loop scattering amplitudes in Yang-Mills. One might think based upon Feynman graph techniques that going to higher-loop order would be a loosing proposition: the number of graph-topologies increases exponentially with loop order. The discovered algebraic structure between graph numerators, however, drastically changes this landscape, rigidly locking the exponentially increasing number of graphs at higher-loops to the correct expression of an incredibly small number of master graphs. Through all known full multi-loop scattering amplitudes in the maximally supersymmetric theory only one non-vanishing graph needs be specified at each order. Additionally, at tree level and one-loop it is clear that only one numerator needs be specified to all multiplicity – the n-point multi-peripheral (half-ladder) diagram for n-point interactions at tree-level, and the n-gon at one loop. All other numerators are given as straightforward linear functions of those "master" numerators. Fantastically, the differences between the maximally supersymmetric theories and the pure QCD Yang-Mills contribution can be projected onto only these master-numerators.

The computational upshot of all of this is that state of the art calculations of only a few years ago, such as three-loop amplitudes in the maximally supersymmetric theory, can now be done on a blackboard. We can now envision using our developed computational resources to explore higher loops and greater multiplicity than previously imagined or attack theories with less supersymmetry – along the way answering certain open questions that have intrigued the community for years.

II. REVIEW OF COLOR-KINEMATIC DUALITY

Modern approaches to calculating scattering amplitudes, much like the analytic S-matrix program of the 1960s, rely heavily on quantum field theoretic consistency – although we do not eschew field-theoretic input when available. For example, a wealth of tree-level amplitudes for interesting theories can be fully constructed through BCFW [8] using only the on-shell three-point amplitude as input. Similarly the generalized unitarity method (unitarity method [9–12] and see recent reviews [5, 13–15]) allows the construction of all loop level contributions using as input only lower-loop sub-amplitudes. Can one further minimize the theory-specific information needed to construct full color-dressed scattering amplitudes? The duality between color and kinematic factors [1, 2] strongly suggests that the answer is yes.

The duality relies upon the cubic-graph representation of gauge theory amplitudes, which we write ,

$$\mathcal{A}_n^{(L)} = i^L g^{n-2+2L} \sum_{i\in\Gamma_3} \int \frac{d^{LD}\ell}{(2\pi)^{LD}} \frac{1}{S_i} \frac{N_i C_i}{p_{i_1}^2 p_{i_2}^2 p_{i_3}^2 \cdots p_{i_m}^2} . \tag{2.1}$$

I have made the dependence on the coupling constant explicit. Note that in general the N_i cannot be uniquely specified; if calculated by Feynman diagrams they are gauge choice dependent. If calculated by other means they are allowed to transform under a generalized gauge transformation [1, 2, 23], which corresponds to the ability to move contact terms between different graphs, while leaving the amplitude invariant. This freedom allows for many different representations of the same amplitude, some of which make special properties of the theory manifest.

One very special property is the duality between color and kinematics. This duality asserts the existence of representations eq. (2.1) where the kinematic numerators N_i obey the same general algebraic properties as the color factors C_i. Specifically, the N_i obey Jacobi relations and have antisymmetric cubic vertices analogous to the color factors, schematically,

$$\begin{aligned} N_i + N_j + N_k = 0 \quad &\Leftrightarrow \quad C_i + C_j + C_k = 0 \ , \qquad \text{(Jacobi identity)} \\ N_i \to -N_i \quad &\Leftrightarrow \quad C_i \to -C_i \qquad \text{(vertex-flip antisymmetry)} \end{aligned} \tag{2.2}$$

where the first line signifies the Jacobi identity valid for specific triplets of graphs in the amplitude, and the second line represents the action of flipping the ordering of a cubic vertex in a graph. In addition, it is often convenient at loop-level to impose the self-symmetries or graph automorphisms of the N_i, similar to the self-symmetries obeyed by the color factors C_i.

The duality between color and kinematics was first proposed by Bern, Johansson and the present author at tree-level [1], and later conjectured to hold to all loop-orders [2] in generic gauge theories. It generalizes a four-point property of tree-amplitudes observed in cuts of multi-loop-amplitudes [1], and also noticed in studies of zeros of scattering amplitudes in the 1980s [16].

We will see that the constraints imposed by the duality allow us to drastically reduce the necessary input from unitarity cuts. (As long as the duality remains a conjecture, it is of course necessary to verify any obtained amplitude on a set of spanning cuts.) We may loosely quantify the amount of theory-specific information required for the construction of an amplitude by counting the number of necessary maximal and near-maximal cuts. Alternatively, one may count the graphs used to represent the full amplitude, and compare this number to the very small number of graph numerators not completely fixed by the kinematic Jacobi relations eq. (2.2). We call the latter ones "master graphs"; from these all other contributing graph numerators can be expressed using linear relations. In section V, the four-point three-loop $\mathcal{N} = 4$ sYM amplitude, comprised of 12 graphs, is constructed using a single master graph. The numerator of this single master graph, after imposing all symmetries and duality relations, is completely fixed by a single maximal cut, illustrating the dramatic simplification involved.

Nontrivial evidence supporting the duality conjecture at loop level is found by studying the four-point amplitude of the $\mathcal{N} = 4$ sYM theory at one [17], two [18], three [2], and four loops [3], and for the five-point amplitude through three loops [4]. Evidence that the duality holds even in nonsupersymmetric theories is provided [2] by the two-loop four-gluon identical-helicity pure Yang-Mills amplitude [19].

We will elaborate on the duality using graph operators, but first it will be instructive to look at an venerable example from the literature where the duality eq. (2.2) is known to hold.

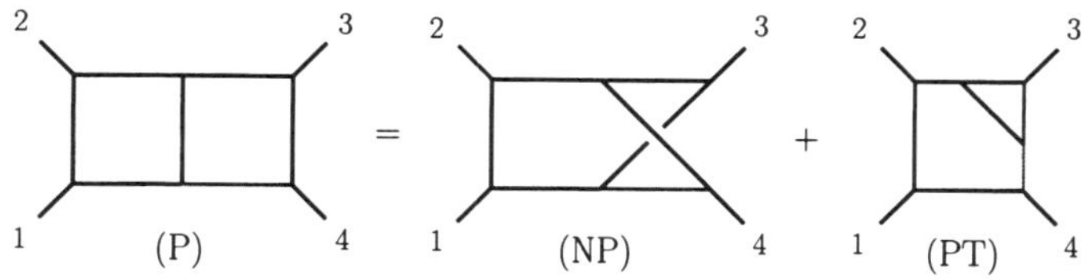

FIG. 1: A Jacobi identity between three cubic four-point two-loop topologies. These three graphs have the same connectivity except for the shaded (pink) edge. Every Jacobi identity and kinematic identity between numerator factors can be graphically seen as a relation local to a set of three edges. For the $\mathcal{N} = 4$ sYM amplitude only graphs (P) and (NP) contribute [18].

III. TWO-LOOP EXAMPLE

An instructive non-trivial example can be found using the four-point two-loop $\mathcal{N} = 4$ sYM amplitude. In the subsequent discussion we will see how such amplitudes can be directly constructed using the duality. The two-loop amplitude is

$$\mathcal{A}_4^{(2)} = -g^6 \Big(N^{(\mathrm{P})} C^{(\mathrm{P})}_{1234} I^{(\mathrm{P})}(s,t) + N^{(\mathrm{NP})} C^{(\mathrm{NP})}_{1234} I^{(\mathrm{NP})}(s,t) \;+ \text{perms} \Big) , \tag{3.1}$$

where the two scalar integrals contributing are drawn in fig. 1, and the "perms" instructs us to sum over all distinct permutations of external labels. For convenience we have pulled the kinematic (and color) factors outside the integral, as they do not depend on the loop momenta. The numerators, $N^{(\mathrm{P})}$ and $N^{(\mathrm{NP})}$ are given by

$$N^{(\mathrm{P})} = N^{(\mathrm{NP})} = \mathcal{K}\,(k_1 + k_2)^2 , \tag{3.2}$$

where $\mathcal{K}$ is the state-dependent prefactor,

$$\mathcal{K}(1,2,3,4) \equiv s_{12}\, s_{23}\, A_4^{\text{tree}}(1,2,3,4) . \tag{3.3}$$

The color factors $C^{(\mathrm{P})}$ and $C^{(\mathrm{NP})}$ are related through the Jacobi relation shown in fig. 1, to a penta-triangle graph $C^{(\mathrm{PT})}$,

$$C^{(\mathrm{P})} = C^{(\mathrm{NP})} + C^{(\mathrm{PT})} . \tag{3.4}$$

If eq. (3.1) satisfies the color–kinematics duality, we should expect a similar relation for the kinematic factors,

$$N^{(\mathrm{P})} = N^{(\mathrm{NP})} + N^{(\mathrm{PT})} . \tag{3.5}$$

This indeed holds; the amplitude eq. (3.1) does not involve any penta-triangle integral, therefore we set $N^{(\mathrm{PT})} = 0$. And thus eq. (3.5) is consistent with the result in eq. (3.2).

IV. GRAPH OPERATORS AND THE COLOR–KINEMATICS DUALITY

In order to phrase the duality precisely, I will now introduce graph operators. This is not an exercise in pedantry. Indeed, precisely defining the duality in terms of graph operators allows for natural automation—especially useful when considering higher-loop contributions that may involve many graph topologies.

The color Jacobi identity directly relates graphs that share entirely the same connectivity except for one edge, *e.g.* the shaded (pink) edge in fig. 1. The language we will use to specify this relation will involve operators that take a graph and a particular edge of that graph, and rearrange that edge's connectivity so as to get another graph. We can specify any internal edge e of a cubic graph in terms of the four edges that touch its bounding nodes. If the order of edges around one node is (a, b, e), and the order of edges around the other node is $(-e, c, d)$, we can specify the edge e as $e = \{(a, b), (c, d)\}$. With this terminology we are ready to consider some mapping $\hat{t}$ to take a graph $\mathcal{G}$ and one of its edges e, to another graph $\mathcal{G}_{t(e)}$,

$$\hat{t}(\mathcal{G}, e) = \mathcal{G}_{t(e)} , \tag{4.1}$$

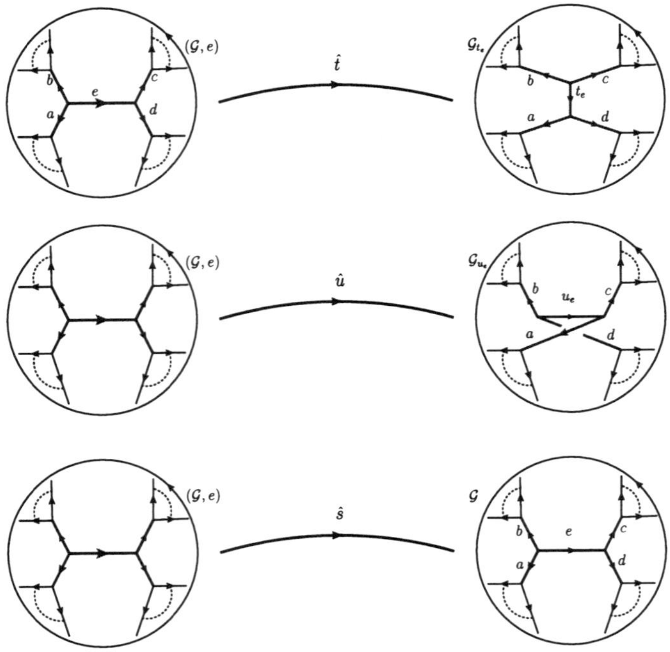

FIG. 2: Three graph operators taking (graph $\mathcal{G}$, edge e) to graphs $\mathcal{G}_{t_e}, \mathcal{G}_{u_e}$ and $\mathcal{G}$. All three graphs have the same connectivity, except for the edge touching edges (a, b, c, d).

where $\mathcal{G}_{t(e)}$ has the same connectivity of $\mathcal{G}$ except the nodes (a, b, e) and $(-e, c, d)$ have been replaced with (d, a, t) and $(-t, b, c)$, thereby defining the edge $t = \{(d, a), (b, c)\}$. We will also use

$$\hat{u}(\mathcal{G},\, e) = \mathcal{G}_{u(e)}\,, \qquad \hat{s}(\mathcal{G},\, e) = \mathcal{G}\,, \tag{4.2}$$

where $\mathcal{G}_{u(e)}$ has the same connectivity of $\mathcal{G}$ except nodes (a, b, e) and $(-e, c, d)$ have been replaced with (d, b, u) and $(-u, c, a)$, with edge $u = \{(c, a), (d, b)\}$. The $\hat{s}$ operator simply returns the original graph unchanged independent of the edge specified. These operations are depicted visually in fig. 2.

We are now ready to concisely specify the color Jacobi identity associated with every edge e of every graph $\mathcal{G}$,

$$\begin{aligned} 0 &= C\left[\hat{s}(\mathcal{G}, e)\right] - C\left[\hat{t}(\mathcal{G}, e)\right] - C\left[\hat{u}(\mathcal{G}, e)\right] \\ &= C[\mathcal{G}] - C[\mathcal{G}_{t(e)}] - C[\mathcal{G}_{u(e)}]\,, \end{aligned} \tag{4.3}$$

where $C[\mathcal{G}]$ gives the color factor arrived at by dressing every node of graph $\mathcal{G}$ with the $\tilde{f}^{abc}$ structure constants. The color–kinematics duality is satisfied in a representation where for all edges e of all graphs $\mathcal{G}$,

$$0 = N\left[\hat{s}(\mathcal{G}, e)\right] - N\left[\hat{t}(\mathcal{G}, e)\right] - N\left[\hat{u}(\mathcal{G}, e)\right]\,, \tag{4.4}$$

where $N[\mathcal{G}]$ gives the kinematic numerator factor associated with the graph $\mathcal{G}$.

Additionally, we may use this formalism to write down some relations that originate from the antisymmetry of the cubic vertices,

$$\begin{aligned} C[\mathcal{G}_{\{(d,a),(b,c)\}}] &= -C[\mathcal{G}_{\{(a,d),(b,c)\}}]\,, \\ N[\mathcal{G}_{\{(d,a),(b,c)\}}] &= -N[\mathcal{G}_{\{(a,d),(b,c)\}}]\,. \end{aligned} \tag{4.5}$$

In the following section we will see that the set of constraints on the numerator factors eq. (4.4) and eq. (4.5) can be understood as a set of functional equations that must hold for the duality to be satisfied.

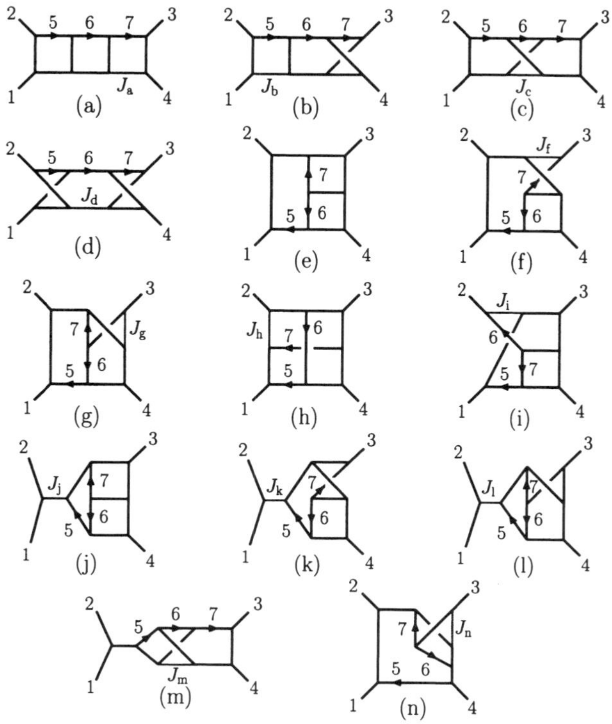

FIG. 3: Three-loop four-point cubic graphs considered in the main text. The external momenta is outgoing and the shaded (pink) edges mark the application of kinematic Jacobi relations used in eq. (5.2). Note that only graphs (a)–(l) contribute to the $\mathcal{N}=4$ sYM amplitude where the duality between color and kinematics is made manifest.

V. THREE-LOOP EXAMPLE

Now we will derive the four-point three-loop $\mathcal{N}=4$ sYM amplitude using the duality between color and kinematics [2]. This amplitude was originally given in [20, 21] in terms of nine cubic diagrams. For this exercise we start by considering a larger set of 25 graphs, which are related to any of the original nine diagrams by a single application of a kinematic Jacobi relation. However, eleven of these diagrams contain triangle subgraphs, which the no-triangle property of $\mathcal{N}=4$ sYM [22] suggests will not contribute. After removing those with one-loop triangle subgraphs we have the 14 graphs depicted in fig. 3. We will see that this set of diagrams is sufficiently large to admit a manifest representation of the duality.

Let us now introduce the kinematic Jacobi relations that the numerators of each diagram must satisfy. Each numerator depends on three independent external momenta and (at most) three independent loop momenta. While in general a graph's numerator factor will depend on the states of all external particles, the four-point amplitude of $\mathcal{N}=4$ sYM is special in that all of its state dependence can be captured in the overall factor $\mathcal{K}$ described earlier. Since this factor is common to all terms and the kinematic Jacobi relations are homogeneous, this factor plays no role in the equations. With this in mind I will suppress all state dependence but make the dependence on momenta explicit,

$$N^{(x)} = N^{(x)}(k_1, k_2, k_3, l_5, l_6, l_7)\,, \tag{5.1}$$

where the canonical labeling for each graph $x \in \{\mathrm{a,b,c,d,e,f,g,h,i,j,k,l,m,n}\}$ is given in fig. 3.

For every internal edge and for every graph there is one functional equation relating the numerator of that graph to the numerator of two other graphs through the kinematic Jacobi relations. The 14 graphs of fig. 3 each have ten propagators, giving 140 relations in total (before accounting for possible overcounts). One can write down many more equations using the graph automorphisms of each numerator – ensuring that numerators are invariant under equivalent ways of labeling symmetric graphs. However, this large set of equations contains much redundancy, the number of independent functional relations is quite small. The following 13 kinematic Jacobi relations, together with all automorphism relations, solve all constraints imposed by the color–kinematics duality:

$$\begin{aligned}
N^{(\mathrm{a})} &= N^{(\mathrm{b})}(k_1,k_2,k_3,l_5,l_6,l_7)\,, && (J_{\mathrm{a}})\\
N^{(\mathrm{b})} &= N^{(\mathrm{d})}(k_1,k_2,k_3,l_5,l_6,l_7)\,, && (J_{\mathrm{b}})\\
N^{(\mathrm{c})} &= N^{(\mathrm{a})}(k_1,k_2,k_3,l_5,l_6,l_7)\,, && (J_{\mathrm{c}})\\
N^{(\mathrm{d})} &= N^{(\mathrm{h})}(k_3,k_1,k_2,l_7,l_6,k_{1,3}-l_5+l_6-l_7)\\
&\quad + N^{(\mathrm{h})}(k_3,k_2,k_1,l_7,l_6,k_{2,3}+l_5-l_7)\,, && (J_{\mathrm{d}})\\
N^{(\mathrm{f})} &= N^{(\mathrm{e})}(k_1,k_2,k_3,l_5,l_6,l_7)\,, && (J_{\mathrm{f}})\\
N^{(\mathrm{g})} &= N^{(\mathrm{e})}(k_1,k_2,k_3,l_5,l_6,l_7)\,, && (J_{\mathrm{g}})\\
N^{(\mathrm{h})} &= -N^{(\mathrm{g})}(k_1,k_2,k_3,l_5,l_6,k_{1,2}-l_5-l_7)\\
&\quad - N^{(\mathrm{i})}(k_4,k_3,k_2,l_6-l_5,l_5-l_6+l_7-k_{1,2},l_6)\,, && (J_{\mathrm{h}})\\
N^{(\mathrm{i})} &= N^{(\mathrm{e})}(k_1,k_2,k_3,l_5,l_7,l_6)\\
&\quad - N^{(\mathrm{e})}(k_3,k_2,k_1,-k_4-l_5-l_6,-l_6-l_7,l_6)\,, && (J_{\mathrm{i}})\\
N^{(\mathrm{j})} &= N^{(\mathrm{e})}(k_1,k_2,k_3,l_5,l_6,l_7) - N^{(\mathrm{e})}(k_2,k_1,k_3,l_5,l_6,l_7)\,, && (J_{\mathrm{j}})\\
N^{(\mathrm{k})} &= N^{(\mathrm{f})}(k_1,k_2,k_3,l_5,l_6,l_7) - N^{(\mathrm{f})}(k_2,k_1,k_3,l_5,l_6,l_7)\,, && (J_{\mathrm{k}})\\
N^{(\mathrm{l})} &= N^{(\mathrm{g})}(k_1,k_2,k_3,l_5,l_6,l_7) - N^{(\mathrm{g})}(k_2,k_1,k_3,l_5,l_6,l_7)\,, && (J_{\mathrm{l}})\\
N^{(\mathrm{m})} &= 0\,, && (J_{\mathrm{m}})\\
N^{(\mathrm{n})} &= 0\,, && (J_{\mathrm{n}})
\end{aligned} \tag{5.2}$$

where $k_4 = -(k_1 + k_2 + k_3)$, $k_{i,j} = k_i + k_j$, and for notational simplicity we have suppressed the arguments of the numerators on the left-hand-sides; these all have the canonical argument $(k_1, k_2, k_3, l_5, l_6, l_7)$. The 13 relations are labeled by (J_x) which correspond to Jacobi relations specified by the marked edges in fig. 3. In the above relations we have immediately set to zero any diagrams that do not feature in fig. 3. This explains why some relations in eq. (5.2) involve fewer than three terms.

Analyzing the result of eq. (5.2) we immediately see that the color–kinematics duality has constrained $N^{(\mathrm{m})}$ and $N^{(\mathrm{n})}$ to vanish, given that we insist on the absence of graphs with one-loop triangles. We see that $N^{(\mathrm{f})}, N^{(\mathrm{g})}, N^{(\mathrm{i})}, N^{(\mathrm{j})}$ are directly expressed as linear combinations of the function $N^{(\mathrm{e})}$. Furthermore, $N^{(\mathrm{k})}, N^{(\mathrm{l})}, N^{(\mathrm{h})}$ are expressed as linear combinations of the functions $N^{(\mathrm{f})}, N^{(\mathrm{g})}, N^{(\mathrm{i})}, N^{(\mathrm{j})}$, and thus they can also be solved in terms of $N^{(\mathrm{e})}$. Finally we note that $N^{(\mathrm{a})} = N^{(\mathrm{b})} = N^{(\mathrm{c})} = N^{(\mathrm{d})}$ and that $N^{(\mathrm{d})}$ can be expressed as a linear combination of $N^{(\mathrm{h})}$. So each of these numerators can also be solved in terms of $N^{(\mathrm{e})}$. Therefore every numerator $N^{(x)}$ is entirely expressible in terms of the planar diagram numerator $N^{(\mathrm{e})}$.

Although the solution to eq. (5.2) is quite constraining, it does not enforce all of the 140 Jacobi-like relations we began with, nor does it enforce all automorphism relations. We obtain the full solution by enforcing the automorphic invariance of each graph numerator. Using the partial solution of eq. (5.2) these relations can be translated to self-constraints on $N^{(\mathrm{e})}$. For example, the fig. 3(e) diagram is invariant under reflections up $\leftrightarrow$ down, thus $N^{(\mathrm{e})}$ must satisfy

$$N^{(\mathrm{e})}(k_1,k_2,k_3,l_5,l_6,l_7) = N^{(\mathrm{e})}(k_2,k_1,k_4,k_{1,2}-l_5,l_7,l_6)\,. \tag{5.3}$$

Similarly, fig. 3(j) is invariant under $1 \leftrightarrow 2$. However, just like color factors, numerators are odd under the reordering of a cubic vertex, therefore we have

$$N^{(\mathrm{j})}(k_1,k_2,k_3,l_5,l_6,l_7) = -N^{(\mathrm{j})}(k_2,k_1,k_3,l_5,l_6,l_7)\,. \tag{5.4}$$

One can easily check that this equation is already enforced by eq. (5.2). As such this particular automorphism relation does not give rise to any new constraints on $N^{(\mathrm{e})}$. Beyond these two examples, every diagram but fig. 3(g) is constrained by automorphism relations, which, through eq. (5.2), translates to possible further constraints on $N^{(\mathrm{e})}$.

One way to efficiently solve the remaining kinematic Jacobi and automorphism relations, is to write down an ansatz for $N^{(\mathrm{e})}$. We will assume, consistent with four-points in the maximally supersymmetric theory, that the state-dependent prefactor is $\mathcal{K} \equiv s_{12}s_{23}A_4^{\mathrm{tree}}(1,2,3,4)$. We write the remaining factors as a local degree-two polynomial in 17 momentum invariants,

$$N^{(\mathrm{e})}_{\mathrm{ansatz}} = \mathcal{K}\,P(s_{12},s_{23},k_i\cdot l_j,l_i\cdot l_j)\,, \tag{5.5}$$

where $k_i = k_1, k_2, k_3$ and $l_i = l_5, l_6, l_7$, and the degree of the polynomial is given by simple dimensional counting. Although we could use various facts about the power counting of $\mathcal{N}=4$ sYM to simplify the ansatz, let us simply proceed with the most general homogeneous degree-two polynomial of 17 variables, which gives $17 \times (17+1)/2 = 153$ parameters to be determined.

Using such an ansatz, only the following four automorphism relations are required in order for all automorphic relations to be satisfied:

$$\begin{aligned} N^{(c)}(k_1,k_2,k_3,l_5,l_6,l_7) &= N^{(c)}(k_1,k_2,k_4,l_5,l_5-l_6,k_{3,4}-l_7)\,, \\ N^{(f)}(k_1,k_2,k_3,l_5,l_6,l_7) &= N^{(f)}(k_2,k_1,k_4,k_{1,2}-l_5,k_4-l_5-l_7,k_4+l_5-l_6)\,, \\ N^{(h)}(k_1,k_2,k_3,l_5,l_6,l_7) &= N^{(h)}(k_1,k_4,k_3,k_1-l_5,l_7,l_6)\,, \\ N^{(h)}(k_1,k_2,k_3,l_5,l_6,l_7) &= N^{(h)}(k_2,k_1,k_4,k_{1,2}-l_5-l_7,-l_6,l_7)\,. \end{aligned} \tag{5.6}$$

Enforcing eq. (5.2) and eq. (5.6) the ansatz is constrained from 153 parameters to only three free parameters, which we label α, β, γ, giving

$$\begin{aligned} N^{(e)}_{\rm sol} = \;& \mathcal{K}\Big[\alpha\, s_{12}s_{23} + \beta\,(s_{12}^2+s_{23}^2) + \gamma\,\Big(s_{12}^2 + 2(s_{12}-s_{23})\,k_1\cdot l_5 \\ & +2(2s_{12}+s_{23})\,k_2\cdot l_5 + 6s_{12}\,k_3\cdot l_5\Big)\Big]\,. \end{aligned} \tag{5.7}$$

After checking that this also solves the original 140 Jacobi relations and all automorphisms, we conclude that eq. (5.7) is a three-fold family of potential solutions that makes the duality between color and kinematics manifest.

One still needs to ensure that these solutions actually describes the $\mathcal{N}=4$ theory. One such additional source of theory-specific data (beyond the form of the ansatz we used) can be a generalized unitarity cut. We consider the maximal cut of graph (e), using the known expression in the literature [18],

$$N^{(e)}\Big|_{\rm cut} = 2\,\mathcal{K}\,s_{12}\,k_4\cdot l_5\,. \tag{5.8}$$

Then we compare to the maximal cut of the solution eq. (5.7),

$$N^{(e)}_{\rm sol}\Big|_{\rm cut} = \mathcal{K}\Big[\alpha\,s_{12}s_{23} + \beta\,(s_{12}^2+s_{23}^2) + \gamma\,(s_{12}s_{23} - 6s_{12}\,k_4\cdot l_5)\Big]\,, \tag{5.9}$$

where we have used the relations $k_1\cdot l_5 = 0$, $2k_2\cdot l_5 = s_{12}$, and $2k_3\cdot l_5 = -s_{12} - 2k_4\cdot l_5$, valid on the maximal cut.

Comparing eq. (5.7) and eq. (5.8), we find that $\alpha = 1/3$, $\beta = 0$, $\gamma = -1/3$, giving the fully solved numerator

$$\begin{aligned} N^{(e)} \;=\;& \frac{1}{3}\mathcal{K}\Big(s_{12}(s_{23}-s_{12}) - 2(s_{12}-s_{23})\,k_1\cdot l_5 \\ & - 2(2s_{12}+s_{23})\,k_2\cdot l_5 - 6s_{12}\,k_3\cdot l_5\Big)\,. \end{aligned} \tag{5.10}$$

All 12 nonvanishing numerators can now be obtained from eq. (5.10), using eq. (5.2). They are explicitly given in [2]. Together, they represent a duality-satisfying expression of the three-point four-loop $\mathcal{N}=4$ sYM amplitude. This was verified on a spanning set of cuts in [2].

VI. GRAVITY

As mentioned in the introduction, one can easily construct gravity amplitudes using the duality; gravity numerators are double copies of gauge theory kinematic factors [1, 2]. Given a n-point L-loop gauge theory amplitude with duality satisfying numerators N_i, a gravity amplitude takes the form

$$\mathcal{M}_n^{(L)} = i^{L+1}\left(\frac{\kappa}{2}\right)^{n-2+2L} \sum_{i\in\Gamma_3} \int \frac{d^{LD}\ell}{(2\pi)^{LD}}\frac{1}{S_i}\frac{N_i\widetilde{N}_i}{p_{i_1}^2 p_{i_2}^2 p_{i_3}^2\cdots p_{i_m}^2}\,, \tag{6.1}$$

where $\widetilde{N}_i$ are a set of kinematic numerators for the amplitude of a possibly different gauge theory (which need not explicitly satisfy the duality), and κ is the gravitational coupling constant. This formula is the generalization of the analogous tree-level ($L=0$) formula [1]. Assuming the color–kinematics duality, the double copy property was proven [23] recursively in n for the explicit case of maximal supersymmetry $[\mathcal{N}=4]\otimes[\mathcal{N}=4]\to[\mathcal{N}=8]$ and similarly for pure Yang-Mills theory going to pure gravity plus axion and dilaton. At the classical level, eq. (6.1) encodes (and arguably explains) the field-theory forms of the string-theoretic Kawai-Lewellen-Tye (KLT) tree-level relations [6] between gravity scattering amplitudes and gauge theory amplitudes. Intriguingly, the double copy form of gravity eq. (6.1) appears to generalize seamlessly to the quantum level, to any loop order [2].

VII. TREE-LEVEL

Consider the formula eq. (2.1) in the tree-level limit $L = 0$. This gives the cubic representation of tree amplitudes introduced in [1],

$$\mathcal{A}_n^{\text{tree}} = g^{n-2} \sum_{i \in \gamma_3} \frac{n_i c_i}{p_{i_1}^2 p_{i_2}^2 p_{i_3}^2 \cdots p_{i_m}^2}, \tag{7.1}$$

where the sum runs over all cubic tree-level graphs γ_3 (including all distinct permutations of external legs), which all have unit symmetry factors. For notational uniformity with [1] we use $n_i = N_i^{\text{tree}}$ and $c_i = C_i^{\text{tree}}$.

As shown in [1], assuming the duality between tree-level color and kinematics leads to novel relations between color ordered tree amplitudes. These decompose any color-ordered partial amplitude into a sum of $(n-3)!$ basis amplitudes, schematically,

$$A_n^{\text{tree}}(\sigma(1), \sigma(2), \ldots, \sigma(n)) = \sum_{\rho \in S_{n-3}} K_\rho^{(\sigma)} A_n^{\text{tree}}(1, 2, 3, \rho(4), \ldots, \rho(n)), \tag{7.2}$$

where σ and ρ are permutations, and $K_\rho^{(\sigma)}$ are some universal (state-independent) non-local kinematic coefficients. We will not give these factors here; they can be deduced using the results in [1]. Similar relations between partial amplitudes were subsequently found in string theory [24] using monodromy of integration on the worldsheet. In the field theory limit these string-relations provided a first proof of eq. (7.2). Using BCFW recursion the same relations were later derived from field theory [25]. In [26] analogous relations was shown to exist in noncommutative Yang-Mills theory.

VIII. DUALITY

As for the duality itself, phrased in terms of color and kinematic numerator factors and corresponding Jacobi relations eq. (2.2), less is known. An investigation in this direction is [27] where the framework of heterotic string theory is used to shed light on the duality. In addition, there are known explicit solutions for tree-level duality-satisfying numerators n_i to all multiplicity [28, 29]. While these solutions demonstrate the existence of numerators satisfying the duality, they take a particularly non-local and asymmetric form (under relabeling of external states). The existence of local or symmetric forms of duality-satisfying tree-level numerators to all mulitplicity remains an open question, although in a recent exploration, Johannes Broedel and I have found nice symmetric forms through six-points [30].

The underlying kinematic algebra responsible for the color-kinematic correspondence, however, is unknown beyond certain sectors in four dimensions [31]. The existence of such an algebra is suggested in general, not only by the kinematic Jacobi relations, but additionally by a trace basis identified by Bern and Dennen in ref. [32]. They present an alternative amplitude representation based on swapping the role of color and kinematics in the traditional color-trace decomposition. The partial-amplitudes in their representation involve the color-factors as numerators, and they introduce kinematic "traces" $\tau(q_1 \ldots q_m)$ in place of the trace over color-generators. It is perhaps worth noting that the numerator functions presented here, for five and six points, each lead to a symmetric τ, i.e. a single function which takes any labeling to the appropriate kinematic contribution.

I close this section by referring the reader to the literature for further application of the duality, the new tree-level amplitude relations, and the gravity double copy property [33];

IX. WHERE TO GO FROM HERE

In this talk I discussed a new approach to calculating gauge-theory amplitudes making use of the conjectured color-kinematic correspondence at loop level. Where do we want to go from here? We do not want to have to write an ansatz. Rather, we would like a direct way to write down master. As an intermediate step, we will be happy with greater control over more fluidly flowing between representations (c.f. recent work with polytopes). I conclude with a list of open future questions:

- Does the duality manifest in higher-genus perturbative string theory?
- What is the connection to recent understanding from higher-spin work – is there a path to string field theory?

- What is the relationship of double-copy construction to holography? Double-copy relationships between Yang-Mills and gravity theories seem to be a weak-weak duality that holds in the gauge theory for any number of colors. In the case of maximal supsersymmmetry in the large-N limit, we see a weak-strong duality between Yang-Mills in $4D$ and supergravity in $5D$.
- What is non-perturbative barrier to ultimately treating gravity as a double-copy of yang-mills when considering questions of information entering black holes, etc?
- What is the relationship of double-copy construction to generalized geometry? It is difficult to imagine that the types of relationships we are seeing cannot be well-expressed in terms of the so called double-field theory, or its generalization to constructions involving branes.

References

---

[1] Z. Bern, J. J. M. Carrasco and H. Johansson, Phys. Rev. D **78**, 085011 (2008) [arXiv:0805.3993 [hep-ph]].
[2] Z. Bern, J. J. M. Carrasco and H. Johansson, Phys. Rev. Lett. **105** (2010) 061602 [arXiv:1004.0476 [hep-th]].
[3] Z Bern, J. J. Carrasco, L. J. Dixon, H. Johansson, and R. Roiban, in preparation.
[4] J. J. M. Carrasco, H. Johansson, [arXiv:1106.4711 [hep-th]].
[5] J. J. M. Carrasco, H. Johansson, to appear in "Scattering Amplitudes in Gauge Theories", special issue of Journal of Physics A, R. Roiban(ed), M. Spradlin(ed), A. Volovich(ed), [arXiv:1103.3298 [hep-th]].
[6] H. Kawai, D. C. Lewellen and S. H. H. Tye, Nucl. Phys. B **269**, 1 (1986);
Z. Bern and A. K. Grant, Phys. Lett. B **457**, 23 (1999) [hep-th/9904026];
Z. Bern, A. De Freitas and H. L. Wong, Phys. Rev. Lett. **84**, 3531 (2000) [hep-th/9912033];
Z. Bern, Living Rev. Rel. **5**, 5 (2002) [gr-qc/0206071];
N. E. J. Bjerrum-Bohr, Phys. Lett. B **560**, 98 (2003) [hep-th/0302131]; Nucl. Phys. B **673**, 41 (2003) [hep-th/0305062];
N. E. J. Bjerrum-Bohr and K. Risager, Phys. Rev. D **70**, 086011 (2004) [hep-th/0407085];
S. Ananth and S. Theisen, Phys. Lett. B **652**, 128 (2007) [arXiv:0706.1778 [hep-th]];
H. Elvang and D. Z. Freedman, JHEP **0805**, 096 (2008) [arXiv:0710.1270 [hep-th]].
[7] Z. Bern, J. J. M. Carrasco and H. Johansson, arXiv:0902.3765 [hep-th];
L. J. Dixon, arXiv:1005.2703 [hep-th];
Z. Bern, J. J. M. Carrasco and H. Johansson, arXiv:1007.4297 [hep-th];
T. Sondergaard, [arXiv:1106.0033 [hep-th]].
[8] R. Britto, F. Cachazo, B. Feng, Nucl. Phys. **B715**, 499-522 (2005). [hep-th/0412308] R. Britto, F. Cachazo, B. Feng and E. Witten, Phys. Rev. Lett. **94**, 181602 (2005) [hep-th/0501052].
[9] Z. Bern, L. J. Dixon, D. C. Dunbar and D. A. Kosower, Nucl. Phys. B **425**, 217 (1994) [hep-ph/9403226]; Nucl. Phys. B **435**, 59 (1995) [hep-ph/9409265].
[10] Z. Bern, L. J. Dixon and D. A. Kosower, Nucl. Phys. B **513**, 3 (1998) [hep-ph/9708239].
[11] Z. Bern, V. Del Duca, L. J. Dixon and D. A. Kosower, Phys. Rev. D **71**, 045006 (2005) [hep-th/0410224].
[12] R. Britto, F. Cachazo and B. Feng, Nucl. Phys. B **725**, 275 (2005) [hep-th/0412103].
[13] Z. Bern and Y. -t. Huang, to appear in "Scattering Amplitudes in Gauge Theories", special issue of Journal of Physics A, R. Roiban(ed), M. Spradlin(ed), A. Volovich(ed), arXiv:1103.1869 [hep-th].
[14] L. J. Dixon, to appear in "Scattering Amplitudes in Gauge Theories", special issue of Journal of Physics A, R. Roiban(ed), M. Spradlin(ed), A. Volovich(ed), arXiv:1105.0771 [hep-th].
[15] R. Britto, to appear in "Scattering Amplitudes in Gauge Theories", special issue of Journal of Physics A, R. Roiban(ed), M. Spradlin(ed), A. Volovich(ed), arXiv:1012.4493 [hep-th].
[16] D. Zhu, Phys. Rev. D **22**, 2266 (1980);
C. J. Goebel, F. Halzen and J. P. Leveille, Phys. Rev. D **23**, 2682 (1981).
[17] M. B. Green, J. H. Schwarz and L. Brink, Nucl. Phys. B **198**, 474 (1982).
[18] Z. Bern, J. S. Rozowsky and B. Yan, Phys. Lett. B **401**, 273 (1997) [hep-ph/9702424].
[19] Z. Bern, L. J. Dixon and D. A. Kosower, JHEP **0001** (2000) 027 [arXiv:hep-ph/0001001].
[20] Z. Bern, J. J. Carrasco, L. J. Dixon, H. Johansson, D. A. Kosower and R. Roiban, Phys. Rev. Lett. **98**, 161303 (2007) [hep-th/0702112].
[21] Z. Bern, J. J. M. Carrasco, L. J. Dixon, H. Johansson and R. Roiban, Phys. Rev. D **78**, 105019 (2008) [arXiv:0808.4112 [hep-th]].
[22] Z. Bern, L. J. Dixon, D. C. Dunbar and D. A. Kosower, Nucl. Phys. B **425**, 217 (1994) [arXiv:hep-ph/9403226].
[23] Z. Bern, T. Dennen, Y. t. Huang and M. Kiermaier, Phys. Rev. D **82**, 065003 (2010) [arXiv:1004.0693 [hep-th]].
[24] N. E. J. Bjerrum-Bohr, P. H. Damgaard and P. Vanhove, Phys. Rev. Lett. **103**, 161602 (2009) [arXiv:0907.1425 [hep-th]];
S. Stieberger, arXiv:0907.2211 [hep-th].

[25] B. Feng, R. Huang and Y. Jia, Phys. Lett. B **695**, 350 (2011) [arXiv:1004.3417 [hep-th]];
Y. X. Chen, Y. J. Du and B. Feng, JHEP **1102**, 112 (2011) [arXiv:1101.0009 [hep-th]].
[26] J. H. Huang, R. Huang and Y. Jia, arXiv:1009.5073 [hep-th].
[27] S. H. Henry Tye and Y. Zhang, JHEP **1006**, 071 (2010) [arXiv:1003.1732 [hep-th]].
[28] M. Kiermaier, Amplitudes 2010, Queen Mary, University of London,
`http://www.strings.ph.qmul.ac.uk/~theory/Amplitudes2010/Talks/MK2010.pdf`
[29] N. E. J. Bjerrum-Bohr, P. H. Damgaard, T. Sondergaard and P. Vanhove, JHEP **1101**, 001 (2011) [arXiv:1010.3933 [hep-th]].
[30] J. Broedel, J. J. M. Carrasco, [arXiv:1107.4802 [hep-th]].
[31] R. Monteiro, D. O'Connell, [arXiv:1105.2565 [hep-th]].
[32] Z. Bern, T. Dennen, [arXiv:1103.0312 [hep-th]].
[33] T. Sondergaard, Nucl. Phys. B **821**, 417 (2009) [arXiv:0903.5453 [hep-th]];
C. R. Mafra, JHEP **1001**, 007 (2010) [arXiv:0909.5206 [hep-th]];
N. E. J. Bjerrum-Bohr, P. H. Damgaard, T. Sondergaard and P. Vanhove, JHEP **1006**, 003 (2010) [arXiv:1003.2403 [hep-th]];
P. Vanhove, arXiv:1004.1392 [hep-th];
Y. Jia, R. Huang and C. Y. Liu, Phys. Rev. D **82**, 065001 (2010) [arXiv:1005.1821 [hep-th]];
N. E. J. Bjerrum-Bohr, P. H. Damgaard, B. Feng and T. Sondergaard, Phys. Rev. D **82**, 107702 (2010) [arXiv:1005.4367 [hep-th]]; Phys. Lett. B **691**, 268 (2010) [arXiv:1006.3214 [hep-th]];
B. Feng and S. He, JHEP **1009**, 043 (2010) [arXiv:1007.0055 [hep-th]];
Z. Bern, J. J. M. Carrasco, L. J. Dixon, H. Johansson and R. Roiban, Phys. Rev. D **82**, 125040 (2010) [arXiv:1008.3327 [hep-th]];
N. E. J. Bjerrum-Bohr, P. H. Damgaard, B. Feng and T. Sondergaard, JHEP **1009**, 067 (2010) [arXiv:1007.3111 [hep-th]];
D. Vaman and Y. P. Yao, JHEP **1011**, 028 (2010) [arXiv:1007.3475 [hep-th]];
C. R. Mafra, arXiv:1007.4999 [hep-th];
B. Feng, S. He, R. Huang and Y. Jia, JHEP **1010**, 109 (2010) [arXiv:1008.1626 [hep-th]];
Y. Abe, Nucl. Phys. B **842**, 475 (2011) [arXiv:1008.2800 [hep-th]];
J. Bjornsson, JHEP **1101**, 002 (2011) [arXiv:1009.5906 [hep-th]];
C. R. Mafra, O. Schlotterer, S. Stieberger and D. Tsimpis, Nucl. Phys. B **846**, 359 (2011) [arXiv:1011.0994 [hep-th]];
H. Nastase and H. J. Schnitzer, JHEP **1101**, 048 (2011) [arXiv:1011.2487 [hep-th]];
J. Broedel, R. Kallosh, JHEP **1106**, 024 (2011). [arXiv:1103.0322 [hep-th]].
N. E. J. Bjerrum-Bohr, P. H. Damgaard, H. Johansson, T. Sondergaard, JHEP **1105**, 039 (2011). [arXiv:1103.6190 [hep-th]].
Y. -t. Huang, [arXiv:1104.2021 [hep-th]].
H. Nastase, H. J. Schnitzer, [arXiv:1104.2752 [hep-th]].
C. R. Mafra, O. Schlotterer, S. Stieberger, JHEP **1107**, 092 (2011). [arXiv:1104.5224 [hep-th]].
S. G. Naculich, H. Nastase, H. J. Schnitzer, [arXiv:1105.3718 [hep-th]].
Y. -J. Du, B. Feng, C. -H. Fu, JHEP **1108**, 129 (2011). [arXiv:1105.3503 [hep-th]].
B. Feng, Y. Jia, R. Huang, Nucl. Phys. **B854**, 243-275 (2012). [arXiv:1105.0334 [hep-ph]].
C. R. Mafra, O. Schlotterer, S. Stieberger, [arXiv:1106.2646 [hep-th]].
C. R. Mafra, O. Schlotterer, S. Stieberger, [arXiv:1106.2645 [hep-th]].
Z. Bern, C. Boucher-Veronneau, H. Johansson, [arXiv:1107.1935 [hep-th]].
P. A. Grassi, L. Sommovigo, [arXiv:1107.3923 [hep-th]].
S. G. Naculich, H. J. Schnitzer, [arXiv:1108.6326 [hep-th]].
Q. Ma, Y. -J. Du, Y. -X. Chen, [arXiv:1109.0685 [hep-th]].
R. H. Boels, R. S. Isermann, [arXiv:1109.5888 [hep-th]].

CP Violation in Charm: a New Method

Gianluca Inguglia
Queen Mary, University of London, Mile End Road, E1 4NS, United Kingdom
g.inguglia@qmul.ac.uk

February 21, 2013

Abstract

We propose for the first time a method to perform analysis of time-dependent CP asymmetries in charm by using both, correlated and un-correlated D^0 mesons. Here we consider the decay channels $D^0 \to K^+K^-$ and $D^0 \to \pi^+\pi^-$. The channel $D^0 \to K^+K^-$ will be used to measure the mixing phase, and the difference between the measured phase $D^0 \to K^+K^-$ and $D^0 \to \pi^+\pi^-$ will open the door to the first measurement of β_c, one of the angle of the charm unitarity triangle. Since in the standard model CP asymmetries in charm are expected to be small, any observation of large time dependent asymmetries or mismatch between predicted and observed value for β_c could signify new physics. We perform and show results of numerical analysis made considering Super$B$ running at charm threshold, Super$B$ running at $\Upsilon(4S)$ and LHCb and find that Super$B$ and LHCb will be able to measure $\beta_{c,eff}$ with a precision of 1.3° and 1.4° respectively. The same analysis shows that ϕ_{MIX} could be measured at Super$B$ with a precision of 1.3°.

1 Introduction

CP violation represents nowadays one of the most intriguing and challenging fields in physics. Assuming the Big Bang theory as the model that describes the evolution of the early Universe, today we should observe a Universe composed in equal quantities by matter and anti-matter. Since the Universe appears to be dominated in its composition only by matter, a mechanism which has favoured matter over anti-matter had to exist during the first instants of its evolution. Considering the discrete symmetries C (charge symmetry), P (parity), and their combination CP, Sakharov in 1967 proposed [1] that the violation of C and CP may have played a very important role in the evolution of the Universe, favoring matter with respect to antimatter. Since the discovery in 1964 of CP violation in the Kaon system [2], the research in the field was intensified leading to the discovery of CP violation also in the B system [3]. But in the charm sector this has not been observed, and this represents a missing piece of information in the understanding of flavor dynamics. As it is shown in the literature, for example in [4], the quantity of CP violation observed in the standard model (SM) of particle physics is not enough to account for the matter anti-matter asymmetry observed in the Universe and new studies are needed in order to look for new sources of CP violation which could explain this asymmetry. This field represents not only a necessary step forward regarding the understanding of elementary particles and flavor dynamics, but it represents a link between particle physics and cosmology.

2 CP violation in the standard model: the CKM Matrix and the Unitarity Triangle(s)

In the SM, CP violation naturally arises from a complex phase appearing in Cabibbo-Kobayashi-Maskawa (CKM) matrix [5] [6]. The matrix is a unitary 3×3 matrix which provides a description of quark mixing in terms of the coupling strengths for up to down quark type transitions, and it may be written as

$$V_{CKM} = \begin{pmatrix} V_{ud} & V_{us} & V_{ub} \\ V_{cd} & V_{cs} & V_{cb} \\ V_{td} & V_{ts} & V_{tb} \end{pmatrix}. \qquad (1)$$

Within this framework the probability to observe a transition between a quark q to a quark q' is proportional to $|V_{qq'}|^2$.

2.1 Wolfenstein and Buras parametrisations of the CKM Matrix

Different parametrisations of the CKM matrix are available, one of them is the Wolfenstein parametrisation [7] shown in Eq. (2) which is an expansion in terms of $\lambda = \sin\theta_c$, A, ρ, and η, where θ_c is the Cabibbo angle. We have chosen a variant of the Wolfenstein parametrisation elaborated by Buras et al. [8] expanded up to and including terms $\mathcal{O}(\lambda^5)$ and shown in Eq. (3)

$$V_{CKM} = \begin{pmatrix} 1-\lambda^2/2 & \lambda & A\lambda^3(\rho - i\eta) \\ -\lambda & 1-\lambda^2/2 & A\lambda^2 \\ A\lambda^3[1-\rho-i\eta] & -A\lambda^2 & 1 \end{pmatrix} + \mathcal{O}(\lambda^4), \quad (2)$$

$$V_{CKM} = \begin{pmatrix} 1-\lambda^2/2-\lambda^4/8 & \lambda & A\lambda^3(\bar\rho - i\bar\eta) + A\lambda^5(\bar\rho - i\bar\eta)/2 \\ -\lambda + A^2\lambda^5[1-2(\bar\rho+i\bar\eta)]/2 & 1-\lambda^2/2-\lambda^4(1+4A^2)/8 & A\lambda^2 \\ A\lambda^3[1-\bar\rho-i\bar\eta] & -A\lambda^2 + A\lambda^4[1-2(\bar\rho+i\bar\eta)]/2 & 1-A^2\lambda^4/2 \end{pmatrix} + \mathcal{O}(\lambda^6), \quad (3)$$

where

$$\bar\rho = \rho[1-\lambda^2/2+\mathcal{O}(\lambda^4)], \quad (4)$$

$$\bar\eta = \eta[1-\lambda^2/2+\mathcal{O}(\lambda^4)]. \quad (5)$$

The choice to consider the CKM matrix with an expansion up to $\mathcal{O}(\lambda^5)$, is due to the fact that while the expansions to $\mathcal{O}(\lambda^3)$, have been sufficient for the B factories era, now it becomes necessary to consider additional terms as we move into the era of LHCb and the Super Flavor Factories (SFF's), not only from a completeness point of view but mainly by another issue: CP violation in charm has to be very small and at $\mathcal{O}(\lambda^3)$ no CP violation would be predicted. The values of the parameters used in the CKM matrix are shown in Tab. (1).

Table 1: Constraints on the Wolfenstein parameters A, λ, ρ, η, $\bar\rho$, and $\bar\eta$ obtained by the UTFit and CKM fitter groups.

Parameter	UTFit	CKM Fitter	Mean Used
λ	0.22545 ± 0.00065	0.22543 ± 0.00077	0.22544 ± 0.00705
A	0.8095 ± 0.0095	$0.812^{+0.013}_{-0.027}$	0.811 ± 0.015
ρ	0.135 ± 0.021	–	–
η	0.367 ± 0.013	–	–
$\bar\rho$	0.132 ± 0.020	0.144 ± 0.025	0.138 ± 0.022
$\bar\eta$	0.358 ± 0.012	$0.342^{+0.016}_{-0.015}$	0.350 ± 0.014

We adopted the convention to write the CKM matrix in terms of $\bar\rho$ and $\bar\eta$ because unitarity of the CKM matrix gives rise to six unitarity triangles in the complex plane and $\bar\rho$ and $\bar\eta$ represent the coordinates of the apex of the bd triangle given in Eq. (6). Since unitarity triangles are mathematically exact, it is very important to measure their angles and sides to verify unitarity. This allows one to check if the CKM mechanism is the ultimate answer to the CP violation problem in the quark sector or if some new physics (NP) effects have still to be considered. In this sense, a recent article [9] has pointed out that the measured value for the $\sin 2\beta$ angle of the beauty-unitarity triangle differs by 3.2 standard deviation from the predicted value. In the paper the authors claim that this discrepancy could be interpreted as the fact that CKM mechanism is breaking down. Two of the six unitarity relationships of the CKM matrix may be written as

$$V_{ud}^* V_{ub} + V_{cd}^* V_{cb} + V_{td}^* V_{tb} = 0, \quad (6)$$

$$V_{ud}^* V_{cd} + V_{us}^* V_{cs} + V_{ub}^* V_{cb} = 0, \quad (7)$$

where Eq. (6) is a well known and studied case referred to as the bd triangle or unitarity triangle (UT), Eq. (7) represents the cu triangle that we will call the *charm* unitarity triangle or simply *charm* triangle.

2.2 The *charm* triangle

In the previous section we have seen that Eq. (7) defines the *charm* unitarity triangle. The angles of the charm unitarity triangle given in Eq. (7) are

$$\begin{aligned} \alpha_c &= \arg\left[-V_{ub}^* V_{cb}/V_{us}^* V_{cs}\right], &(8)\\ \beta_c &= \arg\left[-V_{ud}^* V_{cd}/V_{us}^* V_{cs}\right], &(9)\\ \gamma_c &= \arg\left[-V_{ub}^* V_{cb}/V_{ud}^* V_{cd}\right]. &(10)\end{aligned}$$

Using the averages of CKM Fitter and UTFit values for A, λ, $\bar{\rho}$ and $\bar{\eta}$ and their errors, we predict that, to order λ^6

$$\begin{aligned} \alpha_c &= (111.5 \pm 4.2)^\circ, &(11)\\ \beta_c &= (0.0350 \pm 0.0001)^\circ, &(12)\\ \gamma_c &= (68.4 \pm 0.1)^\circ. &(13)\end{aligned}$$

These predictions for the angles of the charm triangle could, and should, be tested experimentally, either directly (through time-dependent CP asymmetries) or indirectly (through measurements of the sides of the triangle). On comparing Eq. (9) with Eq. (3), one can see that $V_{cd} = V_{cd}e^{i(\beta_c-\pi)}$. The *charm* triangle is shown in Fig. (1), where the sides are not in scale. If one takes the left side of Fig. (1) and divides through by $V_{us}^* V_{cs}$, one obtain the triangle shown

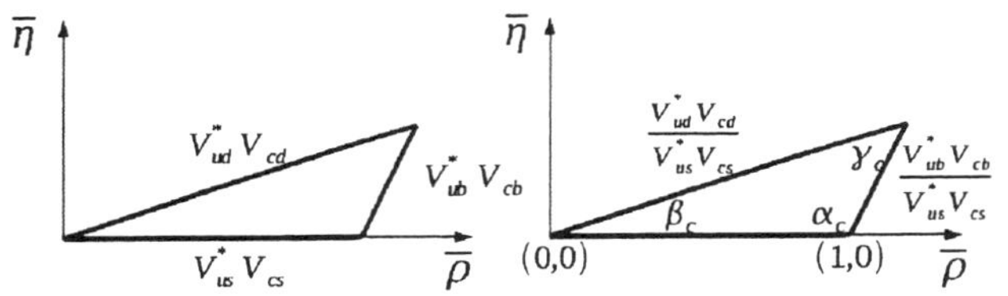

Figure 1: The *charm* triangle before (left) and after (right) baseline normalization.

on the right.

3 Time-dependent formalism

Neutral meson systems $(K, D, B_{d,s})$ do exhibit mixing which may be expressed by

$$i\frac{\partial}{\partial t}\begin{pmatrix} |P^0\rangle \\ |\overline{P}^0\rangle \end{pmatrix} = H_{eff}\begin{pmatrix} |P^0\rangle \\ |\overline{P}^0\rangle \end{pmatrix}, \tag{14}$$

where $|P^0\rangle$ and $|\overline{P}^0\rangle$ are the strong eigenstates of the mesons and the matrix $H_{eff} = M - i\Gamma/2$ appearing on the right-hand side represents the effective Hamiltonian which describes the mixing. The mass eigenstates may be written as

$$|P_{1,2}\rangle = p|P^0\rangle \pm q|\overline{P}^0\rangle, \tag{15}$$

where $q^2 + p^2 = 1$ to normalize the wave function, and

$$\frac{q}{p} = \sqrt{\frac{M_{12}^* - i\Gamma_{12}^*/2}{M_{12} - i\Gamma_{12}/2}}, \tag{16}$$

it is important to mention that if $|q/p| \neq 1$ then CP violation in mixing occurs.

3.1 Time-evolution

Before discussing the time-dependent formalism it is useful to distinguish between two different cases of D^0 meson production: un-correlated and correlated D^0 production. Un-correlated D^0's are produced from the decays of B mesons in electron-positron colliders when particles are collided at a center of mass energy corresponding to the $\Upsilon(4S)$ resonance, or from $c\bar{c}$ continuum, or in hadron machines they can be produced both promptly or as decay products of heavier particles. The correlated D^0 mesons are instead produced in an electron-positron machine running at a center of mass energy corresponding to the $\Psi(3770)$ resonance. We have obtained the time evolution for both situations described above and it can be shown that these are given by [10]

(un-correlated case)

$$\Gamma(P^0 \to f) \propto e^{-\Gamma_1 t}\left[\frac{(1+e^{\Delta\Gamma t})}{2} + \frac{Re(\lambda_f)}{1+|\lambda_f|^2}(1-e^{\Delta\Gamma t}) + e^{\Delta\Gamma t/2}\left(\frac{1-|\lambda_f|^2}{1+|\lambda_f|^2}\cos\Delta M t - \frac{2Im(\lambda_f)}{1+|\lambda_f|^2}\sin\Delta M t\right)\right], \quad (17)$$

$$\Gamma(\overline{P}^0 \to f) \propto e^{-\Gamma_1 t}\left[\frac{(1+e^{\Delta\Gamma t})}{2} + \frac{Re(\lambda_f)}{1+|\lambda_f|^2}(1-e^{\Delta\Gamma t}) + e^{\Delta\Gamma t/2}\left(-\frac{1-|\lambda_f|^2}{1+|\lambda_f|^2}\cos\Delta M t + \frac{2Im(\lambda_f)}{1+|\lambda_f|^2}\sin\Delta M t\right)\right], \quad (18)$$

(correlated case)

$$\Gamma(P^0 \to f) \propto e^{-\Gamma_1|\Delta t|}\left[\frac{h_+}{2} + \frac{Re(\lambda_f)}{1+|\lambda_f|^2}h_- + e^{\Delta\Gamma|\Delta t|/2}\left(\frac{1-|\lambda_f|^2}{1+|\lambda_f|^2}\cos\Delta M\Delta t - \frac{2Im(\lambda_f)}{1+|\lambda_f|^2}\sin\Delta M\Delta t\right)\right], \quad (19)$$

$$\Gamma(\overline{P}^0 \to f) \propto e^{-\Gamma_1|\Delta t|}\left[\frac{h_+}{2} + \frac{Re(\lambda_f)}{1+|\lambda_f|^2}h_- + e^{\Delta\Gamma|\Delta t|/2}\left(-\frac{1-|\lambda_f|^2}{1+|\lambda_f|^2}\cos\Delta M\Delta t + \frac{2Im(\lambda_f)}{1+|\lambda_f|^2}\sin\Delta M\Delta t\right)\right], \quad (20)$$

where

$$h_\pm = 1 \pm e^{\Delta\Gamma|\Delta t|}. \quad (21)$$

and

$$\lambda_f = \frac{q}{p}\frac{\overline{A}}{A}, \quad (22)$$

here q and p are the parameters defining the mixing and A ($\overline{A}$) is the amplitude for the P ($\overline{P}$) decay to a final state f. If $|A|^2 \neq |\overline{A}|^2$ there is direct CP violation (in the decay). The study of λ_f (which should not be confused with the term λ appearing in the CKM matrix) is able to probe the combination of CP violation due to mixing and due to decay, and this form of CP violation is referred to as CP violation in the interference between mixing and decay.

3.2 Time-dependent asymmetries

Considering Eq. (17,18,19,20) we calculated the time-dependent asymmetries associated to the time evolution of the D^0 mesons. The asymmetry is written as follows

$$\mathcal{A}(t) = \frac{\overline{\Gamma}(t) - \Gamma(t)}{\overline{\Gamma}(t) + \Gamma(t)}, \quad (23)$$

we then obtain
(un-correlated case)

$$\mathcal{A}(t) = 2e^{\Delta\Gamma t/2}\frac{(|\lambda_f|^2 - 1)\cos\Delta M t + 2Im\lambda_f \sin\Delta M t}{(1+|\lambda_f|^2)(1+e^{\Delta\Gamma t}) + 2Re\lambda_f(1-e^{\Delta\Gamma t})}, \quad (24)$$

$$(25)$$

in the limit of $\Delta\Gamma = 0$ Eq. (24) becomes

$$\mathcal{A}(t) = -C\cos\Delta M t + S\sin\Delta M t, \quad (26)$$

where

$$S = \frac{2Im\lambda_f}{1+|\lambda_f|^2}, \text{ and } C = \frac{1-|\lambda_f|^2}{1+|\lambda_f|^2}. \quad (27)$$

(correlated case)

In the case of correlated mesons the asymmetry is obtained, as the time evolution, in terms of Δt

$$\mathcal{A}(\Delta t) = \frac{\overline{\Gamma}(\Delta t) - \Gamma(\Delta t)}{\overline{\Gamma}(\Delta t) + \Gamma(\Delta t)} = 2e^{\Delta\Gamma|\Delta t|/2} \frac{(|\lambda_f|^2 - 1)\cos\Delta M \Delta t + 2Im\lambda_f \sin\Delta M \Delta t}{(1+|\lambda_f|^2)h_+ + 2h_- Re\lambda_f}. \quad (28)$$

3.3 Flavor tagging and feasibility of the study in different experimental environments

In this section we define the different ways to tag the D^0 mesons and identify consequently their flavor. The aim of flavor tagging is to identify one (or more) flavor specific final states that can be unambiguously used to determine the flavor of a neutral meson decaying to a CP eigenstate that we want to study. This is a probabilistic assignment and it has consequently an associated probability to be incorrect, which is called mis-tag probability, denoted by ω. This is an important quantity to determine how good a channel is for flavor tagging. A related quantity to determine the power of a channel for flavor tagging is the dilution $D = 1 - 2\omega$: the closer the dilution is to 1, the better the channel is for flavor tagging. Once again we need to distinguish the two cases of un-correlated and correlated production.

(un-correlated case) To flavor tag un-correlated D^0 mesons it is necessary to identify a low momentum pion (slow pion) since the process is driven by $D^{*+} \to D^0\pi^+$ and $D^{*-} \to \overline{D}^0\pi^-$. In this way if the the neutral D meson is produced in association with a π^+ the flavor of the neutral D meson is D^0, while if the neutral D meson is produced in association with a π^- the flavor of the neutral D meson is $\overline{D}^0$.

(correlated case) The flavor tag of D^0 mesons produced in electron-positron collider running at $\Psi(3770)$ is similar to that to flavor tag B^0 mesons produced at $\Upsilon(4S)$. It is possible for example to consider only semi-leptonic decay of the neutral D^0 mesons to tag and understand the flavor (Fig. (2)). The processes $D \to K^{(*)-}l^+\nu$ accounts for 11% of the D decays. The D^0 is associated with a l^+ in the final state, the $\overline{D}^0$ is associated to a l^-: the flavor is unambiguously assigned, with a mistag probability ω which is essentially zero.

To evaluate the feasibility of this study we consider now three possible scenarios: a) LHCb, b) Super$B$ (or Belle II) running at the $\Upsilon(4S)$ or c) Super$B$ at the $\Psi(3770)$. We consider the proven performance of *BaBar* [11] and Belle and the current performance of LHCb [12] to estimate the performance in the scenarios that we are going to examine. For decays to CP eigenstates (i.e. $D^0 \to K^+K^-$, $D^0 \to \pi^+\pi^-$) at LHCb the yield from a sample of 5 fb^{-1} will probably be comparable to that of Super$B$ (Belle II). However since at LHCb hadrons are collided, the levels of background will be higher if compared to those at Super$B$. If one considers the run at charm threshold, then the luminosity will be lower by a factor 10, and the background levels will be lower if compared to those of a run at $\Upsilon(4S)$. To perform a time-dependent analysis the D^0 mesons must to be produced in flight in the laboratory reference frame, and it is then necessary that the flight length of the D^0's exceeds the detector resolution, this will allow one to identify the decay vertex of both D^0 $\overline{D}^0$ and perform the analysis. At LHCb D^0 mesons are produced both promptly and in B decays, with a momentum of hundreds of GeV/c, and it will be important to separate the two different cases. At Super$B$ (Belle II) prompt D^0's from e^+e^- continuum may be easily separated from those from B decays by applying a momentum cut. At charm threshold D^0 mesons are produced in a coherent pair from the decay of the $\Psi(3770)$, and they will have a small momentum, consequently the possibility of the measurement of Δt strongly relies upon the boost. It is important to mention that Super$B$ will be the first asymmetric electron-positron collider running at charm threshold. Certainly a finer granularity in the time-dependent analysis is possible at LHCb than in the other scenarios. Simulations are needed in order to estimate precision achievable in the three scenarios described above.

3.4 Mistag probability and dilution: effects on the asymmetry

We have already mentioned that flavor tagging of neutral mesons is a probabilistic approach to identify the flavor and is affected by mis-tag probability. To account for the mis-tag probabilities one has to consider the physical decay rates instead of the theoretical ones

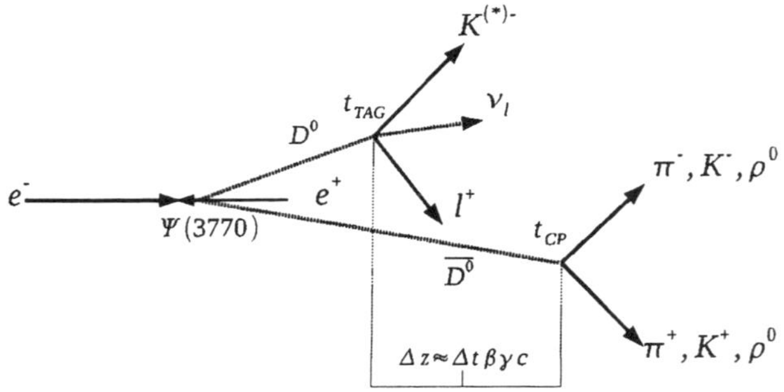

Figure 2: An example of a semi-leptonic tagged event at *charm* threshold.

$$\Gamma^{Phys}(t) = (1-\omega)\Gamma(t) + \overline{\omega}\,\overline{\Gamma}(t), \tag{29}$$

$$\overline{\Gamma}^{Phys}(t) = \omega\Gamma(t) + (1-\overline{\omega})\overline{\Gamma}(t), \tag{30}$$

where $\Gamma(t)$ and $\overline{\Gamma}(t)$ are from Eqns. (17) and (18) and $\overline{\omega}$ represents the mistag probability for anti-particles. Hence for un-correlated mesons we now can write the CP asymmetry

$$\mathcal{A}^{Phys}(t) = \frac{\overline{\Gamma}^{Phys}(t) - \Gamma^{Phys}(t)}{\overline{\Gamma}^{Phys}(t) + \Gamma^{Phys}(t)} = \Delta\omega + \frac{(D-\Delta\omega)e^{\Delta\Gamma t/2}[(|\lambda_f|^2-1)\cos\Delta Mt + 2Im\lambda_f \sin\Delta Mt]}{h_+(1+|\lambda_f|)^2/2 + Re(\lambda_f)h_-}, \tag{31}$$

where $\Delta\omega = \omega - \overline{\omega}$.

With a similar procedure we obtain for the correlated mesons

$$\Gamma^{Phys}(\Delta t) = (1-\overline{\omega})\Gamma(\Delta t) + \omega\,\overline{\Gamma}(\Delta t), \tag{32}$$

$$\overline{\Gamma}^{Phys}(\Delta t) = \overline{\omega}\,\Gamma(\Delta t) + (1-\omega)\overline{\Gamma}(\Delta t), \tag{33}$$

where $\Gamma(\Delta t)$ and $\overline{\Gamma}(\Delta t)$ are now from Eqns. (19) and (20), and

$$\mathcal{A}^{Phys}(\Delta t) = \frac{\overline{\Gamma}^{Phys}(\Delta t) - \Gamma^{Phys}(\Delta t)}{\overline{\Gamma}^{Phys}(\Delta t) + \Gamma^{Phys}(\Delta t)} = -\Delta\omega + \frac{(D+\Delta\omega)e^{\Delta\Gamma|\Delta t|/2}[(|\lambda_f|^2-1)\cos\Delta M\Delta t + 2Im\lambda_f \sin\Delta M\Delta t]}{h_+(1+|\lambda_f|)^2/2 + Re(\lambda_f)h_-}. \tag{34}$$

4 CP eigenstates

In our study we have considered a sample of 35 two and three body CP eigenstate decay channels of D^0 mesons and we determined the CKM matrix elements contributions to the decay amplitudes [10]. From these we evaluated the information on the weak phase that may be extracted by each decay. Here we concentrated on two of these channels: $D^0 \to K^+K^-$ and $D^0 \to \pi^+\pi^-$. We are interested in the value of λ_f and when exploring CP violation, this may be written as

$$\lambda_f = \left|\frac{q}{p}\right| e^{i\phi_{MIX}} \left|\frac{\overline{A}}{A}\right| e^{i\phi_{CP}}, \tag{35}$$

where ϕ_{MIX} is the phase of the mixing between D^0 and $\overline{D}^0$ mesons, ϕ_{CP} the overall CP phase related to decay of the D^0 meson to specific final CP eigenstate f_{CP}, $\overline{A}$ and A represent the amplitude of the decay for $\overline{D}^0$ and D^0. The amplitude A ($\overline{A}$) is determined by the contribution of different decay topologies and consequently it is not always related to one angle of the *charm* triangle. To write the amplitude one has to consider that tree (T), color suppressed tree (CS), weak exchange (W), and penguin (P_q) topologies (amplitudes) may contribute to the decay channel

$$A \;=\; |T|e^{i(\phi_T+\delta_T)} + |CS|e^{i(\phi_{CS}+\delta_{CS})} + |W|e^{i(\phi_W+\delta_W)} + \sum_{q=d,s,b} |P_q|e^{i(\phi_{P_q}+\delta_{P_q})}, \tag{36}$$

where (for $j = T, CS, W, P_q$) ϕ_j are the weak phases (which change sign under CP transformations) of the different topologies, the moduli in front to the exponentials are their magnitude and δ_j are the strong phases (which are invariant under CP transformations) associated to each amplitude.

Table 2: CP eigenstate modes considered in this paper indicating the topologies contributing to each process in terms of the CKM factors associated with T (tree), P_q (penguin where q is a down-type quark), and W_{EX} (W-exchange) transitions. Blank entries in the table denote that a given topology does not contribute to the total amplitude of the decay, and the relative strengths of these amplitudes decrease from left to right (see [10] for the full table).

mode	η_{CP}	T	P_q	W_{EX}
$D^0 \to K^+K^-$	$+1$	$V_{cs}V_{us}^*$	$V_{cq}V_{uq}^*$	
$D^0 \to \pi^+\pi^-$	$+1$	$V_{cd}V_{ud}^*$	$V_{cq}V_{uq}^*$	$V_{cd}V_{ud}^*$

The decays $D^0 \to K^+K^-$ and $D^0 \to \pi^+\pi^-$ are tree dominated and assuming negligible other contributions (this would be adequate for a first rudimentary time-dependent CP asymmetry measurement), it follows that

$$\lambda_f = \left|\frac{q}{p}\right| e^{i\phi_{MIX}} e^{-2i\phi_T^W}, \tag{37}$$

where $|T|$ and the strong phase $e^{i\delta_T}$ cancel in the ratio of $\overline{A}/A$. It is clear that if one wants to relate the CP asymmetry to one of the angles of the *charm* triangle, then the role of the penguin contribution must to be understood and a precision measurement of q/p must be obtained. If now we consider the elements of Table 2 we can evaluate the products of the CKM factors involved in the different processes, and these are:

$$V_{cs}V_{us}^* \;=\; \lambda - \frac{\lambda^3}{2} - \left(\frac{1}{8} + \frac{A^2}{2}\right)\lambda^5, \tag{38}$$

$$V_{cd}V_{ud}^* \;=\; -\lambda + \frac{\lambda^3}{2} + \frac{\lambda^5}{8} + \frac{A^2\lambda^5}{2}[1 - 2(\bar\rho + i\bar\eta)], \tag{39}$$

$$V_{cb}V_{ub}^* \;=\; A^2\lambda^5(\bar\rho + i\bar\eta), \tag{40}$$

Two of these three amplitudes are complex, $V_{cb}V_{ub}^*$ and $V_{cd}V_{ud}^*$. $V_{cb}V_{ub}^*$ has a large phase (γ_c), while $V_{cd}V_{ud}^*$ is related to a small phase (β_c).

4.1 $D^0 \to K^+K^-$

The channel $D^0 \to K^+K^-$ measures the phase of $V_{cd}V_{ud}^*$ only in a sub-dominant penguin transition (of the order of λ^5), and is otherwise dominated by a real tree amplitude with a CKM factor of $V_{cs}V_{us}^*$ (of the order of λ). Hence to first order one would expect to observe an asymmetry consistent with the mixing phase ϕ_{MIX}, with no CKM weak phase contribution. This channel provides measurements of $|q/p|$ and ϕ_{MIX} to complement others that may be available. Given that the SM prediction of the asymmetry in this channel is small, this is also an ideal mode to use when searching for NP. It is interesting to note that V_{cs} is complex at $\mathcal{O}(\lambda^6)$ using the convention of [8]. A measurement of β_c could be possible, this is not the most promising mode to measure the angle.

4.2 $D^0 \to \pi^+\pi^-$

$D^0 \to \pi^+\pi^-$ measures the phase of $V_{cd}V_{ud}^*$ in the leading order tree, one of the penguin amplitudes, and the W exchange topologies. Of the remaining two penguin amplitudes that contribute to this decay, one is completely negligible (mediated by a b quark loop) and the other is of the order of λ. The non-trivial penguin topologies are doubly Cabibbo suppressed loops and proceed at order λ^2, where as the tree amplitude is singly Cabibbo suppressed. A rudimentary measurement of this process could in principle ignore the penguin contribution, in which case $Im\lambda_f \simeq \sin(\phi_{MIX} - 2\beta_c)$. Thus there will be a four-fold ambiguity in any measurement of β_c. However one should note that a more complete analysis would be required in order to extract the weak phase and disentangle the contribution from the $c \to s \to u$ penguin.

5 Time-dependent CP violation: numerical analysis

In this section the three possible experimental scenarios described in Section 3.3 are compared. For the numerical analysis and the extrapolation to the expected precision on β_c we have generated a set of hundred Monte Carlo data samples assuming *SuperB* running at charm threshold (semi-leptonic tagging plus kaon tagging), Super$B$ running at $\Upsilon(4S)$ and the LHCb case, respectively. We generated events according to the time dependences described in Sec. III, and we perform a binned fit to the resulting asymmetry using Eq. (31) and Eq. (34), where $\arg(\lambda_f)$ and $|\lambda_f|^2$ are allowed to vary in the fit to data. This analysis was repeated for different possible values of the phase (i.e. $\arg(\lambda_f) = 0°, 10°$, etc.). The fit results are then expressed in terms of the two parameters and their relative errors and are used to compute the achievable sensitivity on β_c, as described in Sec. 5.1.

(SuperB at $\Psi(3770)$) D meson pairs produced at the $\psi(3770)$ are quantum-correlated, so that the time evolution is given by Eqns. (19) and (20). If one accounts for tagging dilution, then the time-dependent CP asymmetry is given by Eq. (34). On restricting time-dependent analysis to using only semi-leptonic tagged decays, the asymmetry will simplify as there is no dilution, hence both ω and $\Delta\omega$ terms can be neglected, and any systematic uncertainty on the asymmetry arising from $D \simeq 1$ will be small. Furthermore the $e^+e^- \to \psi(3770)$ environment is extremely clean, so that any systematic uncertainty from background contributions will be small and under control. These are important points to stress as we know that the CP phase of interest is small, hence in order to make a precision measurement the systematic uncertainties must be minimized.

With $500\,\text{fb}^{-1}$ of data at charm threshold one can expect to accumulate approximately 1.89×10^9 D meson pairs. We anticipate that Super$B$ could record $158,000$ $X^+e\nu_e$ tagged events, corresponding to 489500 events when using the full set of $K^{(*)}\ell\nu_\ell$ tagged events, $\ell = e, \mu$ [10]. In order to estimate the precision with which one can measure $\arg(\lambda_f)$, we perform unbinned likelihood fits to a number of simulated experiments, each consisting of D^0 and $\overline{D}^0$ tagged signal events, with distributions in Δt given by Eqns. (32) and (33) and asymmetry given by Eq. (34) with $\omega = \Delta\omega = 0$. The average uncertainty $\arg(\lambda_f)$ is taken to represent the typical result of a given experiment. This numerical estimate only includes semi-leptonic tagged events. One can also consider using hadronically tagged events, for example $D^0 \to K^-X$ (K^+X), where X is anything, which correspond to 54% (3%) of all D^0 meson decays. From these modes alone, one would expect $\omega \simeq 0.03$, and that the asymmetry in particle identification of K^+ and K^- in the detector will naturally lead to a small, but non-zero value of $\Delta\omega$. We expect that there would be approximately 2.2 million kaon tagged $D^0 \to \pi^+\pi^-$ events in $500\,\text{fb}^{-1}$ at charm threshold. Using these data alone, one would be able to measure ϕ to a precision of 4°, hence if one combines the results from semi-leptonic and kaon tagged events, a precision of $\sigma_\phi \sim 3.4°$ is achievable.

(SuperB at $\Upsilon(4S)$) The scenario at the $\Upsilon(4S)$ is somewhat more complicated than the situation encountered at the $\psi(3770)$. Firstly in order to remove background from D mesons produced in B meson decay, one restricts the analysis to mesons with high momentum. In addition to non-trivial backgrounds, one also has to consider non-zero tagging dilution, where the asymmetry is that given in Eq. (31). Thus it is not obvious that $\Delta\omega$ can be neglected, and indeed $D \neq 1$. We estimate that one could reconstruct 6.6×10^6 tagged $D^0 \to \pi^+\pi^-$ events in a data sample of $75\,\text{ab}^{-1}$ with a purity of 98%.

(LHCb) The final scenario considered is that of measuring time-dependent asymmetries from uncorrelated D mesons in a hadronic environment, where dilution and background effects will be larger than at an e^+e^- machine. Given that the measurement of λ_f is expected to be dominated by systematic uncertainties, it is not clear what the ultimate precision obtained from LHCb will be. The best way to ascertain this would be to perform the measurement. We estimate that LHCb will collect 7.8×10^6 D^* tagged $D^0 \to \pi^+\pi^-$ decays in $5\,\text{fb}^{-1}$ of data, based on an initial $37\,\text{pb}^{-1}$ of data. Based on the data shown in the reference, we estimate a purity of $\simeq 90\%$ and $\omega \simeq 6\%$.

5.1 Sensitivity on β_c

We report here the summary of the sensitivity estimates made on the basis of our simulations. More accurately, since $D^0 \to K^+K^-$ measures the mixing phase and $D^0 \to \pi^+\pi^-$ measures $\phi_{MIX} + 2\beta_{c,eff}$ the difference between phases of λ_f measured in $D^0 \to K^+K^-$ and $D^0 \to \pi^+\pi^-$ decays is $\phi_{CP} = -2\beta_{c,eff}$. If loop contributions can be well measured and both long-distance and weak exchange contributions are negligible, then this constraint can be translated into a measurement of β_c. The corresponding sensitivity estimates for the different scenarios considered are summarized in Table 3. We estimate that it should be possible to measure ϕ_{CP} to $\sim 2.6°$ using this approach [10]. Assuming that penguin contributions can be measured precisely, then the error on $\beta_{c,eff}$ that could be obtained by Super$B$ would be $\sim 1.3°$. LHCb will require input from Super$B$ on the decay modes with neutral particles in the final state in order to translate a measurement of $\beta_{c,eff}$ to one on β_c. Further work is required to understand how penguins and other suppressed amplitudes affect the translation of $\beta_{c,eff}$ to β_c, however it is clear that there will be a significant contribution from penguins given the size of the $D^0 \to \pi^0\pi^0$ branching fraction.

We determined the phase $\arg(\lambda_f)$, but it is worth mentioning that we are also able to constrain $|\lambda_f|$ with the same set of measurements. If one observes $|\lambda_f| \neq 1$ in data it would constitute a measurement of direct CP violation in a given decay channel. We estimate that it should be possible to measure $|\lambda_f|$ with a statistical uncertainty of $1 - 4\%$ at future experiments such as Super$B$. Something interesting to note is that Super$B$ proponents expect to accumulate $500\,\text{fb}^{-1}$ of data at charm threshold in only three months, whereas $75\,\text{ab}^{-1}$ would require five years of running at nominal luminosity. We also obtain the expected precision for the mixing phase (Table 4) at Super$B$ to be $\pm 1.3°$ when running at center of mass energy corresponding to the $\Upsilon(4S)$ and $\pm 2.1°$ when running at the $\Psi(3770)$ center of mass energy and using semi-leptonic and kaon tagged events.

Table 3: Summary of expected uncertainties from $500\,\text{fb}^{-1}$ of data at charm threshold, $75\,\text{ab}^{-1}$ of data at the $\Upsilon(4S)$, and $5\,\text{fb}^{-1}$ of data from LHCb for $D^0 \to \pi^+\pi^-$ decays. The column marked SL corresponds to semi-leptonic tagged events, and the column SL+K corresponds to semi-leptonic and kaon tagged events at charm threshold.

Parameter	SL	SL + K	$\Upsilon(4S)$	LHCb
$\phi = \arg(\lambda_f)$	8.0°	3.4°	2.2°	2.3°
$\phi_{CP} = \phi_{KK} - \phi_{\pi\pi}$	9.4°	3.9°	2.6°	2.7°
$\beta_{c,eff}$	4.7°	2.0°	1.3°	1.4°

The method proposed here represents a new way to measure the mixing phase, since the current method to measure the mixing phase is via a time-dependent Dalitz Plot analysis of D decays to self conjugate final states.

Table 4: Summary of expected uncertainties from $500\,\text{fb}^{-1}$ of data at charm threshold and $75\,\text{ab}^{-1}$ of data at the $\Upsilon(4S)$ for $D^0 \to K^+K^-$ decays.

Parameter	SL	SL + K	$\Upsilon(4S)$
$\phi_{MIX}(\phi_{KK})$	4.8°	2.1°	1.3°

6 Conclusions

We have explored and defined for the first time the formalism required to perform time-dependent CP asymmetries in charm decays using correlated $D^0\overline{D}^0$ decays as well a D^0 mesons tagged from D^* decays, and we discussed two interesting channels: $D^0 \to \pi^+\pi^-$ and $D^0 \to K^+K^-$. One can use K^+K^- decays to measure the mixing phase precisely and other decays can be used to constrain the angle $\beta_{c,eff}$ which is one of the angles of the *charm* triangle. A data sample of $500\,\text{fb}^{-1}$ collected at charm threshold could be collected at Super$B$ in only three months and it would provide a sufficient test to constrain any potential large NP effects. Similar measurements would be possible using D^* tagged decays at Super$B$, Belle II and LHCb. We expect the statistical precision on the measured phase at Super$B$ to be slightly better than the one achievable at LHCb. The measurements proposed here provide a new set of consistency checks of CKM that can be performed using D decays. Measurements of the sides of the triangles would enable us to perform an indirect cross-check of CKM. Super$B$ has a potential advantage over other experiments as it will be able to collect data at charm threshold with a boosted center of mass, as well as being able to explore effects using neutral mesons from a D^* tagged event. Data from charm threshold will be almost pure, with a mis-tag probability of ~ 0 for semi-leptonic tagged events, which could be advantageous if systematic uncertainties dominate measurements from $\Upsilon(4S)$ data and from LHCb. Within our formalism a measurement of $|\lambda_f| \neq 1$ could also signify direct CP violation.

7 Acknowledgments

The work presented here is a contribution to the proceedings of the *"49th Course of the International School of Subnuclear Physics"* held in Erice (Italy). This study has been supported by Queen Mary, University of London and it is based on a previous study done in collaboration with Dr. A.J. Bevan (Queen Mary, University of London) and Prof. B. Meadows (University of Cincinnati) whom I wish to thank for their teaching and suggestions. I wish to thank the Ettore Majorana Foundation and Centre for Scientific Culture and Prof. A. Zichichi for inviting me at the event and for giving me the possibility to share this work with the community.

References

[1] Sakharov A. D 1967 Pisma Zh. Eksp. Teor. Fiz. 5 32-35.

[2] Christenson J.H. et al.Physical Review Letters 13 (1964), pp. 138-140.

[3] BABAR Collaboration, B. Aubert et al., Phys. Rev. Lett. 93, 131801 (2004); Belle Collaboration, Y. Chao et al., Phys. Rev. Lett. 93, 191802 (2004).

[4] J. McDonald Phys. Rev. D 53, 645654 (1996).

[5] N. Cabibbo, Phys. Rev. Lett. 10, 531 (1963).

[6] M. Kobayashi and T. Maskawa, Prog. Theor. Phys. 49, 652 (1973).

[7] L. Wolfenstein, Phys. Rev. Lett. 51, 1945 (1983).

[8] A. J. Buras, M. E. Lautenbacher, and G. Ostermaier, Phys. Rev. D50, 3433 (1994), hep-ph/9403384.

[9] E. Lunghi and A. Soni (2011), 1104.2117.

[10] A. Bevan, G. Inguglia, B. Meadows (2011), arXiv:1106.5075v2.

[11] B. Aubert et al. (BABAR), Phys. Rev. D79, 072009 (2009), 0902.1708.

[12] LHCb Collaboration LHCb-CONF-2011-023.

Mechanical resonances in the read-out chambers of the ALICE TPC

Mangus Mager for the ALICE TPC collaboration
CERN, Geneva, Switzerland
E-mail: Magnus.Mager@cern.ch

ISSP Erice, July 1st 2011

Abstract

It is reported on a subtle aspect of gated multi-wire proportional chambers (MWPCs) that was recently observed in the read-out chambers of the ALICE TPC: periodic switching of the gating grid causes sizable (order of 10 µm) mechanical oscillations of its wires. Moreover, theses oscillations resemble themselves as a systematic pick-up signal on the data with signal heights comparable to those of minimum-ionising particles.

1 Introduction

1.1 ALICE TPC

One of the biggest challenges in the analysis of heavy-ion collisions as they happen at the Large Hadron Collider (LHC) at CERN is the large number of tracks produced in a single event. At mid-rapidity charged particle multiplicities of $\mathrm{d}N_{\mathrm{ch}}/\mathrm{d}\eta \approx 1600$ were observed for head-on collisions with center of mass energies of $\sqrt{s_{NN}} = 2.76\,\mathrm{ATeV}$ [1]. To track and identify them down to low momenta one needs very light detectors with high resolutions. The dedicated heavy-ion experiment ALICE (A Large Ion Collider Experiment) at LHC uses a time projection chamber (TPC) as its main tracking device for the central barrel (see Fig. 1 and Ref. [2] for a recent review).

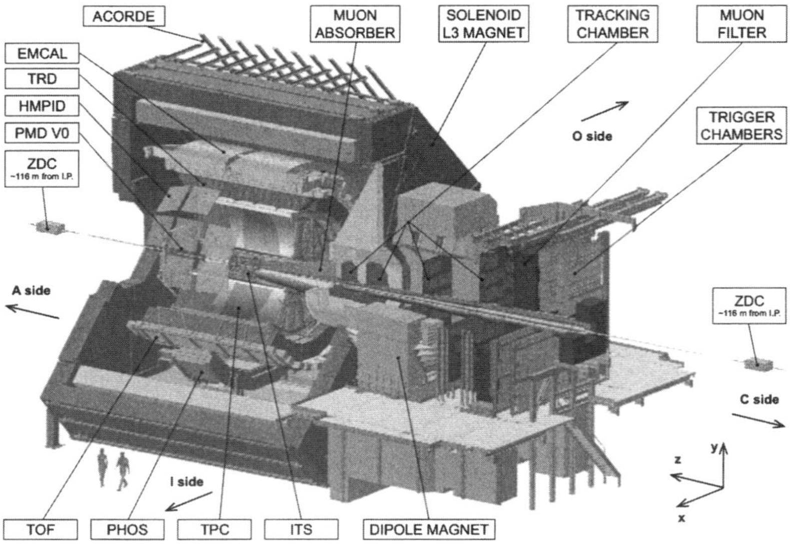

Figure 1: The layout of the ALICE experiment showing the TPC in its central barrel. (Figure taken from [2].)

The TPC is formed as a hollow cylinder with dimensions of $90 < r < 250\,\mathrm{cm}$ times $|z| < 250\,\mathrm{cm}$ (radial times beam direction). It was designed to operate at occupancies of up to 20 000 charged particles within the active volume[1] and to read out up to 200 central PbPb events per second, while keeping the material budget as low as $X/X_0 = 4.1\,\%$ [3].

A central electrode divides the drift volume at $z = 0\,\mathrm{cm}$. It is charged to a potential of $-100\,\mathrm{kV}$, defining the drift field towards the end-plates to $400\,\mathrm{V\,cm^{-1}}$. Charged particles traversing the detector ionise the gas along their tracks. The electric field causes the released ionisation electrons to drift towards the end-plates where their position and arrival time is recorded such that the full three-dimensional image of the collision ca be reconstructed. With the used gas mixture (Ne–CO_2–N_2 [85.7–9.5–4.8][2]) and at nominal conditions, the electrons drift with constant velocity (the accelerating force is compensated by scattering with the gas) of $2.65\,\mathrm{cm/\mu s}$, resulting in a maximal drift time of $94\,\mathrm{\mu s}$.

[1]This corresponds to a multiplicity of $\mathrm{d}N_{\mathrm{ch}}/\mathrm{d}\eta \approx 8000$.

[2]In 2011 the admixture of N_2 was removed, but for the measurements presented here it was still present.

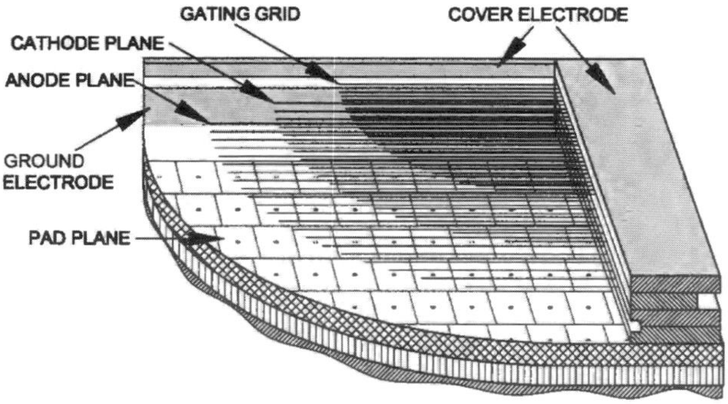

Figure 2: Cut through a read-out chamber indicating the different wire planes. (Figure taken from [2].)

1.1.1 Read-out chambers

Reaching the end-plates the electrons are amplified by multi-wire proportional chambers (MWPCs) and the signal is read-out as induced charge on the finely segmented cathode plate. The TPC is equipped with 72 of these chambers, distributed over 2 sides times 18 sectors with 2 chambers each, an inner read-out chamber (IROC) and an outer read-out chamber (OROC). Properties of these are given in Tab. 1.

The chambers are consist of three wire layers (see Fig. 2):

Anode-wire grid. To enable the gas amplification of drift electrons the anode wires are held at a high potential, creating a high electric field around them.

Cathode-wire grid. The cathode wires define the boundary between drift and amplification regions.

Gating-wire grid. The gating grid is used to make the boundary between amplification and drift either transparent or opaque, both for drift electrons and gas ions created during the gas amplification.

The gating grid is opened when the TPC is triggered and acquires data but it is closed whenever the detector is idle[3]. This is important in order to a)

[3]In spite of being a slow detector in terms of acquisition speed (limited by the drift

Table 1: Properties of the chamber wires [2].

	IROC			OROC		
	Anode	Cathode	Gating	Anode	Cathode	Gating
Material	Au plated W	Cu/Be (98%/2%)		Au plated W	Cu/Be (98%/2%)	
Density [g cm^{-3}]	19.25	8.80		19.25	8.80	
Diameter [µm]	20	75		20	75	
Pitch [mm]	2.5	2.5	1.25	2.5	2.5	1.25
Nom. tension [N]	0.45	0.6		0.45	1.2	
Length [cm]	27 to 44			45 to 84		
Res. frequency [Hz]	310 to 505	200 to 325		162 to 303	105 to 195	
(note)	($E\flat_4$ to B_4)	(G_3 to E_4)		(E_3 to $E\flat_4$)	($A\flat_2$ to G_3)	

prevent unnecessary aging of the chambers, and b) to prevent ions leaking out of the amplification region into the drift volume. Especially the latter is crucial as the ions can potentially distort the drift field, leading to a worsening of spatial resolution [4].

The opening and closure is achieved by applying different potential configurations on its wires. To open it, the wires are held at a potential that corresponds to the potential of the drift field, such that the field lines (which correspond to the electron drift paths) are not distorted. In closed mode, however, wires are biased with alternating potentials, which makes the field lines ending to them and charged particles are intercepted.

Changing the wire potentials implies changing their charges and thus also the charges induced on the pads, which results in an induced signal in the electronics. Since charge is added and removed to the wires in an alternating fashion, the net induced charge should ideally cancel. Slight asymmetries in the switching process, however, yield a non-vanishing pick-up signal (it is the difference of big numbers).

1.1.2 Read-out electronics

Two application specific integrated circuits (ASICs) form the heart of the front-end read-out electronics: A "Pre-Amplifier Shaper" (PASA) [5] and the "ALICE TPC Read-Out Chip" (ALTRO) [6]. They are mounted on 4356 front-end cards (FECs) which are fixed on the detector, only some 10 cm

time to about 100 µs), the TPC cannot be read out faster then a few 100 Hz, i. e. it is idle most of the time.

away from its end-plates. The resulting very dense and close arrangement houses 557 568 complete acquisition channels, which run in parallel.

At nominal conditions the TPC acquires data at 10 MSPS for about 1000 time bins with a resolution of 10 bit. The mean RMS noise is less than 1000 electrons or 0.7 ADC counts, while the average height of a minimum ionising particle (pions at momenta of about 400 MeV/c) is about 10 ADC counts.

Resulting event sizes are of the order of 700 MByte and ask for on detector data reduction in order to be able to ship data off detector at the foreseen rates. This is achieved by applying zero-suppression. To work efficiently it has a strong prerequisite of a equalised signal and the on-detector electronics include a sophisticated digital data processor is used for real time removal of systematic and low frequency variations from the signal. In particular each channel can store a reference acquisition to be subtracted from the signal, which is used to remove any pick-up signal synchronous to the trigger like from the switching of the gating grid.

It is crucial to open the gating grid and start the acquisition as soon as possible after a trigger, as otherwise some part of the TPC active volume is lost (2.65 cm per µs). Therefore, the gating grid was tuned carefully to induce a signal that is small enough to be removed by the ALTRO as described above.

1.2 Wire resonances

Dealing with thin, long wires at high mechanical tensions is one of the delicate design issues for MWPCs [7]. It is known from other wire chambers (not necessarily MWPCs) that oscillations of the wires can build up and lead to degradations of signal or even damages [8]. Amongst the causes for oscillations are external vibrations of the detector and electrical forces occurring due to the ionisation process.

In gated MWPCs, another cause for an exerting force on the wires is the net electrostatic force between open and closed state. This was recently observed and reported in Ref. [9] and forms the main topic of the remaining document.

2 The vibrating string

2.1 The PDE

For small excitations the wires obey the partial differential equation (PDE) of the driven, damped vibrating string:

$$\left(\frac{\partial^2}{\partial t^2} + \frac{2}{\tau}\frac{\partial}{\partial t} - c^2\frac{\partial^2}{\partial y^2}\right) z(y,t) = \frac{\Delta F(y,t)/\Delta y}{\Delta m/\Delta y}\,, \tag{1}$$

where $\Delta F(y,t)/\Delta y$ and $\Delta m/\Delta y$ are the driving force and mass per unit length, c is the velocity of sound, and τ the damping time.

If fixed on both ends and with negligible damping, it in particular yields the eigen-frequencies as:

$$f_n = \frac{n}{2L}c = \frac{n}{2L}\sqrt{\frac{T}{\varrho_w \pi r^2}}\,. \tag{2}$$

Here L is wire length, T the stretching force, ϱ_w the wire density, and r the wire radius. Due to the trapezoidal ("harp-like") shape of the read-out chambers, a range of resonant frequencies appears at the TPC. These are summarised in Tab. 1 together with the resulting musical notes for the ground mode, $n = 1$.

2.2 Driving force

The driving force of the oscillation is the net force originating from the change in charge configuration due to switching of the gating grid potentials. Since forces are acting on charges, the capacitance equation of the system has to be solved for the different potential configurations. It turns out that the anode potential plays a crucial role in the resulting force on the gating grid wires.

The time structure of the force is given by the triggering of the detector.

$$\Delta F(y,t)/\Delta y = \begin{cases} F_0 & t \in [n/f_{\text{trig.}}, n/f_{\text{trig.}} + \Delta T_{\text{GG}}] \\ 0 & \text{otherwise} \end{cases}\,, \tag{3}$$

with $n \in \mathbb{Z}$, the trigger frequency $f_{\text{trig.}}$ and the gating grid opening time ΔT_{GG}. For small gating grid opening times, this represents a Dirac comb, such that the Fourier decomposition of $F(y,t)$ is a Dirac comb as well. This in particular means that $F(y,t)$ has big contributions of integer multiples of $f_{\text{trig.}}$, which increases the probability to hit a resonant frequency.

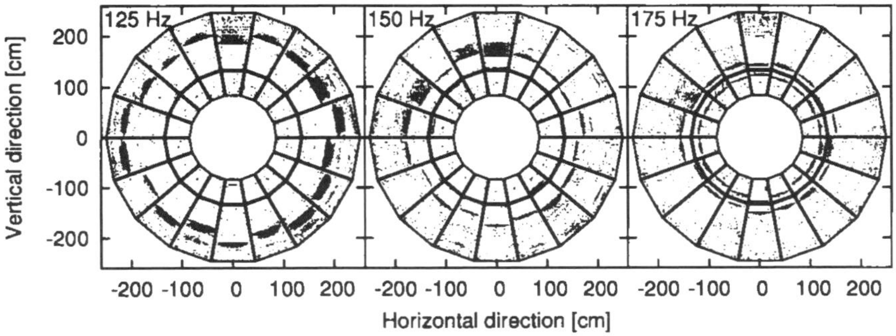

Figure 3: Event display showing the resonance patterns, caused by channels that have a signal above threshold. (Figure taken from [9].)

2.3 Damping

The damping, resulting in the damping time τ, can be attributed to three main sources:

- Air friction
- Energy transfer into the support structure
- Internal friction of the wires

This has been studied previously for musical instruments [10] and it is found there that the internal friction can typically be neglected for thin metal wires, while energy transfer into the support structure depends on the precise attachment and may reach the order of the air friction.

The air friction can be described by Stokes friction. Stokes derived the formula for the friction of cylinders in fluids [11]. He derives a (lengthly) result (Eqs. 102–105 of [11]) and a (short) approximation (last Eq. in paragraph 38 of [11]). It is important to note that while the approximation may to some extend be used to describe musical instruments, it is not applicable to the TPC chamber wires.

3 Measurements

A detailed measurement campaign was carried out to identify and understand the wire oscillations. It has been presented recently in [9] and its most important results are summarised hereunder.

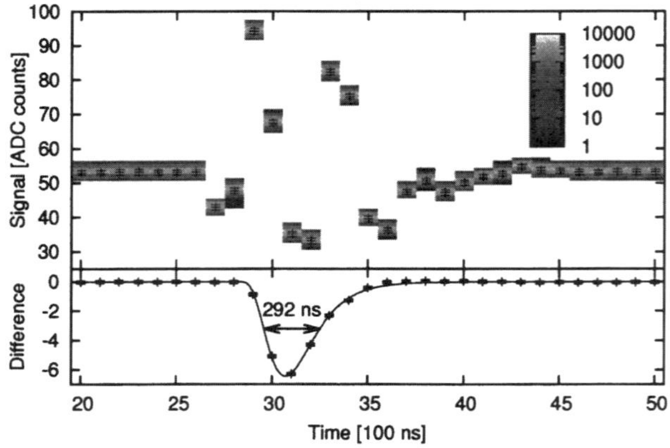

Figure 4: The timestructure of the signal induced upon opening the gating grid. (Figure taken from [9].)

3.1 TPC

The first measurements of the resonances were done during a calibration run—by accident: the trigger-related induced signal resulting from the switching of the gating grid was recorded to be used inside the ALTROs for base line correction. To verify the function of the data processing circuit, the detector was then triggered by at a fixed trigger rate (without beam) and empty events (i. e. without any pick-up signal) were expected. The result (Fig. 3), however, were line structures in the data that depended on the trigger frequency. The frequencies are found to be consistent with the mechanical eigen-frequencies of the cathode and gating wires according to Eq. (2) and Tab. 1.

To further isolate the origin of the signal non zero-suppressed data was taken. Results for a typical channel are shown in Fig. 4, displaying the induced signal upon opening the gating grid (upper plot) as well as how this deviates in case of the resonance (lower plot).

3.2 Laboratory

To characterise the oscillations further a chamber was set up in the laboratory. A microscope was used to identify the amplitude and mode of oscillation of the wires as well as the dependence on various external parameters.

The set-up allows to acquire a time-resolved video of the build-up and

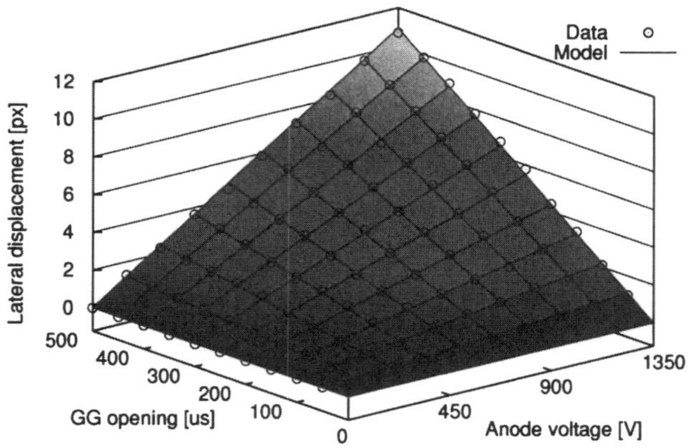

Figure 5: Amplitude dependence of the oscillation, as measured (open points) in the laboratory. The dependence is fitted with a linear function (plane), which is used to extrapolate to TPC operating conditions (1350 V and 100 µs). (Figure taken from [9].)

decay of the resonance[4]. Also does it permit the automatic measurement of the amplitude dependence on different parameters, in particular on anode voltage and gating grid opening time (Fig. 5). From this measurement on may extrapolate[5] the maximal amplitude of oscillation at the TPC operating condition to be about 10 µm.

4 Conclusions

It is shown experimentally that gated MWPCs exhibit resonant behaviour when they are triggered with at the mechanical resonance frequencies of their gating grid wires.

With a maximal amplitude of 10 µm it is assured that resonance will not cause any damage to the ALICE TPC. The pick-up signal induced at the beginning of the acquisition time, however, has implications for the calibration and operation of the detector: firstly it has to be assured that calibration data is acquired without oscillations and secondly the detector has to be op-

[4]The video may be downloaded at: `http://www-linux.gsi.de/~mmager/TIPP/`

[5]There are a few differences (gas composition and electric field configuration) between the laboratory and the experiment, which needed to be taken into account [9]

erated such that no resonance will build up. If these conditions cannot be met, the first some 500 ns (or 1.3 cm of drift) may not be recorded.

Acknowledgements

The author would like to thank Y. Lesenechal for the mechanical preparation of the laboratory set-up and R. Veenhof for his help with simulations.

References

[1] ALICE Collab., K. Aamodt, et al., *Charged-particle multiplicity density at midrapidity in central Pb-Pb collisions at* $\sqrt{s_{NN}} = 2.76\ TeV$, Phys. Rev. Lett. **105** (2010).

[2] ALICE TPC Collab., J. Alme, et al., *The ALICE TPC, a large 3-dimensional tracking device with fast readout for ultra-high multiplicity events*, Nucl. Instr. Meth. A **622** (2010).

[3] ALICE Collab., *Technical design report of the time projection chamber*, CERN/LHCC 2000-001, 2000.

[4] S. Rossegger and J. Thomas, *Space-charge effects in the ALICE TPC: a comparison between expected ALICE performance and current results from the STAR TPC*, ALICE-INT-2010-017, 2010.

[5] H.K. Solveit, *The preamplifier-shaper for the ALICE TPC-detector*, ALICE-INT-2009-039, 2009.

[6] CERN-EP/ED, *ALICE TPC readout chip user manual*, 2002.

[7] J.H. Dieperink, K. Kleinknecht, P. Steffen, J. Steinberger, T. Trippe, and F. Vannucci, *Construction problems with large proportional wire chambers*, CERN NP Internal Report 69-28, 1969.

[8] M. Anderson, J. Godlewski, A Romaniouk, S. Sutchkov, and V. Tikhomirov, *Studies of wire vibrations in TRT straw tubes*, ATL-INDET-2000-018, 2000.

[9] C.J. Horn, A. Junique, C. Lippmann, M. Mager, L. Musa, and T. Richert for the ALICE TPC Collab., *Trigger-induced mechanical resonances of gating grid wires in the multi-wire proportional chambers of the ALICE TPC*, Proceedings of TIPP 2011, to be published in Physics Procedia, 2012.

[10] Neville H. Fletcher and Thomas D. Rossing, *The physics of musical instruments*, 2nd ed. ed., Springer, 1998.

[11] George Gabriel Stokes, *On the effect of the internal friction of fluids on the motion of pendulums*, Trans. Cambridge Phil. Soc. **9** (1850).

First Observation of Positronium Hyperfine Splitting Transition - Particle Physics at a Frequency Frontier -

Akira Miyazaki

Department of Physics, Graduate School of Science
and International Center for Elementary Particle Physics (ICEPP),
The University of Tokyo, 7-3-1 Hongo, Bunkyo-ku, Tokyo, 133-0033, Japan
miyazaki@icepp.s.u-tokyo.ac.jp

Abstract

Positronium is an ideal system for the research of QED in a bound state. The hyperfine splitting of positronium (Ps-HFS: about 203 GHz) is a good tool to test QED, and is also sensitive to new physics beyond the Standard Model. Previous experimental results of Ps-HFS show $3.9\,\sigma$ (15 ppm) discrepancy from the QED prediction up to $\mathcal{O}\left(\alpha^3 \ln 1/\alpha\right)$. A first direct Ps-HFS measurement without one of the systematic uncertainties is proposed in this paper. This measurement needs progress of technology in sub-THz region. The technology for sub-THz and THz is still under development, which is called *terahertz gap*. Technological innovation can create a new probe for *the totally unexpected*. By developing three key devices in sub-THz region, the direct transition between *ortho*-positronium and *para*-positronium was observed for the first time with 5σ. The current status and future prospect are explained.

1 Introduction

Positronium (Ps), the electron-positron bound state, is a purely leptonic system. Since it is free from any hadronic interactions, positronium is a good target to precisely study Quantum Electrodynamics (QED) in a bound state system. The energy difference between *ortho*-positronium (*o*-Ps, $^3S_1$ state) and *para*-positronium (*p*-Ps, $^1S_0$ state) is called hyperfine splitting of positronium (Ps-HFS). The Ps-HFS value is approximately 203 GHz (0.84 meV).

Measurements of the Ps-HFS have been performed in the 70's and 80's [1][2]. The results were consistent with each other, and a combined accuracy of 3.3 ppm was obtained. They were consistent with a $\mathcal{O}\left(\alpha^2\right)$ prediction of the QED calculation available at that time. In 2000, the corrections up to $\mathcal{O}\left(\alpha^3 \ln 1/\alpha\right)$ have been calculated using NonRelativistic QED (NRQED), which is given by

$$\begin{aligned}\Delta^{\rm th}_{\rm HFS} &= \Delta^{\rm th}_{\rm HFS\,0}\left\{1-\frac{\alpha}{\pi}\left(\frac{32}{21}+\frac{6}{7}\ln 2\right)+\frac{5}{14}\alpha^2\ln\frac{1}{\alpha}\right.\\ &\left.+\left(\frac{\alpha}{\pi}\right)^2\left[\frac{1367}{378}-\frac{5197}{2016}\pi^2+\left(\frac{6}{7}+\frac{221}{84}\pi^2\right)\ln 2-\frac{159}{56}\zeta\left(3\right)\right]\right.\end{aligned}$$

$$-\frac{3}{2}\frac{\alpha^3}{\pi}\ln^2\frac{1}{\alpha}+C\frac{\alpha^3}{\pi}\ln\frac{1}{\alpha}+D\left(\frac{\alpha}{\pi}\right)^3+...\Bigg\}, \tag{1}$$

where the coefficient C is calculated as $C=\frac{62}{15}-\frac{68}{7}\ln 2\sim -2.6001$, and D is estimated as an error of the current calulation according to hyperfine splitting of muonium. The new prediction is $\Delta^{\rm th}_{\rm HFS}$ =203.391 69(41) GHz [3]. This calculated value differs from the measured value of $\Delta^{\rm exp}_{\rm HFS}$ =203.388 65(67) GHz by 3.9σ as shown in Fig. 1. This discrepancy may be due to common systematic uncertainties in the previous measurements or to new physics beyond the Standard Model as show in Fig. 2 or Fig. 3.

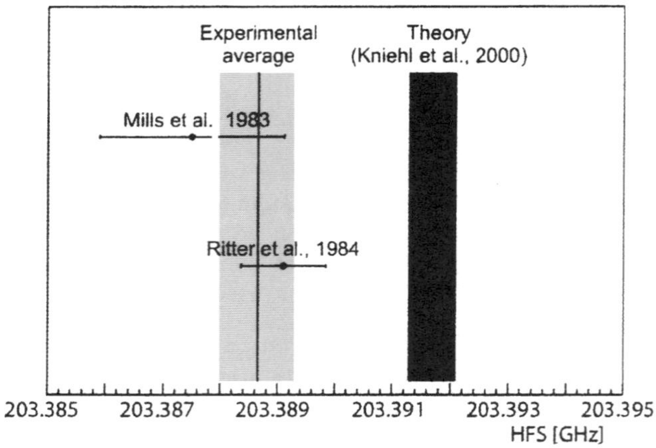

Figure 1: The discrepancy of Ps-HFS value.

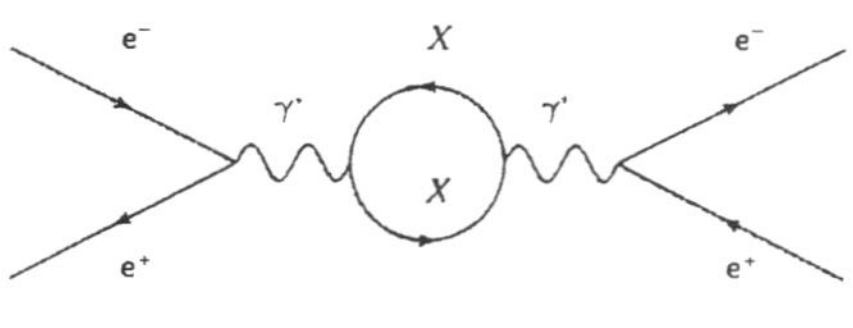

Figure 2: Contribution of unknown fermion X.

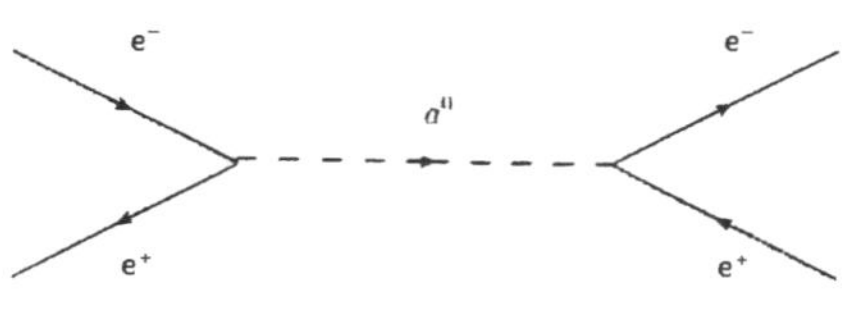

Figure 3: Contribution of unknown pseudo-scalar a_0.

In all previous measurements, the Ps-HFS value was not directly measured, since 203 GHz was too high to be produced and controlled. Zeeman splitting of o-Ps has been measured instead. A static magnetic field makes Zeeman mixing between $m_z = 0$ spin state of o-Ps and p-Ps. As a result, the energy level of $m_z = 0$ state of o-Ps becomes higher than $m_z = \pm 1$ states. This Zeeman splitting, which is approximately proportional to Ps-HFS, is a few GHz under about 1 Tesla magnetic field. A static magnetic field is applied in a RF-cavity where positronium is produced. Zeeman transition from o-Ps of $m_z = \pm 1$ to o-Ps of $m_z = 0$ has been observed. Therefore, the non-uniformity of the magnetic field might be a common systematic uncertainty in all previous experiments.

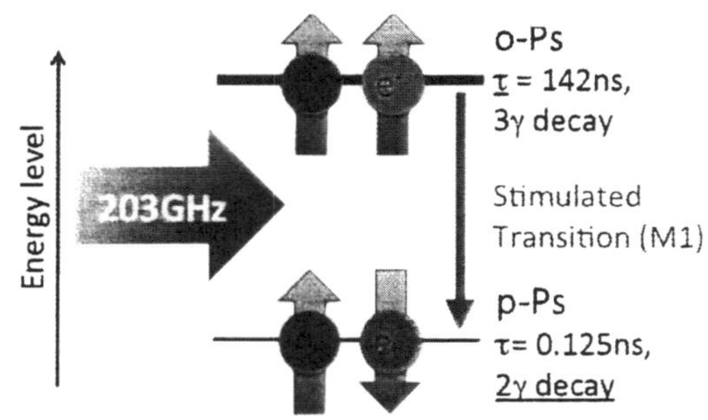

Figure 4: A schematic of the direct transition.

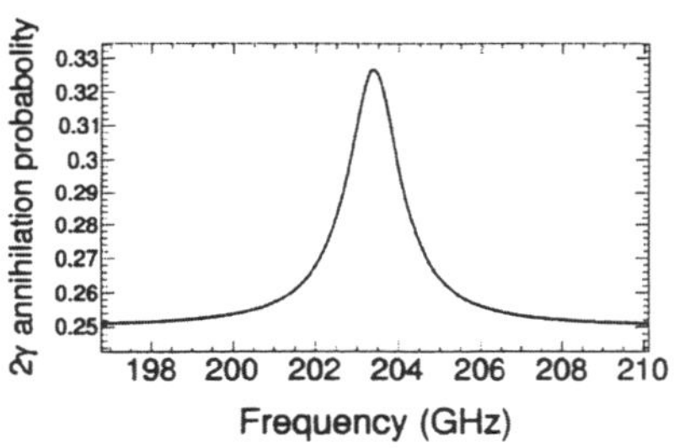

Figure 5: The calculated resonance curve of two γ-ray annihilation without any timing selection. The power of 203 GHz is 10 kW.

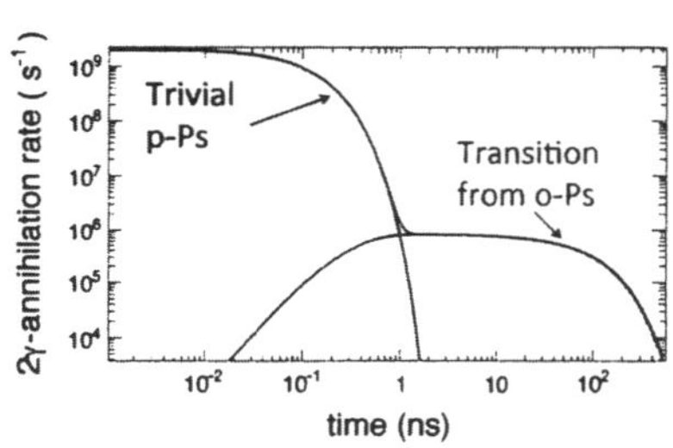

Figure 6: The calculated timing spectra of two γ-ray annihilation. The red line is background p-Ps contribution and the blue one is transited p-Ps. The power of 203 GHz is 10 kW.

We are planning to measure directly the Ps-HFS transition without any magnetic fields. Figure 4 shows a schematic of the direct transition. A very strong RF-field of 203 GHz gives rise to a transition from o-Ps to p-Ps. *Para*-positronium decays into two γ-rays promptly, whereas o-Ps decays into three γ-rays. A signal of two γ-rays indicates the transition. Figure 5 shows the expected resonance of the probability of two γ-decay when the input frequency is swept. Because of spin statistics, 25% of initial positronium is p-Ps, which becomes background. The transition events can be distinguished from such background events by selecting delayed events as shown in Fig. 6.

The direct transition from o-Ps to p-Ps has a very small spontaneous emission rate of 3.37×10^{-8} s^{-1} [4]. Because, this transition is M1-transition, and the Ps-HFS is extremely large. Therefore, a powerful electromagnetic wave of 203 GHz is essential so as to stimulate the direct transition.

The frequency of 203 GHz is just intermediate between optical light and radiowave. There were no high power light sources for spectroscopy in sub-THz region. Our first aim was to observe the direct transition from o-Ps to p-Ps. We developed the following three new sub-THz devices.

1. A sub-THz to THz light source called *gyrotron*,
2. An efficient transportation system of *gaussian converter*,
3. A *Fabry-Pérot cavity* with high-finesse.

These are explained in the next section.

2 Optical system

2.1 Gyrotron

Figure 7: A photograph of gyrotron.

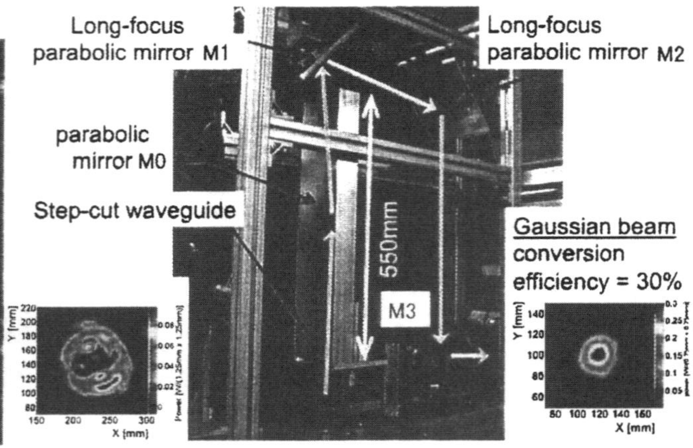

Figure 8: Transportation system.

A gyrotron[5] is a novel high power RF-source for sub-THz to THz frequency region. It has been developed in the field of nuclear fusion, and is dedicated to spectroscopic measurement this time. It contains an electron gun, superconducting magnet, and open RF-cavity. An electron emitted from the gun is accelerated and gyrating inside the cavity. The relativistic cyclotron motion of the electrons coherently stimulates the strong RF-field inside the cavity. The frequency of RF is monochromatic, and is approximately the same as that of the cyclotron motion. This phenomenon is called *cyclotron-maser resonance*.

We developed a gyrotron (Fig. 7) operated at f = 203 GHz with B = 7.34 Tesla, $\gamma \sim 1.04$. The stable power of 250 W is obtained at the output window of the gyrotron. The frequency spread is expected to be less than 1 MHz. The resonant mode is TE_{031}-mode. The transverse mode of in an output section of the gyrotron is azimuthally polarized TE_{03}-mode. The mode inside a Fabry-Pérot cavity is, unfortunately, a linearly polarized *gaussian mode*. The original gyrotron output cannot couple with a Fabry-Pérot cavity. That's why a mode conversion is essential to use gyrotron power efficiently.

2.2 Gaussian Converter

The transportation system is composed of three parabolic mirrors called M0, M1 and M2 as shown in Fig. 8. The first parabolic mirror M0 converts polarization from azimuthal to linear [6]. M1 and M2 change the shape of power distribution from bi-gaussian to gaussian. Then the plain mirror M3 reflects radiation and introduces it into the Fabry-Pérot cavity. The mirror M3 is made of half mirror so as to sample radiation from the gyrotron and reflected radiation from the Fabry-Pérot cavity. This transportation system is the key device which connects microwave technology inside waveguides and optical technology in free space.

The power profile is successfully converted into gaussian mode. The transformation efficiency is 30%, since the output mode of TE_{03}-mode is strongly disturbed by an output taper, a large hole connected to a vacuum pump, and RF-window of gyrotron. The efficiency will become over 80% by optimizing the mode purity in the future.

2.3 Fabry-Pérot cavity

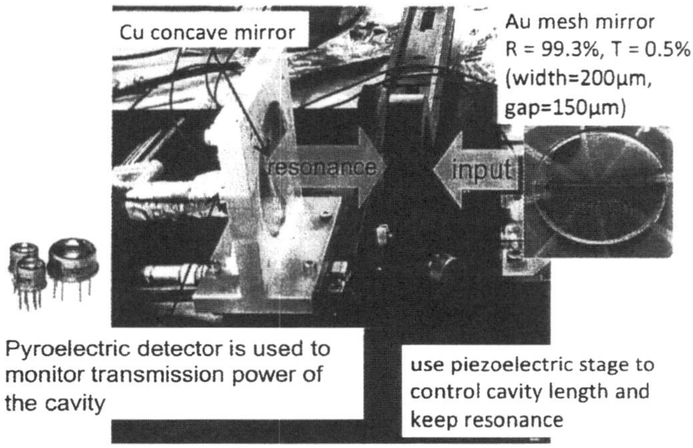

Figure 9: A photograph of the test setup of Fabry-Pérot cavity.

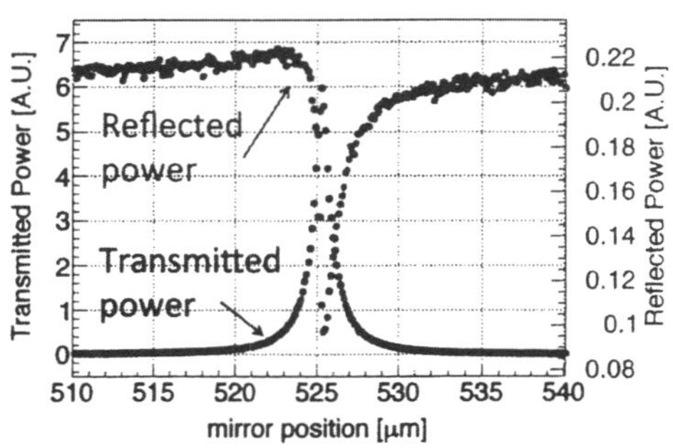

Figure 10: A photograph of the observed resonance of the Fabry-Pérot cavity.

Photons produced at the gyrotron are transported and accumulated in a Fabry-Pérot cavity. The cavity consists of two opposing mirrors as shown in Fig. 9. A golden mesh mirror evaporated on a quartz substrate is used for the input side of the cavity in order to smoothly introduce photons from the gyrotron. A copper concave mirror is used on the other side to transversely stabilize the resonance. The length of the cavity is controlled with a piezo stage under the concave mirror. One can sweep the cavity length around a couple of wavelengths (1.47 mm for 203 GHz), and precisely control it within $\mathcal{O}(100)$ nm as well, since the piezo stage can switch the operating mode between rough-motion and precise-motion.

Transmitted power is measured through a small hole on the concave mirror. Reflected power is monitored from the outside of the cavity, just behind the half mirror M3 in the transportation system. A pyroelectric detector is used as a power monitor. Figure 10 shows a resonance observed in our cavity. A transmitted power increases at the resonance, whereas a reflected power decreases. The performance of the cavity can be estimated by two parameters: finesse $\mathcal{F}$ and input coupling C.

Finesse $\mathcal{F}$ is a parameter of resonance-sharpness, which is given by

$$\mathcal{F} = \frac{\lambda}{2\Gamma}, \tag{2}$$

where Γ is the width of the resonance, and $\lambda = 1.47$ mm is wavelegth of 203 GHz. Γ is measured as 1.1μm, which results in $\mathcal{F} = 650$.

Input coupling C is given by

$$C = 1 - \frac{V_{peak}}{V_{baseline}}. \tag{3}$$

V_{peak} is a voltage at peak decreasing from $V_{baseline}$ which is the voltage of the baseline of the reflected power. An input coupling of 67% is achieved.

The gain G of the cavity can be approximately calculated with these two parameters:

$$G = C\frac{\mathcal{F}}{\pi} \sim 140. \tag{4}$$

The accumulated power and its stability are shown in Fig. 11. The output power of 250 W from the gyrotron reduces to 70 W as propagating through transportation system. The input power to the cavity is stabilized by a feedback control which adjusts heater temperature of an electron gun in the gyrotron. The input power of 70 W times gain G is the accumulated power in the cavity, which is about 10 kW. The achieved power is stabilized by a feedback control of cavity length.

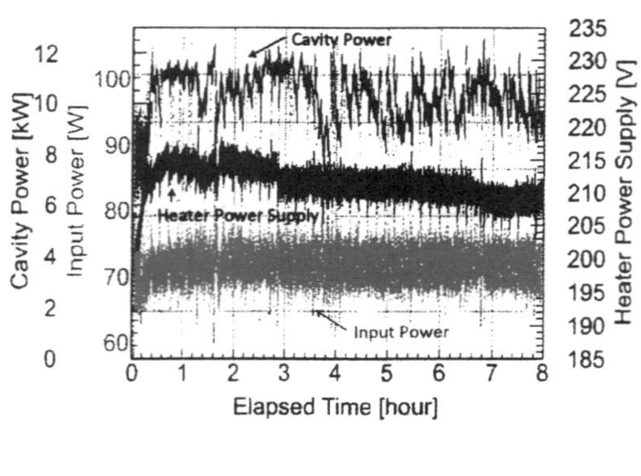

Figure 11: Accumulated power and its stability.

3 Positronium assembly and γ-ray detectors

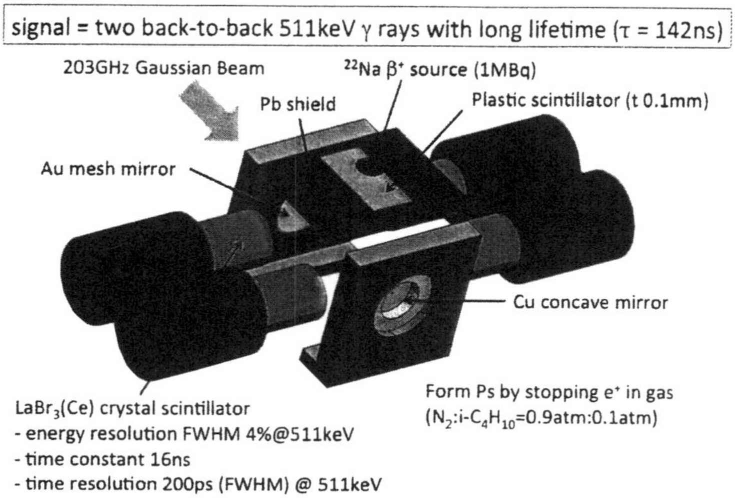

Figure 12: Schematic view of positronium assembly and γ-ray detectors.

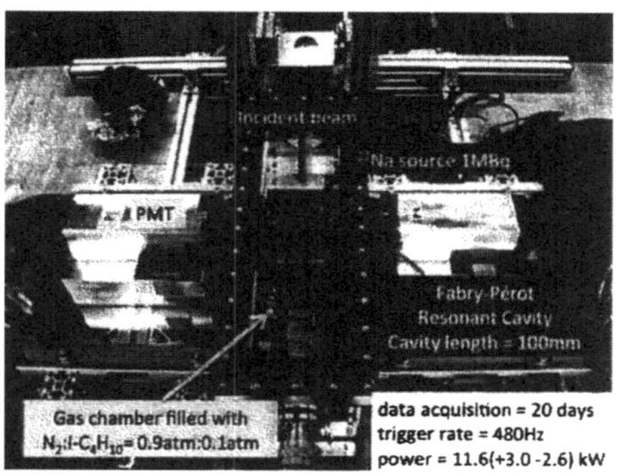

Figure 13: A photograph of the assembly.

Figure 12 and Fig. 13 show a schematic view and a photograph of the positronium assembly and signal detection system, respectively. Gyrotron power is introduced to the cavity via the mesh mirror, and is accumulated inside the Fabry-Pérot cavity. This cavity is placed inside a gas chamber filled with a gas mixture of 0.9 atm nitrogen and 0.1 atm isobutane. The positron is emitted from a ^{22}Na β^+-source, which is located 20 mm above the cavity. In order to generate start timing, the emitted positrons pass through a plastic scintillator, with thickness of 100 μm. A lead collimator is placed under the plastic scintillator so as to select the positrons which go into the cavity. It also works as a shield to protect LaBr$_3$(Ce) scintillators from pileup γ-rays.

A positron forms a positronium with an electron in the gas molecule. *Para*-positronium annihilates into two γ-rays of 511 keV immediately, whereas *o*-Ps remains with lifetime of 142 ns to decay into three γ-rays, whose energy spectra are continuous and less than 511 keV. A signal of the transition from *o*-Ps to *p*-Ps under 203 GHz is a *delayed-two-γ-ray* event. Four $LaBr_3$ crystal scintillators, which have good energy resolution of 4%@511 keV (FWHM), surround the chamber to detect γ-rays. Two γ-decay can be easily separated from three γ-decay with both the energy information and back-to-back topology. The $LaBr_3$ scintillators also have good timing resolution of 200 ps@511 keV (FWHM). Delayed events (i.e. signal of transition) is easily separated from prompt events using delayed coincidence of two signals from $LaBr_3$ and plastic-scintillator.

4 Analysis and Result

Three major selections are applied to improve S/N: (i) Timing selection (ii) Pileup rejection (iii) Energy selection.

(i) Figure 14 shows a typical spectrum of time difference between signals from $LaBr_3$ and plastic-scintillator. A large peak at 0 ns is called *prompt annihilation*, which contains the events that positronium is not formed or that *p*-Ps is formed. They are two γ-decay, and become background. An exponential structure from 50 ns to 400 ns is a contribution of *o*-Ps. Since the direct transition is from *o*-Ps, the events from 100 ns to 320 ns were selected in order to reject the above background.

(ii) The events from 400 ns to 1200 ns are mainly pileup events of plastic-scintillator signal, which extrapolatively exist under *o*-Ps events selected by above timing selection. This background is about 100 times lager than the transition signal. Pileup events is significantly reduced by discarding events in which plastic-scintillator signal synchronizes $LaBr_3$ signal within 50 ns. The result of this selection is shown in Fig. 15, where *o*-Ps events is drastically enhanced.

(iii) The upper side of Fig. 16 shows the energy spectra combined of all $LaBr_3$ detectors. The gyrotron is operated with duty ratio 30% and repetition rate 5 Hz. The events with and without power are normalized by livetime. The lower side of Fig. 16 shows the spectrum subtracted *off* events from *on* events. The excess of power-on events around 511 keV indicates the direct transition. In order to distinguish the transited events from non-transited *o*-Ps events, the events in 511 keV $\pm$ 2 σ are counted to calculate the transition rate. The detection rate is 3.99 $\pm$ 0.71 mHz. The direct transition is firstly observed with significance of 5σ-level. Einstein A coefficient (spontaneous emission rate) is estimated to be 3.9 (+1.8-1.7) $\times\ 10^{-8}\ s^{-1}$, which is consistent with QED calculation of $\mathcal{O}(\alpha)$.

The systematic errors are listed in Table 1. It is estimated that the large systematic error of e^+ tagging rate is due to photons from black-body radiation of the mesh mirror by 10 kW RF-field. Therefore, the plastic scintillator is optically shielded for the next measurement in Sep. 2011. The normalization problem is due to a trouble of one of the CAMAC modules used for data acquisition, and is solved now.

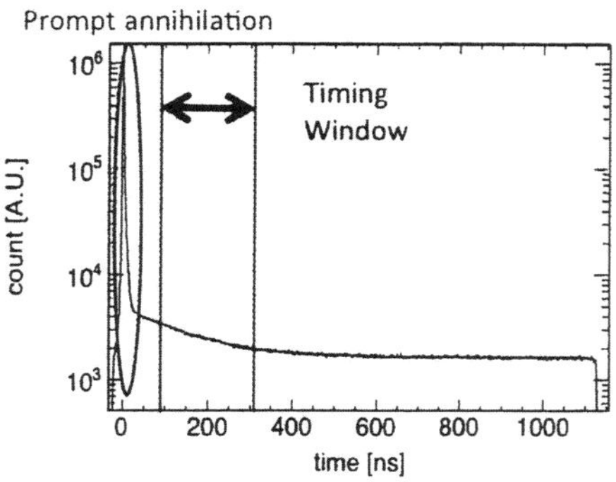

Figure 14: Time spectrum of Positronium decay and timing window.

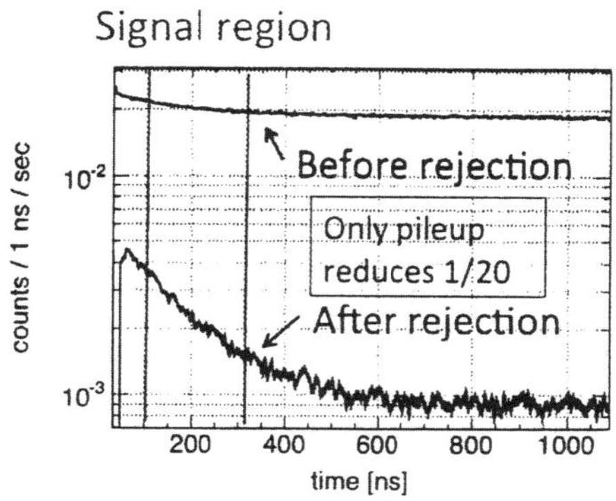

Figure 15: Pileup rejection of plastic-scintillator signal. *Ortho*-Ps events are enhanced after this selection(red histogram)

Table 1: The summary of systematic errors

Source	Systematic error
$LaBr_3$ energy resolution	3.0%
Zero point of the timing spectra	0.3%
e^+ tagging rate	7.7%
normalization	3.1%
Quadrature sum	8.8%

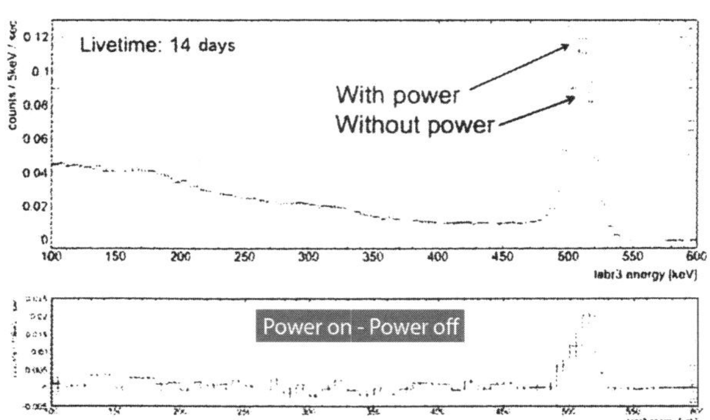

Figure 16: (Upper)Energy spectrum after timing selection and pileup rejection.(Lower) A spectrum subtracted off events from on events.

5 Conclusion and future prospect

There is a large discrepancy between a theory and experiments in the Ps-HFS value. We suspect that non-uniformity of a static magnetic field is one of the common systematic uncertainties in previous experiments. The direct measurement without any magnetic fields is proposed, and a gyrotron, mode converter, and Fabry-Pérot cavity have been developed. We took data for two weeks at fixed frequency of 203 GHz. The direct transition from *o*-Ps to *p*-Ps was for the first time observed with 5σ-level.

In order to measure the whole resonance of the direct transition, the input frequency must be swept within a few GHz range around 203.4 GHz, whereas the gyrotron used in this experiment is monochromatic (called *gyromonotron*). Tunability is an extremely challenging issue in sub-THz region. A new device called reflective gyro-BWO was proposed recently [7]. In terms of the dispersion relation, the resonance occurs at a cross point of cyclotron motion of electrons and resonant frequency of a cavity ($R(z)$ is radius, and $j'_{mn} = 7.016$) as shown in Fig. 17. The output frequency is tuned by changing the magnetic field. We have made the prototype of gyro-BWO as shown in Fig. 18, and testing the basic performance in Sep. 2011.

These experiments are performed in collaboration with T. Yamazaki, T. Suehara, T. Namba, S. Asai, T. Kobayashi, H. Saito, T. Idehara, I. Ogawa and S. Sabchevski.

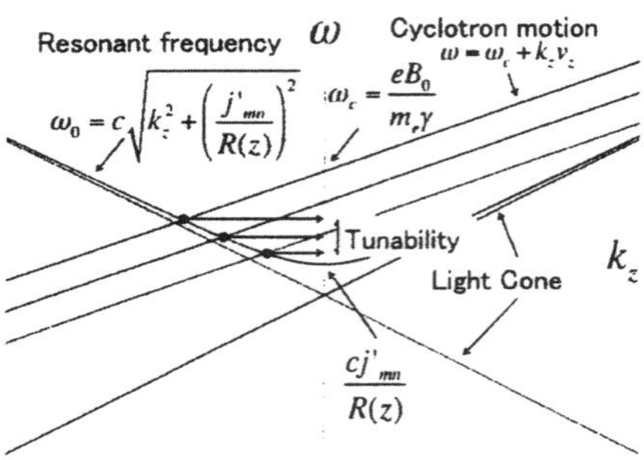

Figure 17: Backward wave interaction.

Figure 18: A photograph of the reflective gyro-BWO.

References

[1] A. P. Mills, Jr. *et al.* Phys. Rev. Lett. **34** (1975) 246; A. P. Mills, Jr. *et al.* Phys. Rev. A **27** (1983) 262.

[2] M. W. Ritter, *et al.* Phys. Rev. A **30** (1984) 1331.

[3] B. A. Kniehl and A. A. Penin, Phys. Rev. Lett. **85** (2000) 5094.

[4] V. V. Burdyuzha,*et al.* Astro. Lett.**25** (1999) 3.

[5] T. Idehara, *et al.* IEEE Trans. Plasma Sci. **27** (1999) 340.

[6] I. Ogawa, *et al.* Int. J. Elec. **83** (1997) 5.

[7] N. C. Chen, *et al.* Appl. Phys. Lett. **96** (2010) 161501.

New Results on Electron Neutrino Appearance in MINOS

Mhair Orchanian – California Institute of Technology

September 30, 2011

1 Introduction and Motivation

Neutrino oscillation is now an established phenomenon, thanks to a variety of experiments [1, 2, 3, 4, 5, 6, 7] which have probed various apparent anomalies in expected neutrino spectra from the sun, atmosphere, nuclear reactors, and accelerator-based beams. The three weak eigenstates – ν_e, ν_μ, ν_τ – are related to the three mass eigenstates – ν_1, ν_2, ν_3 – via the PMNS (for Pontecorvo, Maki, Nakagawa, and Sakata) matrix, a unitary mixing matrix U parametrized by three mixing angles – θ_{12}, θ_{13}, θ_{23} – and three CP-violating phases – δ, α_1, α_2 [8]. No hints exist yet as to the value of δ, and oscillation experiments are not sensitive to the latter two CP-violating phases, which are nonzero only if neutrinos are their own antiparticles (i.e., Majorana particles). So far, two of the mixing angles, θ_{12} and θ_{23}, have been measured [1, 4, 6] and are large. The value of θ_{13}, on the other hand, appears to be quite small. It was first significantly constrained by the CHOOZ experiment in 1999 [9]; the result is shown in Fig. 1. These proceedings report the latest improvements on this constraint by the MINOS experiment and survey the future of research in this area.

The probability of a neutrino of energy E created with flavor α being detected with flavor β after traveling some distance L in vacuum – to wit, the oscillation probability – is given by:

$$P(\nu_\alpha \to \nu_\beta) = \left| \sum_{j=1}^{3} U^*_{\alpha j} \exp\left(-\frac{im_j^2 L}{2E}\right) U_{\beta j} \right|^2 , \tag{1}$$

where m_j is the mass of the j-th neutrino mass eigenstate. The final expression is found to depend not on the absolute squared masses but on two independent squared-mass differences – Δm^2_{21} and Δm^2_{32}. While both the sign and magnitude of the first of these are known [1], only the magnitude of the second squared-mass difference is known [6]. That is, we do not know if ν_3 is more or less massive than ν_1 and ν_2; the first possibility is referred to as the normal mass hierarchy and the latter, the inverted mass hierarchy.

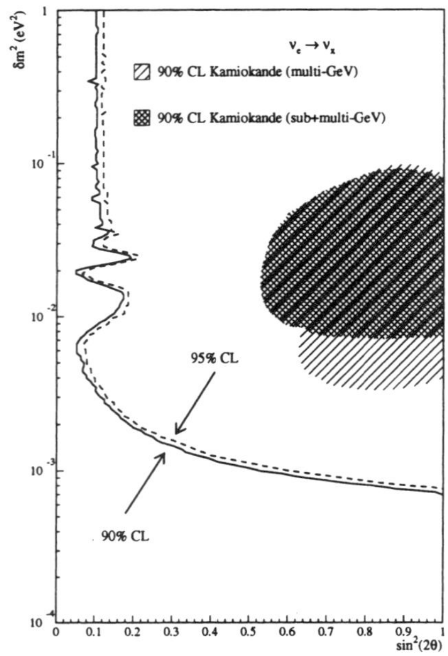

Figure 1: Exclusion contours from the CHOOZ experiment constraining the value of the θ_{13} mixing angle (horizontal axis) as a function of Δm^2_{32} (vertical axis). Taken from Ref. [9].

The oscillation mode of present interest is $\nu_\mu \rightarrow \nu_e$. To leading order:

$$P(\nu_\mu \rightarrow \nu_e) \approx \sin^2(2\theta_{13}) \sin^2(\theta_{23}) \sin^2(1.27 \Delta m^2_{32} L/E), \tag{2}$$

with Δm^2_{32} in units of eV^2, L in km, and E in GeV. When we account for the propagation of the neutrinos through matter, additional, higher order terms appear in Eq. 2 [10]. The full expression depends on the CP-violating phase δ and the mass hierarchy; accordingly, the results presented here will be sensitive to these parameters.

Under certain conditions [11], leptonic CP violation can lead to non-conservation of baryon number, which in turn can help explain the observed matter-antimatter asymmetry in our universe. Access to the value of the leptonic CP-violating phase, namely δ, is controlled by the size of θ_{13}. Should θ_{13} be sufficiently large, one may be able to measure δ with current or next-generation long-baseline neutrino oscillation experiments. Finding, on the other hand, that θ_{13} is zero could point to a new symmetry.

2 The MINOS Experiment

The MINOS experiment consists of three key parts: the Neutrinos at the Main Injector (NuMI) neutrino beam at Fermilab, the Near Detector (ND) at Fermilab, and the Far Detector (FD) in Soudan, MN. The NuMI beam [12] is produced by striking a graphite target with 120-GeV protons from Fermilab's Main Injector, focusing the resulting secondary hadrons (mostly pions, some kaons) using pulsed-current magnetic horns, and allowing these hadrons to decay leptonically in flight. Although the horns' polarities are set to focus π^+ and K^+ and defocus π^- and K^-, a small fraction of negative hadrons travel along the horns' axis, avoid being defocused, and remain in the beam. The primary decay mode for these hadrons is $\mu^+\nu_\mu$ or $\mu^-\bar{\nu}_\mu$; a very small fraction decay to $e^+\nu_e$ or $e^-\bar{\nu}_e$. Ultimately, the resulting neutrino beam is 92% ν_μ and 7% $\bar{\nu}_\mu$; ν_e and $\bar{\nu}_e$ make up the remainder.

The MINOS detectors are functionally identical, magnetized tracking calorimeters. The 980-ton ND is located approximately 1 km downstream of the NuMI target and measures the neutrino beam's initial composition and energy spectrum. The 5.4-kton FD is located 735 km downstream in the Soudan mine in northern Minnesota and measures the oscillated beam's composition and energy spectrum. Both detectors consist of alternating layers of 2.54-cm thick steel and 1-cm thick plastic scintillator. The scintillator is segmented into optically-isolated, 4.1-cm wide strips and serves as the active portion of each detector. Light from the strips is read out via optical fibers and multi-anode photomultiplier tubes. Further details can be found in Ref. [13].

3 Neutrino Interactions in the MINOS Detectors

3.1 Event Types

Though a multi-purpose experiment, MINOS was designed and optimized to study ν_μ disappearance. The event type of interest in such an analysis is a ν_μ charged-current (CC) interaction:

$$\nu_\mu + N \rightarrow \mu^- + X, \tag{3}$$

where N represents a nucleon or a quark inside a nucleon and X represents the resulting hadronic recoil system. The track produced by the muon typically extends far beyond the shower activity of the recoil system (see left of Fig. 2), enabling straightforward identification of such an event. In the analysis at hand, however, the event type of interest is a ν_e CC interaction:

$$\nu_e + N \rightarrow e^- + X. \tag{4}$$

In the MINOS detectors, the electromagnetic shower generated by the electron has a radiation length of 1.76 cm – ~30% less than the thickness of a steel layer

– and a Molière radius of 3.7 cm – ~10% less than the width of a scintillator strip. As a result, the electromagnetic shower becomes fully interspersed with the hadronic shower (see center of Fig. 2). The challenge, then, is to distinguish this type of event from the primary background event type – the neutral-current (NC) interaction (see right of Fig. 2):

$$\nu_x + N \rightarrow \nu_x + X. \tag{5}$$

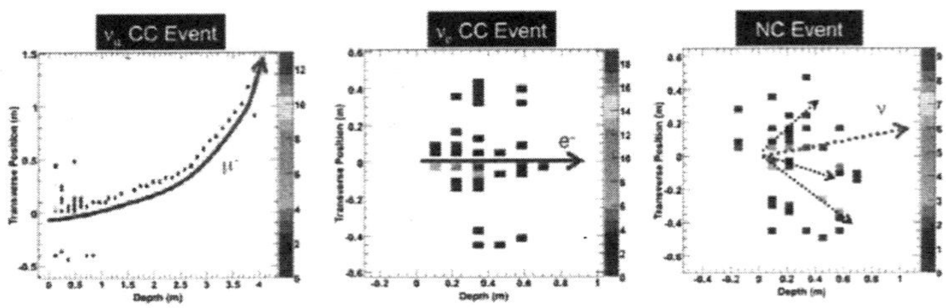

Figure 2: Displays of simulated events in the MINOS detectors. Pulse heights shown are calibrated such that 1 unit is the amount of energy deposited by a minimum ionizing particle.

3.2 Event Identification

This is the third ν_e appearance search conducted by MINOS. In all three analyses, event selection consisted of three general stages. First, we required that the event took place within a fiducial volume. Next, we rejected events that were clearly or likely background. This "pre-selection" eliminated most ν_μ CC events by requiring that any reconstructed track in the event be no more than 24 planes long and extend no more than 15 planes beyond the hadronic shower. We also required that an event have at least five contiguous planes with energy greater than half that deposited on average by a minimum ionizing particle. Events were required to be in-time with and to point in a direction consistent with the neutrino beam pulses. The reconstructed energy of the event had to lie between 1 and 8 GeV, as most events below 1 GeV are from NC interactions and $P(\nu_\mu \rightarrow \nu_e)$ is negligibly small above 8 GeV.

The last stage of the selection involved the use of a sophisticated event identification technique to separate signal from background. In our first two ν_e appearance analyses [14, 15], we used a neural network (named Artificial Neural Network, or ANN) whose input variables were reconstructed quantities characterizing the longitudinal and transverse energy deposition profiles of an event. The present analysis utilizes instead a novel technique named Library Event Matching (LEM) that uses raw energy deposition information instead of reconstructed quantities and is based on a pattern recognition algorithm. We begin by comparing a given candidate event to each of 5×10^7 simulated events (the "library"), of which $\sim 3 \times 10^7$ are background events and $\sim 2 \times 10^7$ are signal events. Intuitively, the comparison determines the extent to which the topologies and strip pulse heights of the candidate and library events are similar. Quantitatively, this is performed using a likelihood:

$$\log \mathcal{L} = \sum_{j=1}^{N_{strips}} \sum_{k=1}^{N_{planes}} \log \left[\int_0^\infty P(n_{cand}^{jk}, \lambda) P(n_{lib}^{jk}, \lambda)\, d\lambda \right], \tag{6}$$

where $P(n, \lambda)$ is the Poisson probability for observing n given mean λ and n_x^{jk} is the charge (in photoelectrons) deposited in strip j of plane k of event x (candidate or library). Once this matching process is completed for a given candidate event, the library events are ranked by their $\log \mathcal{L}$ values. The top 50 of these, designated the "best matches," are summarized by three variables: (1) the fraction that are signal (ν_e CC) events, (2) for these signal events, the average fraction of total event energy in the hadronic recoil system, and (3) for these signal events, the average fraction of deposited charge that overlaps between each of them and the candidate event. These variables, along with the reconstructed energy of the candidate event, are given as inputs to a neural network; the output is the LEM discriminant, shown in Fig. 3.

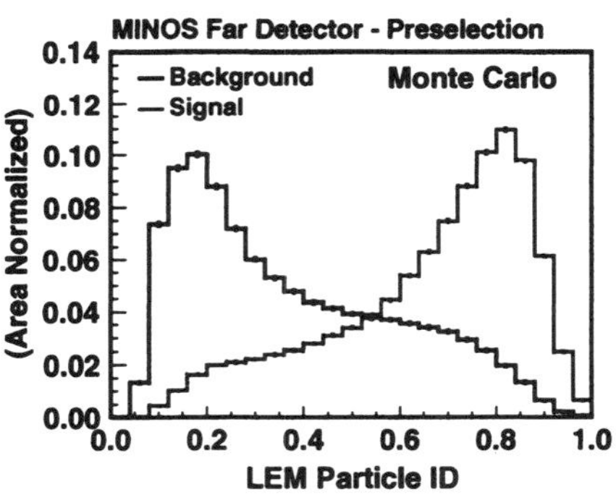

Figure 3: LEM discriminant distributions for signal and background events passing the fiducial and pre-selection cuts.

4 The Analysis

4.1 Overview

The present analysis represents three improvements over the most recent previous analysis, the effects of which are summarized by Fig. 4. An additional exposure of 1.2×10^{20} protons on the NuMI target provides a modest 17% increase in the size of our data set. Using LEM instead of ANN provides a gain in sensitivity to $\theta_{13} = 0$ of about 15%. Finally, instead of cutting at the optimal value of the discriminant (found to be 0.7 for LEM) and comparing the number of events above that value to the expectation (as we did in the previous two analyses), we perform a fit to the LEM discriminant distribution (optimal binning was found to be 0.6-0.7, 0.7-0.8, 0.8-1.0), obtaining an additional 12% increase in sensitivity. The basic approach to making a background prediction, however, remains unchanged – count the events observed in the ND and use the FD and ND simulations to "extrapolate" this to the FD. Further details of this procedure can be found in Ref. [16].

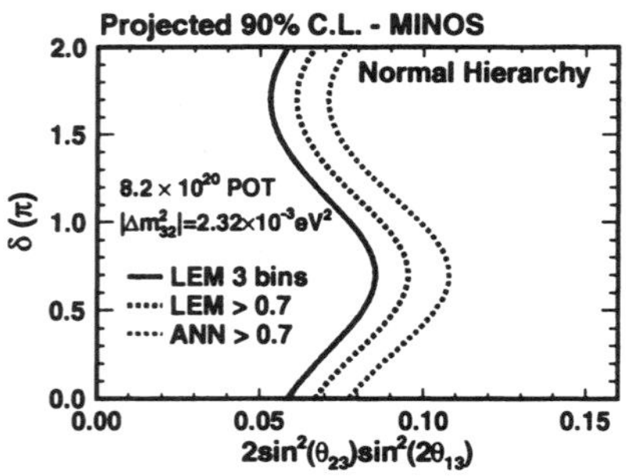

Figure 4: Projected 90% C.L. sensitivities to $\theta_{13} = 0$ assuming the normal mass hierarchy. The contour labeled "ANN > 0.7" is the sensitivity achieved by repeating the last analysis with the additional data included. The "LEM > 0.7" contour is the sensitivity for a counting experiment performed with LEM instead of ANN using the same full data set. Finally, the "LEM 3 bins" sensitivity is obtained by performing a shape fit to the LEM distribution.

In particular, the extrapolation procedure includes the ability to adjust the predictions for different sets of oscillation parameters. To illustrate, Table 1 lists the predicted counts of events with LEM > 0.7 for two reference values of θ_{13}, assuming a commonly used set of values for other oscillation parameters.

4.2 The FD Data

After applying the fiducial and pre-selection cuts to the events in the FD data, we obtain the energy spectrum shown in Fig. 5. Although the pre-selection cuts events with reconstructed energy above 8 GeV, we performed the prediction out to 12 GeV and compared it to the data to test the robustness of our background prediction method.

In the LEM discriminant distribution, we expect negligible signal in the LEM < 0.5 region and thus can use it as a test of the entire analysis chain. We predict 370 ± 19(stat.) background events with LEM < 0.5 and observe 377 – well within statistical error. Sufficiently assured of the soundness of our analysis method, we proceed to unblind the signal-enhanced region of the LEM discriminant distribution. The full spectrum is shown in Fig. 6. For reference only, we can count the number of events with a LEM discriminant above the

Event Class	$\sin^2(2\theta_{13})$	
	0	0.1
NC	34.1	34.1
ν_μ CC	6.7	6.7
ν_e CC	6.4	6.2
ν_τ CC	2.2	2.1
Total Background	**49.4**	**49.1**
$\nu_\mu \rightarrow \nu_e$ **CC**	**0.2**	**19.1**

Table 1: Predicted event counts for events with LEM > 0.7. These numbers were calculated assuming $\delta = 0$, $\theta_{23} = \pi/4$, and $\Delta m^2_{32} = 2.32 \times 10^{-3}\text{eV}^2$ (i.e., normal mass hierarchy). The line labeled "ν_e CC" refers to the intrinsic beam ν_e component. The nonzero signal in the $\theta_{13} = 0$ case is due to Δm^2_{21}-driven oscillations.

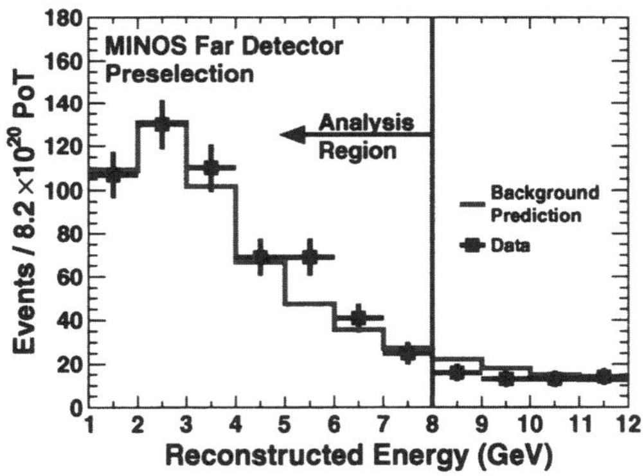

Figure 5: Predicted (assuming $\theta_{13} = 0$) and observed reconstructed energy spectra in the FD.

optimal cut of 0.7; we predict 49.5 ± 2.8(syst.) ± 7.0(stat.) background events and observe 62 events, an excess of approximately 1.7σ.

4.3 Fitting to the Data

The official fit to the data is performed in three LEM bins and five reconstructed energy bins. As an example, Fig. 7 shows the observed reconstructed energy spectra in the three LEM bins superimposed on the best-fit signal and predicted background under the conditions indicated. Our official result, described in Ref. [16], is given in Fig. 8. These contours were calculated according to the Feldman-Cousins prescription [17] and thus provide correct coverage near the physical boundary at $\theta_{13} = 0$. In addition, uncertainties in the other oscillation parameters have been accounted for in generating these contours. Assuming

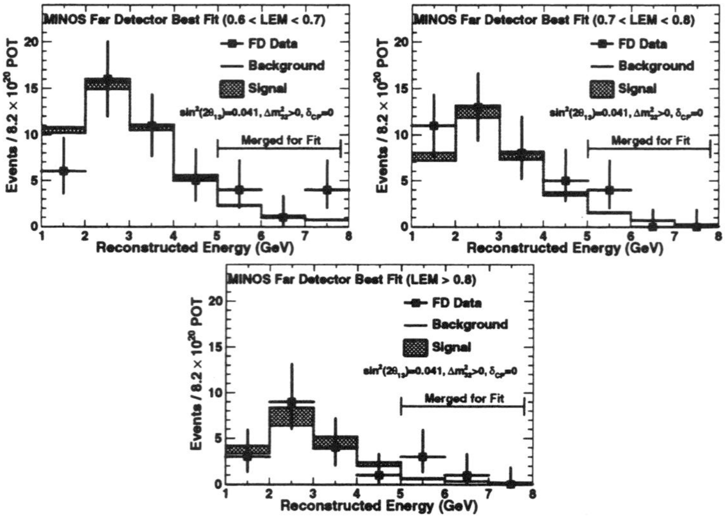

Figure 7: Stacked background prediction and best-fit signal assuming $\delta = 0$, $\theta_{23} = \pi/4$, and $\Delta m^2_{32} = 2.32 \times 10^{-3}\text{eV}^2$, with data superimposed. Note that each reconstructed energy spectrum is for a different bin of the LEM discriminant.

of Athens, Greece; and Brazil's FAPESP, CNPq, and CAPES. We are grateful to the Minnesota Department of Natural Resources, the crew of the Soudan Underground Laboratory, and the staff of Fermilab for their contributions to this effort.

References

[1] B. Aharmim *et al.* (SNO), Phys. Rev. Lett. **101**, 111301 (2008).

[2] T. Araki *et al.* (KamLAND), Phys. Rev. Lett. **94**, 081801 (2005).

[3] Y. Ashie *et al.* (Super-Kamiokande), Phys. Rev. Lett. **93**, 101801 (2004).

[4] Y. Ashie *et al.* (Super-Kamiokande), Phys. Rev. D **71**, 112005 (2005).

[5] K. Abe *et al.* (Super-Kamiokande), Phys. Rev. Lett. **97**, 171801 (2006).

[6] P. Adamson *et al.* (MINOS), Phys. Rev. Lett. **106**, 181801 (2011).

[7] P. Adamson *et al.* (MINOS), Phys. Rev. Lett. **107**, 021801 (2011).

[8] B. Pontecorvo, JETP **34**, 172 (1958); V. N. Gribov and B. Pontecorvo, Phys. Lett. B **28**, 493 (1969); Z. Maki, M. Nakagawa, and S. Sakata, Prog. Theor. Phys. **28**, 870 (1962).

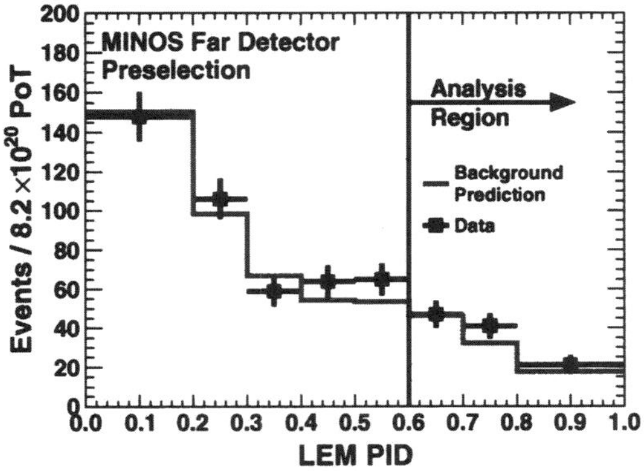

Figure 6: Predicted (assuming $\theta_{13} = 0$) and observed LEM discriminant distributions in the FD.

$\delta = 0$ and $\theta_{23} = \pi/4$, we find that $\sin^2(2\theta_{13}) = 0.041^{+0.047}_{-0.031}$ $(0.079^{+0.071}_{-0.053})$ for the normal (inverted) mass hierarchy. Under the same set of assumptions, $\sin^2(2\theta_{13}) < 0.12\,(0.20)$ at 90% C.L. for the normal (inverted) mass hierarchy. Finally, these results exclude the null hypothesis of $\sin^2(2\theta_{13}) = 0$ at 89% C.L.

5 The Future of θ_{13}

Several next-generation neutrino oscillation experiments are either already online or will be online soon. These include experiments such as Double Chooz, Daya Bay, and RENO that are located near nuclear reactors and will look for $\bar{\nu}_e$ disappearance as well as experiments like NOvA and T2K that are accelerator-based and will look for ν_e appearance in ν_μ beams. As an example, NOvA is the NuMI Off-axis ν_e Appearance experiment – the next-generation NuMI long-baseline experiment succeeding MINOS. It aims to measure θ_{13} or set a much improved limit on it. If θ_{13} happens to be sufficiently large, NOvA might also be able to distinguish between the mass hierarchies and to observe neutrino CP violation (i.e., by measuring δ) – see Fig. 9. Of these experiments, NOvA is the only one that can try to answer all of these questions, thanks mostly to its off-axis location (which exposes the detectors to a narrow-band beam), fine-grained design (which allows for superb ν_e CC event identification), and long baseline (which provides a large matter effect).

6 Acknowledgments

We thank the organizers of the 49th International School of Subnuclear Physics for the opportunity to present this work during the New Talents sessions of the School. These proceedings report work supported by the U.S. DOE; the U.K. STFC; the U.S. NSF; the State and University of Minnesota; the University

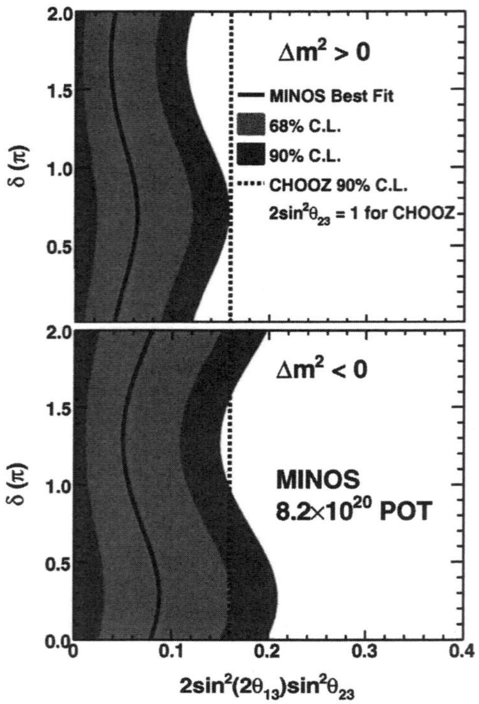

Figure 8: Allowed ranges and best fits for $2\sin^2(\theta_{23})\sin^2(2\theta_{13})$ as a function of δ. The upper (lower) panel assumes the normal (inverted) mass hierarchy. The CHOOZ 90% C.L. upper limit shown assumes $\theta_{23} = \pi/4$ and $\Delta m^2_{32} = 2.32 \times 10^{-3}\text{eV}^2$.

[9] M. Apollonio *et al.* (CHOOZ), Phys. Lett. B **466**, 415 (1999).

[10] E. K. Akhmedov *et al.*, J. High En. Phys. **04**, 078 (2004).

[11] A. D. Sakharov, JETP Lett. **5**, 24 (1967).

[12] K. Anderson et al., FERMILAB-DESIGN-1998-01 (1998).

[13] D.G. Michael et al., Nucl. Inst. & Meth. A **596**, 190 (2008).

[14] P. Adamson *et al.* (MINOS), Phys. Rev. Lett. **103**, 261802 (2009).

[15] P. Adamson *et al.* (MINOS), Phys. Rev. D **82**, 051102 (2010).

[16] P. Adamson *et al.* (MINOS), arXiv:1108.0015v1 [hep-ex] (2011). Accepted by Phys. Rev. Lett.

[17] G. J. Feldman and R. D. Cousins, Phys. Rev. D **57**, 3873 (1998).

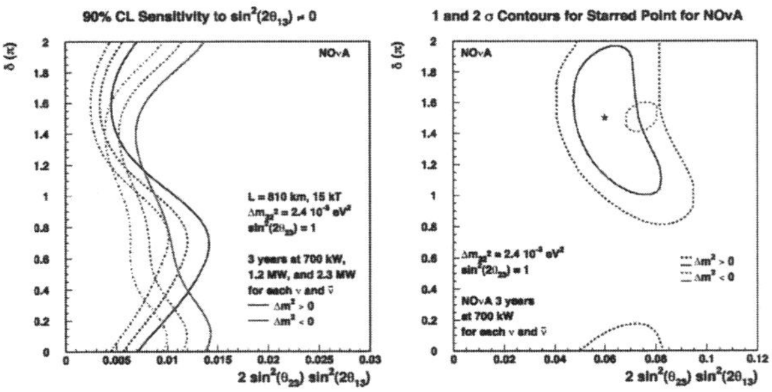

Figure 9: Note that these plots assume six years of running, divided equally between ν and $\bar{\nu}$ beams. (Left) Sensitivity of NOvA to $\theta_{13} = 0$ for different beam intensities. (Right) Expected contours assuming that the true oscillation parameters lie at the starred point. In this case, the mass hierarchy is resolved, and the CP-violating phase is constrained to lie in the upper half-plane. (Taken from Ref. [18].)

[18] http://www-nova.fnal.gov/plots_and_figures/plots_and_figures.html

Measuring $\sin^2 2\theta_{13}$ with the Daya Bay nuclear reactors*

Liangjian Wen
Institute of High Energy Physics
Beijing, 100049, P.R.China
E-mail: wenlj@ihep.ac.cn

The Daya Bay reactor neutrino experiment is designed to precisely measure a new mode of the neutrino oscillation between v_1 and v_3, which is often referred as the neutrino mixing angle θ_{13}. The experiment deploys multiple identical detectors underground at different baselines to reduce the systematic errors and suppress cosmogenic backgrounds. The construction of the experiment is currently close to completion, and the data taking is planned for this summer. A brief overview and the progresses of this experiment will be reported.

1. Introduction

Daya Bay reactor neutrino experiment is located near the Daya Bay nuclear power complexes, in the Shenzhen city in the south of China, 60km away from Hong Kong. The goal of this experiment is to precisely measure the last unknown mixing angle θ_{13}, with a sensitivity of 0.01 or better in $\sin^2 2\theta_{13}$ at 90% C.L.[1]

By comparing the measured number and energy spectrum of neutrinos from reactors at a proper baseline with those of a prediction, the missing probability of electron anti-neutrinos can be obtained and the oscillation parameters can be determined.

The Daya Bay nuclear power complexes have three pairs of reactor cores (Daya Bay, LingAo and LingAo II), with a combined thermal output of 17.4 GW_{th}. In addition, the reactors are near a mountainous region, hence it is convenient to construct underground laboratories to shield cosmic-rays. The layout of Daya Bay experiment is shown in Fig. 1. It uses near and far detector configuration to do a relative measurement, which can cancel the reactor related systematic errors and parts of the detector related systematic errors. The locations of experimental sites are optimized, taking

* http://dayabay.ihep.ac.cn

into account the maximum oscillation baseline, the mountain profile that affects the cosmic-ray flux and the subsequent backgrounds in the underground detectors, as well as the geological conditions for the tunnel and laboratory construction. Detailed discussions can be found in Ref.[2]

Fig. 1. The layout of the Daya Bay experiment.

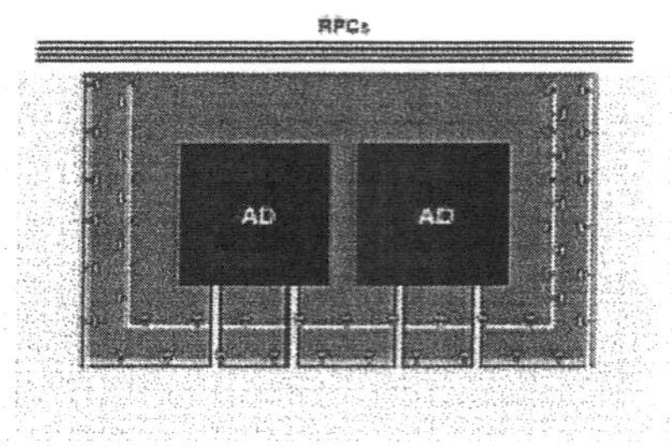

Fig. 2. The layout of the detectors design.

The Daya Bay experiment deploys multiple neutrino detector modules at each site to cross check and reduce un-correlated systematic uncertainties. The detector design is shown in Fig. 2. Anti-neutrino detectors (AD) are immersed in the water pool to be shielded from backgrounds. There are two ADs at each near site and four at far site to allow a side by side check between two ADs at the same site. The un-correlated systematic uncertainties can be reduced by a factor of $\sqrt{N}$ (N is the number of ADs at each site). The water pool is instrumented with photomultipliers (PMT) as a Cherenkov detector to tag cosmic muons for vetoing. On the top of the pool, there are four layers of Resistive Plate Chambers (RPC) to track and tag cosmic-rays. Such a design with multiple AD modules and multiple veto detectors can significantly reduce systematic uncertainties and provide enough redundancy for neutrino detection, muon veto and background subtraction.

With a detailed Monte Carlo simulation of the experiment, as well as a detailed study of various backgrounds, the estimation of signal and backgrounds is summarized in Table. 1. Cosmogenic isotopes ^{9}Li and ^{8}He are one of the most problematic backgrounds, because they have β-neutron cascade decay mode which can mimic a neutrino signal. By a maximum likelihood method, the ^{9}Li/^{8}He backgrounds can be measured *in-situ* by fitting the time between the last muon and the anti-neutrino candidate with the known decay times for these isotopes.[4] The precision of measured background-to-signal ratio using this method can be estimated as $\sigma = \frac{1}{N}\sqrt{(1+\tau R)^2 - 1}$, where N is the number of neutrino candidates, τ is the lifetime of ^{9}Li, and R_μ is the muon rate in the detector.

Table 1. Summary of Daya Bay signal and background rates for an antineutrino detector module at three experimental sites.

	Daya Bay Near	Ling Ao Near	Far Hall
Baseline (m)	363	481 from Ling Ao 526 from Ling Ao II	1985 from Daya Bay 1615 from Ling Ao
Overburden (m)	98	112	350
Radioactivity (Hz)	<50	<50	<50
Muon rate (Hz)	36	22	1.2
Antineutrino Signal (events/day)	840	740	90
Accidental Background/Signal (%)	<0.2	<0.2	<0.1
Fast neutron Background/Signal (%)	0.1	0.1	0.1
^{8}He+^{9}Li Background/Signal (%)	0.3	0.2	0.2

Taking account into all errors from reactor related, detector related and statistics, the requirements on systematic uncertainties for the Daya Bay experiment to reach the design sensitivity of $\sin^2(2\theta_{13})$ are listed in Table.2. With two near sites and one far site, after three years running, the sensitivity of the Daya Bay experiment at 90% C.L is shown in Fig.3.

Table 2. Summary of Daya Bay uncertainties.

Source	Uncertainty
Reactor Power	0.087% (4 cores) 0.13% (6 cores)
Detector (per module)	0.38% (baseline) 0.18% (goal)
Signal Statistics	0.2%

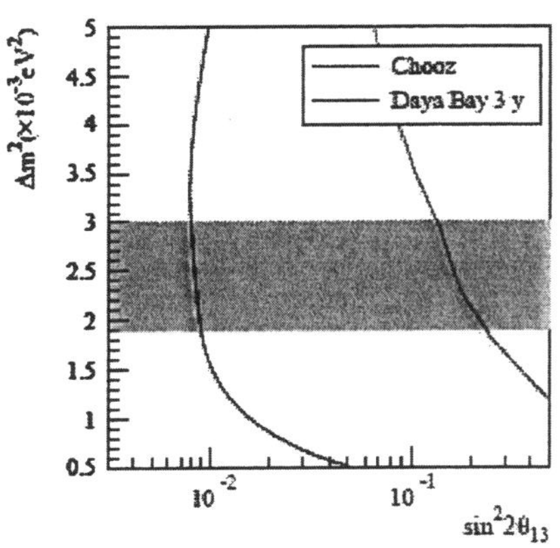

Fig. 3. The Daya Bay sensitivity curve with three years of data.

2. Design and construction of the experiment

Up to now, the major civil construction activities have been completed, and the detector construction is mostly finished. The first two ADs for Daya Bay near hall (hall 1), as well as the muon installation are completed. Hall 1 installation is started and will be ready for physics this summer.

2.1. *Anti-neutrino Detector*

Electron anti-neutrinos are detected via the inverse beta-decay (IBD) reaction $\bar{\nu_e} + p \rightarrow e^+ + n$. e^+s that carrying the $\bar{\nu_e}$ energy information are detected by the liquid scintillator. The neutrons will be captured on Hydrogen or Gadolinium after thermalization, then be detected by the released γ's.

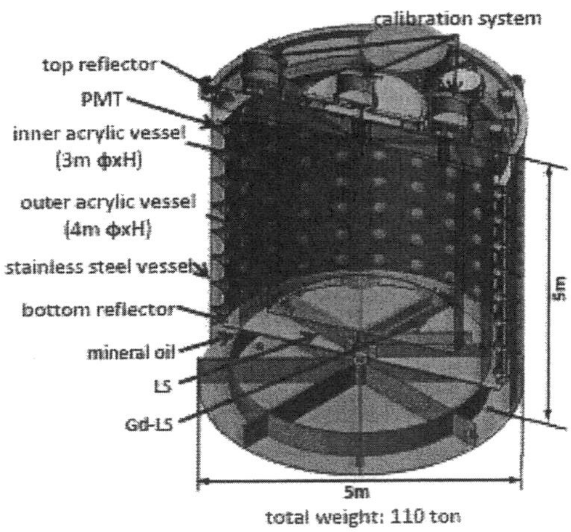

Fig. 4. Schematic drawing of the Daya Bay antineutrino detector module.

Fig. 5. The fabricated reflector.

The Daya Bay antineutrino detector (AD) is a three-zone cylindrical structure, as shown in Fig. 4. The inner-most zone, 3m in diameter and 3m in height, is a 20t Gadolinium doped (0.1%) liquid scintillator (GdLS), which acts as the antineutrino target. The middle volume, with an outer dimension of 4m in diameter and 4m in height, is filled with 20t normal liquid scintillator (LS) as the gamma catcher to contain the energy of γ from neutron capture or positron annihilation. The outer-most volume, is a 40t transparent mineral oil buffer that shields radiation from PMTs and steel entering the target. The three volumes are partitioned by transparent acrylic vessels. PMTs and support structures in the oil buffer are instrumented on a stainless steel vessel (SSV) with a dimension of 5m in diameter and 5m in height. There are three automated calibration units (ACUs) mounted on top of each detector to enable deployment of calibration sources along 3 vertical axes (central, inside the acrylic wall, and in the gamma catcher). LED and radioactive sources can be deployed into the detector to calibrate the PMT and the detector energy response. A manual calibration systems is also designed for a 4π calibration purpose.

An innovative design in the Daya Bay antineutrino detector uses optical reflective panels at the top and bottom of the detector cylinder, close to the γ-catcher vessel, to enlarge photocathode coverage. PMTs are only positioned along the circumference of the cylinder. Usage of reflectors allows reducing the number of PMTs by almost one half compared to a 4π PMT installation. The energy resolution is estimated to be $12\%/\sqrt{E(MeV)}$ and the position resolution is 13cm. This design also simplifies the mechanical structure of such a movable detector. Monte Carlo studies show that reflective panels also improve the energy response uniformity in vertical direction, as shown in Fig. 6. A prototype detector with a scale of 1/3 has been build to test the design principles and construction technologies. Results are in good agreement with the expectations .[3] With extensive technic tests, good quality reflectors are made by ESR film sandwiched between two acrylic panels, as shown in Fig. 5.

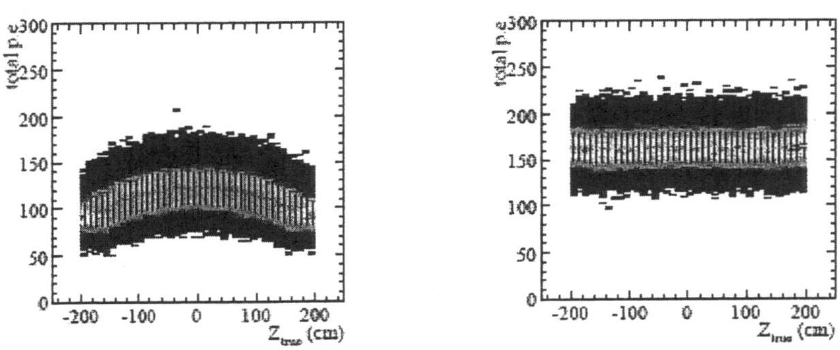

Fig. 6. The simulated detector response along vertical direction, for the situation that without (left) and with (right) reflector.

For a liquid scintillator detector with reflective panels,[5] like Daya Bay, a maximum likelihood (ML) reconstruction algorithm has been developed to determine the vertex and energy from the PMTs' charge pattern. Monte Carlo studies for the prototype detector show that the ML method can reconstruct vertex unbiasedly in the target region and well correct the radial and vertical dependencies of the energy response. This ML method is also tested with ^{137}Cs source data from the prototype detector. Asymmetric energy response caused by the detector geometry and the solid angle effect are well corrected. The resolution of the reconstructed vertex is similar as the intrinsic space smearing of the energy deposit of ^{137}Cs γ source.

Till now, the detector construction has been mostly finished. AD assembly is ongoing at the surface assembly building, and two out of eight ADs have been completed. A full chain test consisting of the first two assembled ADs, automatic calibration units (ACUs), electronics, trigger and DAQ, has been performed without liquid filled. The results show that both ADs are fully functional. Their responses to LED and cosmic-rays are identical and agree with Monte Carlo expectations, as shown in Fig. 7.

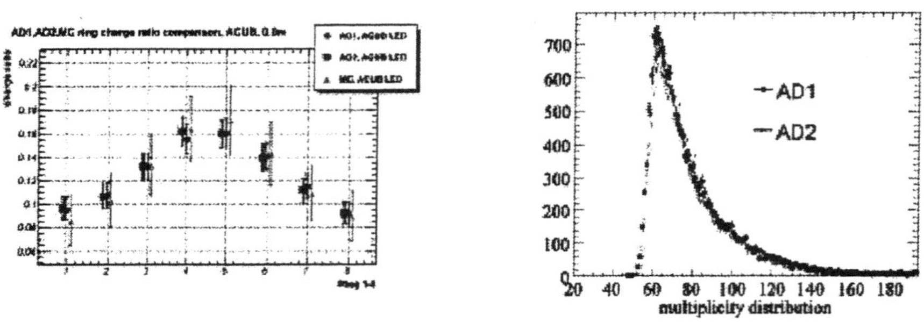

Fig. 7. Results from AD Dry run tests. Left: Charge fraction of each PMT ring from calibration data. Right: Multiplicity distribution of the fired PMTs from cosmic-rays data.

2.2. *Muon Veto Detector*

The muon veto system consists of two parts: the inner and outer water Cerenkov detector and RPCs, as shown in Fig. 8. A minimum of 2.5m water surrounding the ADs can shield gamma-rays from rock and neutrons generated by cosmic-rays in the rock. The tagging efficiency of the water Cherenkov detector is more than 95% and that of RPC is more than 90%, giving a combined inefficiency less than 0.5% with an uncertainty less than 0.25%.

RPCs use a technique invented by the BESIII experiment .[6] Each detector module with a dimension of 2m × 2m, consists of 4 layers of bare RPCs with readout strips in the x- and y- direction alternatively. The concrete water pool is lined by three layers of PermaFlex painting to protect the clean water from pollution. PMTs are mounted on the supporting structure made of stainless steel unistrut which are fixed on the wall.

The water pool PMTs are tested after the installation without water. Results show that all PMTs work well with expectations. In addition, the production system for high purity water is installed and tested. The ADs are being installed in the pool, as shown in Fig. 8. Then the water pool will be filled and the whole detectors will be able to take data.

Fig. 8. AD is being installed into the completed pool. RPCs are at the far side.

3. The prospect of Daya Bay II project

Recently a possible future project after Daya Bay experiment is under discussion. It is a large liquid scintillator detector for the purpose of determine neutrino mass hierarchy.

The survival probability of $\overline{\nu}_e$ from reactors can be expressed as[7] $P_{ee}(L/E) = 1 - P_{21} - P_{31} - P_{32}$, where $P_{21} = \cos^4(\theta_{13})\sin^2(2\theta_{12})\sin^2(\Delta_{21})$, $P_{31} = \cos^2(\theta_{12})\sin^2(2\theta_{13})\sin^2(\Delta_{31})$, $P_{32} = \sin^2(\theta_{12})\sin^2(2\theta_{13})\sin^2(\Delta_{32})$, $\Delta_{ij} = 1.27\Delta m^2_{ij}L/E$, Δm^2_{ij} is the neutrino mass-squared difference $(m_i^2 - m_j^2)$ in eV^2, θ_{ij} is the neutrino mixing angle, L is the baseline from reactor to $\overline{\nu}_e$ detector in meters, and E is the $\overline{\nu}_e$ energy in MeV. $P_{ee}(L/E)$ has three oscillation components, P_{21}, P_{31} and P_{32}, corresponding to three oscillation frequencies in L/E space, which are proportional to $|\Delta m^2_{ij}|$, respectively. The oscillation component $1 - P_{21}$ dominates the P_{ee} oscillation, while P_{31} and P_{32}, which are sensitive to the neutrino mass hierarchy. Thus mass hierarchy in fact can distort the neutrino energy spectrum due to non-zero θ_{13}. An example of observed neutrino spectrum in L/E space is shown in Fig. 9, taking the baseline L to be 60 km and $\sin^2(2\theta_{13}) = 0.1$. Other neutrino mixing parameters used for Fig. 9 are from a global fit, as listed in Ref.[10]

A method using reactor neutrino-based intermediate baseline (40-65 km) experiments has been explored based on precision measurement of distortions of the energy spectrum due to nonzero θ_{13}.[8] A study[9] shows

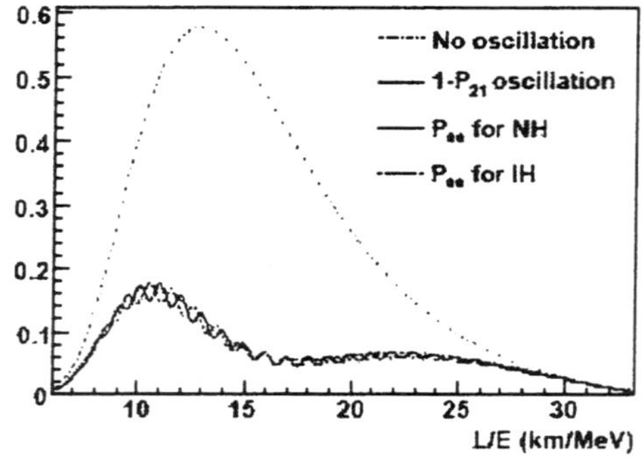

Fig. 9. Reactor neutrino spectra at a baseline of 60 km in L/E space for no oscillation (dashed dotted line), $1 - P_{21}$ oscillation (dotted line) and P_{ee} oscillation in the cases of NH and IH, assuming $\sin^2(2\theta_{13}) = 0.1$.

a new method to distinguish normal or inverted hierarchy after a Fourier transformation of the L/E spectrum of reactor neutrinos. It is observed that the Fourier power spectrum has a small shoulder next to the main peak, and their relative position can be used to determine the mass hierarchy. A filter method is used to improve the sensitivity to the mass hierarchy up to $\sin^2(2\theta_{13}) > 0.02$, if Δm^2_{32} is known *a priori*.

Recently our work shows that by Fourier sine transform (FST) and Fourier cosine transform (FCT), and fully utilizing all information, the capability of determining the neutrino mass hierarchy can be improved without the precise knowledge of Δm^2_{32}.[10] The Fourier transformation used is as following,

$$FST(\omega) = \int_{t_{min}}^{t_{max}} F(t)\sin(\omega t)dt\,, FCT(\omega) = \int_{t_{min}}^{t_{max}} F(t)\cos(\omega t)dt \quad (1)$$

where ω is frequency, $\omega = 2.54\Delta m^2_{ij}$; $t = \frac{L}{E}$ is the variable in L/E space, varying from $t_{min} = \frac{L}{E_{max}}$ to $t_{max} = \frac{L}{E_{min}}$. After applying FCT and FST to the analytical L/E spectrum without considering experimental details, the normal and inverted hierarchy have very different shapes, as shown in Fig. 10.

However, in real experiments with a set of discrete events, the integral can be changed to the summation over all events as in the following.

$$FST(\omega) = \sum_{i=1}^{N} \sin(\omega L/E_i')\,, FCT(\omega) = \sum_{i=1}^{N} \cos(\omega L/E_i') \quad (2)$$

where E_i' is the measured energy of individual events, and N is the total number of events collected.

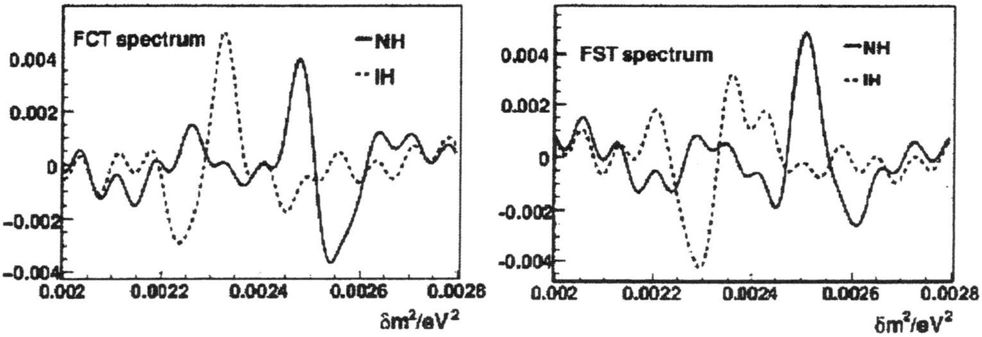

Fig. 10. The FCT and FST spectra and Fourier power spectrum for $\sin^2(2\theta_{13}) = 0.005$. The solid line is for NH and the dashed line is for IH. The FCT and FST spectra have distinctive features to identify the mass hierarchy.

Then in a detailed Monte Carlo study,[11] our work shows that if $\sin^2(2\theta_{13})$ is more than (1-2)%, a (10-50) kt liquid scintillator at a baseline of about 60 km with an energy resolution better than (2-3)% can determine the mass hierarchy at more than 90% C.L. With such a big detector, many other scientific topics can also be studied. The 60km baseline is the ideal baseline to precise measure other mixing parameters like θ_{12}, Δm^2_{12} and Δm^2_{23}. Large liquid scintillator detector is also ideal for supernova neutrinos, because different flavor neutrinos can be detected by liquid scintillator rather than water Cherenkov detectors. Moreover, Geo-neutrinos, atmospheric neutrinos as well as solar neutrinos are also good topics. In fact, there are two main challenges to build such a detector: high efficiency PMT and high transparent liquid scintillator. Now some R&D programs are already started at IHEP.

4. Acknowledgment

This work is supported in part by the Chinese Academy of Sciences, the National Natural Science Foundation of China, the Ministry of Science and Technology of China, the Guangdong provincial government, the Shenzhen Municipal government, the China Guangdong Nuclear Power Group, the Research Grants Council of the Hong Kong Special Administrative Region of China, the United States Department of Energy, the US National Science Foundation, the University of Houston, the University of Wisconsin, the University of Taiwan, and the Ministry of Education, Youth and Sports of the Czech Republic. Acknowledge also the opportunity to report in the Erice School.

References

1. Daya Bay Collaboration, Proposal of Daya Bay experiment, arXiv: hep-ex/0701029, 2007
2. Y. X. Sun *et al.*, Chinese Physics C (HEP&NP) 29, 543 (2005)
3. Z. M . Wang *et al.*, Nucl. Instr. and Meth. A, **602** (2009) 489
4. L. J. Wen *et al.*, Nucl. Instr. and Meth. A, **564** (2006) 471-474
5. L. J. Wen *et al.*, Nucl. Instr. and Meth. A, **629** (2011) 296-302
6. Q. M. Zhang *et al.*, Nucl. Instr. and Meth. A 583, 278 (2007)
7. S. M. Bilenky, D. Nicolo and S. T. Petcov, Phys. Lett. B **538**, 77 (2002) [arXiv:hep-ph/0112216].
8. S. T. Petcov *et al.*, Phys. Lett. B 533, 94 (2002); S. Choubey *et al.*, Phys. Rev. D 68, 113006 (2003)
9. J. Learned *et al.*, Phys. Rev. D 78, 071302 (2008)
10. L. Zhan *et al.*, Phys. Rev. D **78**, 111103 (2008).
11. L. Zhan *et al.*, Phys. Rev. D **79**, 073007 (2009).

DIPLOMAS

Twenty five Diplomas were open for competition among the participants. They have been awarded as follows:

- John S. BELL Diploma to:

  Dmitriy GALAKHOV
  ITEP, Moscow, Russia

- Patrick M.S. BLACKETT Diploma to:

  Magnus MAGER
  CERN, Geneva, Switzerland

- Nicola CABIBBO Diploma to:

  Liagjian WEN
  IHEP, Beijing, China

- James CHADWICK Diploma to:

  Michelle Victoria PREWITT
  Fermilab, Batavia, IL, USA

- Sidney COLEMAN Diploma to:

  Leticia Faria Domingues PALHARES
  Inst.Physique Théorique, Gif-sur-yvette Cedex, France

- Richard H. DALITZ Diploma to:

  Daniel Jesus TAPIA TAKAKI
  IPN – Université de Paris Sud, Orsay, France

- Paul A.M. DIRAC Diploma to:

  Andrea CAMPOLEONI
  Albert Einstein Institute
  Golm, Germany

- Bruno FERRETTI Diploma to:

  Tahsin Cagri SISMAN
  Middle East Techn. Univ., Ankara, Turkey

- Vladimir N. GRIBOV Diploma to:

  Enrico RINALDI
  University of Edinburgh, UK

- Robert HOFSTADTER Diploma to:

  Christine A. AIDALA
  Los Alamos National Laboratory, NM, USA

- Gunnar KÄLLEN Diploma to:

  Philipp BURDA
  ITEP, Moscow, Russia

- Seymour J. LINDENBAUM Diploma to:

  Màtè CSANÀD
  Eötvös Loránd University, Budapest, Hungary

- Yuval NE'EMAN Diploma to:

  Lasma ALBERTE
  Arnold Sommerfeld Center Theor. Physics, Munich, Germany

- Giuseppe P.S. OCCHIALINI Diploma to:

  Elisa MANONI
  INFN - University of Perugia, Italy

- Oreste PICCIONI Diploma to:

  Domenico DI FILIPPO
  LNF - INFN, Frascati, Italy

- Bruno PONTECORVO Diploma to:

  Mhair ORCHANIAN
  California Institute of Technology, Pasadena, CA, USA

- Giampietro PUPPI Diploma to:

  Alexander ROTHKOPF
  The University of Tokyo, Japan

- Isidor I. RABI Diploma to:

  Akira MIYAZAKI
  The University of Tokyo, Japan

- Giulio RACAH Diploma to:

  Pietro BIASSONI
  INFN and University of Milan, Italy

- Bruno ROSSI Diploma to:

  Meike de WITH
  Nikhef, Amsterdam, The Netherlands

- Julian S. SCHWINGER Diploma to:

  John Joseph CARRASCO
  Stanford Institute for Theoretical Physics, CA, USA

- Victor F. WEISSKOPF Diploma to:

  Nils-Erik BOMARK
  University of Bergen, Norway

- Eugene P. WIGNER Diploma to:

  Vasiliy Aleksandrovich ALBA
  ITEP, Moscow, Russia

- Bjorn H. WIIK Diploma to:

  Matthias SCHROEDER
  Universität Hamburg, Germany

- Chien Shiung WU Diploma to:

  Gianluca INGUGLIA
  Queen Mary University of London, UK

AWARDS

The Awards for the best 'New Talent – Speakers' have been attributed to:

i) **Vasiliy Aleksandrovich ALBA**
ITEP, Moscow, Russia
for his original presentation in
MATHEMATICAL PHYSICS

ii) **John Joseph CARRASCO**
Stanford Inst. forTheor. Physics,
CA, USA
for his original presentation in
THEORETICAL PHYSICS

iii) **Magnus MAGER**
CERN, Geneva, Switzerland
for his original presentation in
DETECTOR PHYSICS

iv) **Akira MIYAZAKI**
The University of Tokyo
Tokyo, Japan
for his original presentation in
EXPERIMENTAL PHYSICS

Award for Best Scientific Secretary given to:
Leticia Faria Domingues PALHARES
Institute de Physique Théorique
Gif-sue-Yvette Cedex, France

Award for Best Student given to:
Liagjian WEN
IHEP, Institute of High Energy Physics
Beijing, China

PARTICIPANTS

Andràs AGOCS Hungary	MTA KFKII-RMKI Budapest, Hungary
Andrea AGOSTINELLI Italy	University of Bologna Bologna, Italy
Nazeer Ahmad AHMAD India	Aligarh Muslin University Aligarh, UP, India
Christine A. AIDALA USA	Los Alamos National Laboratory Los Alamos, NM, USA *and* Brookhaven National Laboratory, Upton, NY, USA
Vasiliy Aleksandrovich ALBA Ukraine	ITEP Moscow, Russia
Lasma ALBERTE Latvia	Arnold Sommerfeld Center for Theoretical Physics Munich, Germany
Guido ALTARELLI Italy	CERN Geneva, Switzerland
Duygu ALTIN Turkey	Karadeniz Technical University Trabzon, Turkey
Ignatios ANTONIADIS France	Centre de Physique Théorique - Ecole Polytechnique Palaiseau, France
Anastasia ARISTOVA Russia	ITEP Moscow, Russia
Daniel ARNOLD Switzerland	Albert Einstein Centre for Fundamental Physics Bern, Switzerland
Fatma AYDOGMUS Turkey	Istanbul University Istanbul, Turkey
Yakov AZIMOV Russia	Saint-Petersburg State University Gatchina, Saint-Petersburg, Russia
Marcin BADZIAK Poland	University of Warsaw Warsaw, Poland *and* University of Cambridge, UK
Rinaldo BALDINI Italy	Enrico Fermi Centre Rome, Italy
Viktor BEGUN Ukraine	Johann Wolfgang Goethe University Frankfurt am Main Germany *and* Bogolyubov Inst.Th.Ph., Kiev, Ukraine

Sergio BERTOLUCCI
Italy

CERN
Geneva, Switzerland

Pietro BIASSONI
Italy

INFN and University of Milan
Milano, Italy

Simone BIONDINI
Italy

University of Perugia
Perugia, Italy

Gianluca BLANKENBURG
Italy

University "Roma Tre"
Rome, Italy

Nils-Erik BOMARK
Sweden

University of Bergen
Bergen, Norway

Leron BORSTEN
The Netherland

Imperial College of London, London, UK
& University of Turin, Italy

Raphael BOUSSO
USA

University of California
BERKELEY, CA, US

Massimo BRIDA
Italy

University of Bologna
Bologna, Italy

Philipp BURDA
Russia

ITEP
Moscow, Russia

Xavier CALMET
France

University of Sussex
Brighton, UK

Sheldon CAMPBELL
Canada

Texas A&M University
College Station, TX, USA

Andrea CAMPOLEONI
Italy

Albert Einstein Institute
Golm, Germany

John Joseph CARRASCO
USA / Ireland

Stanford Institute for Theoretical Physics
Stanford, CA, USA *and*
UCLA, Los Angeles, CA, USA

Leonardo CASTELLANI
Italy

University of Piemonte Orientale Amedeo Avogadrc
Alessandria, Italy

Bianca Letizia CERCHIAI
Italy

University of Milano
Milano, Italy

Kirill CHILIKIN
Russia

ITEP
Moscow, Russia

Marcin CHRZASZCZ
Poland

Krakow University of Technology
Krakow, Poland

Fabrizio COCCETTI Italy	Enrico Fermi Centre Rome, Italy
Màtè CSANÀD Hungary	Eötvös Loránd University Budapest, Hungary
Meike de WITH The Netherland	Nikhef Amsterdam, The Netherlands
Mehmet DEMIRCI Turkey	Karadeniz Technical University Trabzon, Turkey
Dmitri DENISOV USA	Fermilab Batavia, IL, USA
Domenico DI FILIPPO Italy	LNF Frascati, Italy
Qader DOROSTI HASANKIADEH Iran	University of Groningen Groningen, The Netherlands
Petr DUNIN-BARKOVSKIY Russia	ITEP Moscow, Russia
Samuel R. EDWARDS Australia	Institute für Kernphysik Darmstadt, Germany
Burcin ENGIN Turkey	Karadeniz Technical University Trabzon, Turkey
Angelo Raffaele FAZIO Italy	Universidad Nacional de Colombia Bogotá, Colombia
Sergio FERRARA Italy	CERN Geneva, Switzerland *and* LNF–INFN, Frascati, Italy
Giancarlo FERRERA Italy	INFN - University of Florence Sesto Fiorentino (Firenze), Italy
Andrea FESTANTI Italy	University of Bologna Bologna, Italy
David-Johannes FISCHER Germany	DESY Hamburg, Germany
Eduardo Souza FRAGA Brazil	Universidade Federal do Rio de Janeiro Rio de Janeiro, Brasil
Matteo FRANCHINI Italy	University of Bologna Bologna, Italy

Dario FRANCIA
Italy
Institute of Physics
Academy of Sciences of the Czech Republic
Prague, Czech Republic

Harald FRITZSCH
Germany
Ludwig-Maximilians-Universitaet
Muenchen, Germany

Ulla GEBBERT
Germany / Finland
Universität Hamburg
Hamburg, Germany

Oleksandr GITULIAR
Ukraine
Institute of Nuclear Physics
Polish Academy of Science
Krakow, Poland

Paolo GIUBELLINO
Italy
CERN
Geneva, Switzerland

Georgy GOLOVANOV
JINR
Dubna, Russian Federation

Barbara GUERZONI
Italy
INFN - University of Bologna
Bologna, Italy

Yuri GUSEV
Russia / Canada
Lebedev Research Center in Physics Institute
Moscow, Russia *and*
Simon Fraser University, Burnaby, BC, Canada

Dieter HAIDT
Germany
DESY
Hamburg, Germany

Rolf-Dieter HEUER
Germany
CERN
Geneva, Switzerland

Gianluca INGUGLIA
Italy
Queen Mary University of London
London, UK

Ayuki KAMADA
Japan
Institute for the Physics and
Mathematics of the Universe (IPMU)
Kashiwa, Japan

Anna Maria KAMINSKA
Italy
University of Warsaw
Warsaw, Poland

Zurab KEPULADZE
Georgia
E. Andronikashvili Institute of Physics
Tbilisi, Georgia

Ahmed Mohamed Saleh Hassanin KHALIL
Egypt
The Egyptian Center for Theoretical Physics
MTI University – Cairo, Egypt

Andrey KHMELNITSKIY
Russia
CERN
Geneva, Switzerland

Martin KIRAKOSYAN
Russia
P.N. Lebedev Institute of Physics
Moscow, Russia

Nadiia KRUPINA
Ukraine
Kharkov Physics and Technology Institute
Kharkov, Ukraine

Evgeny KRYLOV
Russia
ITEP
Moscow, Russia

Maxim KURKOV
Russia
Saint-Petersburg State University
Gatchina, Saint-Petersburg, Russia *and*
Complesso Universitario Monte S. Angelo
Napoli, Italy

Vincenzo LAVORINI
Italy
INFN - University of Calabria
Cosenza, Italy

Paul LAYCOCK
UK
CERN
Geneve, Switzerland *and*
University of Liverpool, UK

Oleg LEMETS
Ukraine
Kharkov Physics and Technology Institute
Kharkov, Ukraine

Heinrich LEUTWYLER
Switzerland
University of Bern
Bern, Switzerland

Ka Ki Carmen LI
Hong Kong
Imperial College London
London, UK

Lev N. LIPATOV
Russia
Saint-Petersburg State University
Gatchina, Saint-Petersburg, Russia *and*
Hamburg University, Germany

Luca LOPEZ
Italy
Scuola Normale Superiore
Pisa, Italy

Narei LORENZO MARTINEZ
Spain
Laboratoire de l'Accélérateur Linéaire
Centre Scientifique d'Orsay
Orsay, France

Árpád László LUKÁCS
Hungary
MTA KFKI RMKI
Budapest, Hungary

Magnus MAGER
Germany
CERN
Geneva, Switzerland *and*
Technische Universitaet Darmstadt, Germany

Elisa MANONI
Italy
INFN - University of Perugia
Perugia, Italy

Alessio MARRANI
CERN

Italy — Geneva, Switzerland

Pierpaolo MASTROLIA
Italy
Max-Planck-Institute for Physics
Munchen, Germany

Annalisa MASTROSERIO
Italy
CERN
Geneva, Switzerland

Alexey MATVEEV
Russia
The University of Tokyo
Tokyo, Japan *and*
Lebedev Physical Institute, Moscow, Russia

Peter MINKOWSKI
Switzerland
University of Bern
Bern, Switzerland

Sergey MIRONOV
Russia
ITEP
Moscow, Russia

Akira MIYAZAKI
Japan
The University of Tokyo
Tokyo, Japan

Piotr MORAWSKI
Poland
Institute of Nuclear Physics PAN
Krakow, Poland

Andrey MOROZOV
Russia
ITEP
Moscow, Russia

Stavros MOUSLOPOULOS
Greece
Sydney Science Tutoring
Sydney, Australia

Alexander V. NESTERENKO
Russia
Joint Institute for Nuclear Research
Dubna, Moscow Region, Russia

Harvey NEWMAN
USA
California Institute of Technology
Pasadena, CA, USA

Matthew NGUYEN
France / USA
CERN
Geneva, Switzerland

Hiromichi NISHIMURA
Japan
Washington University in St. Louis
St. Louis, MO, USA

Francesco NOFERINI
Italy
University of Bologna
Bologna, Italy

Oleg Olegovich NOVIKOV
Russia
Saint-Petersburg State University
Gatchina, Saint-Petersburg, Russia

Emanuele ORAZI
Italy
Politecnico di Torino
Torino, Italy

Mhair ORCHANIAN
USA
California Institute of Technology
Pasadena, CA, USA

Leticia Faria Domingues PALHARES
Brazil

Institute de Physique Théorique
Gif-sue-Yvette Cedex, France

Vito PALLADINO
Italy

LNF
Frascati, Italy

Martina PERANI
Italy

University of Bologna
Bologna, Italy

Antonio POLICICCHIO
Italy

University of Cosenza
Cosenza, Italy

Alexander POPOLITOV
Russia

ITEP
Moscow, Russia

Roberto PREGHENELLA
Italy

Enrico Fermi Centre
Rome, Italy *and* INFN-University of Bologna, Italy

Michelle Victoria PREWITT
USA

Fermilab
Batavia, IL, USA

Olga PTITSYNA
Russia

ITEP
Moscow, Russia

Enrico RINALDI
Italy

University of Edinburgh
Edinburgh, UK

Alexander ROTHKOPF
Germany

The University of Tokyo
Tokyo, Japan

Andrey SADOFYEV
Russia

ITEP
Moscow, Russia

Augusto SAGNOTTI
Italy

Scuola Normale of Pisa
Pisa, Italy

Seyed Sajjad SAHBAEI RAZAVI
Iran

Lund University
Lund, Sweden

Daniela SALVATORE
Italy

University of Cosenza
Cosenza, Italy

Eugenio SCAPPARONE
Italy

INFN – University of Bologna
Bologna, Italy

Robert SCHOEFBECK
Austria

HEPHY, Institute for High Energy Physics
Austrian Academy of Sciences
Vienna, Austria

Matthias SCHROEDER
Germany

Universität Hamburg
Hamburg, Germany

Frederic Paul SCHULLER
Germany

Max Planck Institute for Gravitational Physics
Albert Einstein Institute
Potsdam, Germany

Olga SHEKHOVTSOVA
Ukraine

University of Valencia – CSIC
Valencia, Spain

Sergei SIBIRYAKOV
Russia

Institute for Nuclear Research of RAS
Moscow, Russia

Tahsin Cagri SISMAN
Turkey

Middle East Technical University
Ankara, Turkey

Giacomo SNIDERO
Italy

University of London
London, UK

Elena SOLOVYEVA
Russia

ITEP
Moscow, Russia

Aniello SPIEZIA
Italy

University of Perugia
Perugia, Italy

Michael J. TANNENBAUM
USA

Brookhaven National Laboratory
Upton, NY, USA

Daniel Jesus TAPIA TAKAKI
Mexico

IPN – Université de Paris Sud
Orsay, France

Massimo TARONNA
Italy

Scuola Normale Superiore
Pisa, Italy

Astkhik TER-MINASYAN

RRC "Kurchatov Institute
Moscow, Russia

Gerardus 't HOOFT
The Netherlands

Utrecht University
Utrecht, The Netherlands

Samuel C.C. TING
USA

MIT
CAMBRIDGE, MA, USA

Giuseppe TORRI
Italy

Imperial College
London, UK

Robert VERTESI
Hungary

University of Debrecen
Budapest, Hungary

Alexander VIKMAN
Germany/Ukraine

New York University
NEW YORK, NY, US

Andrey VLASOV
Russia

ITEP
Moscow, Russia

Lucia VOTANO Italy	LNGS L'Aquila, Italy
Liangjian WEN China	IHEP, Institute of High Energy Physics Beijing, China
Horst WENNINGER Germany	CERN Geneva, Switzerland
Ivan YELETSKIKH Ukraine	Joint Institute for Nuclear Research Dubna, Moscow Region, Russia
Armen YERANYAN Armenia	INFN-LNF Frascati, Italy
Danylo YEROKHIN Ukraine	Kharkov Physics and Technology Institute Kharkov, Ukraine
Zeyuan YU China	IHEP, Institute of High Energy Physics Beijing, RPC
Ahmad ZEIN ASSI France	CERN Geneva, Switzerland
Antonino ZICHICHI	CERN Geneva, Switzerland *and* INFN and University of Bologna, Italy *and* Enrico Fermi Centre, Rome, Italy